New Fluorinated Carbons: Fundamentals and Applications

New Fluorinated Carbons: Fundamentals and Applications

Progress in Fluorine Science Series

Edited by

Olga V. Boltalina

Colorado State University, Fort Collins, CO, United States

Tsuyoshi Nakajima

Aichi Institute of Technology, Toyota, Japan

Series Editor

Alain Tressaud

Emeritus Research Director CNRS ICMCB-CNRS, University Bordeaux Pessac, France

ELSEVIER

AMSTERDAM • BOSTON • HEIDELBERG • LONDON
NEW YORK • OXFORD • PARIS • SAN DIEGO
SAN FRANCISCO • SINGAPORE • SYDNEY • TOKYO

Elsevier
Radarweg 29, PO Box 211, 1000 AE Amsterdam, Netherlands
The Boulevard, Langford Lane, Kidlington, Oxford OX5 1GB, United Kingdom
50 Hampshire Street, 5th Floor, Cambridge, MA 02139, United States

Notices
Knowledge and best practice in this field are constantly changing. As new research and experience broaden our understanding, changes in research methods, professional practices, or medical treatment may become necessary.

Practitioners and researchers must always rely on their own experience and knowledge in evaluating and using any information, methods, compounds, or experiments described herein. In using such information or methods they should be mindful of their own safety and the safety of others, including parties for whom they have a professional responsibility.

To the fullest extent of the law, neither the Publisher nor the authors, contributors, or editors, assume any liability for any injury and/or damage to persons or property as a matter of products liability, negligence or otherwise, or from any use or operation of any methods, products, instructions, or ideas contained in the material herein.

Library of Congress Cataloging-in-Publication Data
A catalog record for this book is available from the Library of Congress

British Library Cataloguing-in-Publication Data
A catalogue record for this book is available from the British Library

ISBN: 978-0-12-803479-8

Publisher: John Fedor
Acquisition Editor: Katey Birtcher
Editorial Project Manager: Jill Cetel
Production Project Manager: Paul Prasad Chandramohan
Designer: Mark Rogers

Typeset by TNQ Books and Journals

Contents

List of Contributors

Ph. Bilas Université des Antilles, Pointe à Pitre, France

O.V. Boltalina Colorado State University, Fort Collins, CO, United States

E.V. Bukovsky Colorado State University, Fort Collins, CO, United States

L.G. Bulusheva Nikolaev Institute of Inorganic Chemistry, Siberian Branch of Russian Academy of Sciences, Novosibirsk, Russia

Y.-S. Chen ChemMatCARS University of Chicago, Argonne, IL, United States

T.T. Clikeman Colorado State University, Fort Collins, CO, United States

M. Dubois Université Blaise Pascal, Aubière, France; CNRS, Aubière, France

M. Gola Politecnico di Milano, Milano, Italy; Consorzio INSTM, Firenze, Italy

K. Guérin Université Blaise Pascal, Aubière, France; CNRS, Aubière, France

R. Hagiwara Kyoto University, Kyoto, Japan

A. Hamwi Université Blaise Pascal, Aubière, France; CNRS, Aubière, France

Y. Hattori Shinshu University, Ueda, Japan

K.-I. Ho Chang Gung University, Tao-Yuan, Taiwan

H. Ishikawa Osaka Electro-Communication University, Neyagawa, Osaka, Japan

M. Kawaguchi Osaka Electro-Communication University, Neyagawa, Osaka, Japan

J.H. Kramer University of South Dakota, Vermillion, SD, United States

C.-S. Lai Chang Gung University, Tao-Yuan, Taiwan

B.W. Larson Colorado State University, Fort Collins, CO, United States

Y.-S. Lee Chungnam National University, Daejon, Korea

L. Legras EDF R&D MAI, Les renardières, Moret-sur-Loing, France

J.L. Mansot Université des Antilles, Pointe à Pitre, France

Y. Matsuo University of Hyogo, Himeji, Japan

A. Molza Université des Antilles, Pointe à Pitre, France

T. Nakajima Aichi Institute of Technology, Toyota, Japan

W. Navarrini Politecnico di Milano, Milano, Italy; Consorzio INSTM, Firenze, Italy

A.V. Okotrub Nikolaev Institute of Inorganic Chemistry, Siberian Branch of Russian Academy of Sciences, Novosibirsk, Russia

A.A. Popov Leibniz Institute for Solid State and Materials Research, Dresden, Germany

M. Sansotera Politecnico di Milano, Milano, Italy; Consorzio INSTM, Firenze, Italy

Y. Sato Kyoto University, Kyoto, Japan; National Institute of Advanced Industrial Science and Technology, Tsukuba, Japan

S.H. Strauss Colorado State University, Fort Collins, CO, United States

C.-Y. Su National Central University, Chung-Li, Taiwan

H. Sun University of South Dakota, Vermillion, SD, United States

Ph. Thomas Université des Antilles, Pointe à Pitre, France

H. Touhara Shinshu University, Ueda, Japan

A. Vinogradov St. Petersburg State University, St. Petersburg, Russia

K. Yamada Osaka Electro-Communication University, Neyagawa, Osaka, Japan

Preface

The focus of this second volume in the *Progress in Fluorine Science Series* is molecular and solid-state materials composed principally of carbon and fluorine. The boundary between organic and inorganic chemistry is especially ambiguous when discussing fluorinated forms of carbon allotropes or carbon-rich materials. Is buckminsterfullerene, C_{60}, an inorganic carbon allotrope, like diamond and graphite, or is it an organic carbon allotrope because molecules of C_{60} behave chemically like an electron-poor olefin? Leaving that distinction aside, simply counting the allotropes of carbon can cause dissent, although all scientists agree that many dozens are now known. Debating these issues may be an interesting academic exercise and may even be important to abstracting services and database managers, but the synthesis, characterization, and uses of fluorinated forms of carbon will not depend on what we call them. We believe that the diverse and complex assemblage of materials that are the subject of this book should speak for themselves.

The 1994 volume edited by Nakajima, *Fluorine-Carbon and Fluoride-Carbon Materials: Chemistry, Physics, and Applications*, still serves as an excellent reference on fluorine-containing graphite intercalation compounds. Nevertheless, research on the preparation, characterization, and applications of fluorinated carbon materials has continued unabated, and in fact has accelerated, since that time, justifying the state-of-the-art summaries presented in this book. Chapter 1, by Boltalina, is an overview of current knowledge about the electronic properties of fluorofullerenes and their use as strong electron acceptors in optoelectronic devices. Chapter 2, by Larson, Strauss, and Boltalina, is a survey of trifluoromethylfullerene synthesis. Chapter 3, by Bukovsky, Larson, Clikeman, Chen, Popov, Boltalina and Strauss, is focused on a unique family of 13 structurally-characterized decakis(trifluoromethyl)fullerenes and highlights structure−property relationships relevant to practical applications. Chapter 4, by Popov, and 5, by Sun and Kramer, are accounts of new classes of perfluoroalkylated polycyclic aromatic hydrocarbons, including corannulenes and acenes, with an emphasis on both experimental and theoretical results that bear on their molecular and electronic structures. Chapter 6, by Hattori and Touhara, presents new perspectives on the storage of fluorine by single-wall carbon nanotubes and nanohorns, and Chapter 7, by Lee, is a review of recent advances in the preparation and electrical properties of fluorinated carbon nanotubes and carbon fibers. The functionalization of both molecular and solid-state carbon materials with perfluoropolyethers is discussed by Sansotera, Gola, and Navarrini in Chapter 15. Two other chapters that bridge molecular and solid-state materials are Chapter 8, by Bulusheva and Okotrub, a detailed and up-to-date account of the electronic structure of fluorinated graphene, and Chapter 16, by Ho, Lai, and Su, an introduction to nanoelectronics based on this important new material. The use of NMR spectroscopy to probe the nature of C−F bonds in solid-state fluorinated carbon materials is discussed by Dubois, Guérin, Hamwi, and Vinogradov in Chapter 9, an in-depth summary of their diverse structures is given by Hagiwara and Sato in Chapter 12, and correlations between their structures and macrotribologic properties are discussed by Thomas, Bilas, Molza, Legras, Mansot, Gueren, and Dubois in Chapter 14. The electrochemical synthesis of fluorine- and oxygen-containing graphite and graphene and the fabrication of transparent electrodes are reviewed by Matsuo in Chapter 10, the use of intercalation chemistry for secondary battery applications is summarized by Kawaguchi, Yamada, and Ishikawa in Chapter 11, and the history and fundamentals of lithium-graphite fluoride batteries are presented by Nakajima in Chapter 13. We hope that the structural diversity, unprecedented properties, and practical uses of fluorinated carbon presented herein by authors from eight countries in three continents will inspire further progress in this field for years to come by the next generation of fluorine chemists, physicists, material scientists, and engineers.

We thank Alain Tressaud, the series editor and our longtime friend and colleague, for the invitation to edit this volume and for his constant and invaluable encouragement and support. In closing, we are pleased to acknowledge the creative, dedicated, and impeccably professional Elsevier team that guided and assisted us in the production of this book.

Fort Collins, CO, USA
Toyota, Japan
August 2016

Olga V. Boltalina
Tsuyoshi Nakajima

Electronic Properties and Applications of Fluorofullerenes

1

O.V. Boltalina
Colorado State University, Fort Collins, CO, United States

Chapter Outline

1.1 Introduction

The discovery of new allotropes of carbon, the fullerenes, has been regarded as one of the cornerstone events in chemistry in the end of the last century, which enabled the surge of nanoscience and nanotechnology [1]. Other carbon nanostructures were later synthesized, including carbon nanotubes, nanohorns, nanowhiskers, carbon quantum dots, and, more recently, graphene.

The most abundant fullerene, C_{60}, has a nearly spherical shape, and a diameter of c.7 Å, ie, nearly 1 nm. The C_{60} molecule has 30 double bonds on the cage, and its chemical behavior is similar to that of olefins. Typical chemical reactions include various cycloadditions, or additions of pairs of functional groups to the double bonds. In principle, all 60 carbons of C_{60} can be functionalized to form a $C_{60}X_{60}$ derivative, if the size of the addend small enough, for example, hydrogen or fluorine atoms. Theoretical calculations predict the perhydrogenated molecule $C_{60}H_{60}$ to be thermodynamically stable [2]; however, it has not been isolated, or even detected, experimentally. Only species with 44−46 hydrogen atoms per fullerene cage were observed as the highest known degree of hydrogenation of C_{60} [3]. In contrast, the next smallest addend, the fluorine atom, could be added to C_{60} more readily and in larger numbers per C_{60} cage; even experimental observations of the $C_{60}F_{60-100}$ anionic gas-phase

New Fluorinated Carbons: Fundamentals and Applications. http://dx.doi.org/10.1016/B978-0-12-803479-8.00001-2

species were reported by several research groups [4–6]. A fluorofullerene $C_{60}F_{48}$ is ranked as a single isomer C_{60} derivative with the highest degree of functionalization, in which 48 out of 60 sp^2 carbons are converted into sp^3 carbons upon fluorination, leaving unreacted only six double bonds [7,8]. Furthermore, fluorinated fullerenes were shown to be more thermally stable and chemically inert than hydrogenated fullerenes [3,9–11]. The reason of such drastic difference in properties lies in the relative strengths of the chemical bonds formed as a result of hydrogenation and fluorination: C–F bonds are stronger than C–H bonds.

Fluorine is the most electronegative element in the periodic table, with the electron affinity (EA) of 3.39 eV, and thus its introduction to fullerenes via fluorination and formation of C(cage)–F bonds has the most profound effect on their electronic properties. Noteworthy, derivatives of fullerenes with fluorine-containing groups, such as fluoroalkyl or fluorinated aromatic groups do not belong to the family of fluorofullerenes, as sometimes erroneously mentioned in the literature. Currently, $C_{60}F_{48}$ is regarded as one of the champion electron acceptors among neutral organic molecules, with a gas-phase EA value exceeding 4 eV [12]. Combination of such remarkable electronic properties with exceptional thermal stability distinguishes fluorofullerenes from other classes of fullerene derivatives and explains active research interest in various fields of chemistry and material science. Being molecular forms of fluorinated carbon, fluorofullerenes are particularly suitable for fundamental research and elaboration of structure–property correlations. In contrast, carbon nanotubes or graphene sheets, or other nanoscopic carbon materials typically have rather poorly defined compositions, size distributions, and available structural details, unless significant purification efforts are undertaken.

The goal of this chapter is to provide an outlook at the current state-of-the-art research on fluorinated fullerenes, with the emphasis on the compounds that can be made in commercially viable amounts and with sufficient purity. Emerging applications of fluorofullerenes as strong electron acceptors in new technologies are briefly overviewed, and possible directions for further studies are discussed. Those readers who are interested in the synthetic procedures concerning direct fluorination, isolation, and detailed characterization of their physical and chemical properties can be referred to the reviews published in the past decade [11,13,14].

1.2 Molecular Structures

In early semiempirical theoretical studies of fullerene addition patterns, hydrogen and fluorine have been the most common and preferred types of functional groups studied. First, due to their small size, additions of hydrogen (or fluorine) atoms on neighboring carbon atoms are not sterically hindered, which allows unrestricted range of structures to be considered, with the widest possible range of compositions, in the case of $C_{60}X_{2n}$ derivatives, $2n = 2–60$. Second, formation of the direct C–H (C–F) bonds with the fixed geometry does not require to address conformational isomerism (as it is the case with multiatom functional groups), which considerably lowers the cost of computations. Particularly notable have been the works by Clare and Keppert [15–17], and P. Fowler [18] who have predicted many structures of fluoro- and hydrofullerenes, well before synthetic chemists made their contributions by preparing and structurally characterizing those compounds. It is not surprising that due to the similar sizes of fluorine and hydrogen atoms, the most stable addition patterns were predicted to be similar for a given composition. A particular interest of theorists to hydrogenation and fluorination was also due to very rapid and successful experimental studies of these chemistries: within a few weeks after the publication of the seminal work on the arc synthesis and isolation of fullerenes C_{60} and C_{70} [19], the first reports on the synthesis of hydrofullerenes and fluorinated derivatives were published by several research groups, nearly simultaneously [20–22]. Such an immediate initial success has been a clear indication of the remarkable chemical reactivity of fullerenes, and opening into a promising and fruitful field of chemistry, which has been developing for the past 25 years.

The first synthetic efforts on fluorination included treatment of C_{60} (or a fullerene extract) with elemental fluorine and other strong fluorinating agents. The common observations from these initial studies were that multiple additions of fluorine via C–F bond formation occurred, but all the products were characterized as very complex mixtures of multiple products with the wide range of compositions, as evidenced by broad, featureless ^{19}F NMR spectra and broad "envelopes" of peaks in the mass spectrometry analysis [20,21]. So, no selectivity toward a single product was initially achieved, even when the fullerene was treated with a large excess of a fluorinating agent and for prolonged periods of time, in some cases for several months [23,24]. Several reports appeared in the literature claiming that a highly symmetric $C_{60}F_{60}$ could be present in some of the fluorinated fullerene samples [21,24]. The conclusions were based on the observation of a sharp singlet in the ^{19}F NMR at δ −144 (±5); however, these claims have never been corroborated by further research. Significantly, subsequent experimental detections of the per- and hyperfluorinated ions with the $C_{60}X_{2n \geq 60}$ composition in the gas phase by mass spectrometry have been universally interpreted as evidence of fullerene cage rupture and not as the formation of the $C_{60}F_{60}$ molecules with an intact carbon cage [4,5,25].

The first selective fluorination of C_{60} was reported in 1994: a high-temperature F_2 fluorination of C_{60} powder dispersed in an NaF matrix resulted in the most abundant product, $C_{60}F_{48}$ [8]. Its ^{19}F NMR spectrum predominantly consisted of eight multiplets with equal integrated intensities, pointing to a high-symmetry addition pattern, D_3, as proposed by the authors. Later improvements in the synthesis and reaction scale-up, were followed by vibrational spectroscopy, and single-crystal X-ray crystallography studies, which confirmed the original structural elucidations [7,11,26]. Fig. 1.1 depicts the structure and a Schlegel diagram of D_3-$C_{60}F_{48}$. Out of 30 double bonds on the C_{60} cage, 24 are converted into single C–C bonds, with 48 C atoms bearing F atoms. The remaining six double bonds are nearly evenly distributed on the cage, three on each hemisphere, and their bond lengths are c.1.30 Å. The carbon cage is greatly distorted, indentations in the carbon sphere are due to closer distances of the sp^2 carbon atoms to the geometric center of the molecule than the sp^3 carbon atoms (3.05 and 3.9 Å, respectively) [7].

Further progress in the selective synthesis of fluorofullerenes was achieved two years later, when transition metal–fluoride reactions were discovered as effective fluorination methods for fullerenes [27,28]. In these reactions, two solids, a fullerene powder and a transition metal fluoride, are intimately mixed together and heated to a high which is

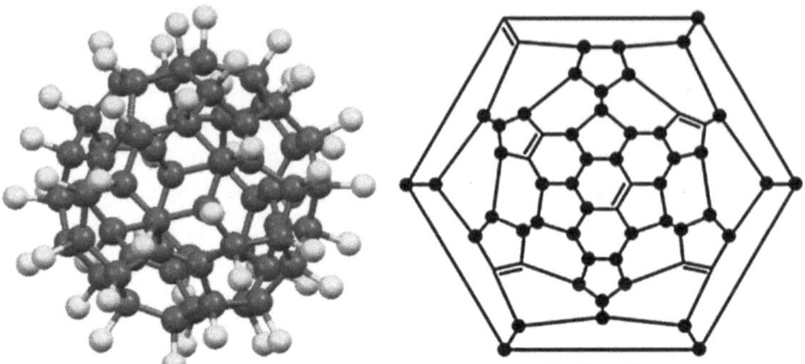

D_3-$C_{60}F_{48}$

Figure 1.1 A drawing of the molecular structure of D_3-$C_{60}F_{48}$ using optimized DFT-calculated coordinates; *yellow balls* (light gray in print versions) represent fluorine atoms, *dark gray balls* represent carbon atoms (left), and a Schlegel diagram illustrating its addition pattern; *black solid circles* represent locations of the fluorine atoms on the cage. *DFT*, density functional theory.

temperature, which is required for thermal activation of the reagents. A fluorinated fullerene product, which has a lower sublimation temperature than the reagents, sublimes out of the hot zone of the reactor, and thus gets separated from the unreacted fullerene, decomposed transition metal fluoride, and other nonvolatile products. The first compound prepared using this approach and MnF_3 as a fluorinating reagent was $C_{60}F_{36}$, the compositional selectivity achieved 90+ mol% (the impurities included species with lower and higher fluorine content, $C_{60}F_{34}$, and $C_{60}F_{38}$, respectively) [28].

In contrast to $C_{60}F_{48}$, which has been synthesized predominantly as a single isomer, $C_{60}F_{36}$ was comprised of three isomers, as followed from the initial spectroscopic analysis, and later confirmed by X-ray crystallography [29]. The most abundant isomer typically comprises 65–70 mol%, it has C_3 symmetry, which is manifested as 12 multiplets in its ^{19}F NMR spectrum. Minor isomers, identified as having T and C_1 symmetry, have characteristic 3- and 36-line ^{19}F NMR spectra, respectively. A drawing of the molecular structure of C_1-$C_{60}F_{36}$ and Schlegel diagrams of three known isomers are shown in Fig. 1.2.

A compound with 18 fluorine atoms per cage, $C_{60}F_{18}$, was made in a selective synthesis by reacting C_{60} with a complex metal fluoride, K_2PtF_6, under similar reaction conditions as used for $C_{60}F_{36}$, albeit at a higher temperature [27]. A single $C_{60}F_{18}$ isomer with C_{3v} symmetry is formed in this case, the structure is shown on Fig. 1.3. Its peculiar addition pattern, in which all 18 fluorine atoms are added in a contiguous loop on one hemisphere results in a highly polar structure, and drastic distortion of the carbon network as determined in the X-ray single crystal study [30].

The fourth fluorinated fullerene compound that was made selectively and in significant quantity has the composition $C_{60}F_{24}$. A fluorofullerene of such composition was first reported as a product of fluorination of the bromo derivative with the same composition, $C_{60}Br_{24}$, using either BrF_3 or XeF_2; however, its molecular structure was not determined [31]. An optimized selective synthesis and structural characterization by ^{19}F NMR and vibrational spectroscopy was published in a later study [32], which

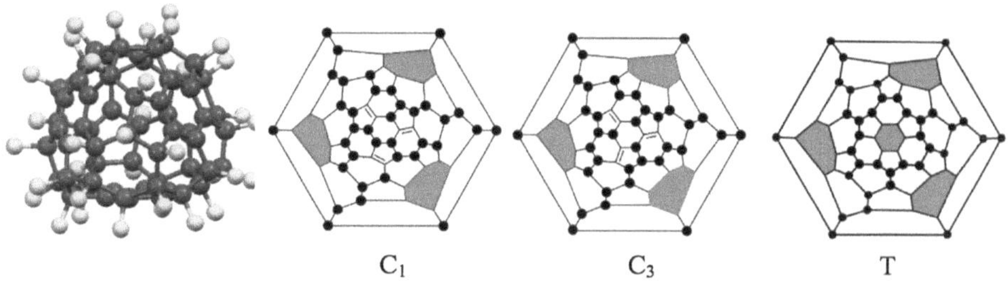

Figure 1.2 A drawing of the molecular structure of C_1-$C_{60}F_{36}$ using X-ray single crystal data; *yellow balls* (light gray in print versions) represent fluorine atoms, *dark gray balls* represent carbon atoms (left), the Schlegel diagrams of C_1, C_3, and T-isomers. *Black solid circles* represent locations of the fluorine atoms on the cage; isolated benzenoid rings are colored with solid blue (gray in print versions).

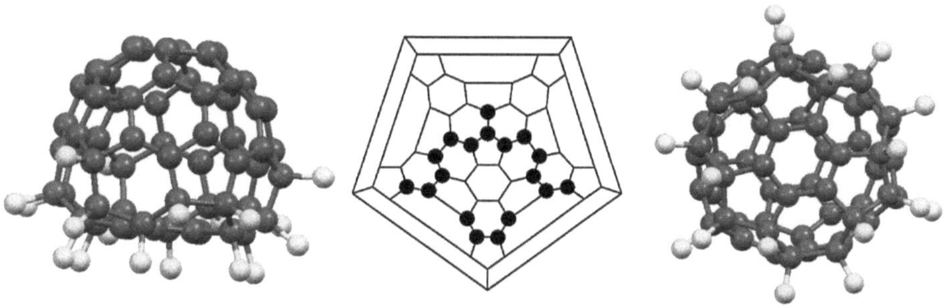

Figure 1.3 A drawing of the molecular structure of C_{3v}-$C_{60}F_{18}$ using X-ray single crystal data; *yellow balls* (light gray in print versions) represent fluorine atoms, *dark gray balls* represent carbon atoms (left), and a Schlegel diagram illustrating its addition pattern; *black solid circles* represent locations of the fluorine atoms on the cage.

determined that fluorine atoms are *ipso* substituted and the fluorofullerene $C_{60}F_{24}$ has the same addition pattern as the bromofullerene substrate; for the latter, its X-ray crystallographic study was carried out to ascertain its molecular structure [33,34]. A drawing of the molecular structure (based on the DFT-optimized coordinates) and the Schlegel diagram for $C_{60}F_{24}$ are presented in Fig. 1.4. In this fluorofullerene molecule, unlike in any other fluorofullerene, fluorine atoms are arranged in a highly symmetrical T_h pattern that does not involve a contiguous addition to neighboring carbon atoms. As a result, its physical and chemical properties are drastically different from the other fluorinated fullerene homologs. In particular, it possesses lower thermal stability than fluorofullerenes prepared under higher temperature conditions. When heated, it partially loses fluorine atoms and rearranges in the $C_{60}F_{18}$ compound; therefore it cannot be sublimed without changing the original structure and composition.

Currently, the fluorofullerenes $C_{60}F_{48}$, $C_{60}F_{36}$, $C_{60}F_{24}$, and $C_{60}F_{18}$ are the only fluorinated derivatives of fullerenes that can be prepared in gram quantities in selective high-yield synthetic procedures employing different types of fluorinating reagents, ranging from elemental fluorine gas, anhydrous HF, xenon difluoride, or high-valent transition metal fluorides. Most of these reagents are very strong oxidizers and need to be handed

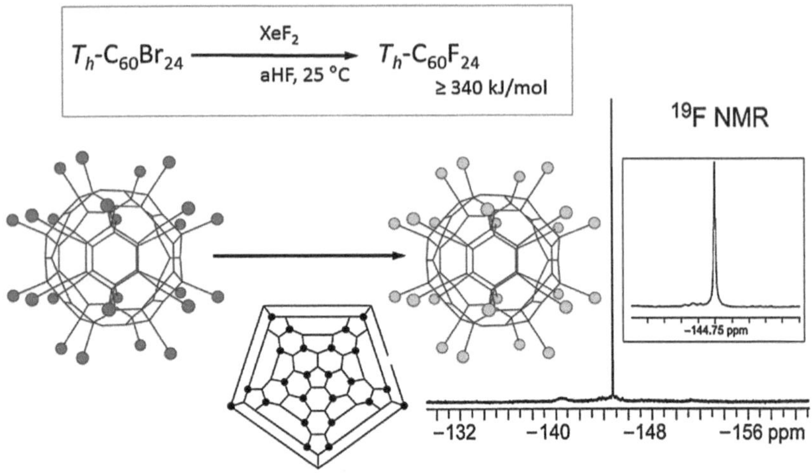

Figure 1.4 A scheme of the reaction used for preparation of $C_{60}F_{24}$ (in the *red box* (gray in print versions)); 3D representations of the molecular structures of $C_{60}F_{24}$ and $C_{60}Br_{24}$ and their Schlegel diagram; ^{19}F NMR spectrum of $C_{60}F_{24}$ in $CDCl_3$.

by specially trained personnel in laboratories equipped to safely use such chemicals. All these compounds can be prepared in one step with fairly good compositional purity of 90+ mol% or higher. If these materials are desired with higher purity, one can employ well-developed high performance liquid chromatography (HPLC) separation methods. In particular, it is advisable to use HPLC for the isolation of highly pure $C_{60}F_{24}$ and $C_{60}F_{18}$ that may contain impurities of compounds with partially substituted bromine atoms or compounds with lower fluorination degree, respectively, in the as-synthesized samples. In the case of $C_{60}F_{36}$, which forms three isomers in the synthesis, the isolation of the pure isomers is also possible using the HPLC method [29,35].

A number of other isomerically pure fluorofullerenes were isolated from the crude reaction products and characterized spectroscopically [36,37]. This group of compounds includes a series $C_{60}F_{2n}$ where $n = 0$, 1, 2, 3, 4, and 10. These lower fluorides typically form as by-products of reactions between complex metal fluorides and C_{60}, with $C_{60}F_{18}$ being a major product. Overall, their combined yield does not exceed single digit mol%, and their isolation as pure isomers involves multistep HPLC processes, involving the use of at least two different types of chromatography columns. For example, the "saturnene" compound $C_{60}F_{20}$ (see its structure and Schlegel diagram in Fig. 1.5) has been isolated by laborious HPLC, its retention time on the COSMOSIL Buckyprep column is more than twice as long as that of $C_{60}F_{18}$ (84 min vs 38 min, respectively). Therefore, the preparation of large quantities of fluorofullerenes with low fluorine content may require significant effort if one uses currently available synthetic and separation methods [38,39].

The isolation of the compounds $C_{60}F_2$, $C_{60}F_4$, $C_{60}F_6$, and two isomers of $C_{60}F_8$ has been reported and their addition patterns were elucidated based on NMR spectra and DFT calculations, see Fig. 1.6 for their Schlegel diagrams [38,39]. It is apparent that

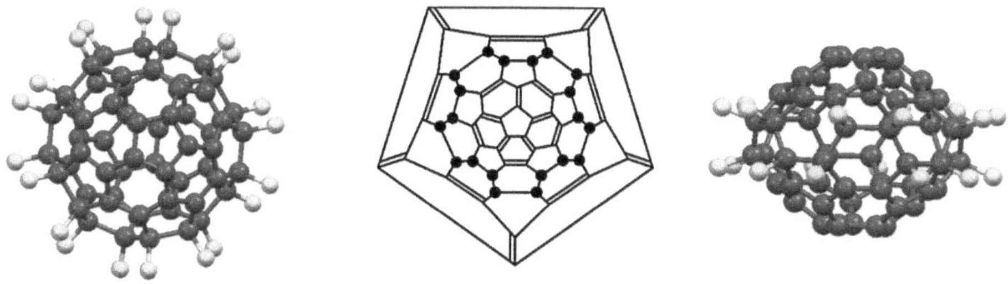

Figure 1.5 A drawing of the molecular structure of D_{5d}-$C_{60}F_{20}$ using X-ray single crystal data; *yellow balls* (light gray in print versions) represent fluorine atoms, *dark gray balls* represent carbon atoms (left), and a Schlegel diagram illustrating its addition pattern; *black solid circles* represent locations of the fluorine atoms on the cage.

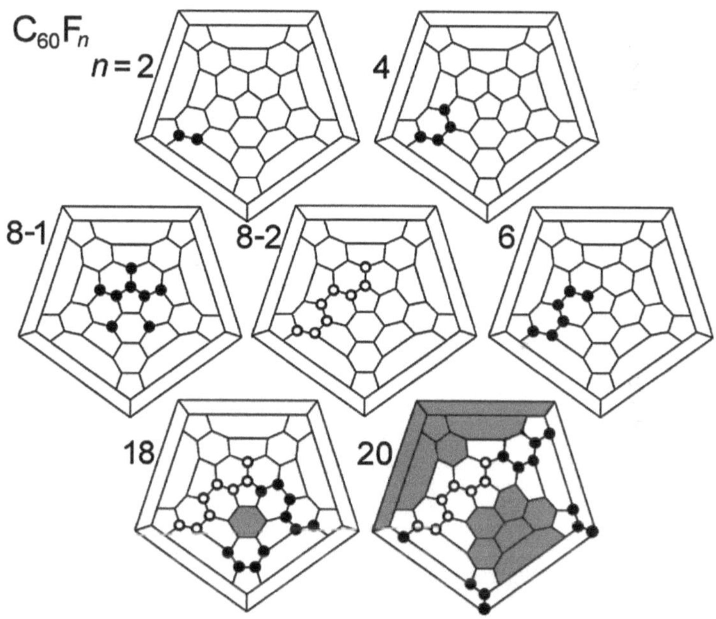

Figure 1.6 A chart showing evolution of the addition patterns in fluorofullerene series from $C_{60}F_2$ to $C_{60}F_{20}$. The common substructures in homologs are shown in *hollow circles*. Aromatic substructures are highlighted in blue (gray in print versions) [40].

consecutive additions of pairs of fluorine atoms occur to neighboring double bonds on the carbon cage in the series from $C_{60}F_2$ to $C_{60}F_8$ (isomer 1). Further additions occur to form either a belt of 20 fluorine atoms around the C_{60} equator, as in $C_{60}F_{20}$ [40], or to form a belt of 15 fluorine atoms and 3 atoms on the outside of the belt around the hexagon on one of the poles of the C_{60} cage as in $C_{60}F_{18}$ [30]. The formation of such specific addition patterns has been rationalized in terms of their relative thermodynamic stability, stabilization due to the formation of aromatic substructures (ie, an isolated benzenoid ring in $C_{60}F_{18}$ or two corannulene units in $C_{60}F_{20}$) [30,40,41].

In summary, fluorination of the fullerene C_{60} has yielded a wealth of chemical derivatives with the widest compositional range among any fullerene substrate, or even any other organic substrate. Among the fluorinated C_{60} fullerenes, four compounds with specific addition patterns can be made in gram amounts: $C_{60}F_{48}$, $C_{60}F_{36}$, $C_{60}F_{24}$, and $C_{60}F_{18}$. These compounds are most likely to have practical importance due to the relative ease of their scale-up. Other fluorofullerenes fall in the category of less accessible materials at this time, until innovative synthetic approaches are found that afford selective preparation of such compounds. For example, it has been speculated that ipso-substitution with fluorine atoms in specific chloro- or bromofullerenes can lead to a selective synthesis of fluorofullerenes with new structures, such as C_s-$C_{60}F_6$ with the characteristic skew pentagonal pyramid addition pattern as in C_s-$C_{60}Cl_6$ or $C_{60}Br_6$ [42]. However, such synthesis has not yet been realized experimentally. Another promising route toward fluorofullerenes with uncommon fluorination degrees is selective defluorination of higher fullerene fluorides, such as $C_{60}F_{48}$ or $C_{60}F_{36}$. The only known successful realization of such an approach is the selective defluorination of $C_{60}F_{48}$ with a ferrocene derivative that yielded a new single-isomer symmetric compound $C_{60}F_{44}$ (see Fig. 1.7 for its molecular structure, Schlegel

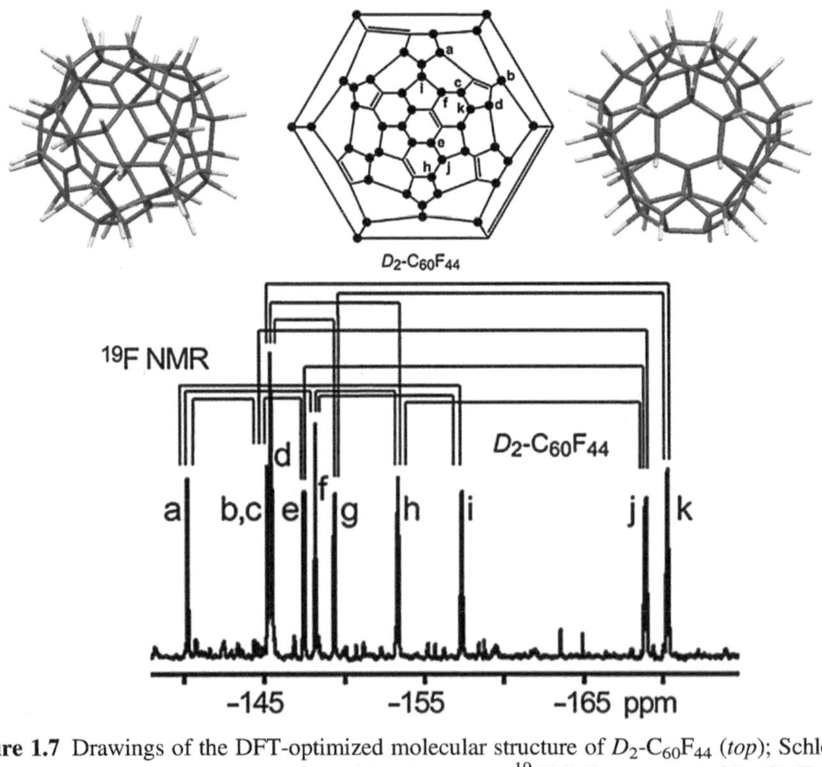

Figure 1.7 Drawings of the DFT-optimized molecular structure of D_2-$C_{60}F_{44}$ (*top*); Schlegel diagram with notations for unique types of fluorine atoms; ^{19}F NMR spectrum of D_2-$C_{60}F_{44}$ in $CDCl_3$ [43].

diagram and ^{19}F NMR spectrum) [43]. So far, this fluorofullerene was made only in a few milligram quantity sufficient for basic characterization, but its scale-up (if larger amounts are desired) does not appear to be difficult, since it does not require sophisticated reaction setup or rare chemicals.

The fluorination of higher fullerenes such as C_{70}, C_{74}, C_{78}, C_{84}, and some others has been studied less extensively than that of C_{60} [44–46]. First, the lower availability (and the high price) of higher fullerenes is known to be an obstacle. For example, even going from C_{60} to C_{70}, the price for the same-purity sample increases 10-fold. Second, since the symmetries of the cages of higher fullerenes are lower than that of C_{60}, the number of theoretically possible isomers of their derivatives increases, even in the cases of the simplest bis-adducts. So far, no selective formation of a single-isomer fluorofullerene of C_{70} has been reported. When strong fluorinating agents were used, such as elemental fluorine or xenon difluoride, the average compositions observed in the crude mixtures were $C_{70}F_{52-56}$ [11,47]. When milder reagents (eg, transition metal fluorides) were applied, lower fluorination degree was observed in the mixed products, eg, $C_{70}F_{36-40}$ [28]. Similar results were reported for other higher fullerenes, such as C_{78} or C_{84}, and no single-isomer products were prepared or isolated in any of these studies, with one exception [48].

At the same time, a surprisingly selective fluorination of C_{74} was observed in the reaction with K_2PtF_6, which produced a single-isomer symmetric product D_3-$C_{74}F_{38}$ [45]. The fullerene C_{74} belongs to a group of small-band-gap carbon clusters, which are not soluble in organic solvents, like other fullerenes, ie, C_{60}, C_{70}, C_{76}, C_{78}, and it possesses enhanced chemical reactivity, including polymerization in the solid state. However, when C_{74} is chemically functionalized, for example, via direct fluorination or trifluoromethylation, drastic improvements in the stability of the products were noted [45,49]. The DFT-calculated energy levels in the parent C_{74} and its fluorinated counterpart, $C_{74}F_{38}$, are shown in Fig. 1.8 [50]. It is apparent that the band gap in

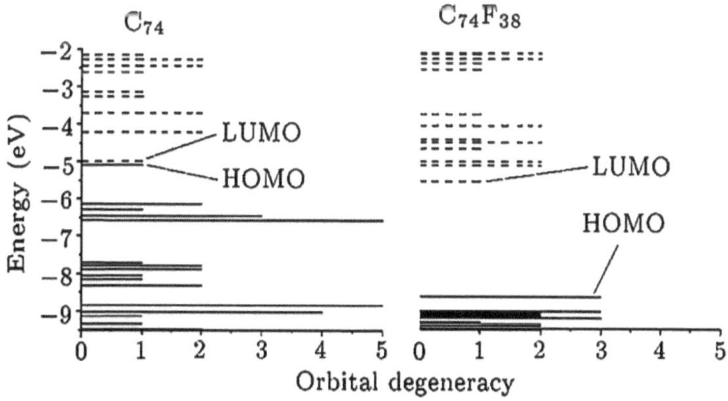

Figure 1.8 DFT-calculated energy level diagrams of C_{74} (left) and $C_{74}F_{38}$ (right). Unoccupied energy levels are shown with *dashed lines*, occupied levels with *solid lines* [50]. *DFT*, density functional theory.

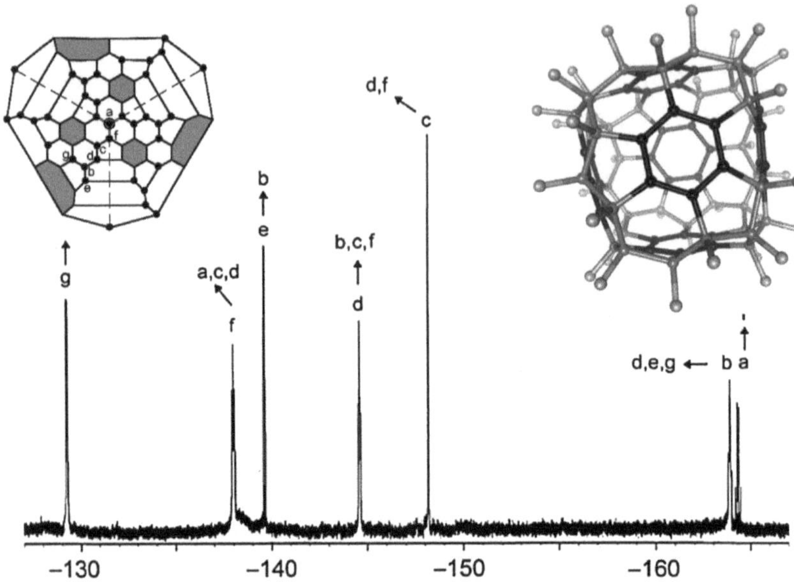

Figure 1.9 Left: Schlegel diagram of D_3-$C_{74}F_{38}$. *Black solid circles* show positions of fluorine atoms on the cage, blue color (gray in print versions) highlights isolated benzenoid rings. Right: DFT-optimized 3D representation of the molecular structure of D_3-$C_{74}F_{38}$ [45]. *DFT*, density functional theory.

$C_{74}F_{38}$ increases drastically compared to the band gap in the parent C_{74}, leading to higher chemical stability.

The molecular structure of this fluorinated derivative was deduced based on ^{19}F NMR data and DFT calculations, as shown in Fig. 1.9 [45,50]. Interestingly, in the proposed structure, six isolated benzenoid moieties are formed due to peculiar contiguous arrangements of fluorine atoms on the C_{74} cage. This proposed most favorable addition pattern in $C_{74}F_{38}$ is a manifestation of the role of aromatic substructures on the carbon core as stabilizing factors in such derivatives, as was also noted for the addition patterns of C_{60} and C_{70} fluorides.

1.3 Electronic Properties

One of the most prominent properties of fullerenes that was immediately recognized, even before fullerenes were isolated as bulk materials is their high electron affinity (EA) and hence the ability to form stable negative ions [51,52]. In the solid state, they are characterized as n-type semiconductors, and have shown some of the highest charge mobilities in organic transistors [53]. Organic molecules with high EA are n-type semiconductors, in which electrons are the main type of charge carriers, whereas for p-type semiconductors, holes are the majority charge carriers.

One of the principles of the molecular design of n-type organic semiconductors is introduction of electron-withdrawing groups in the organic molecule structures. Electronegative atoms such as chlorine, fluorine, or bromine are often used to functionalize organic molecules to enhance their EA. For example, perfluorinated pentacene has been synthesized recently, and it was shown to be a stronger electron acceptor than pentacene itself, and while the parent pentacene is a p-type semiconductor, its perfluorinated analog is an n-type semiconductor with the electron mobility of 0.22 $cm^2/(V\ s)$ [54]. In the case of fullerenes, even underivatized fullerenes are sufficiently strong electron acceptors, with C_{60} having a gas-phase EA = 2.683(18) eV [55], higher fullerenes tend to increase their acceptor strength with the carbon cage size. When chemical functionalization of a fullerene molecule takes place, one or more double bonds get converted into single bonds, and carbon atoms, which form new covalent bonds, get transformed from sp^2-hybridized carbon atoms into sp^3-hybridized carbon atoms. This transformation releases the cage strain, on the one hand, and decreases the fullerene π system, on the other hand. For example, in the case of mono-cycloaddition, the simplest functionalization, the 60-π-electron system of C_{60} becomes a 58-π-electron system. If addition of electron-donating functional groups occurs, the EA of the derivative predictably decreases compared to that of the parent fullerene, and if more such groups are added to the cage, an even lower EA is expected. One well-known example of such a fullerene is a mono-cycloadduct phenyl-C_{61}-butyric acid methyl ester, $PC_{61}BM$, which is the most common acceptor used in organic photovoltaics. The reduction potential of mono $PC_{61}BM$ was measured to be c. 90 mV less positive than that of C_{60}, whereas the bis adduct has an even larger negative shift of 190 mV with respect to C_{60} [56]. If electron-withdrawing groups are added to the fullerene core, it is reasonable to expect the EA to increase, even though the π system shrinks. Indeed, numerous calculations performed for fullerene derivatives established enhanced electron acceptor strength for molecules bearing electron-withdrawing groups such as F, Cl, CN, CF_3, NO_2, and others, ie, they have lower-lying LUMO and higher EA than the parent fullerene. For example, for the molecules 1,7-$C_{60}X_2$, where X is an electron-withdrawing group, the following order of the acceptor strength was determined: X = CN > NO_2 > F > CF_3 [57].

Interestingly, fluorination appears to have a more drastic effect on the electronic properties of fullerenes, than trifluoromethylation, ie, the addition of CF_3 groups. Some of the best electron acceptors among trifluoromethyl derivatives of C_{60} have a positive shift of 0.5−0.7 V in $E_{1/2}$ values [57], whereas the strongest fluorofullerene acceptor, $C_{60}F_{48}$ has a positive shift of 1.38 V compared to the parent fullerene. The situation is opposite for the fluorinated and trifluoromethylated polycyclic aromatic hydrocarbons, such as polyacenes, for which trifluoromethylation causes greater improvements in acceptor strength. This difference can be rationalized in terms of the nature of C−F bonding in both types of molecular substrates. In the case of fluorinated PAHs, fluorine atoms are bonded to sp^2 carbon atoms, and the back π-donation decreases overall electron-withdrawing effect, whereas in fluorofullerenes the fluorine atoms are bonded to sp^3 carbon atoms, and they are not directly connected to the remaining π system of the fullerene, so the back π-donation does not have a detrimental effect on electron-accepting property.

1.3.1 Electron Affinity in the Gas Phase

EA of a molecule A in the gas phase is defined as the energy released upon the attachment of an electron to A to form the molecular anion, A^-, via the reaction:

$$A + e = A^-, \quad EA = \Delta H_0^0 \tag{1.1}$$

If EA is negative, the anion is thermodynamically unstable, whereas the higher the positive EA value the stronger electron acceptor the molecule is. The most electronegative atoms Cl and F have EA values of 3.60 and 3.39 eV, respectively. The fullerenes C_{60} and C_{70} have EA values of 2.683(8) eV [55], and 2.765(10), respectively [58], as determined by low-temperature photoelectron spectroscopy (LT-PES), the best experimental technique for gas-phase EA measurements. EA of a fluorofullerene, $C_{60}F_{48}$ was first reported as 4.06 ± 0.30 eV in 1994 [12]. This value remains the only experimentally derived gas phase EA of this molecule, and thus it is the only cited value in the literature. Noteworthy, the experimental error is 30 times larger than that cited above for C_{60} and C_{70}. In fact, 4.06 eV is not a directly measured energy value for the electron detachment from the $C_{60}F_{48}^-$ anion but rather an estimate derived by the bracketing between the two other molecules with known electron affinities.

In the Fourier transform ion cyclotron mass spectrometry (FTMS) experiment, $C_{60}F_{48}$ species were brought in contact with NO_3^- anions generated from ammonium nitrate in the gas phase. This led to the appearance of $C_{60}F_{48}^-$ and disappearance of the NO_3^- anions, indicating that electron exchange occurred, as shown in reaction (1.2). The literature value of $EA(NO_3) = 3.92$ eV was thus taken as a lower bracket for $EA(C_{60}F_{48})$. Then a gas-phase species with the higher EA, CF_3COO, which was generated from trifluoroacetic acid, was used in a similar experiment, and the electron exchange with $C_{60}F_{48}$ was not observed. Thus, the value of $EA(CF_3COO) = 4.20$ eV was taken as a higher bracketing estimate, so $3.92 < EA(C_{60}F_{48}) < 4.2$ eV. A control experiment with ReO_4^- anions ($EA(ReO_4) = 4.46$ eV) confirmed absence of charge transfer with $C_{60}F_{48}$. The average of the lower and upper bounds was reported as $EA(C_{60}F_{48}) = 4.06 \pm 0.30$ eV.

$$C_{60}F_{48} + NO_3^- = C_{60}F_{48}^- + NO_3 \tag{1.2}$$

Similar experiments were carried out with the gas-phase species $C_{60}F_{44}$, $C_{60}F_{46}$, $C_{70}F_{52}$, and $C_{70}F_{54}$ and their charge transfer behavior was tested within the same brackets, their EAs were thus indistinguishable from $EA(C_{60}F_{48})$ [59]. An experiment involving a reaction of the direct electron exchange between the neutral $C_{60}F_n$ and $C_{70}F_x$ anions was carried out to reveal that no electron exchange occurred, implying that $EA(C_{70}F_{52(54)}) > EA(C_{60}F_{44(46)})$, the difference could not be quantified, but it is probably no greater than a few 100 meV. The molecular anion $C_{60}F_{48}^-$ most likely has a similar structure as the neutral parent, single-isomer D_3-$C_{60}F_{48}$, and thus the EA can be assigned to a well-defined structure. At the same time, the other fluorofullerene anions in this study are most likely mixtures of multiple isomers, and therefore, estimated EA brackets represent an average for all the species with the same molecular mass. Considering the large uncertainty in the bracketing experiment, possible differences between different compositions, eg, $C_{60}F_{44}$ and $C_{60}F_{48}$, could not be determined, let alone variations in the electronic properties of different isomers of the same composition.

A LT-PES study was carried out with three fluorofullerenes whose molecular structures are known and which differ in fluorine content, ie, $C_{60}F_{18}$, $C_{60}F_{36}$, and $C_{60}F_{48}$ [43]. To generate anions in the electrospray ion source and to determine electron photodetachment energy, fresh solutions of fluorofullerenes mixed with the donor compound, tetrakis(dimethylamino)ethylene TDAE, were used. Such solutions contained only fluorofullerene anions with the odd fluorine content, $C_{60}F_{2n-1}^-$, due to partial defluorination, a phenomenon observed in other mass spectrometry studies [43,60,61]. The anion $C_{60}F_{47}^-$ had the highest fluorine content, and it was generated from $C_{60}F_{48}$, whereas the anion with the lowest fluorine content $C_{60}F_{17}^-$ was produced in the solution of $C_{60}F_{18}$ and TDAE. In these experiments, the EA values for the series of open-shell fluorofullerenes were determined with a high precision typical for LT-PES, which allowed one to measure the difference even for the species separated by two fluorine atoms in composition. The graph in Fig. 1.10 reveals a linear

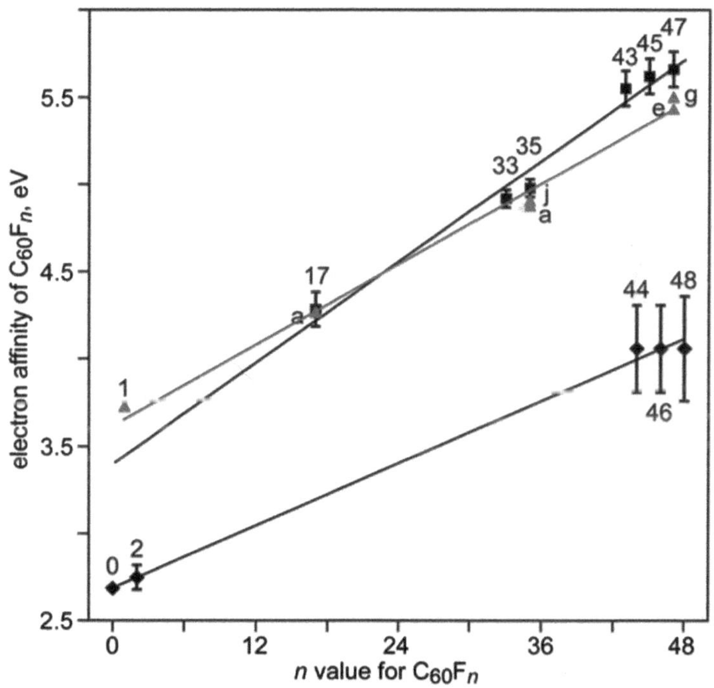

Figure 1.10 Plot of gas-phase electron affinity of fluorofullerenes as a function of fluorine content per molecule. *Black solid squares* are EAs determined by LT-PES. Solid red triangles (gray in print versions) are DFT-calculated EAs for the most probable structures of fluorofullerene radicals. *Solid black rhombs* are gas-phase EA data obtained either by FTMS bracketing method [59], or by Knudsen cell mass spectrometry [63]. *DFT*, density functional theory; *EA*, electron affinity; *FTMS*, fourier transform ion cyclotron mass spectrometry; *LT-PES*, low-temperature photoelectron spectroscopy.

correlation between EA and fluorine content in the studied fluorofullerenes. The absolute EA values of the $C_{60}F_{2n-1}$ are very high, reaching 5.66(1) eV for $C_{60}F_{47}$, this is not unexpected for open-shell species; the DFT calculations predict the EA difference of c.1 eV between the closed-shell and open-shell fluorofullerene species [62]. It is therefore possible that the current EA estimate of 4.06 eV for closed-shell $C_{60}F_{48}$ is somewhat underestimated. The local density functional calculation of EA($C_{60}F_{48}$) yielded a value of 5.07 eV, an overestimate by at least 15–25% [59].

Gas-phase EAs of fluorofullerenes with low fluorine content were also measured by the Knudsen cell mass spectrometry technique [63]. In this method, an effusion cell is loaded with the compounds of interest, and when thermal equilibrium is reached between the neutral and thermally generated charged species, the relative pressures of the neutral and ionic species are measured in the experimentally available temperature interval.

A series of measurements of equilibrium constants of the gas-phase electron exchange reactions were performed for mono- and difluorinated C_{60} and difluorinated C_{70} fullerenes.

$$C_{60}F_2^- + C_{60} = C_{60}^- + C_{60}F_2 \quad (1.3)$$

Fullerene C_{60} was used as the standard due to its well-known EA of 2.683(8) eV [55]. The EAs of difluorofullerenes of C_{60} and C_{70} are slightly higher than the EAs of parent fullerenes: EA($C_{60}F_2$) = 2.76 eV and EA($C_{70}F_2$) = 2.82 eV (these values were recalculated from Ref. [55] using most accurate EA(C_{60}) [55]). The EA estimates for $C_{60}F$ and $C_{70}F$ were reported as >2.93 and >2.99 eV, respectively, in accordance with the open-shell nature of these species.

In a different Knudsen cell mass spectrometry experiment, a mixture of different fullerenes, ranging from C_{60} to C_{84}, was loaded in the effusion cell with potassium fluoride together with a separate platinum ampoule filled with $C_{60}F_{46}$, which was used as the fluorinating agent. Ramping the temperature led to the reaction between the fluorinating reagent and fullerenes yielding difluorinated fullerenes. A representative mass spectrum of the thermally generated negative ions is shown in the left panel of Fig. 1.11. Measuring relative ion intensities of the thermally generated negative ions, for the C_n^- species of interest and a reference compound (ie, C_{60}), enabled determination of the equilibrium constants and Gibbs energies for several fluorine exchange reactions, as shown below.

$$C_n^- + C_{60}F_2^- = C_{60}^- + C_nF_2^- \quad (1.4)$$

Applying second- or third-thermodynamics law, the enthalpies of the reaction (1.4) were determined. Using a known value of fluorine affinity (FA) of C_{60}, the FAs for higher fullerenes were determined. FA is defined as an enthalpy of reaction (1.5), and it characterizes the strength of C–F bonds in the difluorofullerene anions:

$$C_{2n}^- + 2F = C_{2n}F_2^- \quad (1.5)$$

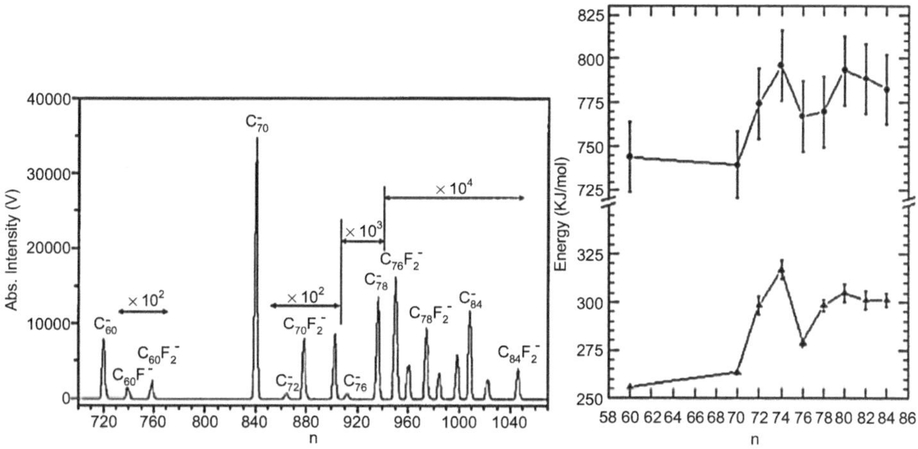

Figure 1.11 Left: A representative mass-spectrum of the fullerene-$C_{60}F_{46}$ system at $T = 975$ K showing thermally generated anions of fullerenes and their difluorinated counterparts. Right: Plot of the fluorine affinity (FA) of fullerene anions (*solid circles*) and the electron affinity (EA) of fullerenes (*solid triangles*) as a function of the number of carbon atoms. The *vertical lines* show total uncertainties in FA and EA values [63].

Fig. 1.11 (right panel) depicts a graph demonstrating a correlation between the change in the EA of the parent fullerenes and the FA of their anions, both determined from gas-phase equilibria studies by Knudsen cell mass spectrometry. Notably, the change in these electronic properties is not linear with the cage size. A sharp increase in the EA of small-band-gap fullerene C_{74} compared to its neighbors is responsible for such a nonlinear trend. This anomalous behavior of C_{74} is consistent with the theoretical calculations that predict a lower LUMO energy for the small-band-gap fullerenes such as C_{74} compared to the more stable wide-band-gap fullerenes. While small-band-gap fullerenes are practically impossible to isolate from the soot due to their high reactivity, their derivatives were shown to be stabilized by chemical derivatization with the electron-withdrawing groups such as fluorine atoms [45], or trifluoromethyl groups [49], as mentioned in the previous section.

Attachment of the two electrons to highly fluorinated fullerenes in the gas phase has been observed in several studies [12,60,64,65]. Furthermore, the measurements of the second EAs for a number of fluorofullerenes showed that doubly charged anions are remarkably stable: EA($C_{60}F_{34}^-$) = 2.50(5) [43] (2.4(1) eV) [66] and EA($C_{60}F_{46}^-$) = 3.30(5) eV [43] (3.2(1) eV) [66]. Large cross sections for electron capture by fluorofullerene anions were found to depend on the fluorine content [64]. Presence of repulsive charges on a single molecule (in the absence of solvent molecules which can stabilize multiply charged anions in solution) explains the fact that multiply charged anions are very rarely observed in the gas phase. In the case of fluorofullerenes, stabilization of the two negative charges is due to the large size of the molecules and the electronegativity of fluorine atoms. Therefore, even triply charged anions, $C_{60}F_{47}^{3-}$ were experimentally observed in the two independent mass spectrometry studies, despite the prediction that the third EA of $C_{60}F_{47}$ has a negative value [65,67].

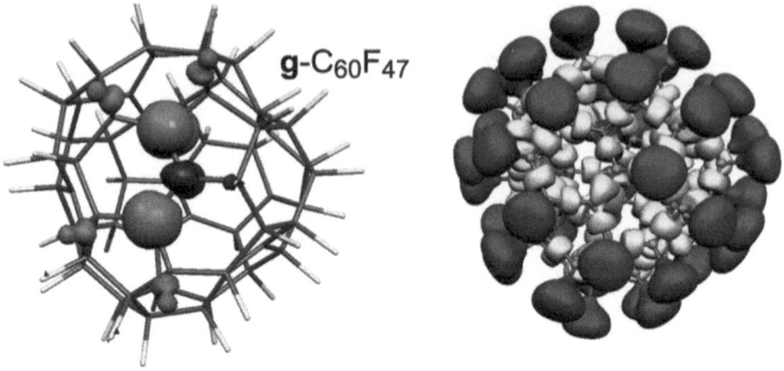

Figure 1.12 Left: The spin density of the g-$C_{60}F_{47}$ radical computed at the B3LYP/6-311G*//PBE/TZ2P level with an isovalue of 0.005 au. Right: The ELF of the g-$C_{60}F_{47}^-$ anion with an isovalue of 0.72 (color codes: yellow (very light gray in print versions), disynaptic V(C,C) basins; dark green (dark gray in print versions), disynaptic V(C,F) and monosynaptic V(F) basins; light green (light gray in print versions), core C(C) basins; red (gray in print versions), monosynaptic valence V(C) basins) [43].

An intriguing insight into the stability of negative ions of fluorofullerenes was obtained in the theoretical study of the excess electron delocalization [43]. For $C_{60}F_{47}^-$, which was found to be particularly stable in the gas phase and in solution, upon chemical reduction, electron delocalization was visualized as the spatial distribution of the electron localization function (ELF) for the anion (Fig. 1.12, right) or as the spin density distribution in the corresponding neutral radical $C_{60}F_{47}$ (Fig. 1.12, left). The "red (gray in print versions)" monosynaptic valence V(C) basin in $C_{60}F_{47}^-$, with a population of 0.98 e^-, was found inside the fullerene cage, in contrast to fluorofullerene anions with lower fluorine content, where V(C) is located outside the cage. The unusual concave regions of $C_{60}F_{47}^-$ make it possible for the "lone pair" of electrons to locate inside the carbon cage and shield the negative charge thereby providing enhanced chemical stability of the anion, not only in the gas phase but also in solution, which was confirmed experimentally [43]. A similar concept has been described in a theoretical study of hypothetical structures $e_2@C_{60}F_{60}$ with two excess electrons encapsulated inside the fullerene cage [68].

1.3.2 Reduction Potentials in Solution

Cyclic voltammetry is used for measurements of electrochemical properties of various materials in solution or in the solid state (as thin films deposited on an electrode). It is a relatively simple and straightforward method that allows one to determine electrochemical potentials relative to an internal standard, as long as the analytes exhibit reversible redox behavior. Results of cyclic voltammetry are the primary source of the information about frontier orbital energy levels of different materials for organic electronics or for battery applications.

Fullerene C_{60} exhibits up to six reversible reductions. Its first reduction potential, $E_{1/2}$ was determined to be −1.06 V versus $FeCp_2^{+/0}$ which indicates that it is a strong electron acceptor, in agreement with the gas-phase data on its EA. Fullerene C_{70} has a slightly more positive $E_{1/2}$ relative to C_{60}, and therefore is a stronger acceptor in solution [69].

Most initial cyclic voltammetry studies reported data for fluorofullerene mixtures with varying average compositions, which were estimated either by elemental analysis or mass spectrometry [24,70,71]. These results should be interpreted with caution due to uncertainty in the molecular structures of the components of such mixtures. Nevertheless, some similarities in the electrochemical behavior of such fluorofullerene samples are worth mentioning: (1) a significant positive shift in the reduction potential compared to the underivatized fullerene, which indicated stronger acceptor properties upon fluorination; (2) electrochemical reduction of fluorofullerenes was not a reversible process, in contrast to the parent fullerene, regardless of scan rate; (3) partial loss of fluorine atoms from the fullerene cage was believed to be responsible for such behavior, in agreement with earlier work. More pronounced acceptor properties of fluorofullerenes compared to the parent fullerene attracted attention of researchers and led to the studies of fluorofullerenes as electrodes for primary solid-state lithium batteries [72].

Several cyclic voltammetry studies were carried out with purified fluorofullerenes of known composition and molecular structure (Table 1.1). The reversible first electrochemical reduction of D_3-$C_{60}F_{48}$ was observed, and a positive shift of 1.38 V in $E_{1/2}$ value relative to C_{60} was determined. Formation of the molecular anion $C_{60}F_{48}^-$ was also observed when $C_{60}F_{48}$ was chemically reduced with I^- to produce I_2. Subsequent reduction steps, either electrochemical, or using chemical reagents, were irreversible, and loss of fluorine as fluoride anions was detected [73].

Electron transfer reactions between the two-electron donor 10,10′-dimethyl-9,9′,10,10′-tetrahydro-9,9′-biacridine $(AcrH)_2$ and the fluorofullerenes $C_{60}F_{18}$ and $C_{60}F_{36}$ were studied [74]. The electron transfer was followed by the cleavage of C−F bond in $C_{60}F_{18}$ leading to the anionic species $C_{60}F_{17}^-$ and $AcrH^+$. The E^0_{red} of $C_{60}F_{18}$ was estimated as 0.47 V more positive than C_{60} [74].

Comparison of the electrochemical properties of the three isomers of $C_{60}F_{36}$ can be found in Ref. [75]. Despite very minor differences in the addition patterns of the $C_{60}F_{36}$ isomers (see their Schlegel diagrams in Fig. 1.2), their electrochemical behavior was different, and reduction potentials differed by more than 100 mV (Table 1.1). The C_3 isomer exhibited reversible reduction, whereas the other two isomers showed some reversibility only at scan rates exceeding 50 V/s. Similar to the other reports, multiple reductions were accompanied by the partial loss of fluorine atoms from the carbon cage.

Cyclic voltammetry of a $C_{60}F_{36}$ sample (a mixture of three isomers) deposited on the ITO substrate from o-dichlorobenzene solution was reported in Ref. [76]. The observed reduction was irreversible, and the onset potential reported was 0.8 V more positive than that of $PC_{61}BM$ (measured by the authors under similar conditions), which corresponds to a 0.7 V shift versus C_{60}.

The compounds $C_{60}F_2$ and $C_{60}F_4$ exhibited cyclic voltammograms with irreversible reductions, therefore only cathodic peak potentials E_{pc} were reported (Table 1.1) [77]. At a scan rate of 0.02 V/s, the cathodic peak for both compounds is accompanied by an

Table 1.1 Electrochemical Potentials of Fluorofullerenes

Compound	$E_{1/2}$, or [E_{pc}] V	Reference	Electrolyte (Conc., M)	Scan Rate, V/s	$E_{1/2}$ or [E_{pc}] vs $C_{60}{}^{0/-}$, V	References
$C_{60}F_{48}$	0.79	SCE	$TBA^+PF_6^-$ (0.08)	0.2	1.38	[73]
C_{60}	−0.59	SCE	$TBA^+PF_6^-$ (0.08)	0.2	0.0	[73]
$C_{60}F_{36}$	−0.05	Ag/Ag^+	$TBA^+PF_6^-$ (0.1)	0.02	0.83	[71]
$C_{60}F_{46}$	0.51	Ag/Ag^+	$TBA^+PF_6^-$ (0.1)	0.02	1.36	[71]
C_{60}	−0.88	Ag/Ag^+	$TBA^+PF_6^-$ (0.1)	0.02	0.0	[71]
$C_{70}F_{54}$	0.72	Ag/Ag^+	$TBA^+PF_6^-$ (0.1)	0.02	1.57	[71]
C_{70}	−0.85	Ag/Ag^+	$TBA^+PF_6^-$ (0.1)	0.02	0.05	[71]
$C_{60}F_{36}$	−0.18	Ag/Ag^+	$TBA^+PF_6^-$ (ACN)	0.1	0.7	[76]
C_1-$C_{60}F_{36}$	0.07	SCE	TBAH (0.05)	10	0.66	[75]
C_3-$C_{60}F_{46}$	0.12	SCE	TBAH (0.05)	1.0	0.71	[75]
T-$C_{60}F_{36}$	0.04	SCE	TBAH (0.05)	50	0.63	[75]
$C_{60}F_{48}$	0.78	SCE	TBAH (0.05)	0.5	1.37	[75]
$C_{60}F_{18}$	[−0.33]	SCE	TBAH (0.05)	5.0	[0.26]	[75]
$C_{60}F_{18}O$	[−0.28]	SCE	TBAH (0.05)	0.5	[0.31]	[75]
$C_{60}F_2$	[0.04]	C_{60}/C_{60}^-	$TBA^+BF_4^-$ (0.1)	0.02	[0.04]	[77]
$C_{60}F_4$	[0.08]	C_{60}/C_{60}^-	$TBA^+BF_4^-$ (0.1)	0.04	[0.08]	[77]
C_{60}	0.0	C_{60}/C_{60}^-	$TBA^+BF_4^-$ (0.1)	0.02	0.0	[77]

additional cathodically shifted peak which appeared to be reversible in most experiments. Similar patterns were observed at all scan rates probed (>5 V/s) except that the intensities of the cathodically shifted peaks varied with scan rate. NIR spectra for $C_{60}F_2$ and $C_{60}F_4$ were recorded at potentials more negative than 0 V versus $C_{60}{}^{0/-}$, which yielded a band at 1080 nm characteristic of the $C_{60}{}^{-}$ radical anion. Additionally, an absorption band at 940 nm, characteristic of the $C_{60}{}^{2-}$ dianion, was observed at more negative potentials. Taken together, these results pointed to the occurrence of electrochemical defluorinations via an ECE (electrochemical–chemical–electrochemical) mechanism, which might involve cleavage of one or more C–F bonds. Such a behavior was also observed for fluorinated fullerenes with higher fluorine content, such as $C_{60}F_{18}$, $C_{60}F_{36}$, and others. Fig. 1.13 presents a plot of available reduction potentials of fluorofullerenes as a function of fluorine content per molecule. Taking into account that for some compounds ($C_{60}F_{18}$, $C_{60}F_2$, and $C_{60}F_4$) the $E_{1/2}$ values could not be determined because of irreversibility (and only E_{pc} values are available for comparison), the general trend clearly points to the correlation between the acceptor strength and the degree of fluorination in fluorofullerenes.

1.3.3 Electronic Properties in the Solid State

Knowledge of the electronic properties in the solid state is a prerequisite for the design of successful material for electronic devices. For example, an optimized offset of the HOMO of the donor material and LUMO of the acceptor molecular material determines, among many other factors, efficient charge transfer and performance of the

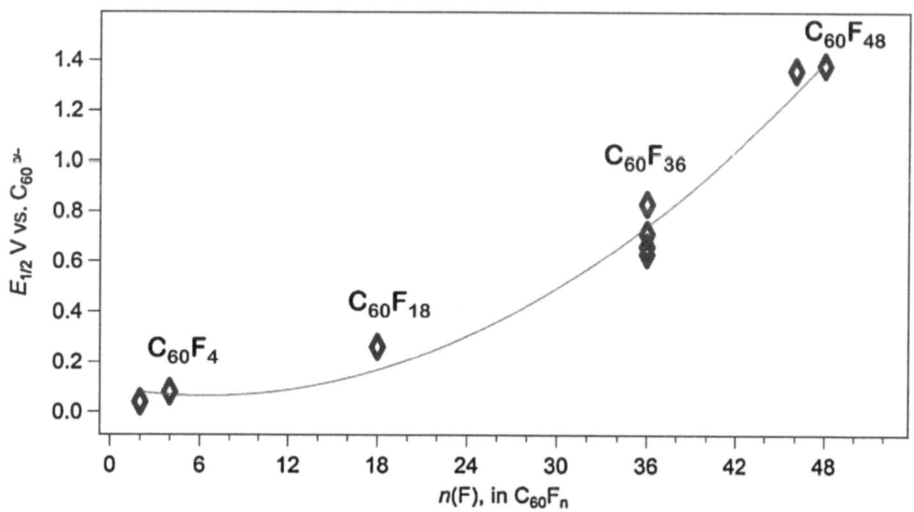

Figure 1.13 Plot of reduction potentials of fluorofullerenes as a function of fluorine content per molecule in $C_{60}F_n$. The *red line* (gray in print versions) is only placed for easy eye guide. Several data points in the case of $C_{60}F_{36}$ are present on the graph because measurements were done for several isomers of $C_{60}F_{36}$.

organic photovoltaic device, or doping efficiency in molecular electronics. Current practice in energy level engineering in molecular electronics utilizes data on the reduction potentials from cyclic voltammetry; the terms E(LUMO) and EA in the solid state are both used in the literature. Such estimates can only be considered as crude approximations as they do not take into account intermolecular interactions in the solid phase. Nevertheless, in the absence of more direct and reliable data for the solid state, the cyclic voltammetry-derived E(LUMO)s are commonly used in the literature, and some data are available for fluorofullerenes. For example, for $C_{60}F_{36}$ E(LUMO) values of 4.4 eV [76] and 4.1 eV [78] were reported. The difference most likely originate from the different conversion methods, a typical issue in the organic semiconductor field [79,80]. This difference, however, does not reflect the difference in electronic properties of $C_{60}F_{36}$ isomers which was found in Ref. [75] as the samples used in both Refs [76,78] were mixtures of isomers. Additional complication with using cyclic voltammetry data of fluorofullerenes for solid-state EA estimations stems from the fact that some fluorofullerenes, including $C_{60}F_{36}$, exhibit irreversible electrochemical behavior. In these cases the onsets of reduction potentials have to be used instead of $E_{1/2}$ values, which make such estimates less reliable.

Another approach involves an indirect method based on estimation of the solid-state EAs as a sum of ionization potential (from PES) and the transport gap. Sometimes the optical gap instead of the transport gap is inaccurately used; the latter can be larger by 0.5−1.5 eV, and hence EA overestimation may occur if the optical gap is used. The most accurate estimate of the transport gap is from combined PES and inverse photoemission spectroscopy (IPES) measurements. If such data are unavailable, the transport gap can be estimated from the empirically derived linear correlation between the transport and optical gaps for organic semicondutors. Such linear fit with a slope of 1.39 ± 0.15 and intercept of 0.46 ± 0.38 eV, and with a correlation coefficient of 0.80 was obtained in Ref. [81] Experimentally measured data for the optical gap (from UV−Vis spectroscopy) and transport gap (difference in the ionization potential and EA in the solid state from PES and IPES, respectively) for 22 organic compounds were used to derive the correlation. This approach was applied to estimate the solid state EA for $C_{60}F_{36}$ [82]. An optical gap of 2.5 ± 0.1 eV was determined from the photoluminescence spectrum of a thin film of $C_{60}F_{36}$. Using the above linear fit, the transport gap was calculated as 3.0 ± 0.3 eV. With the ionization energy from PES of 8.38 eV, this yielded a solid-state EA estimate as 5.38 ± 0.3 eV, which is significantly larger value than E(LUMO) derived from cyclic voltammetry [76].

The E(LUMO) value of 5.0 eV for $C_{60}F_{48}$, a fluorofullerene with higher fluorine content, was estimated in Ref. [83] by adding polarization energy (assumed to be 0.95 eV, the same as that for C_{60}) to the gas-phase value of 4.06 eV. This value is lower than a solid-state electron affinity (EA_s) estimate for $C_{60}F_{36}$, which is reverse to their reduction potentials relations. The ionization threshold energy (HOMO level) determined for fluorofullerene $C_{60}F_{36}$ in Ref. [83] was 8.2 eV, lower than reported in Ref. [82], and hence it would yield a lower solid-state EA_s of 5.2 eV, if the same transport gap is used as in Ref. [82]. It was suggested that this value can be even lower, ie, close to 5.0 eV, as based on the comparison of the p-doping effect of $C_{60}F_{36}$ on the host materials with varying ionization potentials (see further discussion in the next section) [84].

The most direct experimental method to determine the solid-state EA_S of a compound or a material is IPES. In this technique, incident electrons are relaxed to unoccupied levels, and analysis of the emitted photons allows to probe LUMO levels of the materials. Currently, solid-state EA values are known for various organic semiconductors, including several fullerenes and a few of their derivatives, no such data are available yet for fluorofullerenes. When the IPES method is applied to organic materials, the problem of radiation damage may jeopardize results [85−87]. It is particularly relevant to the fluorofullerenes as recent studies have shown their instability toward irradiation with intense electron beams, or high-energy photons [88,89]. Researchers observed also surface charging due to dielectric nature of fluorofullerenes, and reaction of the substrates with fluorine atoms from fluorofullerenes that lead to the covalent bond formation, eg, Si−F bonds on the silicon surface, or C−F bonds on the graphene [90,91]. The most recent advancement in IPES methodology includes the use of ultralow energy beam that allows one to minimize radiation degradation of organic materials, and thus to reliably determine solid-state EA [92]. Such experimental studies are highly desirable for fluorofullerenes, because (1) some of them exhibit irreversible redox behavior in solution, and hence accurate reduction potential are not possible to determine, with a few exceptions; (2) use of low-energy beam instead of intense electron beam may prevent degradation via partial defluorination in some fluorofullerenes; (3) knowledge on the solid-state electronic properties of fluorofullerenes will aid in designing highly functioning electronic devices, and will help in interpretation of the experimental findings concerning fluorofullerene performance in such devices.

1.4 Applications

1.4.1 Surface Doping of Diamond, Silicon, and Graphene

Artificial diamond materials prepared via relatively inexpensive CVD method have been considered promising components for the next-generation electronics due to a number of unique surface properties that silicon and germanium do not possess. At the same time, undoped diamond exhibits much lower electrical conductivity, and significant research efforts to improve it by doping have been made in the last decade, following the discovery of increased surface conductivity of hydrogen-terminated diamond compared to bare diamond surface [93]. Interaction of hydrogen-terminated diamond surface with hydronium ions that ubiquitously physisorb upon exposure to air causes the phenomenon of surface conductivity of diamond [94]. Besides environmentally induced surface conductivity, the concepts and models of molecular surface doping have been developed, which are based on the use of strong electron acceptors as p-dopants for hydrogen-terminated diamond. Fluorofullerenes were chosen as p-dopants for hydrogen-terminated diamond because of their high EA, and thus their potency to accept electron from the diamond surface to form holes and enhance p-type conductivity.

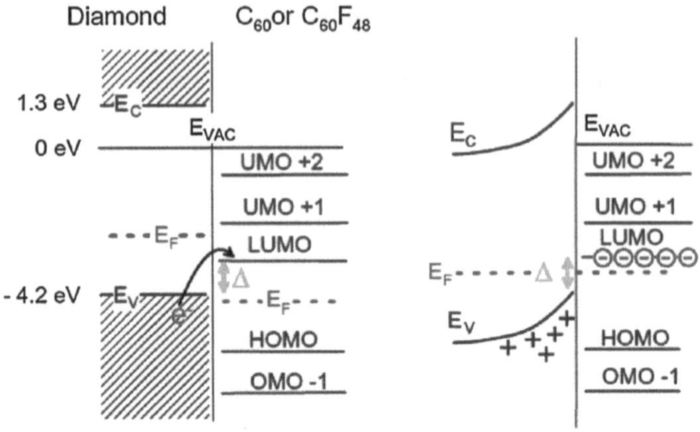

Figure 1.14 Left: The energy diagram of hydrogenated diamond and molecular adsorbate in nonequilibrium state, when E_F, Fermi levels of diamond and molecular adsorbate deviate. Right: The energy diagram of hydrogenated diamond and molecular adsorbate in equilibrium state when Fermi levels of diamond and adsorbate are equalized. E_V is the diamond valence band edge, E_C is diamond conduction band edge; OMO is the occupied molecular orbital; UMO are unoccupied molecular orbitals of the adsorbate molecules relative to the vacuum level. HOMO is the highest occupied molecular orbital, LUMO is the lowest unoccupied molecular orbital. The activation energy Δ is energy required for electrons to be excited from the diamond to the adsorbate system.
Adapted from P. Strobel, M. Riedel, J. Ristein, L. Ley, O. Boltalina, Surface transfer doping of diamond by fullerene, Diamond Relat. Mater. 14 (2005) 451–458.

For the electron transfer between the two materials to occur, the EA of the acceptor should be higher than the ionization potential of the donor (ie, the activation energy to transfer electrons from diamond to the dopant surface need to be overcome, as shown in the energy diagram Fig. 1.14) [95]. Hydrogenated diamond has the lowest ionization energy (IP = 4.2 eV) of any semiconductor, and therefore it is easier to find an acceptor with matching high EA. Interestingly, even underivatized C_{60} was shown to have p-doping effect when several monolayers are deposited on the surface of hydrogenated diamond. Calculations have shown that in going from the gas phase (EA = 2.683 eV [55]) to the bulk van der Waals fullerite solid, its EA increases by more than 1.3 eV, thus the activation energy of the surface electron transfer decreases correspondingly [96,97]. It was experimentally demonstrated [95] and theoretically supported [96] that use of even stronger acceptors than C_{60}, eg, fluorinated derivative $C_{60}F_{36}$, could lower the activation energy for charge transfer and induce greater surface conductivity. Due to the increased EA of $C_{60}F_{36}$ compared to C_{60}, it was predicted that lower surface coverage levels would be needed to achieve comparable doping effect [96].

Indeed, when fluorofullerenes with varying fluorine content were used as surface dopants for hydrogenated diamond, the effects on the diamond conductivity were much greater than that observed for underivatized C_{60}. As shown in Fig. 1.15, left panel, the greatest increase in surface conductivity was achieved for the most fluorinated dopant, $C_{60}F_{48}$. Somewhat smaller effects were observed for $C_{60}F_{36}$ and $C_{60}F_{18}$,

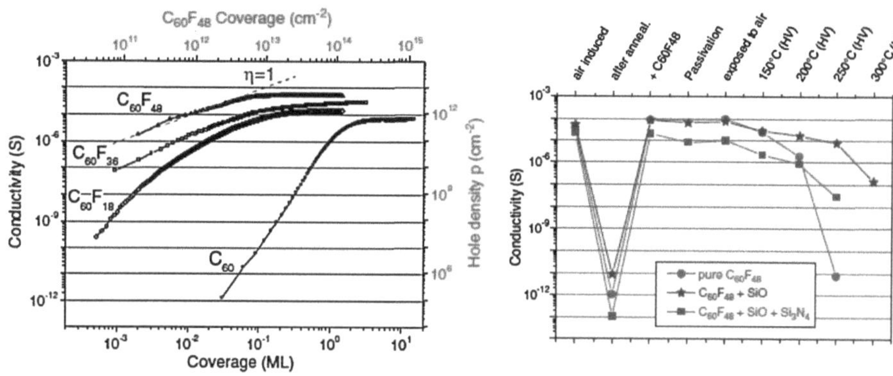

Figure 1.15 Left panel: Comparison of the C_{60}, $C_{60}F_{18}$, $C_{60}F_{36}$ and $C_{60}F_{48}$ induced surface conductivities on hydrogen terminated diamond. The top scale is valid for $C_{60}F_{48}$ only. Right panel: Temperature/treatment profile of surface conductivity of $C_{60}F_{48}$-doped hydrogenated diamond [95].

in accordance with their relative acceptor strengths. It is also evident from the graph, that in the case of the fluorofullerenes, significantly lower doping concentrations are needed, than in the case of C_{60}, to observe drastic increase in surface conductivity. For example, in the case of $C_{60}F_{48}$, even at the level of 2×10^{-3} of a monolayer, the diamond conductance of a few microsiemens was measured. In contrast, the C_{60} dopant needs to be at least at the 10% of a monolayer to observe appreciable conductance increase. For all three fluorofullerene dopants, the saturation is reached when the value of the common Fermi level of diamond and dopant layer becomes lower than the dopant LUMO level [98]. The dopant efficiency of unity estimated for $C_{60}F_{48}$, implies that each dopant molecule extracts one electron from the diamond valence band.

An insight into diamond–fullerene interface formed by the vapor-deposited $C_{60}F_{36}$ onto bare and hydrogenated diamond was obtained from the core-level photoemission spectroscopy and high-resolution electron energy loss spectroscopy [99]. When adsorbed on the hydrogen-terminated diamond, submonolayers of $C_{60}F_{36}$ cause a significant upward band bending of 0.8 eV in the surface region of diamond, which is indicative of the hole accumulation due to transfer doping as shown by energy diagram in Fig. 1.14. When bare diamond was used (with the much higher ionization energy of c. 5.5 eV), deposition of $C_{60}F_{36}$ did not cause such a significant effect, only 0.3 eV binding energy shift in C1s bulk diamond peak was observed; however, this change could not be firmly assigned to transfer doping effect. Notably, direct spectroscopic evidence of chemical interaction between diamond carbon atoms and fluorine atoms from fluorofullerene was reported; after annealing beyond 300°C a new peak due to C(diamond)–F bonds was detected for both hydrogenated and bare diamond [99].

Thermal stability of surface conductivity is of primary concern for practical applications in electronic devices. For example, air-induced conductivity of diamond sharply drops when a sample is thermally heated in vacuo beyond 60°C. Thermal range of surface conductivity of fluorofullerene-doped diamond has been therefore tested to compare with the air-induced surface conductivity. The fluorofullerene dopants exhibited higher thermal stability, for example, when the $C_{60}F_{48}$-doped diamond was heated to 150°C, surface conductivity only dropped by a factor of 10 (Fig. 1.15, right panel). When it was passivated with a SiO protective film, the thermal stability improved noticeably: after heating for 3 days at 250°C, the conductivity decreased by a factor of 10. The underivatized C_{60} deposited on hydrogenated diamond exhibited a superior stability in the same study, both due to higher sublimation temperature and greater chemical inertness than fluorofullerenes, albeit the surface conductivity was much lower than that induced by fluorofullerenes [95].

Charge-transfer doping, with fluorofullerenes, of a semiconductor material, oxidized silicon, was studied aiming at the formation of ultrashallow junctions on Si [100]. Ultrashallow junctions have been viewed as one of the solutions for leakage current in miniaturized metal oxide semiconductors [101]. The fluorofullerene doping effects on the silicon substrate covered with a 2-nm layer of SiO_2 were detected by the change in conductivity in Si. Deposition of $C_{60}F_{36}$ or $C_{60}F_{48}$ caused charge transfer doping, ie, the accumulation of holes in the interfacial ultrathin inversion layer, manifesting the formation of an ultrashallow junction. The band bending of 0.9 eV was measured for Si surface after deposition of $C_{60}F_{36}$. The observed charge (hole) density induced by $C_{60}F_{36}$ (5.2×10^{12} cm^{-1}) was higher than that caused by $C_{60}F_{48}$ doping, in contrast to the presumably reverse acceptor strength order between the two fluorofullerenes [100]. The LUMO level of $C_{60}F_{36}$ on the oxidized silicon was estimated as 5.6–5.7 eV below vacuum, whereas the LUMO level of $C_{60}F_{48}$ was found to be lower, ie, 5.2 eV. This might imply that $C_{60}F_{48}$ is a weaker surface acceptor than $C_{60}F_{36}$ contrasting the relation for the isolated molecules (via gas-phase EA) and in solution (via reduction potentials). Such inconsistencies revealed by comparison of the performance of different fluorofullerenes as dopants calls for in-depth study of their electronic properties in the solid phase, and in the interfaces with different substrates, so that predictable doping performance can be achieved.

Effective p-doping of epitaxial graphene surface was achieved when $C_{60}F_{48}$ was used as a molecular dopant [102]. Carrier density values were experimentally determined as a function of dopant coverage. The change of the carrier density magnitude and carrier sign from n-type to p-type was observed when 0.8 monolayer $C_{60}F_{48}$ coverage was reached. Such control of electrical properties of the epitaxial graphene may become invaluable tool in the design of doped graphene-based electronic devices.

1.4.2 Bulk p-Doping of Organic Semiconductors in OLEDs and OPVs

Many hole-transporting semiconductors have intrinsically low conductivity [103–105]. Therefore application of high operating voltages is needed, which affects device power efficiency. It has been demonstrated that introduction of the dopants to organic semiconductor materials may help alleviate this deficiency [82,84,106–108]. Molecules with electron affinities higher than the ionization potential of the hole-transporting material are necessary for p-doping so that electrons can flow from the semiconducting host to the dopant. Bulk doping (which is more common than surface doping) can be achieved by introducing molecular defects, halogen atoms, oxygen, or molecules with good electron-accepting properties. Bulk molecular doping is a particularly attractive approach since it (1) offers exceptional tunability of electronic properties, and physical characteristics such as molecular weight, molecular structures, solubility, and volatility, (2) it can be introduced after a semiconductor synthesis, (3) provides control over molecular diffusion throughout the material due to a larger molecular weight relative to atomic dopants, (4) and can be controlled during the device fabrication [109].

It has been shown that effective p-doping with high EA materials reduces charge injection barrier at the contacts because the Fermi level of the hole-transport material is shifted toward the charge carrier level. Accumulation of excessive positive charge in the semiconductor at the interface with the p-dopant occurs, whereas negative charge is built up in the p-dopant material is in contact with the semiconductor. A similar description was used for surface doping as shown on the diagram in Fig. 1.14. This process leads to the desirable outcome of increased charge in the semiconductor, higher conductivity, and lower operating voltage.

Bulk molecular p-doping of semiconductors was introduced several decades ago. Metal oxides are commonly used as p-dopants, and the most common organic p-dopants are TCNQ and its derivatives. Fluorofullerenes attracted attention as bulk p-dopants for semiconductors relatively recently, in 2008 [78], after they have already been proven as efficient surface p-dopants for hydrogenated diamond [95]. Important advantages of fluorofullerenes as bulk dopants include good thermal stability allowing for vapor deposition fabrication, large molecular weight preventing diffusion through the bulk of the organic semiconductor, and relatively low volatility. The latter property is essential as other commonly used p-dopants based on TCNQ and their fluorinated analogs have extremely high volatility (causing severe vacuum chamber contamination), high cost, diffusion in the solid host material over time (F4-TCNQ has molecular weight of 276 vs 1404 of $C_{60}F_{36}$), and as a consequence, short-lifetime devices [82].

In one experimental study, when 10 wt% $C_{60}F_{36}$ was added to P3HT, a common hole-transporting conjugated polymer, a new feature in the absorption spectrum ascribed to charge-polaron absorption was observed [76]. Photoluminescence quenching of P3HT with $C_{60}F_{36}$ was faster and over a longer range than with C_{60}, suggesting more efficient charge transfer, in accordance with higher EA of the fluorofullerene. Luminescence quenching was significant even with 0.1 wt% doping level of $C_{60}F_{36}$. Thin-film transisitors (TFTs) were then fabricated from 0.1 and 10 wt% $C_{60}F_{36}$-doped P3HT to test doping effect on the electrical conductivity and to compare it to that of the undoped P3HT and C_{60}-doped P3HT. A conductivity increase of two orders of magnitude relative to the undoped P3HT TFT was recorded for the 10 wt% $C_{60}F_{36}$-doped TFT. High electrical conductivity is desirable for the hole-injection layers in the organic light-emitting devices (OLEDs), and therefore $C_{60}F_{36}$-doped hole-transporting OLED material was examined by the authors. The solution-processed film consisting of poly-(9-vinylcarbazole) with cinnamate pendant groups, P3HT and $C_{60}F_{36}$, was prepared and the conductivity was measured as a function of time. The initial remarkably high conductivity of $C_{60}F_{36}$-doped layer of nearly 0.1 S/cm was recorded. It dropped by one order of magnitude to 10^{-2} S/cm over the span of 120 h and continuing heating in vacuum at 110°C, this final measured value was still nearly three orders of magnitude higher than that of the commercially used hole-injection layer PEDOT:PSS. Interestingly, when other conjugated polymer (polyTPD, *N,N′*-diphenyl-*N,N′*-bis(3-methylphenyl)-(1,1′-biphenyl)-4,′4-diamine) was tested for doping with $C_{60}F_{36}$ the measured conductivity was three orders of magnitude lower than in the experiment with P3HT. As a possible reason for such a difference in electrical doping, a lower-lying HOMO of polyTPD (5.3 eV) than that of P3HT (5.1 eV) was proposed. The light-emitting diode (LED) was also made using $C_{60}F_{36}$-doped polyTPD hole-transporting layer, with the following LED characteristics: turn-on voltage of 5 V, luminance of 1000 cd/m^2 at 12 V, and efficiency of 5 cd/A. According to the authors, such p-doping was not only beneficial with respect to the improved brightness and turn-on voltage, but it also did not compromise the electron blocking layer function in the devices [78]. Despite the overall modest OLED performance, and unknown lifetime of the device, this work provided the first example of the operational p-doping with fluorofullerenes in organic electronic devices, and paved the way to further explorations.

The p-doping efficiency of F4-TCNQ and $C_{60}F_{36}$ in the hole-transporting layers in OLED and organic photovoltaic cell devices was compared. In the OLED study, identical device architectures were used, except for hole-transporting layers as shown on Fig. 1.16 left panel [82].

With both dopants, highly efficient OLED devices were fabricated, with some characteristics being slightly better for the $C_{60}F_{36}$ dopant than for F4-TCNQ. A particular advantage of using fluorofullerene is its higher thermal stability, lower volatility, and larger molecular weight. These properties are beneficial for operation at elevated temperatures, longer lifetime due to decreased diffusion within the device, and ease of fabrication due to better control over evaporation.

In a comparative study of doping performance by $C_{60}F_{36}$ and F6-TCNNQ, conductivity and Seebeck studies were carried out with two amorphous hole-transporting materials: MeO-TPD (*N,N,N′,N′*-tetrakis 4-methoxyphenyl-benzidine) and BF-DPB (*N,N*-bis (9,9-dimethyl-fluoren-2-yl)-*N,N′*-diphenyl-benzidine) [84]. These semiconductors have different ionization potentials (5.07 eV for MeO-TPD and 5.23 eV for

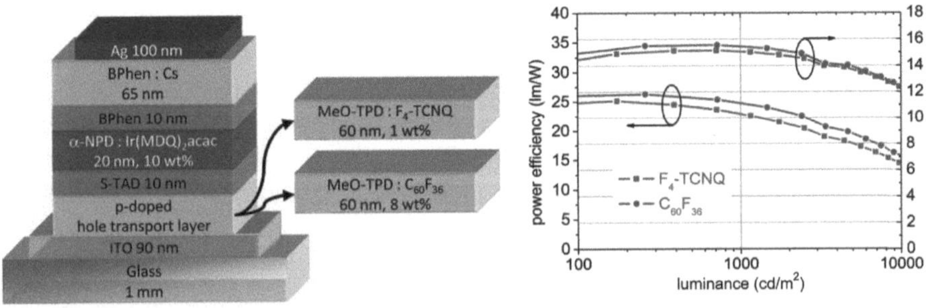

Figure 1.16 Left: OLED stack architecture used for the comparison of the F_4-TCNQ and $C_{60}F_{36}$ dopant performances. Right: Power efficiency and external quantum efficiency of the p-i-n OLEDs with F_4-TCNQ and $C_{60}F_{36}$ dopants [82].

BF-DPB) and both have been used as effective hole-transport materials in OLEDs and OPV (organic photovoltaics). Available cyclic voltammetry-derived data on E(LUMO) values of the dopants indicated that F6-TCNNQ may be a weaker acceptor than $C_{60}F_{36}$ [82]. Therefore the initial assumption was that $C_{60}F_{36}$ can perform better in both host materials due to a better match of the energy levels for charge transfer. Both compounds were efficient dopants and improved the conductivity of the hosts by several orders of magnitude, indicating that charge transfer takes place and hole accumulation in the semiconductor was effective. Annealing at mild temperatures did not have a detrimental effect on conductivity figures of merit. The relative performance of $C_{60}F_{36}$ in the host with a lower ionization potential, MeO-TPD, was much better than in BF-DPB, in accordance with their energy level relation. However, when performance of $C_{60}F_{36}$ was compared to that of F6-TCNNQ, the latter showed a larger effect on conductivity, pointing to a stronger acceptor properties, which contradicted with literature data on their E(LUMO). These observation made the authors to question that the reported E(LUMO) of $C_{60}F_{36}$ could be as high as 5.38 eV [82]. In fact, the authors suggested that it can be as low as 5.0−5.1 eV, ie, smaller than the ionization potential of MeO-TPD [84]. Such inconsistency in the fundamental dopant characteristic of the solid $C_{60}F_{36}$ calls for further elaboration of its electronic properties, which will not only assist in the interpreting existing results, but also in the future design of electronics with $C_{60}F_{36}$-semiconductors.

1.5 Summary and Outlook

Nearly three decades of fundamental research in the field of fluorination of fullerenes have resulted in a basic understanding of the reactivity of various carbon clusters toward fluorine. In several cases, efficient, selective, and scalable synthetic procedures were developed that afford gram amounts of fluorofullerene compounds with the specific compositions and structures (Fig. 1.17). These advancements made fluorofullerenes commercially viable and introduced a new type of fluorocarbon material that possesses many attractive properties for diverse applications.

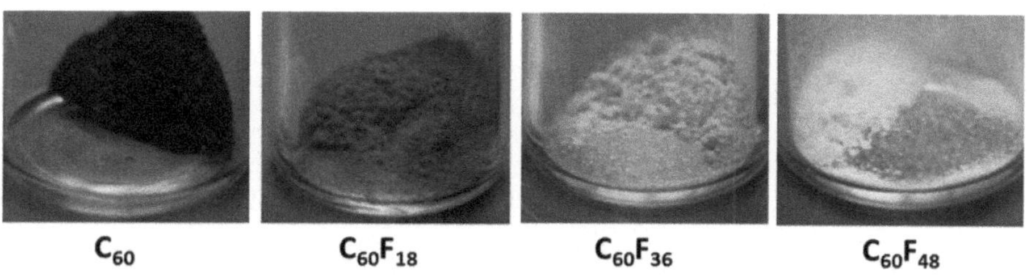

Figure 1.17 Photographs of the research samples of fluorofullerenes prepared in the author's labs. For comparison, a photograph with the sample of 99.5% C_{60} is shown. The as-synthesized sample of $C_{60}F_{18}$ shown of the photograph has a *brownish tint* (dark gray in print versions) due to the presence of small impurity of C_{60} in it. Purified $C_{60}F_{18}$ has a bright yellow color (gray in print versions).

One particular property—the ability of fluorofullerenes to effectively attach electron(s) and form stable anions—has become the reason of the recent surge in the research and their practical applications as electrical molecular dopants in modern organic electronics. Theoretical studies and experimental data collected for various fullerene substrates suggest that both increase in the size of the carbon cluster and the number of fluorine atoms per cage increase EA of fluorofullerenes. Such predictive behavior provides a valuable tool for fine-tuning energy levels of fluorofullerenes for specific applications. For example, in the case of C_{60} fluorides, nearly unlimited number of materials with gradual change of fluorine content (and hence, gradual change in the EA or E(LUMO)) can be currently prepared for the composition range from $n = 36$ to $n = 48$ in $C_{60}F_n$. Taking into account a nearly 1 eV difference in EA between $C_{60}F_{36}$ and $C_{60}F_{48}$ molecules, nearly continuous variations in the respective energy levels of the fluorofullerene materials can be achieved synthetically. It has not yet been proven experimentally that this trend observed for the isolated molecules in the gas phase, will be repeated in the solid state; however, the possibility to have such a uniquely tunable material warrants further research in this direction.

Less evidence has been obtained to understand the effect of the relative positions of fluorine atoms on the cage, ie, addition pattern, on the EA for molecules with the same compositions. Partly, a unique selectivity in high-temperature fluorinations leading to the single-isomer fluorination products is responsible for the sparsity of such experimental data. When alternative synthetic methods of fullerene fluorination become available in the future, eg, low-temperature halogen substitution reactions, this gap in our knowledge can be filled.

Significant body of data has been obtained on the electronic properties of fluorofullerenes in the gas phase, and many fluorofullerene compounds were studied by electrochemical methods. This provided some valuable information of the trends in the electronic properties in these media. For advancing applied science with these materials, it is imperative to obtain reliable data on their electronic properties in the solid state as neat solid films, and in interfaces with different substrates. The low-energy IPES method developed recently [86, 87] will be a method of choice for such studies. Several technological questions concerning environmental and thermal stability, lifetime of the devices which incorporate fluorofullerenes, will need to be addressed as commercialization of such devices becomes a possibility.

References

[1] H.W. Kroto, J.R. Heath, S.C. O'Brien, R.F. Curl, R.E. Smalley, C_{60}: buckminsterfullerene, Nature 318 (1985) 162–163.

[2] B. Dunlap, D. Brenner, J. Mintmire, R. Mowrey, C. White, Geometric and electronic structures of $C_{60}H_{60}$, $C_{60}F_{60}$, and $C_{60}H_{36}$, J. Phys. Chem. 95 (1991) 5763–5768.

[3] A. Lobach, A. Perov, A. Rebrov, O. Roshchupkina, V. Tkacheva, A. Stepanov, Preparation and study of hydrides of fullerenes C_{60} and C_{70}, Russ. Chem. Bull. 46 (1997) 641–648.

[4] O.V. Boltalina, A.K. Abdul-Sada, R. Taylor, Hyperfluorination of [60] fullerene by krypton difluoride, J. Chem. Soc. Perkin Trans. 2 (1995) 981–985.

[5] A.A. Tuinman, A.A. Gakh, J.L. Adcock, R.N. Compton, Hyperfluorination of buckminsterfullerene: cracking the sphere, J. Am. Chem. Soc. 115 (1993) 5885–5886.
[6] O. Boltalina, A.Y. Lukonin, V. Pavlovich, L. Sidorov, R. Taylor, A. Abdul-Sada, Reaction of [60] fullerene with terbium (IV) fluoride, Fullerene Sci. Technol. 6 (1998) 469–479.
[7] S.I. Troyanov, P.A. Troshin, O.V. Boltalina, I.N. Ioffe, L.N. Sidorov, E. Kemnitz, Two isomers of $C_{60}F_{48}$: an indented fullerene, Angew. Chem. Int. Ed. 40 (2001) 2285–2287.
[8] A.A. Gakh, A.A. Tuinman, J.L. Adcock, R.A. Sachleben, R.N. Compton, Selective synthesis and structure determination of $C_{60}F_{48}$, J. Am. Chem. Soc. 116 (1994) 819–820.
[9] G. Gigli, G. Balducci, V.Y. Markov, O.V. Boltalina, A.A. Goryunkov, L.N. Sidorov, R. Taylor, Saturated vapor pressure and sublimation enthalpy of $C_{60}F_{18}$, J. Chem. Thermodyn. 34 (2002) 57–61.
[10] G. Gigli, Thermodynamic functions of fullerenes and fluorofullerenes, Proc. Electrochem. Soc. 2000–2012 (2000) 100–108.
[11] O.V. Boltalina, N.A. Galeva, Direct fluorination of fullerenes, Russ. Chem. Rev. 69 (2000) 609–621.
[12] C. Jin, R.L. Hettich, R.N. Compton, A. Tuinman, A. Derecskei-Kovacs, D.S. Marynick, B.I. Dunlap, Attachment of two electrons of $C_{60}F_{48}$: coulomb barriers in doubly charged anions, Phys. Rev. Lett. 73 (1994) 2821–2824.
[13] E.Y. Misochko, A.V. Akimov, V.A. Belov, D.A. Tyurin, D.N. Laikov, High resolution EPR spectroscopy of $C_{60}F$ and $C_{70}F$ in solid argon: reassignment of $C_{70}F$ regioisomers, J. Chem. Phys. 127 (2007) 084301/1–084301/9.
[14] O.V. Boltalina, Fluorination of fullerenes and their derivatives, J. Fluorine Chem. 101 (2000) 273–278.
[15] B.W. Clare, D.L. Kepert, The structures of $C_{60}F_{36}$ and new possible structures for $C_{60}H_{36}$, J. Mol. Struct. THEOCHEM 466 (1999) 177–186.
[16] B.W. Clare, D.L. Kepert, Early stages in the addition to C_{60} to form $C_{60}X_n$, X = H, F, Cl, Br, CH_3, C_4H_9, J. Mol. Struct. THEOCHEM 621 (2003) 211–231.
[17] B.W. Clare, D.L. Kepert, Stereochemical patterns in $C_{60}X_n$, J. Phys. Chem. Solids 58 (1997) 1815–1821.
[18] J.P. Sandall, P.W. Fowler, The energies of some isomers of $C_{60}F_8$: the use of experimental and theoretical considerations to limit candidate structures, Org. Biomol. Chem. 1 (2003) 1061–1066.
[19] W. Krätschmer, L.D. Lamb, K. Fostiropoulos, D.R. Huffman, C_{60}: a new form of carbon, Nature 347 (1990) 354–358.
[20] H. Selig, C. Lifshitz, T. Peres, J. Fischer, A. McGhie, W. Romanow, J. McCauley Jr., A. Smith III, Fluorinated fullerenes, J. Am. Chem. Soc. 113 (1991) 5475–5476.
[21] J.H. Holloway, E.G. Hope, R. Taylor, G.J. Langley, A.G. Avent, T.J. Dennis, J.P. Hare, H.W. Kroto, D.R.M. Walton, Fluorination of buckminsterfullerene, J. Chem. Soc. Chem. Commun. (1991) 966–969.
[22] R.E. Haufler, J. Conceicao, L.P.F. Chibante, Y. Chai, N.E. Byrne, S. Flanagan, M.M. Haley, S.C. Obrien, C. Pan, Z. Xiao, W.E. Billups, M.A. Ciufolini, R.H. Hauge, J.L. Margrave, L.J. Wilson, R.F. Curl, R.E. Smalley, Efficient production of C_{60} (buckminsterfullerene), $C_{60}H_{36}$, and the solvated buckide ion, J. Phys. Chem. 94 (1990) 8634–8636.
[23] H. Selig, K. Kniaz, G. Vaughan, J. Fischer, A. Smith, Fluorinated fullerenes: synthesis and characterization, in: Macromolecular Symposia, 1994 (Wiley Online Library).
[24] A. Hamwi, C. Fabre, P. Chaurand, S. Della-Negra, C. Ciot, D. Djurado, J. Dupuis, A. Rassat, Preparation and characterization of fluorinated fullerenes, fullerenes, nanotubes, Carbon Nanostruct. 1 (1993) 499–535.

[25] S.K. Chowdhury, S.D. Cameron, D.M. Cox, K. Kniaz, R.A. Strongin, M.A. Cichy, J.E. Fischer, A.B. Smith, Mass spectrometric investigation of polyfluorinated C_{60}. Evidence for the existence of $C_{60}F_{2n}$ ($n = 0-30$), Org. Mass Spectrom. 28 (1993) 860−866.

[26] A.A. Popov, V.M. Senyavin, O.V. Boltalina, K. Seppelt, J. Spandl, C.S. Feigerle, R.N. Compton, Infrared, Raman, and DFT vibrational spectroscopic studies of $C_{60}F_{36}$ and $C_{60}F_{48}$, J. Phys. Chem. A 110 (2006) 8645−8652.

[27] O.V. Boltalina, V.Y. Markov, R. Taylor, M.P. Waugh, Preparation and characterization of $C_{60}F_{18}$, Chem. Commun. (1996) 2549−2550.

[28] O.V. Boltalina, A.Y. Borschevskii, L.N. Sidorov, J.M. Street, R. Taylor, Preparation of $C_{60}F_{36}$ and $C_{70}F_{36/38/40}$, Chem. Commun. (1996) 529−530.

[29] A.G. Avent, B.W. Clare, P.B. Hitchcock, D.L. Kepert, R. Taylor, $C_{60}F_{36}$: there is a third isomer and it has C_1 symmetry, Chem. Commun. (2002) 2370−2371.

[30] I.S. Neretin, K.A. Lyssenko, M.Y. Antipin, Y.L. Slovokhotov, O.V. Boltalina, P.A. Troshin, A.Y. Lukonin, L.N. Sidorov, R. Taylor, $C_{60}F_{18}$, a flattened fullerene: alias a hexa-substituted benzene, Angew. Chem. Int. Ed. 39 (2000) 3273−3276.

[31] N. Yudanov, A. Okotrub, L. Bulusheva, I. Asanov, V. Lisoivan, Y.V. Shevtsov, Synthesis and electron structure of $C_{60}F_{24}$, Mol. Cryst. Liq. Cryst. Sci. Technol. Sect. C 7 (1996) 127−132.

[32] N.I. Denisenko, S.I. Troyanov, A.A. Popov, I.V. Kuvychko, B. Zemva, E. Kemnitz, S.H. Strauss, O.V. Boltalina, T_h-$C_{60}F_{24}$, J. Am. Chem. Soc. 126 (2004) 1618−1619.

[33] S.I. Troyanov, P.A. Troshin, O.V. Boltalina, E. Kemnitz, Bromination of [60] fullerene. II. Crystal and molecular structure of [60] fullerene bromides, $C_{60}Br_6$, $C_{60}Br_8$, and $C_{60}Br_{24}$, fullerenes, nanotubes, Carbon Nanostruct. 11 (2003) 61−77.

[34] F.N. Tebbe, R.L. Harlow, D.B. Chase, D.L. Thorn, G.C. Campbell, J.C. Calabrese, N. Herron, R.J. Young, E. Wasserman, Synthesis and single-crystal x-ray structure of a highly symmetrical C-60 derivative, $C_{60}BR_{24}$, Science 256 (1992) 822−825.

[35] O.V. Boltalina, J.M. Street, R. Taylor, $C_{60}F_{36}$ consists of two isomers having T and C_3 symmetry, J. Chem. Soc. Perkin Trans. 2 (1998) 649−654.

[36] O.V. Boltalina, A.Y. Lukonin, A.G. Avent, J.M. Street, R. Taylor, Formation and characterisation of $C_{60}F_8O$, $C_{60}F_6O$ and $C_{60}F_4O$; the sequential pathway for addition to fullerenes, Perkin 2 (2000) 683−686.

[37] A.G. Avent, R. Taylor, O.V. Boltalina, A.Y. Lukonin, B. De La Vaissiere, P.W. Fowler, J.M. Street, Isolation of some fluorofullerenes of low fluorine content; the contiguous addition pathway for fullerenes, Proc. − Electrochem. Soc. 2000−2011 (2000) 109−116.

[38] A.A. Goryunkov, I.E. Kareev, I.N. Ioffe, A.A. Popov, I.V. Kuvychko, V.Y. Markov, I.V. Goldt, A.S. Pimenova, M.G. Serov, S.M. Avdoshenko, P.A. Khavrel, L.N. Sidorov, S.F. Lebedkin, Z. Mazej, B. Zemva, S.H. Strauss, O.V. Boltalina, Reaction of C_{60} with $KMnF_4$: isolation and characterization of a new isomer of $C_{60}F_7(CF_3)$ and the known isomer of $C_{60}F_8$, J. Fluorine Chem. 127 (2006) 1423−1435.

[39] A.A. Popov, A.A. Goryunkov, I.V. Goldt, I.E. Kareev, I.V. Kuvychko, W.-D. Hunnius, K. Seppelt, S.H. Strauss, O.V. Boltalina, Raman, infrared, and theoretical studies of fluorofullerene $C_{60}F_{20}$, J. Phys. Chem. A 108 (2004) 11449−11456.

[40] N.B. Shustova, Z. Mazej, Y.S. Chen, A.A. Popov, S.H. Strauss, O.V. Boltalina, Saturnene revealed: x-ray crystal structure of D_{5d}-$C_{60}F_{20}$ formed in reactions of C_{60} with a_xMF_y fluorinating agents (a = Alkali metal; M = 3d metal), Angew. Chem. Intern. Ed. 49 (2010) 812−815.

[41] O.V. Boltalina, V.Y. Markov, P.A. Troshin, A.D. Darwish, J.M. Street, R. Taylor, $C_{60}F_{20}$: "saturnene", an extraordinary squashed fullerene, Angew. Chem. Int. Ed. 40 (2001) 787−789.

[42] F. Zhou, X. Wei, H. Su, Q. Fu, Q. Li, F.Y. Boey, Electronic and steric mechanisms in mono- and double Fluorination of $C_sC_{60}Cl_6$, Mod. Phys. Lett. B 22 (2008) 2727−2738.

[43] X.-B. Wang, C. Chi, M. Zhou, I.V. Kuvychko, K. Seppelt, A.A. Popov, S.H. Strauss, O.V. Boltalina, L.-S. Wang, Photoelectron spectroscopy of $C_{60}F_n^-$ and $C_{60}F_m^{2-}$ ($n = 17$, 33, 35, 43, 45, 47; m = 34, 46) in the gas phase and the generation and characterization of C_1-$C_{60}F_{47}^-$ and D_2-$C_{60}F_{44}$ in solution, J. Phys. Chem. A 114 (2010) 1756−1765.

[44] O.V. Boltalina, T.V. Avakyan, V.Y. Markov, T.J.S. Dennis, A.K. Abdul-Sada, R. Taylor, formation of $C_{76}F_{38}$, $C_{78}F_{38}$, $C_{82}F_{44}$, $C_{84}F_{40}$, and $C_{84}F_{44}$, J. Phys. Chem. B 103 (1999) 8189−8191.

[45] A.A. Goryunkov, V.Y. Markov, I.N. Ioffe, R.D. Bolskar, M.D. Diener, I.V. Kuvychko, S.H. Strauss, O.V. Boltalina, $C_{74}F_{38}$: an exohedral derivative of a small-bandgap fullerene with D3 symmetry, Angew. Chem. Int. Ed. 116 (2004) 1015−1018.

[46] O.V. Boltalina, A.A. Goryunkov, V.Y. Markov, I.N. Ioffe, L.N. Sidorov, In situ synthesis and characterization of fullerene derivatives by Knudsen-cell mass spectrometry, Int. J. Mass Spectrom. 228 (2003) 807−824.

[47] A.A. Tuinman, P. Mukherjee, J.L. Adcock, R.L. Hettich, R.N. Compton, Characterization and stability of highly fluorinated fullerenes, J. Phys. Chem. 96 (1992) 7584−7589.

[48] A.D. Darwish, N. Martsinovich, J.M. Street, R. Taylor, C_2 isomers of $C_{84}F_{40}$ and $C_{84}F_{44}$ are cuboid and contain benzenoid and naphthalenoid aromatic patches, Chem. Eur. J. 11 (2005) 5377−5380.

[49] N.B. Shustova, B.S. Newell, S.M. Miller, O.P. Anderson, R.D. Bolskar, K. Seppelt, A.A. Popov, O.V. Boltalina, S.H. Strauss, Discovering and verifying elusive fullerene cage isomers: structures of C_2-p11-(C_{74}-D_{3h})$(CF_3)_{12}$ and C_2-p11-(C_{78}-D_{3h}(5))$(CF_3)_{12}$, Angew. Chem. Ed. 119 (2007) 4189−4192.

[50] H.-P. Wu, G.-L. Lu, Y.-B. Yuan, K.-M. Deng, Y.-Z. Liu, J.-L. Yang, Geometric and electronic properties of highly fluorinated fullerene $C_{74}F_{38}$, Chin. Phys. Lett. 23 (2006) 2563−2565.

[51] L.S. Wang, J. Conceicao, C.M. Jin, R.E. Smalley, Threshold photodetachment of cold C_{60}, Chem. Phys. Lett. 182 (1991) 5−11.

[52] S.H. Yang, C.L. Pettiette, J. Conceicao, O. Cheshnovsky, R.E. Smalley, UPS of buckminsterfullerene and other large clusters of carbon, Chem. Phys. Lett. 139 (1987) 233−238.

[53] T.D. Anthopoulos, B. Singh, N. Marjanovic, N. Saricifici, A.M. Ramil, H. Sitter, M. Colle, D.M. de Leeuw, High performance n-channel organic field-effect transistors and ring oscillators based on C_{60} fullerene films, Appl. Phys. Lett. 89 (2006) 213504.

[54] Y. Inoue, Y. Sakamoto, T. Suzuki, M. Kobayashi, Y. Gao, S. Tokito, Organic thin-film transistors with high electron mobility based on perfluoropentacene, Jpn. J. Appl. Phys. 44 (2005) 3663.

[55] X.B. Wang, H.K. Woo, L.S. Wang, Vibrational cooling in a cold ion trap: vibrationally resolved photoelectron spectroscopy of cold C_{60}^- anions, J. Chem. Phys. 123 (2005) 051106/1.

[56] M. Lenes, G.J.A. Wetzelaer, F.B. Kooistra, S.C. Veenstra, J.C. Hummelen, P.W. Blom, Fullerene bisadducts for enhanced open-circuit voltages and efficiencies in polymer solar cells, Adv. Mater. 20 (2008) 2116−2119.

[57] A.A. Popov, I.E. Kareev, N.B. Shustova, E.B. Stukalin, S.F. Lebedkin, K. Seppelt, S.H. Strauss, O.V. Boltalina, L. Dunsch, Electrochemical, spectroscopic, and DFT study of $C_{60}(CF_3)_n$ frontier orbitals ($n = 2$−18): the link between double bonds in pentagons and reduction potentials, J. Am. Chem. Soc. 129 (2007) 11551−11568.

[58] X.-B. Wang, H.-K. Woo, X. Huang, M.M. Kappes, L.-S. Wang, Direct experimental probe of the on-site coulomb repulsion in the doubly charged fullerene anion C_{70}^{2-}, Phys. Rev. Lett. 96 (2006) 143002.
[59] R. Hettich, C.M. Jin, R. Compton, Determination of the electron-affinities of fluorinated fullerenes ($C_{60}F_{44,46}$, $C_{70}F_{52,54}$) by fourier-transform mass-spectrometry, Int. J. Mass Spectrom. 138 (1994) 263–274.
[60] T. Drewello, H. Frauendorf, R. Herzschuh, A.A. Goryunkov, S.H. Strauss, O.V. Boltalina, The formation of long-lived fluorofullerene dianions by direct electrospray ionization, Chem. Phys. Lett. 405 (2005) 93–96.
[61] R. Cozzolino, O. Belgacem, T. Drewello, L. Kaseberg, R. Herzschuh, S. Suslov, O.V. Boltalina, Laser desorption/ionization of fluorinated fullerenes, Eur. Mass Spectrom. 3 (1997) 407–414.
[62] I.N. Ioffe, A.A. Goryunkov, O.V. Boltalina, A.Y. Borschevsky, L.N. Sidorov, Computational study of structure and thermochemistry of some endo- and exohedral fullerene derivatives, fullerenes, nanotubes, Carbon Nanostruct. 12 (2004) 169–173.
[63] O.V. Boltalina, D.B. Ponomarev, A.Y. Borschevskii, L.N. Sidorov, Thermochemistry of the gas phase reactions of fluorofullerene anions, J. Phys. Chem. A 101 (1997) 2574–2577.
[64] O.V. Boltalina, P. Hvelplund, T.J.D. Jorgensen, M.C. Larsen, M.O. Larsson, D.A. Sharoitchenko, Electron capture by fluorinated fullerene anions in collisions with Xe atoms, Phys. Rev. A: Mol. Opt. Phys. 62 (2000) 023202/1–023202/7.
[65] R.W. Kirschbaum, M. Hausmann, O.V. Boltalina, S.H. Strauss, T. Drewello, Energy-dependent gas-phase fragmentation of fluorofullerene multiply charged anions (MCAs), Phys. Chem. Chem. Phys. 17 (2015) 23052–23058.
[66] I. Ioffe, S. Avdoshenko, O. Boltalina, L. Sidorov, K. Berndt, J. Weber, Mass spectrometry, photoelectron spectroscopy, and quantum chemical studies of fluorofullerene dianions, Int. J. Mass Spectrom. 243 (2005) 223–230.
[67] O.V. Boltalina, A.V. Streletskii, I.N. Ioffe, P. Hvelplund, B. Liu, S.B. Nielsen, S. Tomita, Formation of long-lived fluorofullerene trianions in collisions with Na, J. Chem. Phys. 122 (2005) 021102/1–021102/3.
[68] Y.F. Wang, W. Chen, G.T. Yu, Z.R. Li, D. Wu, C.C. Sun, Evolution of lone pair of excess electrons inside molecular cages with the deformation of the cage in e2@ $C_{60}F_{60}$ systems, J. Comput. Chem. 32 (2011) 2012–2021.
[69] L. Echegoyen, L.E. Echegoyen, Electrochemistry of fullerenes and their derivatives, Acc. Chem. Res. 31 (1998) 593–601.
[70] D. Claves, J. Giraudet, A. Hamwi, R. Benoit, Structural, bonding, and electrochemical properties of perfluorinated fullerene C_{70}, J. Phys. Chem. B 105 (2001) 1739–1742.
[71] N. Liu, H. Touhara, Y. Morio, D. Komichi, F. Okino, S. Kawasaki, Electrochemical behavior of fluorinated fullerenes $C_{60}F_x$ and $C_{70}F_x$, J. Electrochem. Soc. 143 (1996) L214–L217.
[72] N. Liu, H. Touhara, F. Okino, S. Kawasaki, Y. Nakacho, Solid-state lithium cells based on fluorinated fullerene cathodes, J. Electrochem. Soc. 143 (1996) 2267–2272.
[73] F. Zhou, G.J. Van Berkel, B.T. Donovan, Electron-transfer reactions of fluorofullerene $C_{60}F_{48}$, J. Am. Chem. Soc. 116 (1994) 5485–5486.
[74] K. Ohkubo, R. Taylor, O.V. Boltalina, S. Fukuzumi, Electron transfer reduction of fluorofullerenes, $C_{60}F_{18}$ and $C_3C_{60}F_{36}$, Proc. – Electrochem. Soc. 2003–2015 (2003) 20–30.
[75] D. Paolucci, F. Paolucci, M. Marcaccio, M. Carano, R. Taylor, Electrochemistry of perfluorinated fullerenes: the case of three isomers of $C_{60}F_{36}$, Chem. Phys. Lett. 400 (2004) 389–393.

[76] O. Solomeshch, Y.J. Yu, A.A. Goryunkov, L.N. Sidorov, R.F. Tuktarov, D.H. Choi, J.I. Jin, N. Tessler, Ground-state interaction and electrical doping of fluorinated C_{60} in conjugated polymers, Adv. Mater. 21 (2009) 4456–4460.
[77] A.A. Popov, J. Tarábek, I.E. Kareev, S.F. Lebedkin, S.H. Strauss, O.V. Boltalina, L. Dunsch, Poly(trifluoromethyl)fullerene radical anions. an ESR/Vis-NIR spectroelectrochemical study of $C_{60}F_{2,4}$ and $C_{60}(CF_3)_{2,10}$, J. Phys. Chem. A 109 (2005) 9709–9711.
[78] Y.-J. Yu, O. Solomeshch, H. Chechik, A. Goryunkov, R. Tuktarov, D.H. Choi, J.-I. Jin, Y. Eichen, N. Tessler, p-type doping in organic light emitting diodes based on fluorinated C_{60}, J. Appl. Phys. 104 (2008) 124505.
[79] C.M. Cardona, W. Li, A.E. Kaifer, D. Stockdale, G.C. Bazan, Electrochemical considerations for determining absolute frontier orbital energy levels of conjugated polymers for solar cell applications, Adv. Mater. 23 (2011) 2367–2371.
[80] B.W. Larson, J.B. Whitaker, X.B. Wang, A.A. Popov, G. Rumbles, N. Kopidakis, S.H. Strauss, O.V. Boltalina, Electron affinity of phenyl-C-61-butyric acid methyl ester (PCBM), J. Phys. Chem. C 117 (2013) 14958–14964.
[81] P.I. Djurovich, E.I. Mayo, S.R. Forrest, M.E. Thompson, Measurement of the lowest unoccupied molecular orbital energies of molecular organic semiconductors, Org. Electron. 10 (2009) 515–520.
[82] R. Meerheim, S. Olthof, M. Hermenau, S. Scholz, A. Petrich, N. Tessler, O. Solomeshch, B. Lüssem, M. Riede, K. Leo, Investigation of $C_{60}F_{36}$ as low-volatility p-dopant in organic optoelectronic devices, J. Appl. Phys. 109 (2011) 103102.
[83] R. Mitsumoto, T. Araki, E. Ito, Y. Ouchi, K. Seki, K. Kikuchi, Y. Achiba, H. Kurosaki, T. Sonoda, H. Kobayashi, O.V. Boltalina, V.K. Pavlovich, L.N. Sidorov, Y. Hattori, N. Liu, S. Yajima, S. Kawasaki, F. Okino, H. Touhara, Electronic structures and chemical bonding of fluorinated fullerenes studied by NEXAFS, UPS, and vacuum-UV absorption spectroscopies, J. Phys. Chem. A 102 (1998) 552–560.
[84] T. Menke, D. Ray, H. Kleemann, M.P. Hein, K. Leo, M. Riede, Highly efficient p-dopants in amorphous hosts, Org. Electron. 15 (2014) 365–371.
[85] H. Yoshida, K. Yoshizaki, Electron affinities of organic materials used for organic light-emitting diodes: a low-energy inverse photoemission study, Org. Electron. 20 (2015) 24–30.
[86] H. Yoshida, K. Yamada, J. Tsutsumi, N. Sato, Complete description of ionization energy and electron affinity in organic solids: determining contributions from electronic polarization, energy band dispersion, and molecular orientation, Phys. Rev. B 92 (2015).
[87] H. Yoshida, Measuring the electron affinity of organic solids: an indispensable new tool for organic electronics, Anal. Bioanal. Chem. 406 (2014) 2231–2237.
[88] N.Y.U. Svetchnikov, A. Kolmakov, V. Stankevitch, Photoelectron spectra of submonolayer $C_{60}F_{48}$ films on d-metal substrates, Surf. Rev. Lett. 9 (2002) 1269–1274.
[89] V. Shnitov, V. Mikoushkin, Y.S. Gordeev, O. Boltalina, I. Gol'dt, Single and collective electron excitations in the solid $C_{60}F_{18}$, fullerenes, nanotubes, Carbon Nanostruct. 14 (2006) 297–301.
[90] S.L. Wong, H. Huang, Y. Wang, L. Cao, D. Qi, I. Santoso, W. Chen, A.T.S. Wee, Quasi-free-standing epitaxial graphene on SiC (0001) by fluorine intercalation from a molecular source, ACS Nano 5 (2011) 7662–7668.
[91] Y. Fujikawa, J.T. Sadowski, K.F. Kelly, K.S. Nakayama, E.T. Mickelson, R.H. Hauge, J.L. Margrave, T. Sakurai, Adsorption of fluorinated C_{60} on the Si (111)-(7 $\times$ 7) surface studied by scanning tunneling microscopy and high-resolution electron energy loss spectroscopy, Jpn. J. Appl. Phys. 41 (2002) 245.

[92] H. Yoshida, Low-energy inverse photoemission study on the electron affinities of fullerene derivatives for organic photovoltaic cells, J. Phys. Chem. C 118 (2014) 24377−24382.

[93] M. Landstrass, K. Ravi, Resistivity of chemical vapor deposited diamond films, Appl. Phys. Lett. 55 (1989) 975−977.

[94] F. Maier, M. Riedel, B. Mantel, J. Ristein, L. Ley, Origin of surface conductivity in diamond, Phys. Rev. Lett. 85 (2000) 3472.

[95] P. Strobel, M. Riedel, J. Ristein, L. Ley, O. Boltalina, Surface transfer doping of diamond by fullerene, Diamond Relat. Mater. 14 (2005) 451−458.

[96] S. Sque, R. Jones, J. Goss, P. Briddon, S. Öberg, First-principles study of C_{60} and $C_{60}F_{36}$ as transfer dopants for p-type diamond, J. Phys. Condensed Matt. 17 (2005) L21.

[97] J. Goss, B. Hourahine, R. Jones, M. Heggie, P. Briddon, p-type surface doping of diamond: a first-principles study, J. Phys. Condensed Matt. 13 (2001) 8973.

[98] P. Strobel, J. Ristein, L. Ley, K. Seppelt, I. Goldt, O. Boltalina, Surface conductivity induced by fullerenes on diamond: passivation and thermal stability, Diamond Relat. Mater. 15 (2006) 720−724.

[99] T. Ouyang, K.P. Loh, D. Qi, A.T.S. Wee, M. Nesladek, Chemical bonding of fullerene and fluorinated fullerene on bare and hydrogenated diamond, ChemPhysChem 9 (2008) 1286−1293.

[100] T. Tada, N. Uchida, T. Kanayama, H. Hiura, K. Kimoto, Charge-transfer doping by fullerenes on oxidized Si surfaces, J. Appl. Phys. 102 (2007) 074504/1−074504/5.

[101] S.M. Sze, Semiconductor Devices: Physics and Technology, John Wiley & Sons, 2008.

[102] A. Tadich, M.T. Edmonds, L. Ley, F. Fromm, Y. Smets, Z. Mazej, J. Riley, C.I. Pakes, T. Seyller, M. Wanke, Tuning the charge carriers in epitaxial graphene on SiC(0001) from electron to hole via molecular doping with $C_{60}F_{48}$, Appl. Phys. Lett. 102 (2013) 241601/1−241601/5.

[103] S. Reineke, F. Lindner, G. Schwartz, N. Seidler, K. Walzer, B. Lussem, K. Leo, White organic light-emitting diodes with fluorescent tube efficiency, Nature 459 (2009) 234−238.

[104] M. Riede, C. Uhrich, J. Widmer, R. Timmreck, D. Wynands, G. Schwartz, W.M. Gnehr, D. Hildebrandt, A. Weiss, J. Hwang, S. Sundarraj, P. Erk, M. Pfeiffer, K. Leo, Efficient organic tandem solar cells based on small molecules, Adv. Funct. Mater. 21 (2011) 3019−3028.

[105] F. Ante, D. Kalblein, U. Zschieschang, T.W. Canzler, A. Werner, K. Takimiya, M. Ikeda, T. Sekitani, T. Someya, H. Klauk, Contact doping and ultrathin gate dielectrics for nanoscale organic thin-film transistors, Small 7 (2011) 1186−1191.

[106] W. Gao, A. Kahn, Controlled p-doping of zinc phthalocyanine by coevaporation with tetrafluorotetracyanoquinodimethane: a direct and inverse photoemission study, Appl. Phys. Lett. 79 (2001) 4040−4042.

[107] N. El-Khatib, B. Boudjema, G. Guillaud, M. Maitrot, H. Chermette, Theoretical and experimental doping of molecular materials: P and N doping of zinc phthalocyanine, J. Less Common Met. 143 (1988) 101−112.

[108] M.L. Tietze, L. Burtone, M. Riede, B. Lüssem, K. Leo, Fermi level shift and doping efficiency in p-doped small molecule organic semiconductors: a photoelectron spectroscopy and theoretical study, Phys. Rev. B 86 (2012) 035320.

[109] S. Braun, W.R. Salaneck, M. Fahlman, Energy-level alignment at organic/metal and organic/organic interfaces, Adv. Mater. 21 (2009) 1450−1472.

Synthesis and Isolation of Trifluoromethylfullerenes

2

B.W. Larson, S.H. Strauss, O.V. Boltalina
Colorado State University, Fort Collins, CO, United States

Chapter Outline

2.1 Introduction

Since the early 2000s, fullerenes and their derivatives have matured from lab-scale curiosities to multimillion dollar industrial revenue streams and have proven useful in a plethora of established and emerging technologies, thanks to the truly vast range of optical, electronic, thermal, and chemical properties that can be accessed through derivatization of the parent fullerene. While the library of fullerene derivatives expands daily through the multidisciplinary efforts of researchers around the globe, this chapter will focus exclusively on the largest and most fully characterized family of fullerene derivatives, trifluoromethylfullerenes (TMFs). At the time of this writing, over 100 isomers of $C_{60,70}(CF_3)_n$ have been structurally characterized by single crystal X-ray diffraction or ^{19}F NMR, and more than 60 others have been identified for the higher fullerenes, C_{74-96} [1]. While trifluoromethylated derivatives of higher fullerenes, heterofullerenes, and fullerenes encapsulating metal atoms, clusters of atoms, and small molecules, known as endohedral fullerenes, further expand and enrich the family of TMFs [1,2], significant challenges and opportunities must be met in regard

New Fluorinated Carbons: Fundamentals and Applications. http://dx.doi.org/10.1016/B978-0-12-803479-8.00002-4

to the production and availability of the respective parent fullerenes before their perfluoroalkyl derivatives can realize true commercial potential. At the time of this writing, the most current anthology of literature related to fluorinated alkyl (R_F) cage substituents on a 3D molecular carbon nanosubstrate was reviewed by Boltalina et al., which succinctly details the full gamut of physical, chemical, and electronic properties of fullerene(R_F) compounds [1]. This chapter should serve as a practical guide to efficiently prepare and isolate appreciable quantities of TMFs.

2.2 Synthetic Methodologies

2.2.1 General Remarks on Trifluoromethylfullerene Formation

Several synthetic routes toward TMF production have been developed over the years and are discussed in detail in the following sections of this chapter. Fig. 2.1 shows a general scheme that should aid in the following discussions about TMF formation. It depicts the reaction processes known to occur between bare fullerene, CF_3 radicals, and also TMFs. In this figure, the area inside the box represents a general reaction environment where conditions are suitable for TMF formation, ie, the reactive zone. Reactive zone conditions may include but are not limited to sufficiently high temperature or UV exposure, as well as presence of any reactive CF_3 source. Within this zone, four different trifluoromethylation reactions are shown: (1) addition of CF_3 radicals, (2) loss of CF_3 radicals, (3) rearrangement of CF_3 radicals, and (4) transfer of CF_3 radicals

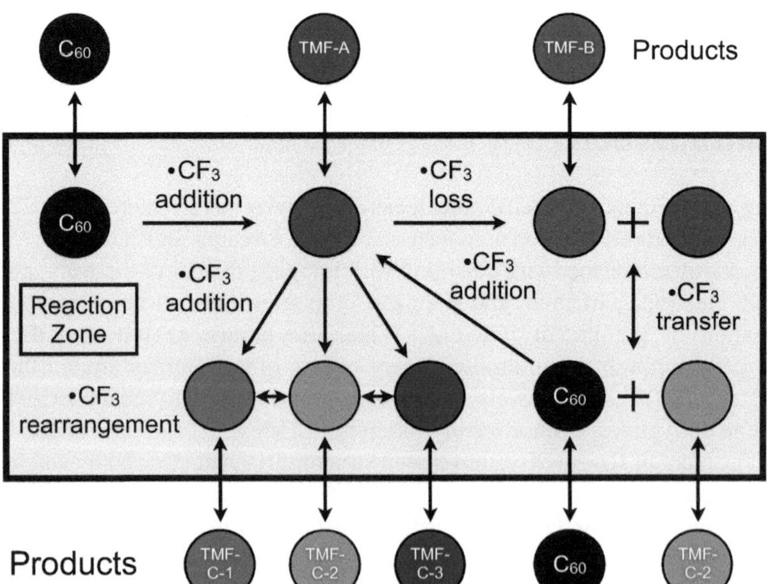

Figure 2.1 A general scheme for TMF formation is depicted, including the reactive zone (within the box) and the nonreactive zone (outside the box). *TMF*, trifluoromethylfullerene.

from one fullerene to another. The species depicted outside the box are considered to be in an unreactive environment, and therefore represent the reaction products. Note that TMF products can reenter the reactive zone or be used as starting materials in subsequent reactions, and can undergo the same trifluoromethylation reactions as bare fullerene.

In general, fullerene trifluoromethylation is not a selective process; however, the composition of the product mixture can be more or less tuned by precisely controlling the specific parameters that define the reactive zone environment. Such parameters include concentration and molar ratio of reactive species, presence of side product scavengers, hetero- or homogeneity of reaction phase, presence of gaseous or liquid buffer molecules, reaction pressure, temperature, UV exposure, and reaction medium. Each of these reaction parameters has a unique effect on the composition of the TMF product mixture and many have been independently studied, as highlighted in the following sections. In addition to the parameters of the reactive zone, the residence time within the reactive zone has a major impact on the composition of product mixture. In the scheme, the reaction rates are not depicted, since the composition of the product mixture strongly depends on *both* the reactive zone conditions *and* the residence time within the reactive zone.

In general, the composition of the product mixture can be tuned toward higher TMFs (ie, $C_{60}(CF_3)_n$ where $n > 6$) or lower TMFs (where $n \leq 6$) by choosing suitable reactive zone conditions. According to the scheme, a longer residence time coupled with a large stoichiometric excess of a CF_3 source would favor formation of higher TMFs, while a shorter residence time and a lower molar ratio of CF_3 to fullerene would favor formation of lower TMFs. In addition to producing either higher or lower TMFs, isomeric selectivity within the targeted product mixture can also be achieved. This level of product selectivity becomes especially important for higher TMFs, where the total number of possible isomers increases exponentially. To tune isomeric distributions in the product mixture, the reactive zone conditions and the residence time are balanced in such a way that either kinetic or thermodynamic products are favored. Thermodynamic products are typically favored with a longer residence time within the reactive zone assuming otherwise constant parameters that target either higher or lower TMFs, as observed for S_6-$C_{60}(CF_3)_{12}$ and C_1-$C_{70}(CF_3)_{10}$ [3,4]. By tuning the reactive zone parameters to favor higher TMFs and allowing only a short residence time, more than 30 isomers of $C_{60}(CF_3)_{10}$ were synthesized in a single reaction. These examples, as well as other literature reports that collectively support this scheme will now be presented.

2.2.2 Solution Reactions

In 1993, Fagan and coworkers carried out the very first fullerene trifluoromethylation reactions by generating CF_3 radicals from trifluoromethyl iodide in saturated solutions of C_{60} or C_{70} either at elevated temperature (175–200°C) or under UV irradiation at room temperature [5]. These reactions (typically 24 h long) produced extremely complex mixtures of TMFs; analysis by mass spectrometry revealed compositions with 2–14 CF_3 groups. When the reaction took place in benzene or trichlorobenzene, substantial quantities of hydrogenated TMFs (ie, $C_{60}(CF_3)_nH_m$) due to H atom transfer from solvent molecules were also observed in the product mixture. However, when suspensions of C_{60} or C_{70} in either C_6F_6 or $CClF_2CCl_2F$ (Freon 113) were reacted with CF_3 radicals, hydrogenation was completely suppressed. Although multigram quantities of TMFs were prepared, no attempts to isolate pure compounds were reported. Around the same time, Yoshida and coworkers reported a different trifluoromethylation approach via reaction of C_{60} with diacyl peroxide $(CF_3CO_2)_2$ in chlorobenzene to give the trifluoromethylfullerol $C_{60}(CF_3)OH$, which was isolated by size exclusion chromatography [6]. Reduction of $C_{60}(CF_3)OH$ with $(n\text{-Bu})_3SnH$ in benzene yielded the simplest hydrogenated TMF $C_{60}(CF_3)H$, which was separated by flash chromatography [7]. It should be noted that hydrogenated TMFs are less stable than normal TMFs and readily decompose at elevated temperatures (see Section 2.2.3).

Nearly two decades later, in 2010, Shustova et al. reevaluated the UV irradiation reaction of CF_3I and C_{60} in C_6F_6 [8]. After 2–7 days of continuous irradiation, the crude product mixture was shown to contain TMFs with up to 23 CF_3 groups as evidenced by electron-impact (EI)–mass spectrometry (MS). After fractionating the crude product mixture by high-performance liquid chromatography (HPLC), one fraction, containing $C_{60}(CF_3)_{8-12}$, was further reacted with CF_3I in C_6F_6 under UV irradiation for 55 h in attempt to push the products to a higher degree of trifluoromethylation. Interestingly, EI–MS analysis of the product revealed only a slight increase in the degree of trifluoromethylation (16 was the highest number of CF_3 groups observed, up from 12), yet a significant number of TMFs with *fewer* than eight CF_3 groups were also observed. This was the first time that UV-induced rearrangement and transfer of CF_3 groups on the fullerene cage was experimentally shown, and indicates CF_3 groups may be more labile than previously thought. Thermally induced rearrangement of fullerene cage-CF_3 groups has been studied and is discussed later in the chapter, but UV-induced rearrangement mechanisms and reactions are yet to be further explored. More information about this type of CF_3 rearrangement would be especially interesting for potential photovoltaic applications, a field where fullerenes are so prevalent, and of course exposed to solar irradiation. Vorobiev et al. UV irradiated a mixture of TMFs suspended directly in liquid CF_3I for 24 h to enrich the content of stable radical $C_{60}(CF_3)_n$ (where $n = 11-17$, odd) compounds, yet no structural information with insights about CF_3 rearrangement could be gleaned from the electron spin resonance, HPLC, and matrix-assisted laser desorption ionization-MS characterization data reported [9].

The only other reported solution phase trifluoromethylation reaction involved heating a mixture of C_{60} and 18 eq. of CF_3I in 1,2-dichlorobenzene at 180°C in the presence of copper powder for 24 h [10]. Copper has been shown to promote CF_3I dissociation, and also acts as a scavenger of iodine by forming insoluble CuI [10,11]. Although the method proved successful for the preparation of a variety of bis-perfluoroalkylfullerenes, minimal conversion of C_{60} to TMF was observed. After several hours of heating it was noted that the typically deep magenta colored C_{60} solution appeared faint pink along with the presence of black precipitate on the walls of the ampule; the solution returned to deep magenta upon cooling at the completion of the 24-h reaction period. These observations were consistent with the experimentally studied anomalous temperature dependence of C_{60} solubility in organic solvents such as 1,2-dichlorobenzene [12,13]; however, no mention of this issue was reported in the earlier work by Fagan and coworkers. One possible explanation is that a far greater molar excess of CF_3I was used in the earlier work, which shifted the reaction equilibrium toward a faster rate of TMF formation than C_{60} precipitation.

In summary, solution phase reactions provide a simple stir and heat/irradiate method to produce appreciable quantities of TMFs with minimal equipment. However, the reaction conditions must be carefully chosen to minimize side product formation, such as hydrogenated TMFs. Additionally, these reactions have not been successfully optimized to selectively target isomerically or compositionally enriched product mixtures. Product workup methodologies, such as those discussed in Section 2.3.1, do accommodate separation of complex product mixtures, though at the cost of purchasing expensive preparative HPLC equipment that can handle large volume injections and high throughput. At present, this synthetic route is best suited for producing complex bulk mixtures of TMFs with minimal cost. When appropriate separation equipment is available, this route is also recommended for capture of large isomeric families of TMFs.

2.2.3 Solid–Solid Reactions

Solid–solid fullerene trifluoromethylation reactions involve intimate mixing (ie, grinding) of fullerene and trifluoromethylating reagent and either a high-temperature heating step or UV irradiation of the powder mixture. Metal trifluoroacetates (TFAs) have been extensively used as a source of trifluoromethyl radicals in such solid–solid reactions. Their use became popular after it was discovered that a mixed fluorofullerene–TMF (ie, $C_{60}F_x(CF_3)_y$) was formed as a by-product during the solid–solid reaction of C_{60} with K_2PtF_6 [14]. In that case, the authors hypothesized that extensive fluorination of the fullerene resulted in cage rupture, providing a CF_3 source for the isolated compound, $C_{60}F_{17}CF_3$. In 2001, a series of transition metal TFAs (AgTFA, $Cu(TFA)_2$, $Pd(TFA)_2$, and $Cr(TFA)_2$) were examined in trifluoromethylation reactions [15]. Finely ground mixtures of either C_{60} or C_{70} with a metal TFA were heated at 300–400°C for 6 h, and the resulting product mixtures were analyzed by EI and laser desorption ionization MS. Interestingly, the extent of fullerene trifluoromethylation could be controlled based on which metal TFA was used; metal TFAs with higher decomposition temperatures (and therefore higher reaction temperatures) produced TMFs carrying a lower number CF_3 groups for both C_{60} and C_{70}. For example, C_{60} mixtures with AgTFA ($T_{decomp} = 230°C$) heated at 300°C showed as many as 20 CF_3 groups on the cage, while the same reaction using $Cr(TFA)_2$ ($T_{decomp} = 310°C$) heated at 400°C yielded TMFs carrying only up to four CF_3 groups. This fact has not since been exploited to control reactions where a lower degree of trifluoromethylation is desired. Further investigation of trifluoromethylation by other metal TFAs (AgTFA has been used exclusively since the first report) may unveil new reaction insights and provide access to new TMFs or better product selectivity.

Much is now known about the trifluoromethylation of fullerenes using AgTFA. In 2003 and 2004, Darwish et al. produced three publications in which a finely ground fullerene/AgTFA mixture was heated at 300°C under dynamic vacuum for 1 h to synthesize TMFs of C_{60}, C_{70}, and their mixtures [16–18]. Numerous TMFs were produced (more than 60 from C_{60} and 46 from C_{70}) in addition to significant amounts of hydrogenated TMFs, yet some of the isolated compounds were sufficiently pure to be characterized by UV–vis, ^{1}H, ^{13}C, ^{19}F NMR spectroscopy. While this method did allow for high conversion of starting fullerene, it also came with several drawbacks such as loss of unreacted sublimed AgTFA, a highly complex TMF/hydrogenated TMF product mixture, and formation of unstable soluble TMF-Ag complexes. The latter was seriously problematic; the authors reported it caused irreparable damage to the expensive HPLC columns used in the product workup.

Goryunkov et al. solved these problems by improving the reaction vessel and adding a pre-HPLC crude product workup step [19]. Instead of heating the AgTFA/fullerene mixture under dynamic vacuum, they loaded the finely ground mixture into an open-ended glass ampule, which was then sealed inside a metal pipe (usually copper). The use of closed reaction environment mitigated loss of unreacted AgTFA and allowed for more precise control over reaction stoichiometry. To solve the more serious problem of irreversible clogging of the HPLC column by TMF-Ag complexes in the product mixture, they also implemented one or more high temperature (c.400–600°C) sublimations of the crude product [19]. The sublimate was found to contain neither hydrogenated TMFs nor TMF-Ag complexes, and therefore no column damage occurred during HPLC workup. The simplification of the product mixture during sublimation can be explained by many kinetic isomers rearranging to fewer, more thermodynamically stable, isomers. Thermally induced rearrangement of CF_3 groups on the cage (or product simplification) is discussed in more detail in Section 2.3.3.

The improved AgTFA approach has since been employed in numerous fullerene trifluoromethylation reactions [20–24], but it is not a good candidate for the economical production of appreciable quantities of TMFs for several reasons. First, to increase the reaction scale, larger pipe reactors are needed, which in turn must be able to handle the pressure generated by release of CO_2 upon decomposition of TFAs in the sealed reaction vessel. Second, this process involves an extra sublimation step, which is required to remove unstable TMF-Ag complexes from the product mixture, but also initiates CF_3 loss and rearrangement and limits the final TMF composition to only the most thermodynamically stable. As a result, the tunability of this trifluoromethylation approach is also limited such that kinetic isomers cannot be targeted. Nonetheless, there is still unexplored reaction parameter space within the solid–solid approach that may hold useful information for the overall improvement of TMF preparation.

2.2.4 Gas–Solid Reactions

The most chemically simplistic approach to synthesizing TMFs involves the reaction of gaseous CF_3I with solid fullerene. It is the most commonly employed method to date, and greatly reduces the rate of formation of non-TMF side products. In only one case has an iodine–fullerene bond been reported; however, the authors noted that even in air at room temperature the compound $C_{60}(OO\text{-}t\text{-}Bu)_4(OH)I$ decomposed via loss of iodine to form both the bis-hydroxyl and epoxide analogs [25]. At elevated

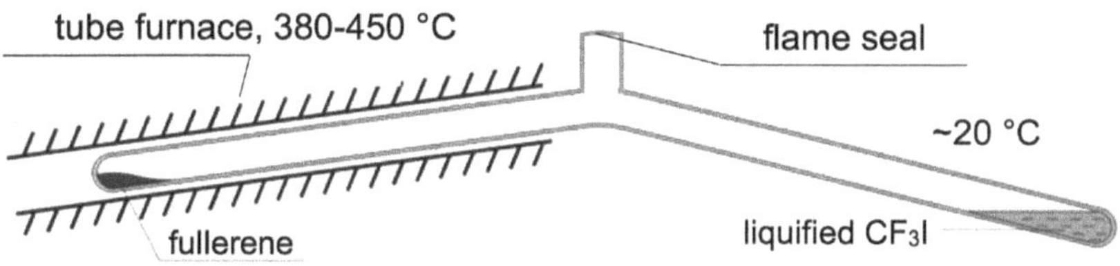

Figure 2.2 A general design schematic is shown for fullerene trifluoromethylation reactions using the sealed ampule method.

temperatures, iodine has never been shown to form stable bonds with fullerenes, and therefore can be removed as I_2 or scavenged as CuI in the presence of copper powder in TMF reactions. There have been three general reactor designs for gas–solid fullerene trifluoromethylation: (1) sealed ampule, (2) flow tube, and (3) gradient-temperature gas–solid (GTGS) reactions. Experiments from each of these reactor designs have provided valuable information on how to tune the reactive zone conditions in order to prepare particular TMFs.

The basic reactor design for a sealed ampule reaction is shown in Fig. 2.2. In a typical experiment, fullerene is loaded into one end of a glass ampule, evacuated, and then an excess of liquid CF_3I is condensed into the other end. The ampule is then flame sealed and the section containing fullerene is placed into a tube furnace, heated at 380–450°C. The end containing liquid CF_3I is kept at room temperature and maintains c.5 atm of CF_3I pressure within the ampule. Caution: the highest care should be taken when heating sealed glass ampules that contain CF_3I or any volatile compounds. The pressure within the ampoule can reach as high as 30 atm due to formation of C_2F_6, which can cause violent detonation of the ampule if it is not properly sealed or specified to handle such pressures.

During the reaction, CF_3I molecules homolytically cleave to generate CF_3 radicals and atomic iodine, which condenses outside the reactive zone as I_2. TMFs are formed as CF_3 radicals attach to exposed fullerene cage, and the latter can also form gaseous C_2F_6. TMF products sublime out of the hot zone and condense on the relatively cooler walls of the ampule. Ampule trifluoromethylation reactions typically achieve high conversion of starting fullerene, but also require reaction times on the order of days. Also, these reactions typically produce product mixtures containing mostly higher TMFs (especially carrying 10 to 14 groups), likely because of the high pressure of CF_3I and longer reaction times. In order for fresh solid fullerene to be exposed, the outer layer of the pile must be trifluoromethylated to the extent that it can sublime away. TMFs sublime at lower temperatures than their parent fullerene, and in general their sublimation temperature is inversely related the number of CF_3 groups on the cage. For example, the sublimation temperature of C_{60} under dynamic vacuum was measured to be 500°C, yet under the same conditions the sublimation temperature of $C_{60}(CF_3)_{10}$-3 was measured at c.250°C; a similar sublimation temperature difference was observed for C_{70} and $C_{70}(CF_3)_{10}$-1. Simply, this can be understood by a weakening of intermolecular forces between cages due to an incrementally diminished pi system and also an increased average cage–cage distance as CF_3 groups are added.

There is evidence that the reactive zone conditions in the ampule synthesis are not well suited for preparation of lower TMFs. In an example where the reaction time was shortened from days to 6 h, the main difference in the product mixture was a much lower conversion of C_{60}; very few lower TMFs were present. These observations do not support an alternate hypothesis that involves a mechanism where the bulk (as opposed to the surface) of the fullerene is trifluoromethylated. Other variable reactive zone parameters such as the length of the hot end of the ampule (thus residence time of TMFs in the gas phase) and temperature dependence of the CF_3I end (thus CF_3I pressure and larger hot to cold zone temperature gradient) have not been explored. Many TMFs have been isolated in high isomeric purity using this method; however, the nature of the reactor design makes it extremely difficult and risky to scale up (the burst pressure of a sealed glass ampule is inversely proportional to its diameter). Also, using larger quantities of fullerene in thick-walled smaller diameter ampules has caused clogging of the ampule in cooler sections where TMF products and I_2 condense.

The risk of explosion is overcome using a flow tube reactor, which is depicted in Fig. 2.3. A flow tube reaction is typically performed as follows. A glass or fused silica tube is loaded with finely ground fullerene and connected to a gas handling system equipped with Ar or N_2 and CF_3I gas inlets. The location of where the fullerene is loaded within the glass tube can be varied. The outlet of the tube is generally connected to a mineral oil bubbler, which keeps the relative pressure of the system slightly positive (to atmospheric pressure) and prevents backflow of air into the reaction. Once connected, the system is thoroughly flushed with an inert atmosphere. Then, the portion of the tube containing fullerene is heated using a tube furnace. Once the reaction temperature is reached (400–550°C), CF_3I gas is very slowly flowed through the system. Iodine vapor can be seen forming when the CF_3I gas reaches the hot reactive zone in the tube. Volatile TMF products as well as I_2 sublime out of the hot zone and condense on the cooler walls of the tube outside of the furnace. Reaction times for $C_{60,70}$ are shorter than ampule reactions at 1–2 h, but some have been reported for up to 4 h [4]. The reactor should be located in a well-ventilated fume hood to avoid inhalation exposure to any CF_3I decomposition vapors.

Tube furnace reactions have produced TMFs carrying from 2 to 18 CF_3 groups. Typically, the compositions of fullerene(CF_3) products are narrowly distributed, but both higher and lower TMFs can be targeted. For example, C_{60} trifluoromethylation reactions carried out at 460°C yield full conversion of C_{60} and mostly higher TMFs (ie, $C_{60}(CF_3)_{8-12}$), while reactions carried out at 600°C produce mostly $C_{60}(CF_3)_{2-6}$ as well as a significant amount of unreacted C_{60} [26,27]. The competing processes of

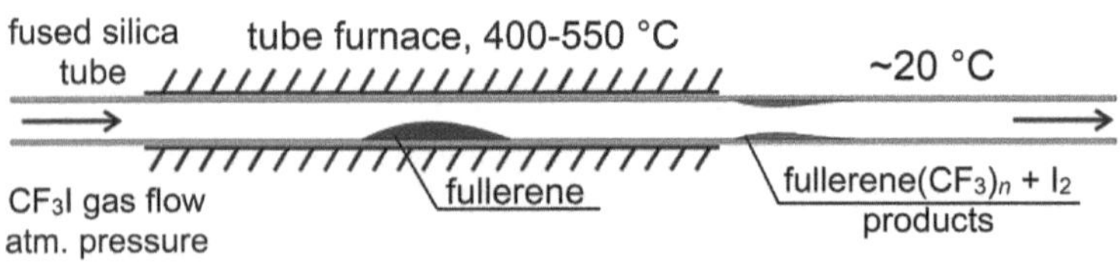

Figure 2.3 A general design schematic is shown for fullerene trifluoromethylation reactions using the flow tube method.

reaction zone residence time and sublimation temperature can explain this inverse relationship between temperature and product distribution. At higher temperatures, more fullerene(CF_3) molecules are subliming (the sublimation temperature for fullerene$(CF_3)_n$ decreases with increasing n), and therefore exiting the hot zone, which shortens the residence time. Experimental evidence suggests that the reactive zone processes depicted in Fig. 2.1 occur more rapidly in the gas phase than in the solid phase. When lower reaction temperatures are used, two key points favor formation of higher TMFs: (1) more CF_3 groups must accumulate on the fullerene to promote sublimation (more CF_3 groups are required to drive the sublimation temperature down to the reaction temperature) and (2) the residence time in the gas phase is longer due to a smaller temperature differential between the hot and cold zone. According to this hypothesis, this would also mean that for a given temperature, the length of the hot zone could be used to directly tune the gas phase residence time. Some evidence already supports this supposition, as in fact longer hot zones do favor higher TMF formation [4,27–29].

The presence of copper powder intimately mixed with fullerene in the hot zone was shown to promote CF_3I dissociation at lower temperatures, and was used to presumably generate CF_3 radicals in closer proximity to fullerene [26]. Using this approach, the synthesis of $C_{70}(CF_3)_{10}$ (abbreviated as 70-10-1, similar notations are used for TMFs throughout the text; see Appendix 1 for Schlegel diagrams of structures) was improved from 27% isolated yield to as high as 55 mol% yield at 90% purity [4,30]. Overall, the flow tube reactor provides a safer alternative to ampule reactions. Partial tunability of the product distribution can be achieved using the flow tube reactor, although it is generally better suited toward production of higher TMFs. While the flow tube reactor is safer, simpler, and more scalable than the ampule method, a major drawback is continuous (albeit low—0.5 mmol/min) flow of CF_3I gas. For scaling up, this issue can become quite costly and wasteful.

The third reactor for gas—solid reactions was conceived at Colorado State University and is known as the GTGS reactor, which is depicted in Fig. 2.4. It is equipped with unique design features that allow for precise control over the temperature of the hot zone, length of the hot zone, pressure of CF_3I, hot to cold zone temperature gradient, and is capable of multigram reaction batches. A general GTGS reaction is performed as follows. Finely ground fullerene powder is loaded into a quartz thimble and connected to a gas handling line with vacuum capabilities, such as a Schlenk line. The system is then evacuated and the quartz thimble is placed into a bored out brick furnace; the length of the hot zone can be controlled by the depth of the bore. A recirculating cold finger can also be positioned inside the thimble to control the length of the hot zone. It is also needed in order to effectively collect iodine vapor during larger reaction batches. The thimble is heated to 400—550°C over about 30 min and then a specified pressure of CF_3I gas (10^{-2} up to 1.05 atm) is introduced into the system. During the course of the reaction (0.5—8 h), the pressure is monitored and can be adjusted by adding fresh CF_3I gas. At the completion of the reaction, the brick furnace is lowered to allow the thimble to cool, and then the system is evacuated to remove residual CF_3I and any C_2F_6 byproducts. Reaction products are collected from the thimble by rinsing with a minimal volume of organic solvent such as toluene or 1,2-dichlorobenzene.

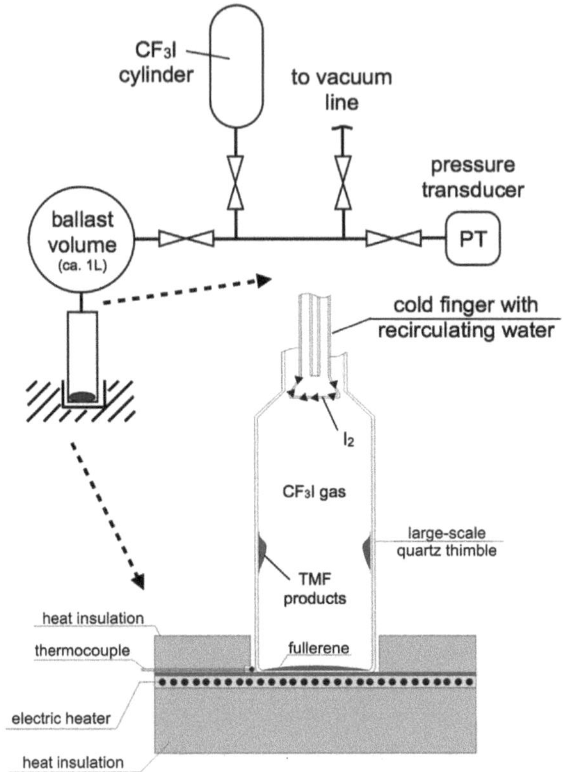

Figure 2.4 A general design schematic is shown for fullerene trifluoromethylation reactions using the gradient-temperature gas–solid method. *TMF*, trifluoromethylfullerene.

A major advantage of the GTGS reactor over ampule or flow tube reactors is that it offers more control over the reaction zone conditions and therefore the product distribution. Kuvychko et al. reported the detailed effects of each reaction zone parameter on fullerene trifluoromethylation [11]. For otherwise identical reaction conditions, an increase in CF_3I pressure results in higher conversion of fullerene and product compositions containing more CF_3 groups. Increasing the reaction temperature for a given pressure also increases the content of higher TMFs. When only the length of the hot zone is increased, the overall conversion increases and higher TMFs are favored. When the reaction time increased, the product distribution is mostly unchanged, but the conversion increases. For large-scale reactions, a cold finger in the reaction thimble is necessary to condense iodine vapor, which suppresses the reaction by buffering contact of CF_3I with fullerene. The synthesis of lower TMFs in high yield was achieved by using a short reaction zone, high temperature (c.500°C), and a lower pressure of CF_3I gas (10–30 torr) [11,31].

In addition to compositional control over the product mixture, the GTGS reactor allows for more kinetic products to be targeted as well, to an extent that cannot be accomplished with either ampule or flow tube methods. Access to large isomeric sets of TMFs can greatly benefit certain research goals, such as structure–property relationships, DFT validation, and even applied charge transfer studies [4,24,32–35]. Separation of these mixtures of isomers is no longer problematic (see Section 2.3.1 for more), so designing synthetic routes toward complex product mixtures is actually desirable. Recently, 27 isomers of $C_{60}(CF_3)_{10}$ were isolated from a mixture of higher TMFs and 8 of them yielded crystals that were studied by single-crystal X-ray diffraction [36]. Previous attempts to synthesize a large number of $C_{60}(CF_3)_{10}$ compounds using a flow tube reactor yielded only six isomers [32]. Fig. 2.5 shows two HPLC chromatograms that compare the formation of higher TMFs using either a flow tube reactor or a GTGS reactor. Fewer, sharper peaks are seen in the flow tube product chromatogram in contrast to many more and broader peaks in the GTGS product chromatogram, which is due to the presence of many isomers that have similar retention times (more about retention times in Section 2.3.1). While both reactions target similar compositional products, the GTGS reactor produces far more kinetic products. Again, controlling the reactive zone conditions is the key to controlling TMF product composition. For this GTGS reaction, a higher pressure of CF_3I gas was used in addition to a shorter hot zone and a high reaction temperature. By creating an environment with a high concentration of reactive species and a short residence time, many kinetic products exit the reaction zone before annealing into thermodynamic products. For this reason, the flow tube product chromatogram shows a narrower distribution of higher TMFs and less isomeric diversity.

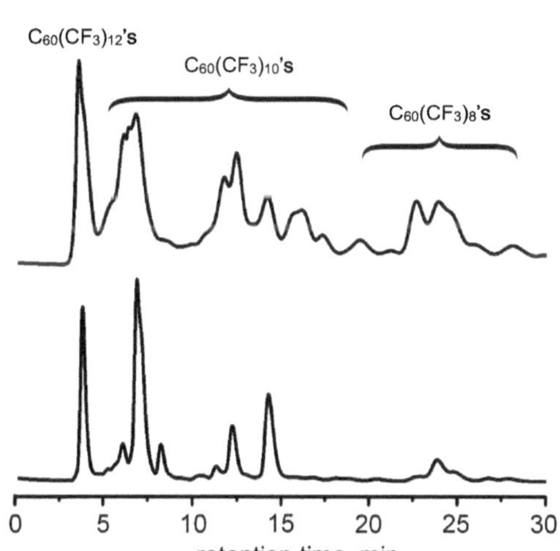

Figure 2.5 HPLC traces are shown for a mixture of higher TMFs synthesized using reaction conditions favoring kinetic products (GTGS, top) versus conditions favoring thermodynamic products (flow tube, bottom). The HPLC conditions were 10 × 250 mm Buckyprep column, 5 mL/min flow rate, 20/80 v/v toluene/heptane, 300 nm detection. *GTGS*, gradient-temperature gas–solid; *HPLC*, high-performance liquid chromatography; *TMFs*, trifluoromethylfullerenes.

Several options exist for fullerene trifluoromethylation reactions between solid fullerene and gaseous CF_3I. Of the three discussed here, the flow tube and GTGS methods are best suited for scaling up, even though considerable design changes should be made to make CF_3I gas usage more economical in the case of the flow tube. Both ampule and flow tube reactions generally produce narrow compositional product mixtures of higher TMFs that contain mostly the most stable thermodynamic products. The GTGS reactor offers the most control over product composition and gives the greatest access to kinetic products because it has so many reactive zone variable parameters, but with the downside that it is a specialized custom built reactor. Most importantly, all three reactors have collectively contributed to the knowledge that will inevitably lead to the next generation TMF reactor design.

2.3 Trifluoromethylfullerene Isolation Methodologies

2.3.1 High-Performance Liquid Chromatography

Besides the few rare examples described above, fullerene trifluoromethylation is not an isomerically selective process. Compositional selectivity can be achieved with a moderately narrow distribution (ie, fullerene$(CF_3)_n$ where $n = 2-6$, $8-12$, $10-14$, etc.); however, complex mixtures of isomers may still be present within a compositionally narrow product mixture, which further complicates the task of separation. Clearly, the need for powerful separation methodologies is evident when dealing with such potentially complex product mixtures. HPLC has been the most commonly employed separation technique for TMFs to date. Many aspects of an HPLC separation method can be varied, including the column size and packing material, flow rate, composition of the mobile phase, and temperature. More about these will be discussed below, but for now the general procedure is described. First, the crude product mixture is dissolved in the solvent that will be used as the mobile phase for separation, and then subjected to submicron filtration prior to injection. Since the product mixture can contain many TMFs as well as unreacted starting fullerene, a multiple-stage separation strategy is often implemented. In general, TMFs have shorter retention times than their parent fullerene (and higher TMFs elute faster than lower TMFs), so the first stage of separation is usually designed to crudely fractionate the product mixture and recover any unreacted starting fullerene. Each fraction is then dried and redissolved in the solvent (or mixture) that will be used as the mobile phase in the next stage of separation. Conditions for subsequent HPLC stages are chosen to intentionally lengthen the retention times compared to the previous stage. In some cases, this allows for baseline or near baseline collection of peaks consisting of isomerically pure TMFs. An example of this strategy is shown in Fig. 2.6 for the isolation of three lower TMFs: 60-2-1, 60-4-3, and 60-4-3-O.

The top panel in the figure shows an HPLC chromatogram of the products from a GTGS reaction designed to produce lower TMFs. In the first stage of separation, 100% toluene was used as the mobile phase, and the fraction eluting between 6.5 and 7.8 min was collected for second stage separation. Note that under first stage conditions, the

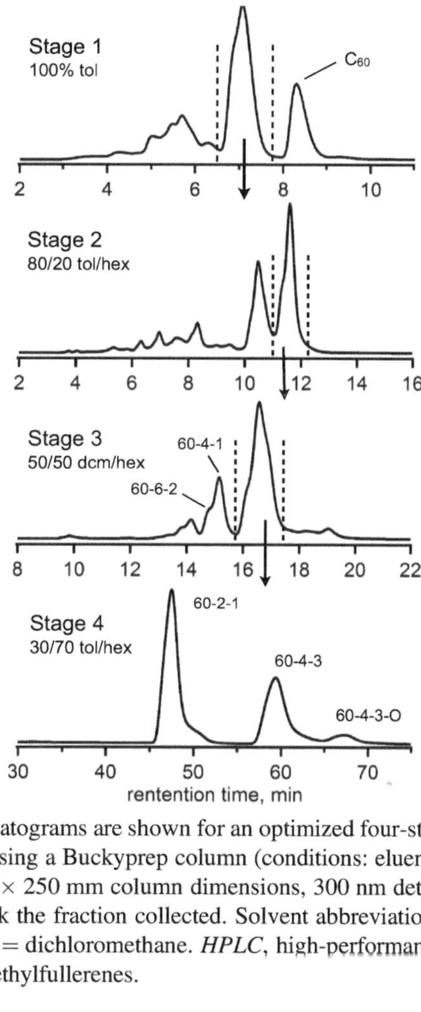

Figure 2.6 HPLC chromatograms are shown for an optimized four-stage preparative separation scheme of lower TMFs using a Buckyprep column (conditions: eluent noted in figure legend, 16 mL/min flow rate, 25 × 250 mm column dimensions, 300 nm detection wavelength). The vertical dashed lines mark the fraction collected. Solvent abbreviations are tol = toluene, hex = hexanes, and dcm = dichloromethane. *HPLC*, high-performance liquid chromatography; *TMFs*, trifluoromethylfullerenes.

unreacted C_{60} (eluting between 8 and 9 min) is excluded from the collected fraction. The weaker solvent conditions (80/20 v/v toluene/hexanes) used in the second stage of separation lengthened the retention time compared to the first stage, and expanded the single first-stage peak into two main peaks, the second of which (t_r = 11−12.2 min) was collected for a third stage of separation. In the third stage, a 50/50 dichloromethane/hexanes mobile phase was used to further resolve the fraction and increase t_r. Conditions for the fourth and final stage of separation allowed for baseline separation of the three TMFs. Note that the 30/70 toluene/hexanes eluent in this stage of separation resulted in nearly 70 min of retention.

As illustrated in the example above, the composition of the mobile phase can have a dramatic impact on TMF retention times. HPLC separation relies on the interplay between the affinity of the analyte toward the stationary phase versus the solvation strength of the mobile phase. "Strong" solvents, meaning the TMF has more affinity for the mobile phase than the stationary phase, result in shorter retention times. These solvents are typically aromatic or olefinic hydrocarbons, but also include singly, doubly, or triply substituted halo-benzenes, toluene, and derivatives of naphthalene. Weaker solvents have an opposite effect on retention times because the analyte has more affinity for the stationary phase and thus takes longer to pass through the column; these solvents include normal chain alkanes, cycloalkanes, and even some primary alcohols such as hexyl, heptyl, and octyl alcohol. As also shown in the above example, mixtures of strong and weak solvents can be used to fine-tune the separation of compounds with similar retention times. Other TMF solvents that can be used alone or in mixtures include chloroform, dichloromethane, and acetonitrile, which have different effects on retention time depending on the analyte mixture and specific stationary phase material in the HPLC column. Caution is urged when experimenting with solvents to ensure chemical compatibility with the HPLC column and equipment in order to avoid causing costly irreversible damage. It should be noted that even though the use of a weaker mobile phase allows for isolation of isomerically pure compounds, it may come at a considerable twofold expense. In addition to long retention times (see Fig. 2.6, Stage 4), a weaker mobile phase also reduces the overall solubility of the TMF mixture, and can therefore require substantially more injections (and time) for preparative separation.

Different columns can also be used to tune the separation capabilities for TMFs. At present, the most efficient reverse phase columns for HPLC separation of fullerene derivatives are commercially available from Nacalai Tesque Inc., Japan. Each contains high purity spherical silica beads, but differ in the surface modification of those beads. The chemical structures of the stationary phase modifiers are shown for five different fullerene HPLC columns in Fig. 2.7. Each column offers unique separation capabilities because each stationary phase interacts differently with TMFs. In nearly every case of HPLC isolation of TMFs in the literature, a Buckyprep column was used, as it is advertised as the premiere column for fullerene derivative separation. For the Buckyprep column, the retention times of different TMFs can be understood by the

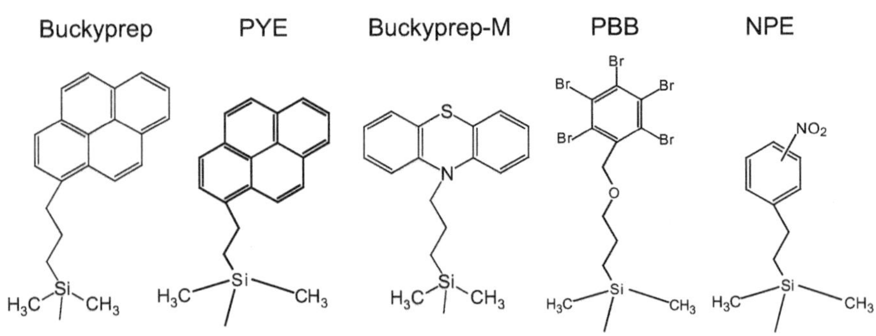

Figure 2.7 Molecular structures are shown for five types of modified silica packing material in columns advertised for fullerene separation by Nacalai Tesque Inc.

fundamentally pi—pi interaction between the pyrenylpropyl-coated silica packing material and the fullerene cage pi system; the latter is determined by the number and addition pattern of CF_3 groups on the cage. The observed result is that higher TMFs, which have fewer remaining cage double bonds and therefore less pi—pi interactions, have shorter retention times compared to lower TMFs and the bare parent fullerene. As shown above, this mode of resolving power has been successfully exploited to obtain isomerically pure compounds and is quite useful for separation of lower TMFs and fractionation of compositional mixtures.

Higher TMFs, and especially isomeric mixtures, are considerably more problematic to separate using a Buckyprep column because there is less fullerene pi character and structure differentiation. A different column, Buckyprep-M, was found to be an excellent alternative to Buckyprep for separation of TMFs. It contains phenothiazine as the active component in its stationary phase, which has an aromatic component for pi—pi retention as well as atoms with lone electron pairs, sulfur in this case, that can also interact with molecules in the mobile phase. Buckyprep-M columns were designed and are advertised for separation of endohedral metallofullerenes, which are zwitterionic compounds (the encapsulated metal atoms or clusters formally donate electrons to the cage); and hence they possess unique electronic properties. Clearly, having this additional form of interaction between the stationary phase and analyte owes to their successful use in endohedral metallofullerene chromatography, and may also explain why isomeric mixtures of TMFs are better resolved with this column. Fig. 2.8 shows a comparison of HPLC chromatograms for a mixture of $C_{60}(CF_3)_{10}$ isomers analyzed using either a Buckyprep or Buckyprep-M column.

The only difference between the two analyses was the diameter of the column, which was accounted for by adjusting the flow rate. Two major differences in the chromatograms are immediately noticeable: (1) the sample elutes between 12 and 47 min on the Buckyprep column and 4—30 min on the Buckyprep-M column, and (2) the top chromatogram shows 4 main peaks while the bottom reveals at least 13. An interesting observation is that some of the $C_{60}(CF_3)_{10}$ isomers were retained for at least 25 min on the Buckyprep column, but eluted almost immediately (shortly after dead volume elution) from the Buckyprep-M column, meaning that these compounds have much greater affinity for pyrene than for phenothiazine. Many of the sharp peaks in the Buckyprep-M chromatogram represent pure compounds.

It is important to reiterate that the effect of the mobile phase on TMF retention for one column will not necessarily produce the same effect with another column. This concept is presented in Fig. 2.9. From a mixture of $C_{60}(CF_3)_{10}$ compounds, 60-10-14 can be isolated using either a Buckyprep or Buckyprep-M column. When a Buckyprep column is used, the fraction containing 60-10-14 contains a significant amount of other isomers, as evidenced by the ^{19}F NMR spectrum of that fraction. Note that 60-10-14 is the last major peak to elute, at approximately 27 min. When a Buckyprep-M column is used, many other smaller peaks are resolved, but also, both the retention order (now second) *and* retention time (approximately 9 min) for 60-10-14 changes. In both cases, 60-10-14 can be isolated in sufficient purity to grow single crystals, yet the ^{19}F NMR spectrum reveals much higher purity was achieved in the latter case.

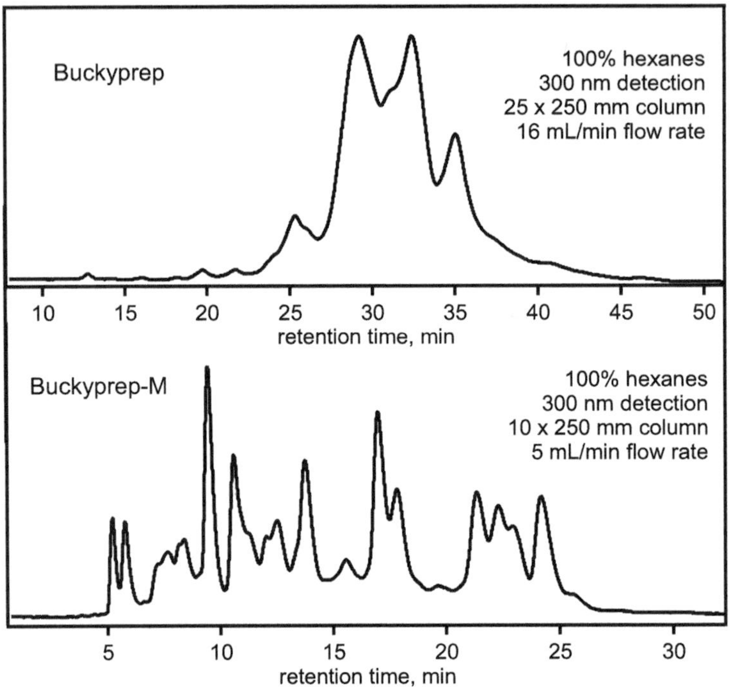

Figure 2.8 High-performance liquid chromatography chromatograms are shown for a sample containing $C_{60}(CF_3)_{10}$ compounds analyzed using either a Buckyprep (top) or Buckyprep-M (bottom) column.

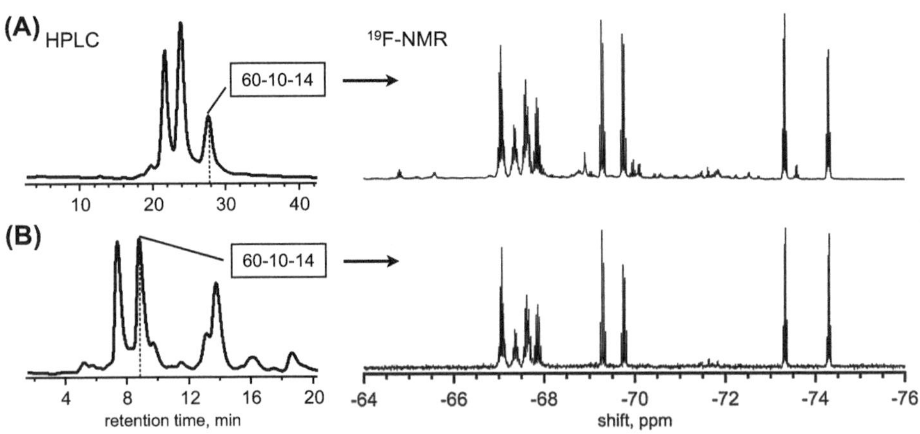

Figure 2.9 High-performance liquid chromatography chromatograms (left) are shown for the analysis of a mixture of $C_{60}(CF_3)_{10}$ compounds using either a Buckyprep (A) or Buckyprep-M (B) column with 100% hexanes eluent and 300 nm detection. ^{19}F NMR spectra (right) highlight the purity of $C_{60}(CF_3)_{10}$-14 isolated from the mixture using each column. The spectra were recorded in $CDCl_3$ at 376.48 MHz using C_6F_6 as an internal standard (δ−164.9).

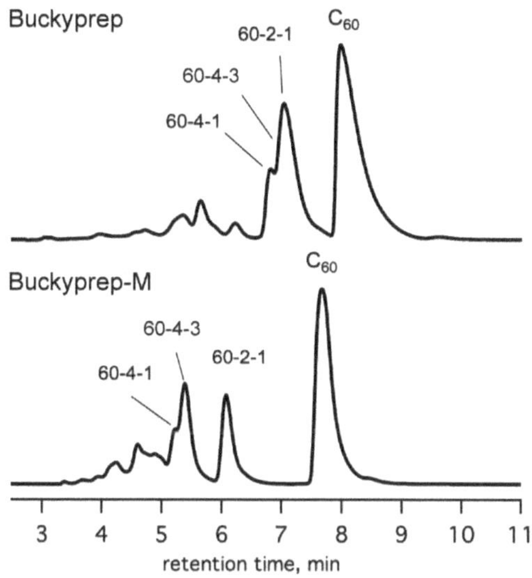

Figure 2.10 High-performance liquid chromatography chromatograms are shown for a crude product mixture of lower trifluoromethylfullerenes (TMFs) analyzed using either a Buckyprep (top) or Buckyprep-M (bottom) column with 100% toluene eluent and 300 nm detection. Peaks corresponding to lower TMFs are labeled.

In addition to isomeric separation of higher TMFs, the Buckyprep-M column can also be used to achieve excellent compositional separation of lower TMFs. In fact, 60-2-1 can be isolated in high purity using a single stage of HPLC separation in less than 10 min. This compares to four stages totaling 120 min of retention (refer to Fig. 2.6) using a Buckyprep column. Fig. 2.10 shows the HPLC chromatograms of a crude product mixture of lower TMFs for the two columns under otherwise identical conditions. In the top chromatogram (Buckyprep), 60-2-1 coelutes with at least two other TMFs whereas baseline peak resolution is achieved using Buckyprep-M in a single stage. Another advantage is that 100% strong eluent (toluene) is used, which increases the mass loading of each injection due to higher solubility, further decreasing the total separation time by decreasing the overall number of injections required to process a given amount of material. This HPLC method, along with the improved synthesis of 60-2-1 on a gram plus scale, provides a reasonable pathway toward its availability for applied research efforts.

The use of a weaker mobile phase further improves the separation of lower TMFs on a Buckyprep-M column, too. When a 50/50 mixture of toluene/hexanes is used, several other TMFs (in addition to 60-2-1) can be also isolated in a single stage, as shown in Fig. 2.11. Even with this weaker mobile phase, the retention time of C_{60} is still only about 16 min, much shorter than it is under the same conditions with a Buckyprep column. This could be due to the fact that phenothiazine has a smaller surface area of conjugation compared to pyrene, which means less physical space for pi–pi interactions and gives rise to shorter retention times. As the library of TMFs expands, experiments can be designed to better understand the relationship

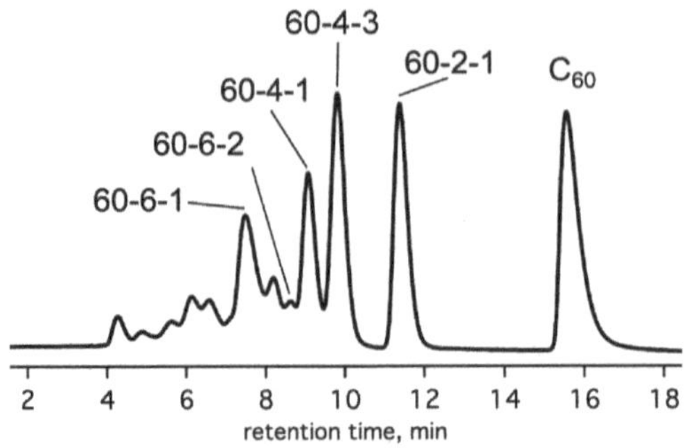

Figure 2.11 A high-performance liquid chromatography chromatogram is shown for the single-stage separation of lower trifluoromethylfullerenes (conditions: 50/50 v/v toluene/hexane eluent, 5 mL/min flow rate, 10 × 250 mm Buckyprep-M column, 300 nm detector wavelength).

between specific electronic and structural properties and the interaction with a particular stationary phase. Such findings would be greatly beneficial to not only TMF separation, but also fullerene derivative HPLC methods in general. One set of compounds that would be particularly well-suited for that kind of study would be the 13 crystallographically proven isomers of $C_{60}(CF_3)_{10}$ [36].

In summary, HPLC is a powerful technique for efficient isolation of isomerically pure TMFs. By varying the mobile phase composition, column type, or both, complex compositional and isomeric mixtures can be readily separated. The recent use of a Buckyprep-M column for separation of TMFs has produced dozens of new compounds from previously inseparable mixtures, and serves as an example that the use or development of other new columns may result in even better separation capabilities. HPLC separation of TMFs is no longer as laborious and time-consuming as it once was, but the upfront cost of the equipment is still appreciable. For large-scale preparatory purposes, though, HPLC remains the most versatile and effective option.

2.3.2 Solvent Extraction and Selective Crystallization

TMFs exhibit an impressive spectrum of physical and chemical properties, and these properties can be exploited to easily separate compounds that would otherwise require significant chromatographic workup. One property that can be exploited is solubility; TMFs with a higher degree of substitution are more soluble in solvents that are poor for bare C_{60} or C_{60} carrying only two CF_3 groups. Therefore, selective fractionation of the product mixture can be implemented to greatly shorten later stages of workup that would otherwise require HPLC. Fig. 2.12 shows the HPLC analyses of the same complex product mixture of TMFs after extraction by several "poor" solvents for bare fullerenes. In the experiment, a crude TMF product mixture was divided into four portions and dried under vacuum. To each portion, a different poor fullerene solvent (isooctane, hexane, pentane, or 1-propanol) was added and stirred for 20 min. The samples were then filtered, yielding an extract and an insoluble material. The insoluble material was washed with the corresponding solvent, and each fraction was dried and easily redissolved in toluene, then analyzed by HPLC. As seen in Fig. 2.12, a significant amount of higher TMFs

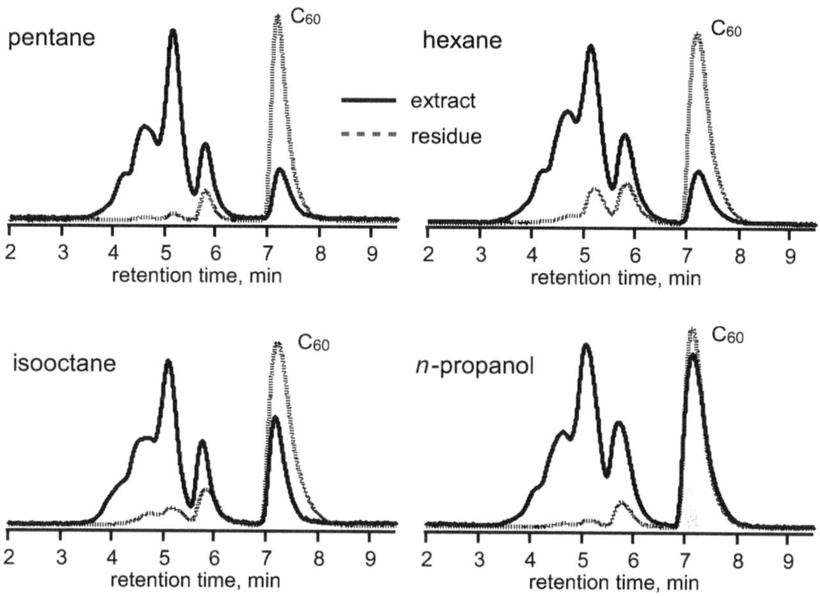

Figure 2.12 High-performance liquid chromatography (HPLC) chromatograms are shown for a crude trifluoromethylfullerene product mixture fractionated by poor fullerene solvents (denoted in each inset) into extract (solid trace) and insoluble residue (dotted trace). HPLC conditions: 100% toluene eluent, 5 mL/min flow rate, 10 × 250 mm Cosmosil Buckyprep-M, 300 nm detection.

can be removed from the product mixture by solvent fractionation, and pentane is a good solvent for extracting TMFs from residual C_{60} in the product mixture.

Two lower TMFs, $C_{60}(CF_3)_4$-1 and $C_{60}(CF_3)_6$-2, have similar HPLC retention times under optimized solubility-to-return conditions. This term refers to the mass loading in a poor solvent versus the retention time and resolution of peaks. In other words, how weak of a mobile phase can be used before the TMFs reach such a low solubility that an unreasonable number of injections are required for separation? One option for isolating 60-4-1 in high purity after one fractional HPLC stage (refer to the HPLC conditions shown in Fig. 2.11) is an HPLC-free second step. Simply, the fraction containing a mixture of 60-4-1 and 60-6-2 (from the first stage of HPLC separation) is adjusted to 30/70 toluene/hexane. In this solvent mixture, 60-4-1 has much lower solubility than 60-6-2 and upon drying, they phase separate: 60-4-1 forms dense black crystals and 60-6-2 (and any other trace TMF compounds) forms orange semicrystalline solids. The degree of crystallinity depends on the presence of other TMF impurities in the fraction, which coprecipitate together with the orange 60-6-2 solids. Then, the 60-6-2 and other impurities if present are filtered or decanted away by rinsing the crystals first with hexanes and then briefly with toluene; 60-4-1 crystals remain intact. This selective precipitation followed by solvent washing method yields highly pure 60-4-1 crystals and does not require additional HPLC purification. It is noted that the 60-4-1 crystals do require gentle heating and stirring to be fully re-dissolved later.

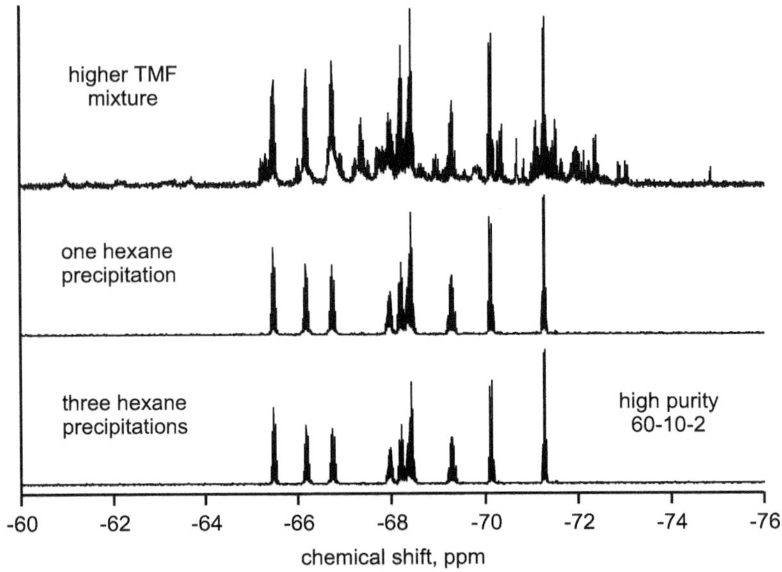

Figure 2.13 ^{19}F NMR spectra are shown following the process of selective precipitation of 60-10-2 (bottom two) from a TMF mixture (top) using hexane. The spectra were recorded in $CDCl_3$ at 376.48 MHz using C_6F_6 as an internal standard ($\delta -164.9$). *TMF*, trifluoromethylfullerenes.

The compound 60-12-1 is a higher TMF that can also be separated easilyby selective precipitation [3]. It forms orange crystals with a dense lock-and-key packing motif and infinite linear chains that result in crystallization even in the presence of other TMF compounds. The tightly packed crystals exhibit low solubility under gentle washing conditions, allowing other TMFs formed in the reaction to be filtered or decanted away. Another higher TMF, 60-10-2, is highly soluble in toluene, but scarcely soluble in hexane. Many other higher TMFs are soluble in hexane, especially if present as mixtures. Therefore, 60-10-2 can be selectively precipitated from a mixture of $C_{60}(CF_3)_{10}$ compounds in hexane by allowing the solution to rest for 24 h. During this time, tiny crystals of 60-10-2 precipitate and settle, which can be collected by filtering and washing with hexane. The purity can reach +95 mol% if the solid is subsequently dissolved in a minimal volume of toluene and reprecipitated, filtered, and washed with hexane. ^{19}F NMR spectra of a higher TMF mixture and the resulting 60-10-2 precipitate are shown in Fig. 2.13 for this method.

2.3.3 Sublimation and Thermal Annealing

Some research purposes may benefit from fewer, more thermodynamically stable, TMFs in the crude product mixture. Regardless of the synthetic route taken, the product mixture can be reworked by thermal annealing or high temperature sublimation. This notion was first conceived to simplify the product workup in metal TFA trifluoromethylation reactions [19]. Temperatures above 400°C (or in c.100°C excess of the

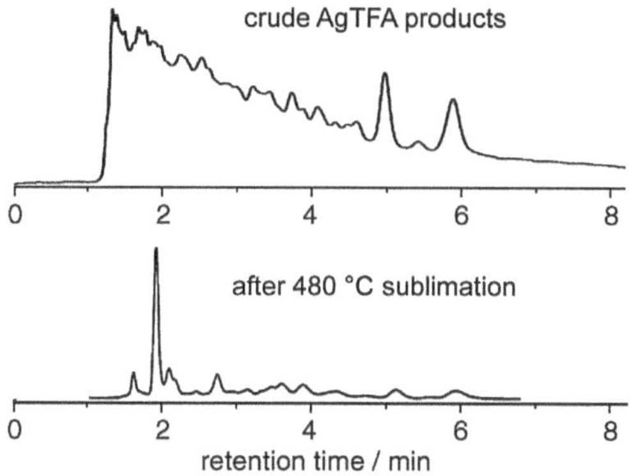

Figure 2.14 High-performance liquid chromatography chromatograms are shown for the crude product of an AgTFA reaction with C_{70} before (top) and after (bottom) subliming at 480°C. Figure reproduced and modified from E.I. Dorozhkin, et al., Synthesis, characterization, and theoretical study of stable isomers of C70(CF3)*n* (*n* = 2, 4, 6, 8, 10), Chem. Eur. J. 12 (2006) 3876–3889.

sublimation temperature of the TMF) should be used for product simplification/rearrangement, as lower temperatures have been shown not to cause TMF rearrangement [19,26]. Representative HPLC traces are shown in Fig. 2.14 for the simplification of the product mixture from a metal TFA reaction with C_{70}. In this case, other hydrogenated TMFs as well as soluble metal TMF complexes are decomposed, however, the distribution of TMF products simplified to mostly 70-8-1 [22]. In the presence of bare fullerene, TMFs of C_{60} and C_{70} can rearrange and transfer CF_3 groups at high temperatures, leading to formation of the most thermodynamically stable isomers [37–40]. Care should be taken when utilizing this method, since elevated temperatures can lead to formation of unwanted side products, such as CF_2-linked TMF dimers due to fragmentation of cage CF_3 groups [41].

2.4 Conclusions and Outlook

The first TMFs were synthesized over 20 years ago, and yet this library of compounds continues to grow, along with valuable examples of the practical uses of these molecules. TMFs have proven to be a rich and diverse family of nanocarbon, and their study has directly produced innumerable scientific contributions to fullerene chemistry and beyond. Researchers have achieved a greater mechanistic understanding of TMF formation in the past two decades, and this knowledge has enabled specific isomers to be targeted by choosing the appropriate reaction conditions. At the same time, new HPLC products and separation methodologies have greatly increased the efficiency of isolating pure compounds from complex TMF mixtures. For example, the reactions directed toward producing kinetic isomers (ie, such as those in the original 1993 work by Fagan et al.) yielded practically inseparable mixtures as recent as only a few years ago, but now can be divided with relative ease into the wealth of individual compounds within. In terms of realizing TMFs in commercializable quantities, the bottleneck problem of separation has significantly eased; however, many opportunities for improvement still exist. Methods of improving fullerene production are steadily driving down starting material costs, and a higher global demand is further fueling those improvements. The need for physically and chemically robust organic molecules with a wide range of electronic properties grows with each day, and it is only a matter of time before TMFs are readily available enough to rise to the occasion.

Appendix

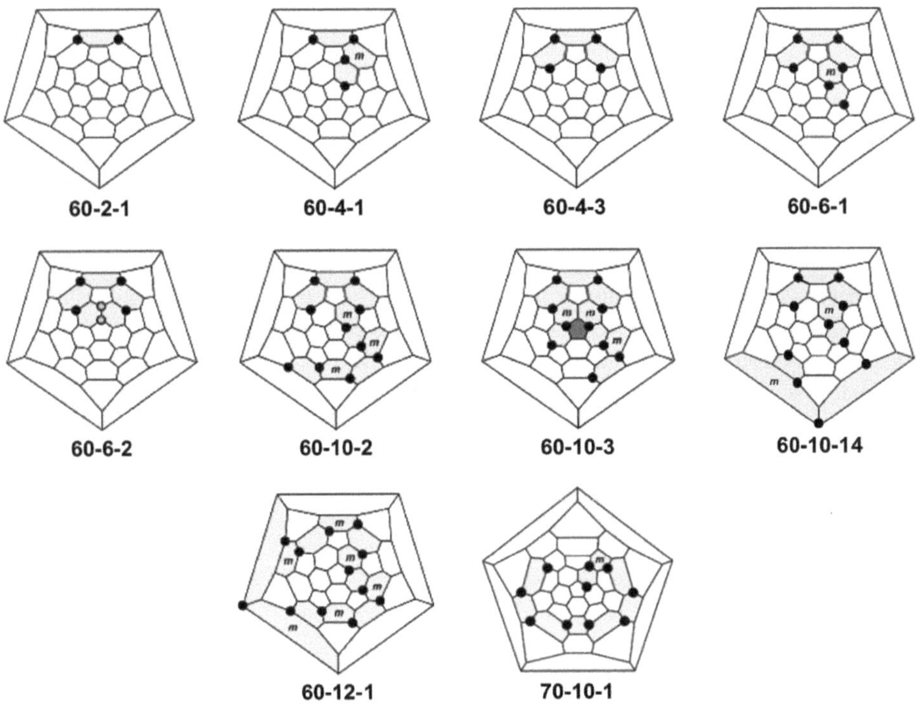

Figure A2.1 Schlegel diagrams [1] for specific trifluoromethylfullerenes discussed in this chapter.

Acknowledgment

Authors are thankful for partial financial support from National Science Foundation (grants CHE-1362302, CHE-1012468).

References

[1] O.V. Boltalina, A.A. Popov, I.V. Kuvychko, N.B. Shustova, S.H. Strauss, Perfluoroalkylfullerenes, Chem. Rev. 115 (2015) 1051−1105.

[2] A.A. Popov, S. Yang, L. Dunsch, Endohedral fullerenes, Chem. Rev. 113 (2013) 5989–6113.
[3] S.I. Troyanov, A. Dimitrov, E. Kemnitz, Selective synthesis of a trifluoromethylated fullerene and the crystal structure of $C_{60}(CF_3)_{12}$, Angew. Chem. Int. Ed. 45 (2006) 1971–1974.
[4] A.A. Popov, et al., Synthesis, spectroscopic and electrochemical characterization, and DFT study of seventeen $C_{70}(CF_3)_n$ derivatives (n = 2, 4, 6, 8, 10, 12), Chem. Eur. J. 14 (2007) 107–121.
[5] P.J. Fagan, et al., Production of perfluoroalkylated nanospheres from buckminsterfullerene, Science 262 (1993) 404–407.
[6] M. Yoshida, et al., Reaction of C_{60} with diacyl peroxides containing perfluoroalkyl groups. The first example of electron transfer reaction via $C_{60}^{+}\cdot$ in solution, Tetrahedron Lett. 34 (1993) 7629–7632.
[7] M. Yoshida, A. Morishima, Y. Morinaga, M. Iyoda, Reactions of fullerols and fullerene dimer containing perfluoroalkyl groups with tributyltin hydride, Tetrahedron Lett. 35 (1994) 9045–9046.
[8] N.B. Shustova, et al., High-temperature and photochemical syntheses of C_{60} and C_{70} fullerene derivatives with linear perfluoroalkyl chains, J. Fluorine Chem. 131 (2010) 1198–1212.
[9] A.K. Vorobiev, et al., Stable trifluoromethylated fullerene radicals $C_{60}(CF_3)_{15}$ and $C_{60}(CF_3)_{17}$, Mendeleev Commun. 20 (2010) 7–9.
[10] I.V. Kuvychko, S.H. Strauss, O.V. Boltalina, Journal of fluorine chemistry, J. Fluorine Chem. 143 (2012) 103–108.
[11] I.V. Kuvychko, et al., Pressure effect on heterogeneous trifluoromethylation of fullerenes and its application, J. Fluorine Chem. 132 (2011) 679–685.
[12] R.J. Doome, S. Dermaut, A. Fonseca, M. Hammida, J.B. Nagy, New evidences for the anomalous temperature-dependent solubility of C_{60} and C_{70} fullerenes in various solvents, Fullerene Sci. Technol. 5 (1997) 1593–1606.
[13] A. Demortier, R. Doome, A. Fonseca, J.B. Nagy, in: Magnetic Resonance in Colloid and Interface Science, Springer, Netherlands, 2002, pp. 513–518, http://dx.doi.org/10.1007/978-94-010-0534-0_47.
[14] O.V. Boltalina, P.B. Hitchcock, P.A. Troshin, J.M. Street, R. Taylor, Isolation and spectroscopic characterisation of $C_{60}F_{17}CF_2CF_3$ and isomers of $C_{60}F_{17}CF_3$; insertion of CF_2 into fluorofullerene C–F bonds, J. Chem. Soc. Perkin Trans. 2 (2000) 2410–2414, http://dx.doi.org/10.1039/b006890j.
[15] I.S. Uzkikh, E.I. Dorozhkin, O.V. Boltalina, A.I.A. Boltalin, New method for the synthesis of perfluoroalkyl derivatives of fullerenes, Doklady Chem. 379 (2001) 204–207.
[16] A.D. Darwish, A.G. Avent, A.K. Abdul-Sada, R. Taylor, [60]- and [70] Fullerenes are trifluoromethylated across 5:6-bonds, Chem. Commun. (2003) 1374, http://dx.doi.org/10.1039/b303496h.
[17] A.D. Darwish, et al., Unusual addition patterns in trifluoromethylation of [60] fullerene, Org. Biomol. Chem. 1 (2003) 3102.
[18] A.D. Darwish, et al., Novel addition in trifluoromethylation of [70] fullerene, J. Fluorine Chem. 125 (2004) 1383–1391.
[19] A.A. Goryunkov, et al., Isolation of $C_{60}(CF_3)_n$ ($n = 2, 4, 6, 8, 10$) with high compositional purity, J. Fluorine Chem. 124 (2003) 61–64.
[20] A.A. Goryunkov, et al., Crystal and molecular structures of $C_{70}(CF_3)_8\cdot$PhMe, Mendeleev Commun. 15 (2005) 225–227.
[21] V.Y. Markov, et al., Mass spectrometric studies of trifluoromethylated fullerenes, Int. J. Mass Spectrom. 251 (2006) 16–22.

[22] E.I. Dorozhkin, et al., Synthesis, characterization, and theoretical study of stable isomers of $C_{70}(CF_3)_n$ ($n = 2, 4, 6, 8, 10$), Chem. Eur. J. 12 (2006) 3876–3889.
[23] E.I. Dorozhkin, et al., Structure of 1,4,10,19,25,41-$C_{70}(CF_3)_6$, isomer with unique arrangement of addends, J. Fluorine Chem. 127 (2006) 1344–1348.
[24] E.I. Dorozhkin, et al., Synthesis, structure, and theoretical study of lower trifluoromethyl derivatives of [60]Fullerene, Eur. J. Org. Chem. 2007 (2007) 5082–5094.
[25] S. Huang, et al., Preparation of covalent-bound iodofullerene through selective opening of fullerene epoxide to form halohydrin fullerene derivatives, Synlett 2006 (2006) 1266–1268.
[26] I.E. Kareev, et al., Thermally stable perfluoroalkylfullerenes with the Skew-Pentagonal-Pyramid pattern: $C_{60}(C_2F_5)_4O$, $C_{60}(CF_3)_4O$, and $C_{60}(CF_3)_6$, J. Am. Chem. Soc. 128 (2006) 12268–12280.
[27] I.E. Kareev, et al., Synthesis, structure, and 19F NMR spectra of 1,3,7,10,14,17,23,28,31,40-$C_{60}(CF_3)_{10}$, J. Am. Chem. Soc. 127 (2005) 8362–8375.
[28] N.B. Shustova, D.V. Peryshkov, I.E. Kareev, O.V. Boltalina, S.H. Strauss, 1,4,7,11,18,21,24,31,35,39,51,58,61,64-Tetradecakis(trifluoromethyl)-1,4,7,11,18,21,24,31,35,39,51,58,61,64-tetradecahydro(C_{70}–D5 h)[5,6]fullerene p-xylene trisolvate, Acta Crystallogr. E Struct. Rep. Online 63 (2007) o3928–o3929.
[29] N.B. Shustova, O.P. Anderson, O.V. Boltalina, S.H. Strauss, I.E. Kareev, 1,3,7,10,14,17,21,28,31,42,52,55-Dodecakis(trifluoromethyl)- 1,3,7,10,14,17,21,28,31,42,52,55-dodecahydro(C_{60}–I_h)[5,6]fullerene, Acta Crystallogr. E Struct. Rep. Online 64 (2008) o159.
[30] Y. Takano, et al., Electron donor-acceptor interactions in regioselectively synthesized exTTF2-$C_{70}(CF_3)_{10}$ dyads, Chem. Eur. J. 16 (2010) 5343–5353.
[31] I.V. Kuvychko, et al., Substituent effects in a series of 1,7-$C_{60}(R_F)_2$ compounds ($R_F = CF_3$, C_2F_5, n-C_3F_7, i-C_3F_7, n-C_4F_9, s-C_4F_9, n-C_8F_{17}): electron affinities, reduction potentials and E(LUMO) values are not always correlated, Chem. Sci. 3 (2012) 1399–1407.
[32] A.A. Popov, et al., Electrochemical, spectroscopic, and DFT study of $C_{60}(CF_3)_n$ frontier orbitals ($n = 2$–18): the link between double bonds in pentagons and reduction potentials, J. Am. Chem. Soc. 129 (2007) 11551–11568.
[33] D.V. Ignat'eva, et al., Isolation and structural characterization of the most highly trifluoromethylated C_{70} fullerenes: $C_{70}(CF_3)_{18}$ and $C_{70}(CF_3)_{20}$, New J. Chem. 37 (2013) 299–302.
[34] D.C. Coffey, et al., An optimal driving force for converting excitons into free carriers in excitonic solar cells, J. Phys. Chem. C 116 (2012) 8916–8923.
[35] D.V. Ignat'eva, A.A. Goryunkov, I.N. Ioffe, L.N. Sidorov, Trifluoromethylation of fullerenes: kinetic and thermodynamic control, J. Phys. Chem. A 117 (2013) 13009–13017.
[36] E.V. Bukovsky, et al., Structures and structure-related electronic properties of new $C_{60}(CF_3)_{10}$ isomers, J. Fluorine Chem. (2016), http://dx.doi.org/10.1016/j.jfluchem.2016.02.007.
[37] N.A. Romanova, et al., S6 Isomer of $C_{60}(CF_3)_{12}$: synthesis, properties and thermodynamic functions, J. Chem. Thermodyn. 66 (2013) 59–64.
[38] N.M. Belov, et al., Transalkylation of higher trifluoromethylated Fullerenes with C_{70}: a pathway to new addition patterns of $C_{70}(CF_3)_8$, Chem. Eur. J. 20 (2013) 1126–1133.
[39] T. Mutig, E. Kemnitz, S.I. Troyanov, Trifluoromethyl derivatives of fullerene C_{70}, $C_{70}(CF_3)_2$, $C_{70}(CF_3)_8$ and $C_{70}(CF_3)_{14}$, Mendeleev Commun. 19 (2009) 30–31.
[40] T. Mutig, I.N. Ioffe, E. Kemnitz, S.I. Troyanov, Crystal and molecular structures of C2-$C_{70}(CF_3)_8 \cdot 1.5$PhMe, Mendeleev Commun. 18 (2008) 73–75.
[41] V.Y. Markov, et al., Mass spectrometric identification of multicage fullerene cycloadducts, J. Anal. Chem. 65 (2010) 1495–1503.

Thirteen Decakis(trifluoromethyl) decahydro(C_{60}-I_h)[5,6]fullerenes ($C_{60}(CF_3)_{10}$): Structures and Structure-Related Properties of the Largest Set of Fullerene$(X)_n$ Isomers

3

E.V. Bukovsky [1], B.W. Larson [1], T.T. Clikeman [1], Y.-S. Chen [2], A.A. Popov [3], O.V. Boltalina [1], S.H. Strauss [1]

[1]Colorado State University, Fort Collins, CO, United States; [2]ChemMatCARS University of Chicago, Argonne, IL, United States; [3]Leibniz Institute for Solid State and Materials Research, Dresden, Germany

Chapter Outline

3.1 Introduction

One of the most important properties of the hollow nanospheroidal molecules known as fullerenes is that they are excellent electron acceptors, as evidenced by their gas-phase electron affinities (EAs; eg, 2.689(8), 2.77(1), and 2.98(1) eV for C_{60} [1], C_{70} [2], and C_{76}-$D_2(1)$ [3], respectively), and by their ability to undergo up to six reversible one-electron reductions in solution [4,5]. It is therefore no surprise that fullerenes have

New Fluorinated Carbons: Fundamentals and Applications. http://dx.doi.org/10.1016/B978-0-12-803479-8.00003-6

been extensively studied as electron acceptors for energy storage, photovoltaic energy conversion, and other molecular electronic applications [6–8]. Ever since the discovery of fullerenes [9] and their availability in macroscopic quantities [10], chemists have prepared and studied fluorofullerenes (FFs) with the expectation that they would be even stronger electron acceptors [11–14]. Although FFs do have EAs in excess of 4 eV (eg, 4.06(30) eV for $C_{60}F_{48}$ [15]), their usefulness in electronic devices is limited because they readily undergo reductive defluorination in condensed phases [16–20] and because at least some of them are unstable towards hydrolysis [21].

Complex mixtures of C_{60} derivatives with multiple electron-withdrawing perfluoroalkyl (R_F) groups were first prepared by Fagan et al. in 1993 [22]. Although the mixtures were not separated into individual compounds, the authors clearly demonstrated that $C_{60}(R_F)_n$ derivatives do not hydrolyze in contact with water and can be heated to at least 270°C without decomposition. Since that time, hundreds of well-characterized, HPLC-purified compounds collectively known as perfluoroalkylfullerenes (PFAFs) have been reported, with fullerenes from C_{60} to C_{90}, with heterofullerenes and endohedral metallofullerenes, and with straight-chain and branched R_F groups from CF_3 to C_8F_{17} [23,24]. PFAFs are now one of the largest classes of fullerene derivatives. For example, consider just one fullerene, C_{60}, and one type of R_F group, CF_3. There are more than 60 well-characterized $C_{60}(CF_3)_n$ compounds, with even n values from 2 to 18 (obviously there are multiple isomers known for most values of n; an exception is that 1,7-$C_{60}(CF_3)_2$ is the only isomer with this composition) [24,25]. Some PFAFs are stable at temperatures exceeding 350°C, most are significantly stronger electron acceptors than their respective parent fullerene, and many have been shown to have one-electron-reduced radical anions that are stable indefinitely in solutions of organic solvents at 25°C [24].

The abbreviations we shall use to denote specific $C_{60}(CF_3)_n$ derivatives consist of three numbers. As an example, the two known isomers of $C_{60}(CF_3)_6$ are denoted 60-6-1 and 60-6-2. The first number indicates that the compound is a derivative of C_{60}; the second number specifies the number of CF_3 groups (even numbers from 2 to 18); the third number is an arbitrary number denoting a particular isomer of that composition, that is to say, a particular addition pattern. In 2007, we published a spectroscopic, electrochemical, and DFT study of the 18 $C_{60}(CF_3)_n$ derivatives shown in Fig. 3.1 [26]. At the time, many of the compounds had not been structurally characterized by X-ray crystallography, but a combination of through-space Fermi-contact J_{FF} values in ^{19}F NMR spectra and DFT calculations allowed all of the addition patterns to be reliably determined (all fullerene$(CF_3)_n$ compounds exhibit ^{19}F NMR through-space J_{FF} couplings for CF_3 groups that share the same hexagon or pentagon [26–33]). With few exceptions, the addition patterns of $C_{60}(CF_3)_n$ derivatives with $n \leq 12$ were found to follow three guidelines, not only for the 18 compounds in Fig. 3.1 but also for two additional isomers of $C_{60}(CF_3)_4$ [34,35] and for seven recently reported structures of $C_{60}(CF_3)_{10}$ (see later).

The first guideline is that the CF_3 groups are arranged on one or more *para* (*p*) $C_6(CF_3)_2$ hexagons and/or on one or more ribbons or loops of edge-sharing *meta*-(*m*-) and/or *p*-$C_6(CF_3)_2$ hexagons (each shared edge is a $C(sp^3)$–$C(sp^2)$ bond). The *para-meta-para* (*pmp*) ribbon region in 60-4-1 is also shown in Fig. 3.1 [34]. The

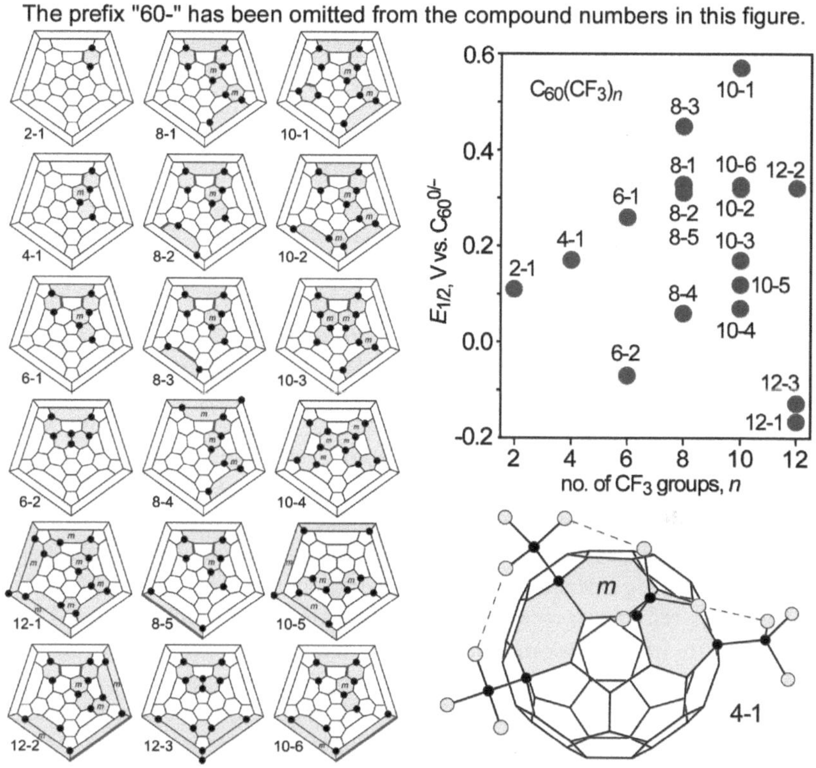

Figure 3.1 Left: Schlegel diagrams for the 18 $C_{60}(CF_3)_n$ derivatives reported in Ref. [26] ($n = 2{-}12$). The *black circles* indicate the cage C atoms to which the CF_3 groups are attached. The ribbons and loops of edge-sharing *p*- and *m*-$C_6(CF_3)_2$ hexagons and isolated *p*-$C_6(CF_3)_2$ hexagons are highlighted in yellow (light gray in print versions); the *m*-$C_6(CF_3)_2$ hexagons are indicated with the letter *m*. The bonds highlighted in red (gray in print versions) are *nonterminal* double bonds-in-pentagons (*nt*-DBIPs; see text for details). Upper right: Experimental $E_{1/2}(0/-)$ values determined by cyclic voltammetry (0.1 M $N(n\text{-}Bu)_4^+BF_4^-$ in CH_2Cl_2) and reported in Ref. [26] for these 18 $C_{60}(CF_3)_n$ derivatives versus the number of CF_3 groups (on this scale $E_{1/2}(0/-)$ for C_{60} is 0.0 V). Note that the range of $E_{1/2}(0/-)$ values for the six isomers of $C_{60}(CF_3)_{10}$ is 0.5 V. Note also that three of these compounds, 60-6-2, 60-12-1, and 60-12-3, have $E_{1/2}(0/-)$ values more negative than C_{60}. Lower right: The *pmp* ribbon addition pattern of 60-4-1 (Ref. [34]). Three F···F distances shorter than 2.7 Å are indicated with *dotted lines*. These interactions give rise to through-space Fermi-contact spin–spin couplings in the ^{19}F NMR spectra of all fullerene $(CF_3)_n$ compounds [26-33], including 60-4-1 [27].

compound 60-6-1 has a p^3mp ribbon, and 60-8-2, 60-8-3, and 60-8-5 have a p^3mp ribbon plus a different isolated *p*-$C_{60}(CF_3)_2$ hexagon, so their addition patterns are denoted p^3mp,p. The first three isomers of $C_{60}(CF_3)_{10}$ have the addition patterns p^3mpmp,p (60-10-1), $p^3mpmpmp$ (60-10-2), and pmp^3mpmp (60-10-3). The second guideline is that there is only one CF_3 group per pentagon, undoubtedly for steric reasons. The only exceptions are 60-10-3 and 60-12-2. The third guideline is that CF_3

groups do not occupy adjacent cage $C(sp^3)$ atoms, undoubtedly also for steric reasons, and the only exceptions are 60-6-2 and 60-12-3. These addition-pattern guidelines are in sharp contrast to the earlier prediction by Taylor that the CF_3 groups in $C_{60}(CF_3)_n$ derivatives occupy a string of contiguous cage $C(sp^3)$ atoms [36,37].

Finally, Fig. 3.1 also shows a plot of the first reduction potentials of the 18 $C_{60}(CF_3)_n$ derivatives versus the number of CF_3 groups [26]. The range of $E_{1/2}(0/-)$ values is more than 0.7 V. For the six $C_{60}(CF_3)_{10}$ isomers, the range of $E_{1/2}(0/-)$ values is 0.5 V. This large variation in reduction potentials for different isomers [26] was, and still is, three times greater than for any other set of $C_{60}(X)_n$ derivatives. For example, the next largest reported $E_{1/2}(0/-)$ ranges are 0.14 and 0.16 V for eight isomers of $C_{60}(c\text{-}CH_2N(R)CH_2)_2$ [38] and seven isomers of $C_{60}(C(CO_2Et)_2)_3$ [39], respectively. The electrochemical behavior of these 18 $C_{60}(CF_3)_n$ derivatives demonstrated that the addition pattern is as important, if not more important, than the number of substituents or the electronic properties of the substituents. Although CF_3 groups are strong electron-withdrawing substituents, note that 60-6-2, 60-12-1, and 60-12-3 were found to be *harder* to reduce than C_{60} itself (eg, by as much as 0.17 V for 60-12-1) [26].

Some of the cage C—C bonds in the Schlegel diagrams in Fig. 3.1 are highlighted in red (gray in print versions). These are pentagon—hexagon edges that are relatively short (and therefore have significant double-bond character) and have *two* $C(sp^2)$ neighbors. These bonds are referred to as *nonterminal* double-bonds-in-pentagons (*nt*-DBIPs) to distinguish them from *terminal* double-bonds-in-pentagons (*t*-DBIPs), which have only one $C(sp^2)$ neighbor. In 2007, it was found that the number and positions of *nt*-DBIPs in $C_{60}(CF_3)_n$ derivatives, and by inference in other $C_{60}(X)_n$ derivatives, are correlated with the positions, extents of delocalization, and relative energies of their lowest unoccupied molecular orbitals (LUMOs), and hence with their $E_{1/2}(0/-)$ values [26].

Fig. 3.2 shows this effect in an unambiguous and dramatic way. The DFT-predicted LUMOs of 60-10-1, 60-10-2, and 60-10-3 all include large contributions from their *nt*-DBIP carbon atoms. In all three cases, the two *nt*-DBIPs associated with the p^3 portion of their addition-pattern ribbon form a fulvene-like fragment of the π system. The fulvene fragment in 60-10-3 is isolated from the rest of the π system, and the LUMO is essentially confined to this fragment. In 60-10-2, the fulvene fragment is not isolated from the rest of the π system, and the LUMO is delocalized over several $C(sp^2)$ atoms on adjacent hexagons. As a result, 60-10-2 has a lower DFT-predicted LUMO energy than 60-10-3 and is 0.22 V easier to reduce than 60-10-3. The compound 60-10-1 has a third *nt*-DBIP, and its position relative to the fulvene-like fragment leads to a much more delocalized LUMO than in 60-10-2 or 60-10-3. This dramatic difference in LUMO delocalization results in 60-10-1 being 0.27 V easier to reduce than 60-10-2 and 0.49 V easier to reduce than 60-10-3 [26,40]. Equally dramatic is the fact that the three $C_{60}(CF_3)_n$ compounds that are harder to reduce than C_{60}, 60-6-2, 60-12-1, and 60-12-3, have no *nt*-DBIPs.

In summary, $C_{60}(X)_n$ derivatives with multiple *nt*-DBIPs, and especially with *nt*-DBIPs in close proximity, have extensively delocalized and low-energy LUMOs and are the best electron acceptors. In an effort to provide a wider variety of $C_{60}(CF)_n$ addition patterns, 21 new isomers of $C_{60}(CF_3)_{10}$ were prepared [40,41] and seven had their addition patterns determined by X-ray crystallography. The structural, electrochemical, and DFT characterization of these seven isomers, and

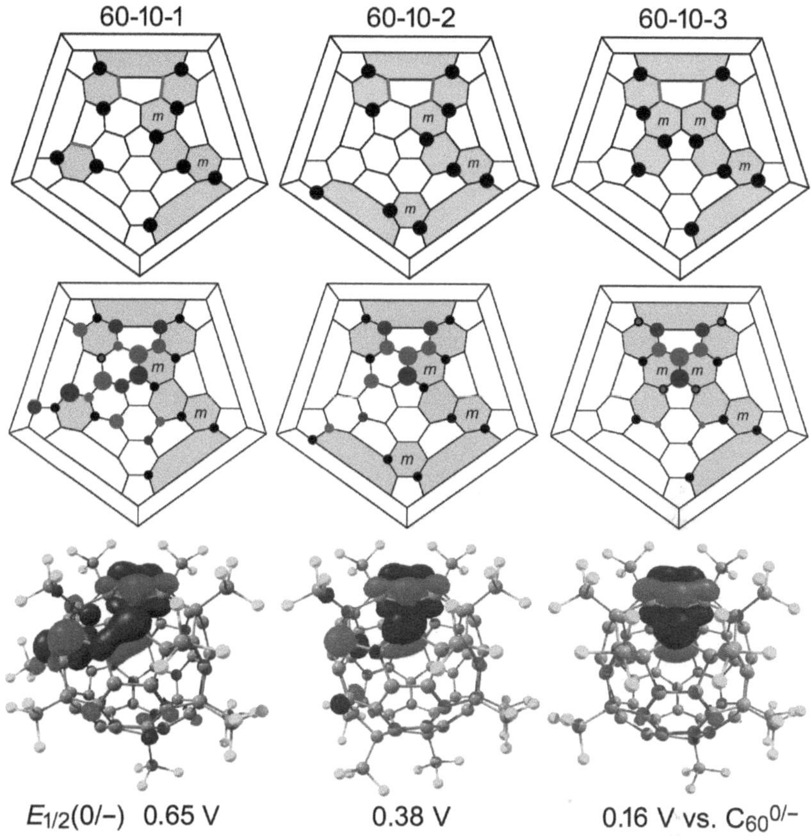

Figure 3.2 Schlegel diagrams, DFT-predicted LUMOs, and $E_{1/2}$(0/−) values for the $C_{60}(CF_3)_{10}$ isomers 60-10-1, 60-10-2, and 60-10-3, based on the results reported in Refs. [26,40]. The *black circles* indicate the cage C(sp^3) atoms to which the CF_3 groups are attached. The bonds highlighted in red (gray in print versions) are *nt*-DBIPs (see Fig. 3.1 caption for more details). The *blue* (darker gray in print versions) (+) and *green* (dark gray in print versions) (−) *circles* in the LUMO-Schlegel diagrams in the middle represent the exohedral lobes of the p-π atomic orbitals for cage C(sp^2) atoms. The diameters of the circles are scaled approximately to the contributions of those orbitals to the LUMO. *LUMOs*, lowest unoccupied molecular orbitals; *nt-DBIPs*, *nonterminal* double-bonds-in-pentagons.

comparisons with the six previously studied isomers shown in Fig. 3.3, are the subject of this chapter. The synthesis and HPLC purification of the seven new isomers have already been described [40,41] and will not be discussed in this chapter.

3.2 The 13 Isomers of $C_{60}(CF_3)_{10}$

The structurally characterized isomers of $C_{60}(CF_3)_{10}$ now number 13. Table 3.1 lists their abbreviations, addition patterns, point-group symmetries, IUPAC lowest locants, cyclic-voltammetric $E_{1/2}$(0/−) values, and DFT-predicted relative energies and

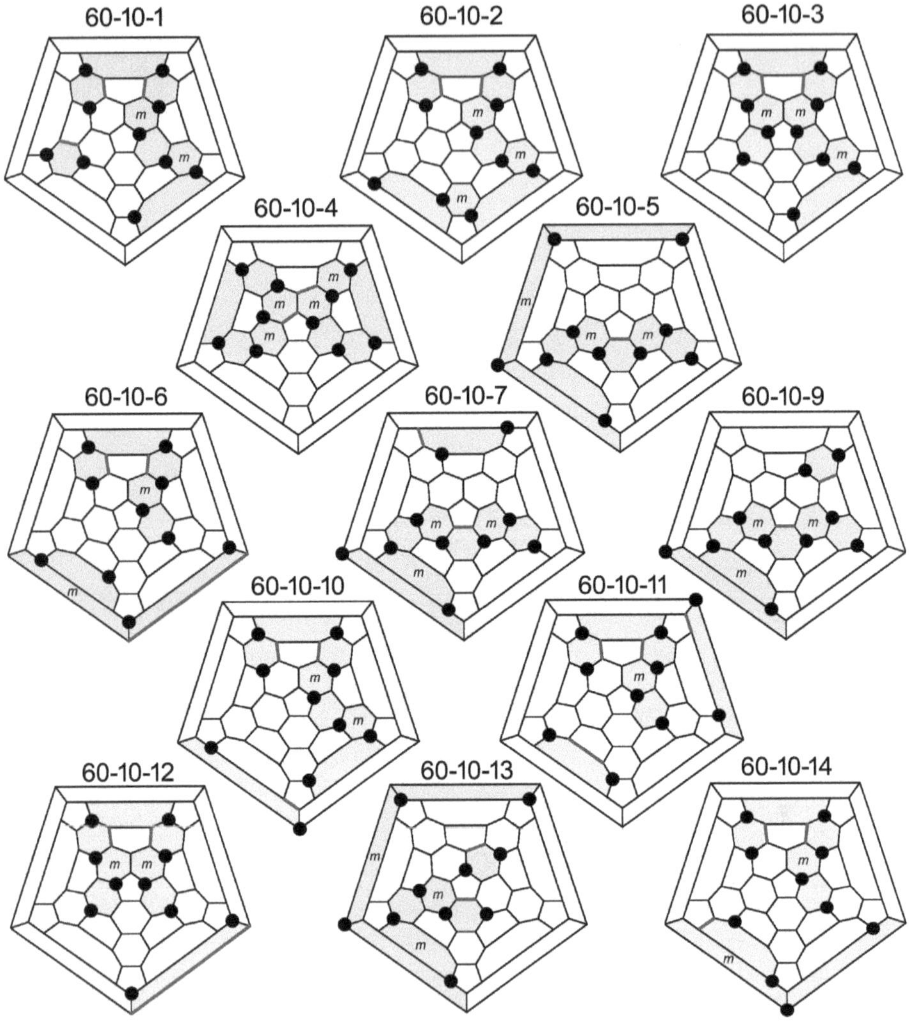

Figure 3.3 Schlegel diagrams for the 13 structurally characterized isomers of $C_{60}(CF_3)_{10}$. The *nt*-DBIPs are highlighted in red (gray in print versions) (see text). A hypothetical addition pattern denoted 60-10-8 (not shown) was discussed in Ref. [24] and references therein. *nt-DBIPs, nonterminal* double-bonds-in-pentagons.

E(LUMO) values [26,28,34,40–45]. Five compounds either did not have an electrochemically quasi-reversible first reduction (60-10-12) or else a sufficient amount of compound was not available for a cyclic-voltammetric experiment (60-10-4, 60-10-7, 60-10-9, and 60-10-11), and for these compounds $E_{1/2}(0/-)$ values in 1,2-$C_6H_4Cl_2$ are not listed (note that $E_{1/2}(0/-)$ values for 60-10-1 through 60-10-6 in CH_2Cl_2 were previously reported [26]). Schlegel diagrams for all 13 isomers are shown in Fig. 3.3. Crystal data, refinement parameters, densities, and fullerene centroid···centroid (⊙···⊙) distances for all 13 structures are listed in Tables 3.2–3.4 [46,47]. Table 3.4

Table 3.1 Structurally Characterized $C_{60}(CF_3)_{10}$ Isomers[a]

Abbreviation[b]	Addition-Pattern[c]	IUPAC Lowest Locants[d]	Synthesis and Isolation Refs.	X-ray Refs.	^{19}F NMR Refs.	$E_{1/2}(0/-)$,[e] V vs. $C_{60}^{0/-}$	DFT Predictions[f]	
							Rel. Energy (kJ mol^{-1})	E(LUMO) (eV)
60-10-1	C_1-p^3mpmp,p	1,6,11,16,18,24,27,36,41,57	[28]	[42]	[28]	0.65 (0.57)	9.22	−5.129
60-10-2	C_1-$p^3mpmpmp$	1,6,11,16,18,24,27,36,54,60	[28]	[34,43]	[28]	0.38 (0.32)	0.06	−4.894
60-10-3	C_1-pmp^3mpmp	1,3,7,10,14,17,23,28,31,40	[28]	[28]	[28]	0.16 (0.17)	4.56	−4.754
60-10-4	C_2-$(p^3m^2\text{-loop})^2$	1,6,12,15,18,23,25,41,45,57	[26]	[44]	[26]	(0.07)	7.60	−4.529
60-10-5	C_1-$pmpmpmpmp$	1,6,11,16,18,26,36,41,44,57	[26]	[26]	[26]	0.12 (0.12)	1.25	−4.638
60-10-6	C_1-p^3mp,pmp	1,6,11,18,24,27,33,51,54,60	[26]	[45]	[26]	0.41 (0.33)	0.00	−4.922
60-10-7	C_1-$pmpmpmp,p$	1,6,11,16,18,28,31,36,42,56	[40,41]	[42]	[40,41]	–	10.10	−4.816
60-10-9	C_1-$pmpmpmp,p$	1,6,11,16,18,28,31,36,44,58	[40,41]	[42]	[40,41]	–	17.36	−4.829
60-10-10	C_1-p^3mpmp,p	1,6,11,16,18,24,27,36,50,60	[40,41]	[42]	[40,41]	–	20.55	−4.924
60-10-11	C_1-p^3mp,p,p	1,6,11,14,18,24,27,31,53,56	[40,41]	[42]	[40,41]	–	29.60	−5.090
60-10-12	C_1-pmp^3mp,p	1,3,7,10,14,17,28,31,43,55	[40,41]	[42]	[40,41]	0.37	15.41	−4.867
60-10-13	C_1-$pmpmpmp,p$	1,6,11,16,18,25,36,45,54,60	[40,41]	[42]	[40,41]	0.42	18.46	−4.938
60-10-14	C_1-p^3mp,pmp	1,6,11,18,24,27,34,50,52,55	[40,41]	[42]	[40,41]	0.47	16.03	−4.983

IUPAC, International Union of Pure and Applied Chemistry.

[a]The $C_{60}(CF_3)_{10}$ isomers 60-10-2 through 60-10-6 were structurally characterized prior to the work reported in Ref. [42]. A hypothetical addition pattern denoted 60-10-8 was discussed in Ref. [24] and references therein.

[b]The numbers 60 and 10 in the abbreviations refer to the fullerene (C_{60}) and number of CF_3 groups, respectively. The third number is an arbitrary index denoting the specific isomer.

[c]The addition pattern abbreviations refer to the ribbons, loops, and/or isolated edge-sharing *meta-* (*m*) and *para*-$C_6(CF_3)_2$ (*p*) hexagons (each shared edge is a C(sp^3)–C(sp^2) bond) and also indicate the point-group symmetry.

[d]See Schlegel diagram in Fig. 3.2. See also Refs. [64] and [65] for IUPAC-recommended nomenclature rules for C_{60}, C_{70}, and higher fullerenes and their derivatives.

[e]The $E_{1/2}(0/-)$ potentials not in braces are from Refs. [40] and [41] and were measured in 1,2-$C_6H_4Cl_2$ containing 0.1 M N(*n*-Bu)$_4$PF$_6$ and Fe(Cp)$_2$ as an internal standard ($E_{1/2}(\text{Fe(Cp)}_2^{+/0}$ = 1.08 V vs. $C_{60}^{0/-}$). The $E_{1/2}(0/-)$ potentials in braces are from Ref. [26] and were measured in CH_2Cl_2 containing 0.1 M N(*n*-Bu)$_4$BF$_4$ and Fe(Cp)$_2$ as an internal standard.

[f]The DFT-predicted results are from Refs. [40] and [41] except for 60-10-1 through 60-10-6, which were reported in Ref. [26]. The methodology used for all DFT calculations is described in several papers by Popov, Boltalina, Strauss et al. including Refs. [23,26,30].

Table 3.2 Crystal Data, Refinement Parameters, Density, and Fullerene Centroid···Centroid (⊙···⊙) Distances for the X-ray Structures of $C_{60}(CF_3)_{10}$ Isomers 60-10-1 Through 60-10-6[a]

Compound Abbreviation	60-10-1 [42][b]	60-10-2·0.5C_6H_6 [43][c]	60-10-3 [28][b]	60-10-4 [44][b]	60-10-5·$C_6H_4(CH_3)_2$ [26][c]	60-10-6 [45][b]
Crystal system	Triclinic	Triclinic	Triclinic	Orthorhombic	Triclinic	Triclinic
Space group (*Z*)	$P\bar{1}$, 4	$P\bar{1}$, 2	$P\bar{1}$, 2	*Pbcn*, 4	$P\bar{1}$, 2	$P\bar{1}$, 2
a (Å)	12.2718(6)	11.0290(15)	12.1433(7)	18.0971(11)	12.310(3)	11.0257(4)
b (Å)	20.2702(11)	13.7614(18)	12.3153(7)	16.2516(10)	14.197(3)	11.4172(4)
c (Å)	21.0715(11)	16.443(2)	15.6682(9)	15.3735(9)	16.002(3)	20.4527(7)
α (deg)	113.4700(10)	87.785(6)	82.2470(10)	90	90.100(5)	82.369(2)
β (deg)	101.5240(10)	85.082(7)	86.6060(10)	90	107.630(5)	77.010(2)
γ (deg)	90.0410(10)	88.850(6)	81.9760(10)	90	98.136(5)	65.543(2)
V (Å)^3	4692.2(4)	2484.2(6)	2297.1(2)	4521.5(5)	2636.2(10)	2281.36(14)
ρ_{calc} (g cm^{-3})	1.997	1.938	2.040	2.072	1.914	2.054
T (K)	100(2)	100(1)	173(2)	100(1)	173(2)	100(2)
$R(F)$ $(I > 2\sigma(I))^d$	0.0567	0.045	0.0529	0.047	0.0507	0.051
$wR(F^2)$ (all data)^d	0.1420	0.130	0.1292	0.129	0.1129	0.122
GOF	1.002	1.03	1.037	1.03	1.010	1.03
Idealized lattice packing	Not closely packed	Not closely packed	HCP	HCP	Not closely packed	Not closely packed
⊙···⊙ distances (Å)	10.05–14.56	9.88–14.17	10.17–14.17	9.93–12.59	9.88–14.20	10.95–14.47
Mean ⊙···⊙ distance (Å) (σ)	12.57 (1.47)	12.25 (1.41)	11.90 (1.00)	11.81 (0.90)	12.36 (1.55)	12.26 (1.27)
Median ⊙···⊙ distance (Å)	12.31	11.85	12.03	12.09	12.52	11.87

HCP, hexagonal close-packed.
[a]Literature reference given next to compound.
[b]The formula and formula weight for 60-10-1, 60-10-3, 60-10-4, and 60-10-6 are $C_{70}F_{30}$ and 1410.70 g mol^{-1}, respectively.
[c]The formulas for 60-10-2·0.5C_6H_6 and 60-10-5 $C_6H_4(CH_3)_2$ are $C_{73}H_3F_{30}$ and $C_{78}H_{10}F_{30}$, respectively; their formula weights are 1456.81 and 1516.87 g mol^{-1}, respectively.
[d]$R(F) = \sum||F_o| - |F_c||/\sum|F_o|$; $wR(F^2) = \left(\sum\left[w(F_o^2 - F_c^2)^2\right]/\sum\left[w(F_o^2)^2\right]\right)^{1/2}$.

Table 3.3 **Crystal Data, Refinement Parameters, Density, and Fullerene Centroid···Centroid (⊙···⊙) Distances for the X-ray Structures of $C_{60}(CF_3)_{10}$ Isomers 60-10-7 and 60-10-9 Through 60-10-13[a]**

Compound Abbreviation	60-10-7	60-10-9[b]	60-10-10[b]	60-10-11·$CDCl_3$[c]	60-10-12[b]	60-10-13[b]
Crystal system	Monoclinic	Triclinic	Monoclinic	Triclinic	Monoclinic	Orthorhombic
Space group (*Z*)	$P2_1/c$, 4	$P\bar{1}$, 2	$C2/c$, 8	$P2_1/c$, 4	$P2_1/n$, 4	$P2_12_12_1$, 4
a (Å)	20.5157(8)	12.1441(6)	46.403(6)	12.253(1)	17.794(1)	11.521(1)
b (Å)	12.1614(5)	12.3855(7)	12.115(2)	24.771(3)	13.1545(8)	19.777(2)
c (Å)	20.1763(8)	17.3822(9)	17.249(3)	17.001(2)	20.108(1)	20.524(2)
α (deg)	90	93.614(3)	90	90	90	90
β (deg)	114.3613(8)	96.249(3)	104.164(7)	106.035(3)	102.099(2)	90
γ (deg)	90	114.298(2)	90	90	90	90
V (Å)^3	4585.8(3)	2351.9(2)	9402(2)	4959.2(8)	4602.0(5)	4676.5(7)
ρ_{calc} (g cm^{-3})	2.043	1.992	1.993	2.051	2.036	2.004
T (K)	100(2)	120(2)	120(2)	100(2)	100(2)	120(2)
$R(F)$ $(I > 2\sigma(I))$[d]	0.0523	0.0520	0.0458	0.0716	0.0445	0.0644
$wR(F^2)$ (all data)[d]	0.1535	0.1556	0.1436	0.2217	0.1123	0.1651
GOF	1.029	1.036	1.057	0.985	1.017	0.975
Idealized lattice packing	HCP	HCP	HCP	Not close-packed	CCP	Not close-packed
⊙···⊙ distances (Å)	10.86–12.97	9.70–13.74	10.07–14.35	10.08–14.66	10.20–13.96	10.13–14.41
Mean ⊙···⊙ distance (Å) (σ)	11.85 (0.74)	12.01 (1.15)	12.12 (1.47)	12.24 (1.27)	11.97 (1.35)	12.63 (1.38)
Median ⊙···⊙ distance (Å)	11.87	12.14	12.12	12.33	11.95	11.88

CCP, cubic close packed; *HCP*, hexagonal close-packed.
[a]All structures from Ref. [42].
[b]The formula and formula weight for $C_{60}(CF_3)_{10}$ isomers 60-10-7, 60-10-9, 60-10-10, 60-10-12, and 60-10-13 are $C_{70}F_{30}$ and 1410.70 g mol^{-1}, respectively.
[c]The formula and formula weight for 60-10-11·$CDCl_3$ is $C_{71}DF_{30}Cl_3$ and 1531.08 g mol^{-1}, respectively.
[d]$R(F) = \sum||F_o| - |F_c||/\sum|F_o|$; $wR(F^2) = \left(\sum\left[w(F_o^2 - F_c^2)^2\right]/\sum\left[w(F_o^2)^2\right]\right)^{1/2}$.

Table 3.4 **Crystal Data, Refinement Parameters, Density, and Fullerene Centroid···Centroid (⊙···⊙) Distances for the X-ray Structures of $C_{60}(CF_3)_{10}$ Isomer 60-10-14, Faux-Hawk Fullerene, Solvent-Free PCBM, and Solvent-Free 60-10-2[a]**

Compound Abbreviation	60-10-14·$CHCl_3$ [42][b]	Faux-Hawk Fullerene [46][c]	PCBM [47][d]	60-10-2 (Solvent-Free) [34][b]
Crystal system	Triclinic	Orthorhombic	Monoclinic	Monoclinic
Space group (*Z*)	$P\bar{1}$, 2	*Pbca*, 8	$P2_1/n$, 4	*C*2/*c*, 8
a (Å)	11.532(1)	9.9998(6)	13.47310(10)	21.1725(2)
b (Å)	12.088(1)	20.6538(12)	15.13760(10)	11.17780(10)
c (Å)	19.854(2)	31.3512(2)	19.00509(12)	39.4419(5)
α (deg)	72.791(4)	90	90	90
β (deg)	74.569(5)	90	106.9000(7)	99.0521(6)
γ (deg)	85.305(5)	90	90	90
V (Å)^3	2548.3(4)	6475.1(7)	3708.72(4)	9218.1(2)
ρ_{calc} (g cm^{-3})	1.994	1.885	1.631	2.033
T (K)	120(2)	15(2)	100(2)	100(2)
$R(F)$ ($I > 2\sigma(I)$)[e]	0.0628	0.0732	0.0512	0.0449
$wR(F^2)$ (all data)[e]	0.2285	0.1633	0.1503	0.1234
GOF	0.975	1.121	1.058	1.223
Idealized lattice packing	Not close-packed	Not close-packed	Not close-packed	CCP
⊙···⊙ Distances (Å)	9.85–16.01	9.74–10.34	9.95–10.28	9.95–14.52
Mean ⊙···⊙ distance (Å) (σ)	11.94 (1.15)	10.09 (0.20)	10.17 (0.13)	11.87 (1.06)
Median ⊙···⊙ distance (Å)	11.93	10.03	10.24	11.97

CCP, cubic close-packed; *PCBM*, phenyl-C61-butyric acid methyl ester.

[a]Literature reference given next to compound abbreviation except as otherwise noted.

[b]The formula and formula weight for 60-10-14·$CHCl_3$ are $C_{71}HF_{30}Cl_3$ and 1530.08 g mol^{-1}.

[c]Faux-hawk fullerene is 1,9-C_{60}(*cyclo*-CF_2(2-C_6F_4)) (formula and formula weight $C_{67}F_6$ and 918.67 g mol^{-1}; the structure was modeled as a three-component twin with total residue disorder).

[d]PCBM is phenyl-C_{61}-butyric acid methyl ester (formula and formula weight $C_{72}H_{14}O_2$ and 910.88 g mol^{-1}).

[e]$R(F) = \sum||F_o| - |F_c||/\sum|F_o|$; $wR(F^2) = \left(\sum\left[w(F_o^2 - F_c^2)^2\right]/\sum\left[w(F_o^2)^2\right]\right)^{1/2}$.

also includes results for a solvent-free structure of 60-10-2 [34] and for two other fullerene structures that will be compared with the $C_{60}(CF_3)_{10}$ structures later in this chapter.

The first fullerene derivative with more than one CF_3 group to be structurally characterized by X-ray crystallography was 60-10-3 [28]. Its structure is shown in Fig. 3.4. Its 90 cage C–C bond distances are shown pictorially in Fig. 3.4 and graphically in Fig. 3.5. They span a broad range of distances from 1.339(3) Å for the shortest $C(sp^2)–C(sp^2)$ bond to 1.553(3) for the longest $C(sp^2)–C(sp^3)$ bond. The nearly continuous range of cage $C(sp^2)–C(sp^2)$ bond distances in 60-10 isomers, which blurs the distinction between "single" and "double" bonds, can be fully appreciated in a plot of X-ray versus DFT cage C–C distances in 60-10-9, one of the most precise of the 13 $C_{60}(CF_3)_{10}$ structures (the X-ray $C(sp^2)–C(sp^2)$ distances range from 1.346(3) to 1.474(3) Å in 59 increments that average 0.002 Å, which is smaller than the standard error (σ) for the individual distances, 0.003 Å) [42]. This plot, shown in Fig. 3.6, also shows how reliably the DFT code we used duplicates the observed cage C–C bond distances in the X-ray structure at the $\pm 3\sigma$ level of precision except for a few of the shortest bonds [42].

In contrast, the 90 $C(sp^2)–C(sp^2)$ bond distances in DFT-optimized C_{60} are either 1.399 Å or 1.453 Å. This was verified by the most precise X-ray structure of C_{60}, which is in a co-crystal with a platinum porphyrin [48]. The C_{60} molecule is on a general lattice position, making all 90 C–C bonds unique, which allows the small variations in their distances due to packing forces to be observed. Fig. 3.6 shows that the 90 bonds are divided into 30 short hexagon–hexagon edges, 1.379(3)–1.396(3) Å, and 60 long hexagon–pentagon edges, 1.440(3)–1.461(3) Å. The hex–hex and hex–pent edges in fullerenes are usually referred to as "double" and "single" bonds, respectively, although this is an oversimplification because all 90 $C(sp^2)–C(sp^2)$ bonds are conjugated with one another and therefore they all have some degree of double-bond character. The short bond/long bond discontinuity in C_{60} and theoretical results for $C_{60}(X)_n$ derivatives [49,50] have led to the oft-quoted "rule" that double bonds in pentagons (DBIPs) in fullerene derivatives are destabilizing, presumably because of angle strain [51]. Nevertheless, a *short* hex–pent edge, obviously with significant double-bond character, is formed whenever substituents are added to the *para* positions of a C_{60} hexagon, and five *p*-$C_6(CF_3)_2$ hexagons are joined into the 60-10-3 *pmp*^{3}*mpmp* ribbon. Accordingly, five of the six shortest bonds in 60-10-3 are DBIPs, as shown in Fig. 3.3. In fact, all $C_{60}(CF_3)_{10}$ isomers except 60-10-4 have five DBIPs (the two loops in 60-10-4 are not made with five *p*-$C_6(CF_3)_2$ hexagons). The extreme examples are 60-10-11, with four *nt*-DBIPs and only one *t*-DBIP, and 60-10-5, with four *t*-DBIPs and only one *nt*-DBIP.

3.3 Enumerating $C_{60}(CF_3)_{10}$ Addition Patterns That Meet the Guidelines

There are millions of possible isomers of $C_{60}(CF_3)_{10}$, but only a limited number that have DFT-predicted relative energies $\leq$30 kJ mol^{-1} (the DFT-predicted highest

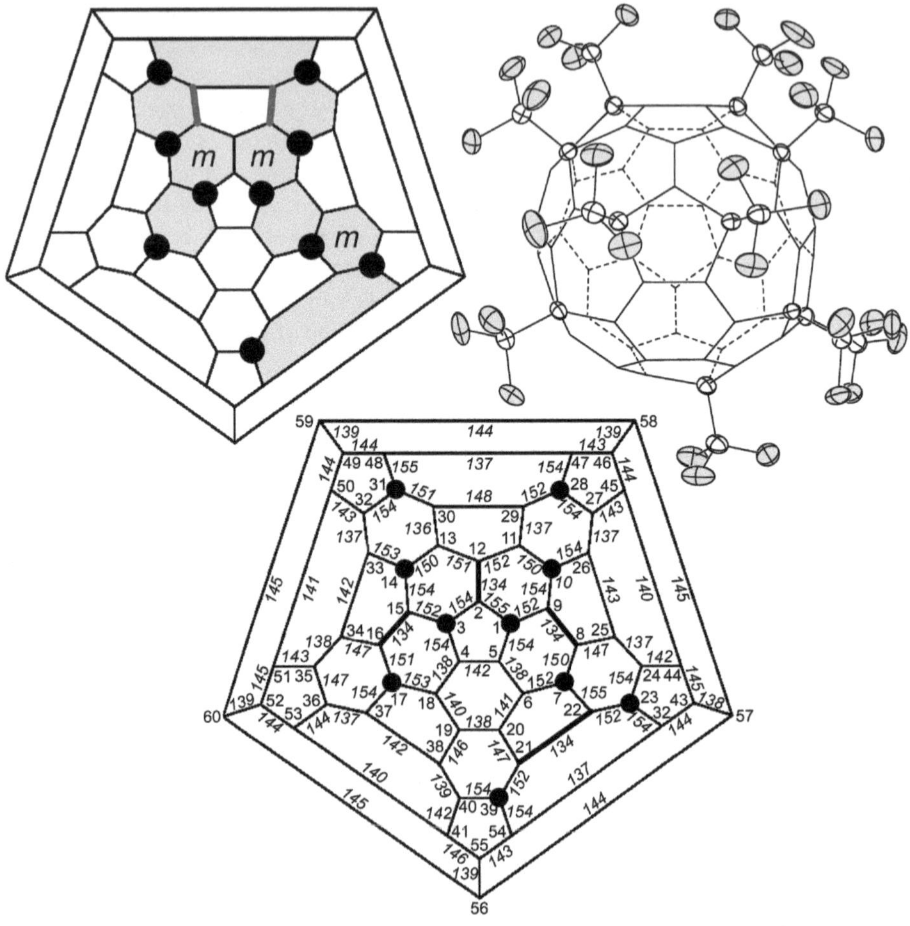

Figure 3.4 Results of the single crystal X-ray structure of the 60-10-3 isomer of $C_{60}(CF_3)_{10}$ reported in Ref. [28]. Top left: Schlegel diagram showing the cage C atoms to which the CF_3 groups are attached as *black circles*. The *pmp*^{3}*mpmp* ribbon of edge-sharing *p*- and *m*-$C_6(CF_3)_2$ hexagons is highlighted in yellow (light gray in print versions) and *m*-$C_6(CF_3)_2$ hexagons are indicated with the letter *m*. The two *nt*-DBIPs, C11−C29 and C13−C31, are highlighted in red (gray in print versions). Top right: Molecular structure showing the CF_3 atoms and the cage C(sp^3) atoms as 50% probability ellipsoids, with the F atoms highlighted in yellow (light gray in print versions). Bottom: Schlegel diagram showing the IUPAC locants for cage C atoms (one- and two-digit numbers) and cage C−C bond distances in pm (three-digit numbers). The standard errors (σ) for the C−C bond distances in this structure are 0.3 pm and therefore ±3σ is c. ±1 pm. The four shortest cage C−C bonds, 134(1) pm, are highlighted as *thicker lines*. Three of these bonds, C8−C9, C15−C16, and C21−C22 are *t*-DBIPs. The next shortest bonds, C11−C29 (137(1) pm) and C13−C31 (136(1) pm), are the two *nt*-DBIPs. See text for details. *nt*-DBIPs, *nonterminal* double-bonds-in-pentagons; *t*-DBIPs, *terminal* double-bonds-in-pentagons; *IUPAC*, International Union of Pure and Applied Chemistry.

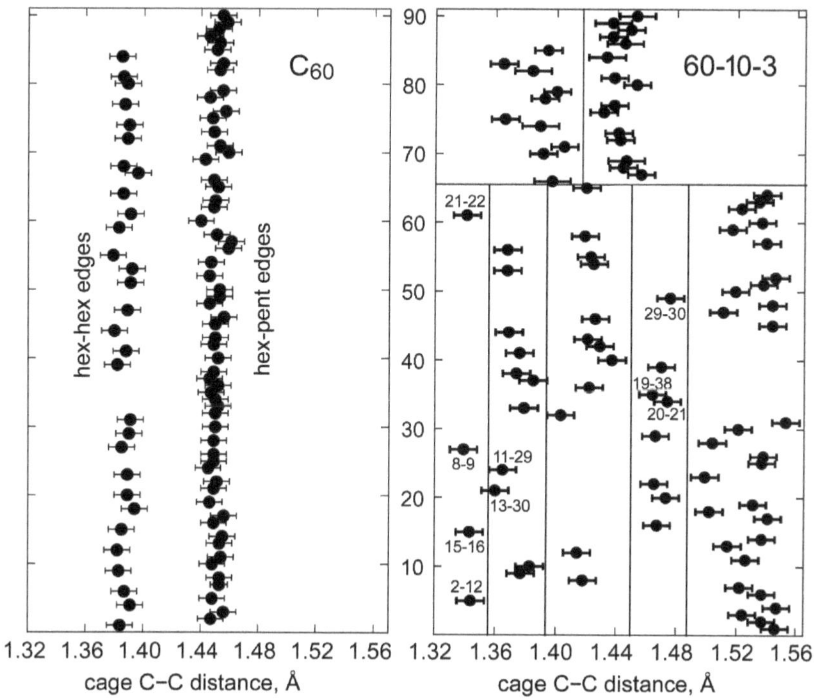

Figure 3.5 The 90 cage C–C distances determined by X-ray crystallography in the C_{60} molecule in $C_{60}\cdot Pt(OEP)\cdot 2C_6H_6$ (Ref. [48]; *OEP*, the dianion of octaethylporphyrin) and in the $C_{60}(CF_3)_{10}$ isomer 60-10-3 (Ref. [28]). The *error bars* shown in both graphs are $\pm 3\sigma$. In both structures the fullerene molecule is on a general position and therefore all 90 cage C–C bonds are unique. In the structure of 60-10-3, C8–C9, C15–C16, and C21–C22 are *t*-DBIPs and C11–C29 and C13–C30 are *nt*-DBIPs (see text for details). *nt*-DBIPs, *nonterminal* double-bonds-in-pentagons; *t-DBIPs*, *terminal* double-bonds-in-pentagons.

energy of the 13 structurally characterized isomers is 29.60 kJ mol^{-1} for 60-10-11). The supporting information in Ref. [26] lists 36 low-energy $C_{60}(CF_3)_{10}$ isomers based on ribbons or loops of edge-sharing *p*- and *m*-$C_6(CF_3)_2$ hexagons (ie, based on the $C_{60}(CF_3)_n$ guidelines discussed above for $n \leq 12$).

There are only three possible single-ribbon addition patterns with five edge-sharing *p*- and *m*-$C_6(CF_3)_2$ hexagons and no 1,3-$C_5(CF_3)_2$ pentagons. The compounds 60-10-2 (*p*^{3}*mpmpmp*) and 60-10-5 (*pmpmpmpmp*) are two of these, and they are within 1.25 kJ mol^{-1} of the lowest-energy isomer, which is 60-10-6 (*p*^{3}*mp,pmp* addition pattern). The third is a different *pmpmpmpmp* isomer yet to be identified, with a relative energy of 8.01 kJ mol^{-1} [26]. There is only one possible single-ribbon isomer that includes one, and only one, 1,3-$C_5(CF_3)_2$ pentagon, and that is 60-10-3 (*pmp*^{3}*mpmp*). No single-ribbon isomers with *p*^{4} addition-pattern fragments are possible for C_{60}, because four consecutive edge-sharing *p*-$C_6(CF_3)_2$ hexagons on C_{60} produces a *p*^{4} loop like the ones that are part of the addition patterns of 60-6-2 and 60-12-3, not a *p*^{4} ribbon. (Note that higher-fullerene CF_3 derivatives *can* have *p*^{n} ribbons with $n \geq 4$; some examples

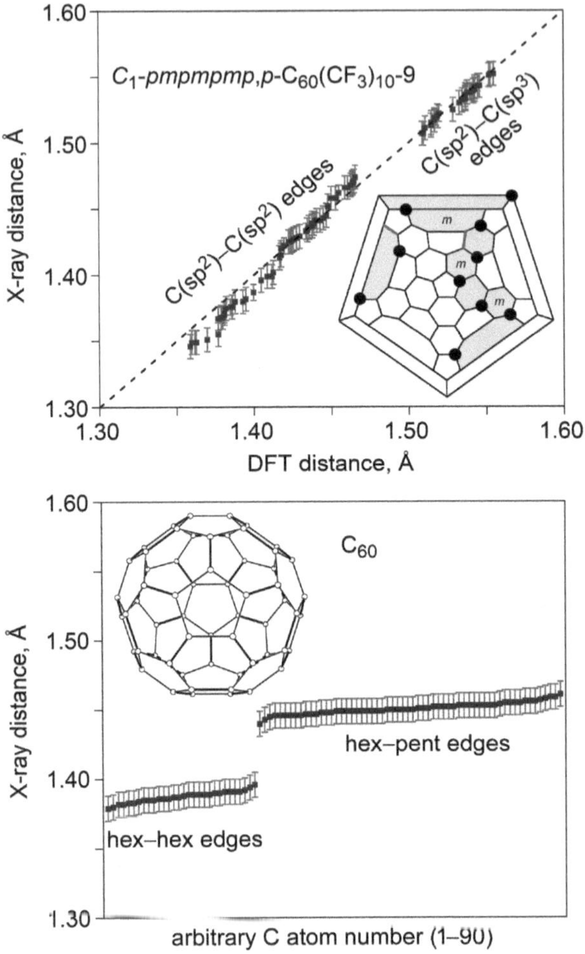

Figure 3.6 Top: X-ray versus DFT cage C—C distances in the $C_{60}(CF_3)_{10}$ isomer 60-10-9 (Ref. [42]). Bottom: The 90 cage C—C distances determined by X-ray crystallography in the C_{60} molecule in $C_{60}\cdot Pt(OEP)\cdot 2C_6H_6$ (Ref. [48]; *OEP*, the dianion of octaethylporphyrin). The error bars for the X-ray distances in both plots are $\pm 3\sigma$. In both structures the fullerene molecule is on a general position and therefore all 90 cage C—C bonds are unique.

are p^7mp-$C_{70}(CF_3)_{10}$ [52], p^9-(C_{78}-C_{2v}(2))$(CF_3)_{10}$ [30], p^5,p^5-(C_{78}-D_3(1))$(CF_3)_{10}$ [30], p^{11}-(C_{78}-D_{3h}(5))$(CF_3)_{12}$ [53], and p^6,p^2,p-(C_{84}-C_2(11))$(CF_3)_{12}$ [54].)

There are 11 "8 + 2" addition patterns among the low-energy isomers, either p^3mpmp,p pmp^3mp,p or *pmpmpmp,p*, and six of them are now known with certainty: 60-10-1, 60-10-7, 60-10-9, 60-10-10, 60-10-12, and 60-10-13. There is one additional p^3mpmp,p isomer (21.01 kJ mol^{-1}, very similar to 60-10-10) and four additional *pmpmpmp,p* isomers (10.05, 16.67, 18.71, and 18.78 kJ mol^{-1}) [26].

There are 17 "6 + 4" addition patterns among the low-energy isomers [26], and only two, 60-10-6 and 60-10-14, with p^3mp,pmp addition patterns, are known with certainty. As of this writing none of the following addition patterns have been confirmed by X-ray crystallography: p^3mp,p^3; *pmpmp,pmp*; *pmpmp*,p^3; SPP,*pmp*; and SPP,p^3 (the SPP addition-pattern fragment is the one exhibited by 60-6-2 [34] (see Fig. 3.1)).

The highest energy addition pattern that follows the guidelines with a relative energy ≤ 30 kJ mol^{-1} is the *pmpmp,p,p* "6 + 2 + 2" isomer 60-10-11. Interestingly, this isomer was not listed in the supporting information for Ref. [26]. And, surprisingly, there is an as-yet-unknown p^3mp,p,p isomer predicted to have a relative energy of only 14.08 kJ mol^{-1} [26]. It is likely that the "as-yet-unknown" or "missing" isomers referred to above are among the other 14 new isomers of $C_{60}(CF_3)_{10}$ that have not yet been characterized by X-ray crystallography [40,41].

Finally, we note that many of the 13 isomers have ribbon fragments in common. The most obvious is that 60-10-3 and 60-10-12 both have a pmp^3mp ribbon. Isomers 60-10-5, 60-10-7, 60-10-9, and 60-10-13 all have a *pmpmpmp* ribbon. Eight isomers have a p^3mp ribbon (60-10-1, 60-10-2, 60-10-3, 60-10-6, 60-10-10, 60-10-11, 60-10-12, and 60-10-14), and four of these have a p^3mpmp ribbon (60-10-1, 60-10-2, 60-10-3, and 60-10-10).

3.4 The Molecular Structures of the Seven Recently Reported $C_{60}(CF_3)_{10}$ Isomers

These seven structures were reported in Ref. [42]. A thermal ellipsoid plot of the structure of 60-10-12 is shown in Fig. 3.7. The other six thermal ellipsoid plots are similar [42]. The asymmetric unit in all seven structures is a single asymmetric $C_{60}(CF_3)_{10}$ molecule, with 90 unique cage C—C bonds, plus a solvent molecule in the cases of 60-10-11 and 60-10-14. All seven structures contain an equal number of both $C_{60}(CF_3)_{10}$ enantiomers, and none of the seven structures exhibits molecular disorder (ie, the enantiomers are not randomly superimposed on a given site in any of the structures). In addition, none of the CF_3 groups in any of the structures is disordered (ie, none exhibit multiple conformations due to rotations about their respective F_3C—C(cage) single bonds). The C—F distances and F—C—F angles within the CF_3 groups are normal (eg, 1.301(3)—1.336(2) Å and 106.1(2)—108.7(2)°, respectively, in 60-10-9), as are the F_3C—C(cage) distances (eg, 1.539(3)—1.553(3) Å in 60-10-9).

In compounds such as *m*- and *p*-$C_6H_4(CF_3)_2$, the F_3C—$C(sp^2)$ bonds lie in the same plane as the aromatic C_6 hexagon. This is not the case for the *m*- and *p*-$C_6(CF_3)_2$ hexagons in $C_{60}(CF_3)_n$ compounds due to the curvature of the fullerene cage. For example, in 60-10-7 the angles between the F_3C—$C(sp^3)$ bonds and the least-square plane of the four $C(sp^2)$ atoms in the $C_6(CF_3)_2$ hexagon are $154 \pm 1°$ for the isolated *p*-$C_6(CF_3)_2$ hexagon.

In general, the CF_3 groups in the structures of 60-10-7 and 60-10-9 through 60-10-14 are staggered with respect to their three underlying cage $C(sp^3)$—$C(sp^2)$ bonds. Two exceptions are the pair of CF_3 groups on the 1,3-$C_5(CF_3)_2$ pentagon in 60-10-12, as shown in Fig. 3.7. The proximity of CF_3 groups on the 1,3-$C_5(CF_3)_2$ pentagons in

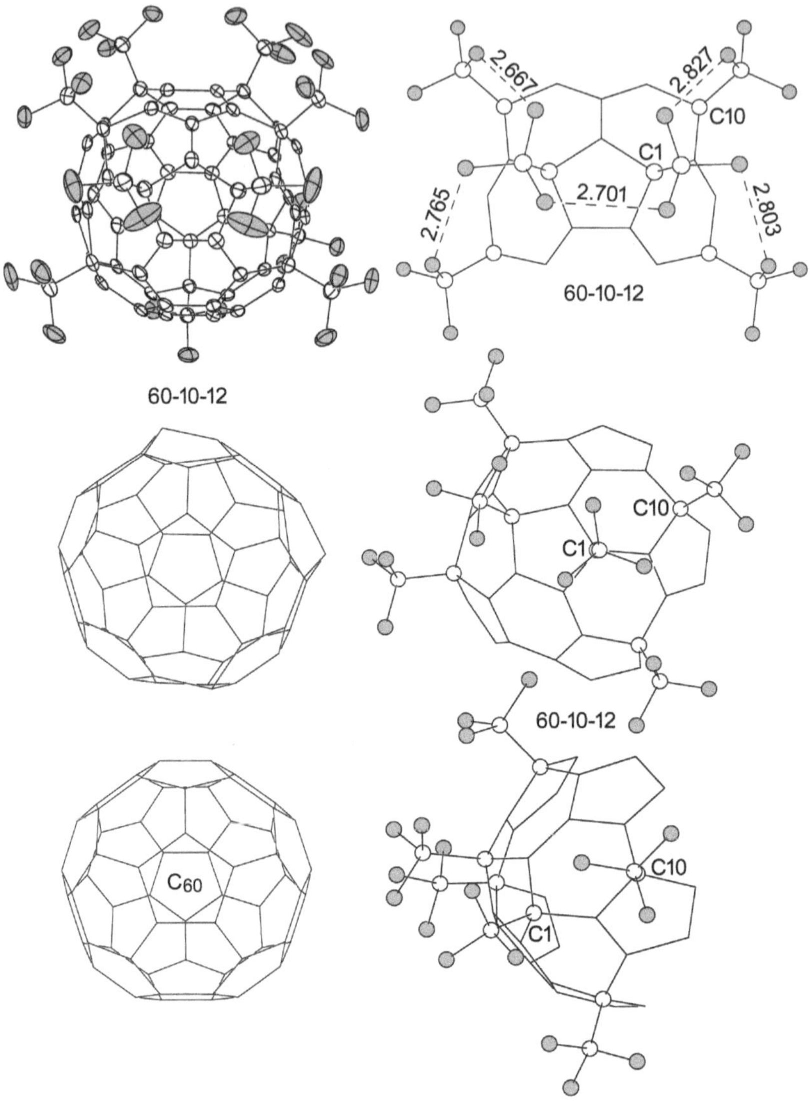

Figure 3.7 Drawings of the X-ray structures of C_{60} (lower left; Ref. [48]) and 60-10-12, including a 50% probability thermal ellipsoid plot (top left; F atoms highlighted in yellow (light gray in print versions)). Portions of the 60-10-12 structure on the right show F···F distances between proximal CF_3 groups and the conformations of the CF_3 groups attached to C1 and C10. The smallest C–C–C–F torsion angles for these two CF_3 groups are 12.1° (ie, significantly eclipsed) and 58.8° (ie, staggered), respectively. The orientation of the 60-10-12 cage (middle left, CF_3 groups not shown) is not the same as the orientation in the thermal ellipsoid plot.

60-10-12, 60-10-3, and 60-12-2 (see Fig. 3.1) and the *m*- and *p*-$C_6(CF_3)_2$ hexagons in all $C_{60}(CF_3)_n$ derivatives results in F···F contacts ≤2.9 Å over the shared hexagons or pentagons, and these contacts in turn result in through-space Fermi-contact $^{6,7}J_{FF}$ nuclear spin−spin couplings observed in their ^{19}F NMR spectra [24,28,30,40,41,55]. The instantaneous $^{6,7}J_{FF}$ values range from 70 to 180 Hz, but only time-averaged 8−20 Hz splittings are observed for fullerene $(CF_3)_n$ derivatives due to rapid rotation of the CF_3 groups on the NMR timescale. When rotational averaging does not occur, as in *p*-$C_6(i\text{-}C_3F_7)_2$ hexagons in the compounds C_2-$C_{60}(i\text{-}C_3F_7)_4$ [56] and C_3-$C_{60}((i\text{-}C_3F_7)_6$ [57], the $^7J_{FF}$ values observed for the $(CF_3)_2C$−F···F−$C(CF_3)_2$ interactions of 2.61 and 2.66 Å, respectively, were found to be 172 and 160 Hz, respectively [24].

The curvature of C_{60} also requires that its C(sp^2) atoms cannot be coplanar with their three nearest neighbors. Each C_{60} vertex is common to two hexagons and one pentagon [51,58], unlike the rigorously planar C(sp^2) atoms in graphite, which are at the junction of three hexagons (the interior angles for the cage C(sp^2) atoms in the $C_{60}(CF_3)_n$ structures are 109 ± 1° for their pentagons and 120 ± 2° for their hexagons). The distortion from planarity introduces significant nonplanar strain energy into C_{60}, believed to be c. 33 kJ mol^{-1} C-atom^{-1} [59]. The exohedral derivatization of fullerenes is driven, in large measure, by the relief of this strain as substituents (such as CF_3 groups) are added to some cage C atoms, changing their hybridization from nominally C(sp^2) to C(sp^3) [60].

The degree to which fullerene cage atoms are nonplanar can be defined by a vector through each cage C atom that makes equal angles, φ, to its three nearest neighbors [51]. A pyramidalization figure of merit, θ, is defined as $\theta = \varphi - 90°$ (ie, for an ideal planar C(sp^2) atom $\theta = 0°$ (ie, 90°−90°) and for an ideal C(sp^3) atom $\theta = 19.5°$ (ie, 109.5°−90°)) [61]. The ideal value of θ for the cage C atoms in DFT-optimized icosahedral C_{60} is 11.64°; for the X-ray structure of C_{60} discussed above the values of θ ranged from 11.4° to 11.9° [48].

The distortion of the C_{60} cage from icosahedral symmetry caused by the addition of 10 CF_3 groups can be seen in Fig. 3.7 for 60-10-12. The 10 cage C(sp^3) atoms are clearly more pyramidal than the 50 remaining cage C(sp^2) atoms. Their θ angles range from 22.0° to 23.5°. Since fullerenes and their derivatives are *closed* carbon cages, the average θ angle for all 60 cage atoms must be essentially constant regardless of the degree of derivatization. As a consequence, most of the remaining cage C(sp^2) atoms in fullerene derivatives must be less pyramidal than in C_{60} itself. In 60-10-12, the average θ angle for the 50 remaining C(sp^2) atoms is 9.4° (they range from 5.1° to 12.4°), and the average θ angle for all 60 cage C atoms is 11.6°. Therefore, the addition of substituents to C_{60} makes some cage atoms more pyramidal because they have become sp^3 hybridized, and makes the remaining C(sp^2) atoms, on average, less pyramidal, both of which are stabilizing changes with respect to the nonplanar strain in underivatized C_{60}. The ranges and average values of C(sp^3) and C(sp^2) θ angles for all $C_{60}(CF_3)_{10}$ isomers are essentially the same as those just enumerated for 60-10-12.

As discussed above, the addition of substituents to *para* positions of C_{60} hexagons results in a significant shortening of a C(sp^2)−C(sp^2) bond in a neighboring pentagon, which is either a terminal or a nonterminal double bond in that pentagon (*t*-DBIP or *nt*-DBIP, respectively). Although the 60 cage C(sp^2)−C(sp^2) bonds that form the pentagons in the X-ray structure of C_{60} have distances ≥1.44 Å [48] (the DFT-optimized value is 1.453 Å [26]), the DBIPs in $C_{60}(CF_3)_{10}$ isomers are ≤1.38 Å, shorter than most of the other cage C−C bonds. This can be seen in Fig. 3.8, which shows the 11 shortest cage C−C bonds in the X-ray structures of 60-10-2 and 60-10-9, which are two of the most precise $C_{60}(CF_3)_{10}$ structures. In both structures, the five DBIPs are among the eight shortest bonds. In the structure of 60-10-2, for example, the five DBIPs range from 1.3545(14)−1.3664(16) Å [43]. All of the remaining $C(sp^2)$−$C(sp^2)$ bonds in pentagons in that structure are longer than 1.4228(16) Å.

3.5 The Links Between Molecular and Electronic Structures of $C_{60}(CF_3)_{10}$ Isomers

With the seven new addition patterns published in Ref. [42], it was possible to further demonstrate the connections between *nt*-DBIPs, LUMO energies, and $E_{1/2}(0/-)$ reduction potentials. The new results [42] were found to be consistent with the correlations first reported in 2007 [26]. Fig. 3.9 is a plot of first reduction potentials in 1,2-$C_6H_4Cl_2$ ($E_{1/2}(0/-)$) for eight of the 13 structurally characterized isomers versus their

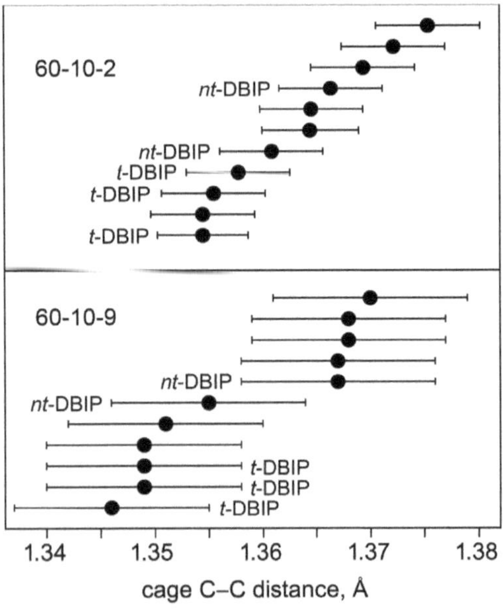

Figure 3.8 The 11 shortest cage C−C bonds in the X-ray structures of 60-10-2 and 60-10-9 (*error bars* are shown as ±3σ). The abbreviations *t*-DBIP and *nt*-DBIP denote *terminal* and *nonterminal* double-bonds-in-pentagons, respectively. The bonds that are not labeled are hexagon−hexagon edges. After the five DBIPs shown in the figure, the next shortest pentagon−hexagon $C(sp^2)$−$C(sp^2)$ bond in 60-10-2 is 1.4228(16) Å, fully 0.056 Å longer than the longest DBIP, 1.3664(16) Å.

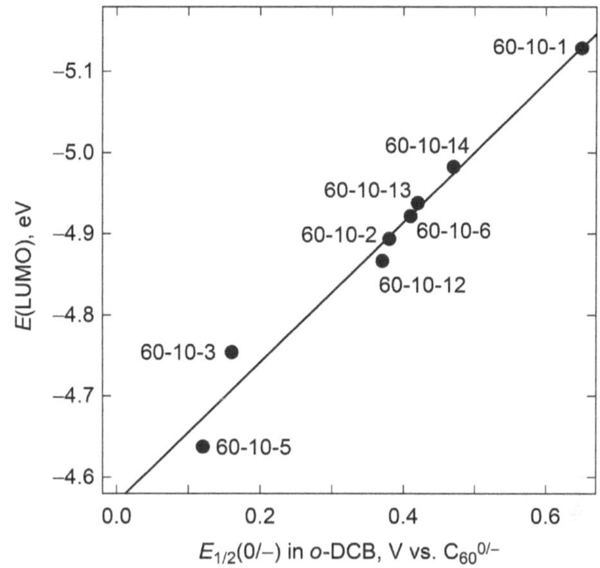

Figure 3.9 First reduction potentials ($E_{1/2}(0/-)$ values) measured in 1,2-$C_6H_4Cl_2$ (*o*-DCB) for eight $C_{60}(CF_3)_{10}$ isomers versus their DFT-predicted LUMO energies (*E*(LUMO) values). The line is a linear least-squares fit to the data. *LUMO*, lowest unoccupied molecular orbital. The data are from B.W. Larson, Synthesis and Characterization of Fluorine-Containing C_{60} Derivatives and Their Charge Carrier Photophysics in Organic Photovoltaics (Ph.D. dissertation), Colorado State University (2013); B.W. Larson, A.A. Popov, S.H. Strauss, O.V. Boltalina, Synthesis and characterization of Twenty-one new isomers of $C_{60}(CF_3)_{10}$, J. Fluorine Chem. (2016) (submitted for publication); and E.V. Bukovsky, B.W. Larson, T.T. Clikeman, Y.-S. Chen, A.A. Popov, O.V. Boltalina, S.H. Strauss, Structures and structure-related electronic properties of $C_{60}(CF_3)_{10}$ isomers, J. Fluorine Chem. 185 (2016) 103–117 and are listed in Table 3.1.

DFT-predicted LUMO energies (*E*(LUMO)) [40,41]. Several observations are noteworthy. The two lowest $E_{1/2}(0/-)$ values are for 60-10-5 and 60-10-3, and both compounds have a single-ribbon addition pattern. Furthermore, the $E_{1/2}(0/-)$ value for 60-10-5, an isomer with only one *nt*-DBIP, is lower than for 60-10-3, which has two *nt*-DBIPs. The four highest $E_{1/2}(0/-)$ values are, in ascending order, for 60-10-6, 60-10-13, 60-10-14, and 60-10-1, all of which have a two-ribbon addition pattern (60-10-6 and 60-10-14) or a ribbon plus an isolated *p*-$C_6(CF_3)_2$ hexagon (60-10-1 and 60-10-13). Furthermore, three of these four high-potential isomers have three *nt*-DBIPs. The isomer 60-10-13 has only two *nt*-DBIPs, but as discussed below these two *nt*-DBIPs are in close proximity and lead to a relatively low-energy LUMO.

Three new examples of the connections between *nt*-DBIPs and the positions, extents of delocalization, and energies of $C_{60}(CF_3)_{10}$ LUMOs are shown in Fig. 3.10. The two LUMO-Schlegel diagrams on the left show that the consequence of the isolated *p*-$C_6(CF_3)_2$ hexagon in 60-10-12 is to give this isomer a third *nt*-DBIP to go along with the two contained in the isolated-fulvene fragment. The isomer

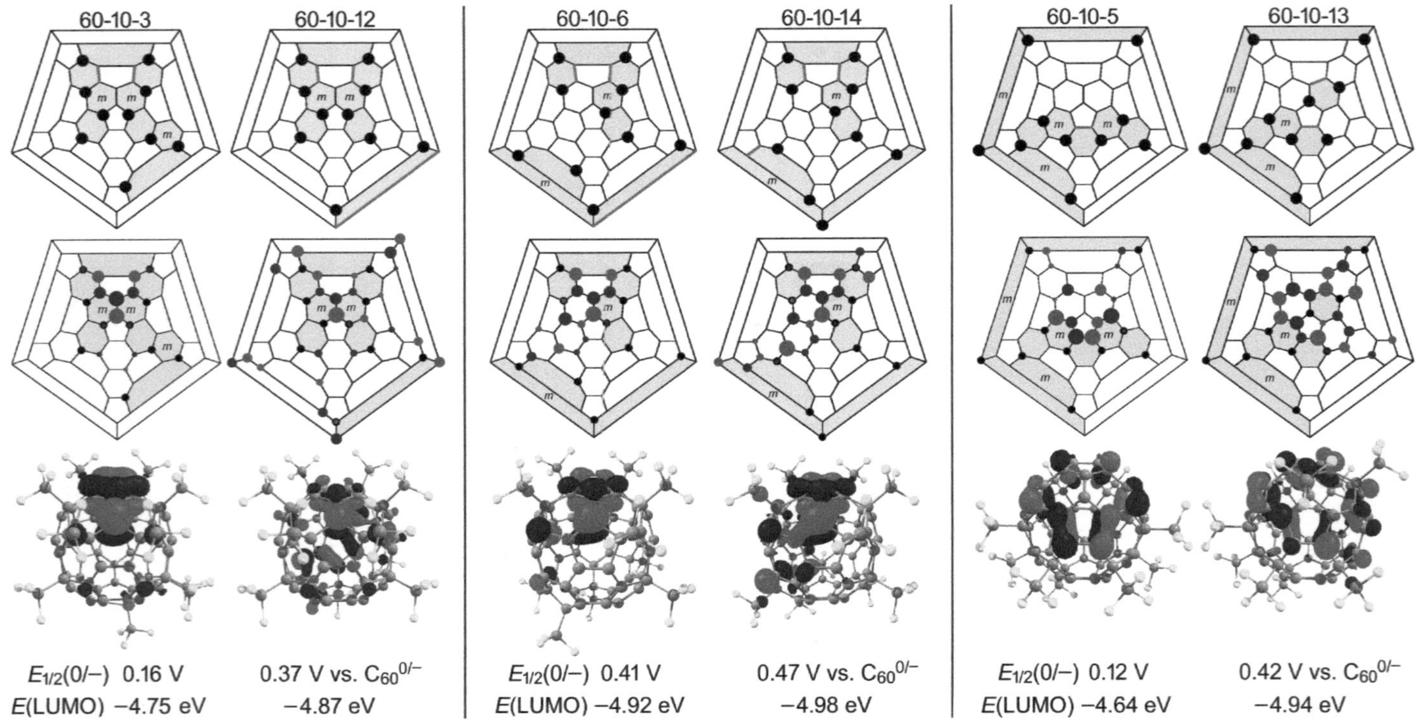

Figure 3.10 Schlegel diagrams with *nt*-DBIPs highlighted in red (gray in print versions), LUMO-Schlegel diagrams, and representations of the LUMOs for six of the 13 isomers of $C_{60}(CF_3)_{10}$, including three of the structurally characterized isomers recently reported in Ref. [42]. The *blue* (darker gray in print versions) (+) and *green* (dark gray in print versions) (−) *circles* in the LUMO-Schlegel diagrams represent the exohedral lobes of the p-π atomic orbitals for cage C(sp^2) atoms. The diameters of the circles are scaled approximately to the contributions of those orbitals to the LUMO. *LUMO*, lowest unoccupied molecular orbital; *nt-DBIPs*, *nonterminal* double-bonds-in-pentagons.

60-10-3 has only the two isolated-fulvene *nt*-DBIPs. For these reasons, the 60-10-12 LUMO is much more extensively delocalized than the 60-10-3 LUMO, and as a result 60-10-12 is easier to reduce than 60-10-3 by 0.21 V. The two isomers 60-10-6 and 60-10-8 have very similar p^3mp,pmp addition patterns that involve the same eight C_{60} hexagons. Nevertheless, the two LUMO-Schlegel diagrams in the middle of Fig. 3.10 demonstrate that the *nt*-DBIP formed by the *pmp* ribbon in 60-10-14 is closer to, and is more extensively delocalized with, the other two *nt*-DBIPs in its LUMO than is the case in 60-10-6. As a result, 60-10-14 is easier to reduce than 60-10-6, in this case by 0.06 V. Finally, the *pmpmpmpmp* and *pmpmpmp,p* addition patterns of 60-10-5 and 60-10-13 are very similar, having the positions of eight of their 10 CF_3 groups in common. The 60-10-5 isomer has only one *nt*-DBIP. The presence of two *nt*-DBIPs in 60-10-13, and their close proximity, results in a very delocalized LUMO relative to the 60-10-5 LUMO, as shown by the two LUMO-Schlegel diagrams on the right in Fig. 3.10. The seemingly minor change in the addition pattern results in a major change in the LUMO and makes 60-10-13 significantly easier to reduce, by 0.3 V, than 60-10-5.

One final link between the molecular and electronic structures of $C_{60}(CF_3)_{10}$ isomers that is not yet fully understood is the correlation between the HOMO−LUMO gap (ie, the absolute value of *E*(HOMO) − *E*(LUMO)) and the number of *nt*-DBIPs, which is shown in Fig. 3.11 [42]. The plot includes a hypothetical p^3mp,pmp isomer denoted 60-10-8, which is related to the addition patterns of 60-10-6 and 60-10-14, and a hypothetical p^3mp,p,p isomer denoted 60-10-15, which is related to the addition pattern of 60-10-1. Schlegel diagrams for 60-10-8 and 60-10-15, which have three and four *nt*-DBIPs, respectively, are also shown in Fig. 3.11. Although the ranges of HOMO−LUMO gaps overlap, there is a clear trend. Isomers with more *nt*-DBIPs tend to have smaller HOMO−LUMO gaps. This may be a general result for $C_{60}(CF_3)_n$ isomers, not only for isomers with $n = 10$, and this correlation should be investigated in more detail in the future.

3.6 The Solid-State Packing of $C_{60}(CF_3)_{10}$ Isomers

It is widely believed that the number of electronically coupled nearest-neighbor fullerene derivative electron acceptors and their three-dimensional arrangement are among the key factors that determine charge transport properties in the fullerene domains in Type II heterojunction solar cells (see Ref. [46] and references therein). Accordingly, the determination of the number of nearest-neighbor fullerene derivatives is important, and it depends on the choice of the maximum relevant fullerene centroid···centroid (⊙···⊙) distance beyond which fullerene−fullerene electronic coupling is probably negligible.

Before discussing the solid-state packing of the 10 solvent-free $C_{60}(CF_3)_{10}$ structures, it is instructive to review the solid-state packing of the solvent-free X-ray structures of two C_{60} derivatives with only a single cycloadduct substituent and therefore with only two cage C(sp^3) atoms, 1,9-C_{60}(*cyclo*-CF_2(2-C_6F_4)) (hereinafter known as

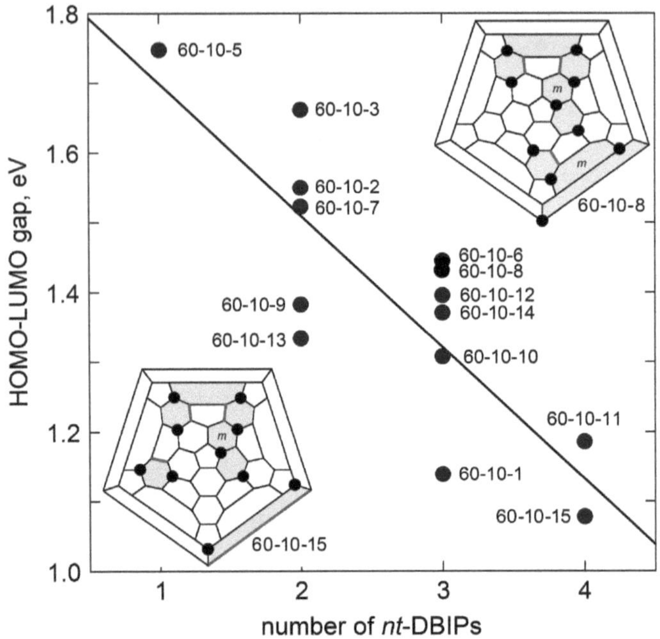

Figure 3.11 Plot of the DFT-predicted HOMO−LUMO gap (ie, the DFT-predicted absolute value of *E*(HOMO) − *E*(LUMO)) versus the number of nonterminal double bonds in pentagons (*nt*-DBIPs) for the 13 structurally characterized $C_{60}(CF_3)_{10}$ isomers plus the hypothetical isomers labeled 60-10-8, with three *nt*-DBIPs, and 60-10-15, with four *nt*-DBIPs. The Schlegel diagrams for 60-10-8 and 60-10-15 are shown at the upper right and lower left corners of the plot, respectively. Their *nt*-DBIPs are highlighted in red (gray in print versions). See Ref. [24] and references therein for information about the 60-10-8 addition pattern. See Ref. [42] for information about the 60-10-15 addition pattern. *nt-DBIPs*, *nonterminal* double-bonds-in-pentagons; *HOMO*, highest occupied molecular orbital; *LUMO*, lowest unoccupied molecular orbital.

the faux-hawk fullerene, or FHF) [46] and phenyl-C_{61}-butyric acid methyl ester (hereinafter known as PCBM) [47], both of which are shown in Fig. 3.12 (note that the two cage $C(sp^3)$ atoms are adjacent to one another). The substituents in these two derivatives have nearly the same number of non-hydrogen atoms, 13 for FHF and 14 for PCBM, but the FHF substituent is clearly more compact. There are only seven nearest-neighbor fullerene molecules in crystalline solvent-free PCBM, with ⊙···⊙ distances of 9.95−10.28 Å (the mean and median distances are 10.17 and 10.24 Å, respectively). On the other hand, there are 10 nearest neighbors in the structure of FHF, with ⊙···⊙ distances of 9.74−10.34 Å (the mean and median distances are 10.09 and 10.05 Å, respectively). The result is that the 1.885 g cm^{-3} density of crystalline solvent-free FHF is 15.6% higher than the 1.631 g cm^{-3} density of crystalline solvent-free PCBM, even though the molar masses of the two compounds, 918.67 g mol^{-1} for FHF and 910.83 g mol^{-1} for PCBM, differ by only 1.1%. This comparison shows that fullerene derivatives that appear to be sterically similar can

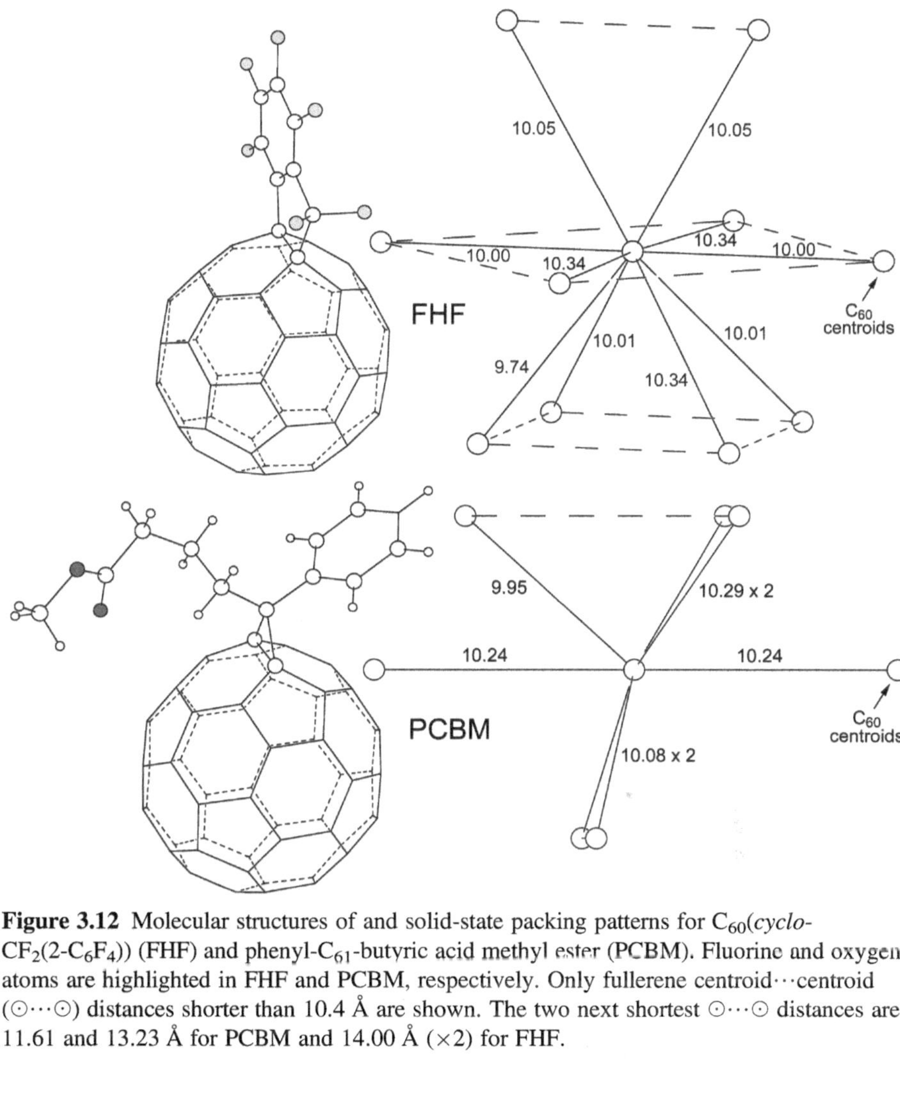

Figure 3.12 Molecular structures of and solid-state packing patterns for C_{60}(*cyclo*-CF_2(2-C_6F_4)) (FHF) and phenyl-C_{61}-butyric acid methyl ester (PCBM). Fluorine and oxygen atoms are highlighted in FHF and PCBM, respectively. Only fullerene centroid···centroid (⊙···⊙) distances shorter than 10.4 Å are shown. The two next shortest ⊙···⊙ distances are 11.61 and 13.23 Å for PCBM and 14.00 Å (×2) for FHF.

have very different packing patterns in the solid-state (and presumably in thin films as well), including the number of closely approaching nearest neighbors.

After the seven closest nearest neighbors in the structure of PCBM, the two next shortest ⊙···⊙ distances are 11.61 and 13.23 Å. On what basis was the 11.61 Å molecule excluded as an electronically relevant nearest neighbor? The criterion used in Ref. [46] is as follows. The 11.61 Å distance was considered to be too long for effective electronic coupling because the closest $C_{cage}\cdots C_{cage}$ distance between these two PCBM molecules is 5.03 Å. In contrast, the closest $C_{cage}\cdots C_{cage}$ distances for PCBM molecules with ⊙···⊙ separations of 10.28 and 10.24 Å are 3.23 and 3.32 Å, respectively, approximately the same as the 3.35 Å interplanar separation in graphite

[62]. For FHF, the closest C(cage)···C(cage) distance between two molecules with a separation of 10.34 Å is 3.38 Å, and the next closest ⊙···⊙ distances are 14.00 Å. Finally, note that one of the ⊙···⊙ separations in FHF, 9.74 Å, is actually shorter than the ⊙···⊙ distance of 9.94 Å in the cubic close-packed structure of pure crystalline C_{60} [63].

The isomers of $C_{60}(CF_3)_{10}$ are composed of 100 non-hydrogen atoms, and as such would not be expected to have multiple short ⊙···⊙ separations as in C_{60}, FHF, and PCBM. Nevertheless, the distances listed in Tables 3.2–3.4 show that the X-ray structures of 11 of the 13 isomers have one or two distances shorter than 10.4 Å (only 60-10-6 and 60-10-7 do not), and five of them have a ⊙···⊙ separation shorter than 9.94 Å.

If we consider only the solvent-free $C_{60}(CF_3)_{10}$ structures (ie, excluding 60-10-5, 60-10-11, and 60-10-14), some surprising observations can be made. The ⊙···⊙ distances for these 10 structures are shown graphically in Fig. 3.13. Seven of these 10 structures have distorted close-packed packing patterns, with 12 nearest neighbors if ⊙···⊙ separations out to 14.6 Å are included. This distance corresponds to a pair of

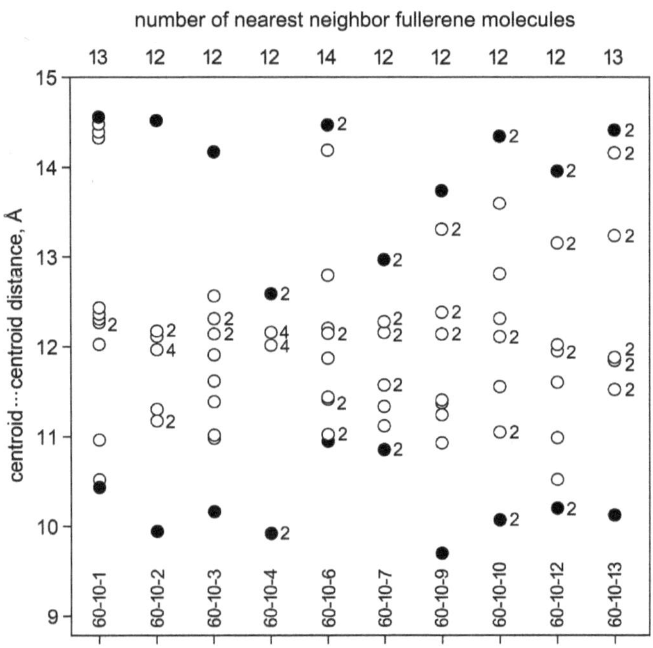

Figure 3.13 The fullerene centroid···centroid (⊙···⊙) distances in the X-ray structures of 10 solvent-free isomers of $C_{60}(CF_3)_{10}$. The shortest and longest ⊙···⊙ distances are highlighted as *black circles* for clarity. The numbers 2 and 4 next to some of the data points indicated that two or four identical ⊙···⊙ distances, respectively, are exhibited by that structure. The numbers at the top of the figure indicate the number of nearest neighbors for each fullerene molecule in each structure. The packing of molecules in the structures of 60-10-3, 60-10-4, 60-10-7, 60-10-9, 60-10-10, and 60-10-12 are distorted cubic or hexagonal close-packed arrays.

$C_{60}(CF_3)_{10}$ molecules with a CF_3 group on each molecule oriented along the ⊙···⊙ vector and with intermolecular F···F distances of c. 3 Å. For these seven structures, the ranges of the 12 ⊙···⊙ distances are very different, almost by a factor of two. The narrowest range, 2.67 Å, is for 60-10-4 (from 9.93 to 12.59 Å); the widest range, 4.57 Å, is for 60-10-2 (from 9.95 to 14.52 Å). The mean ⊙···⊙ distances vary from 11.81 Å for 60-10-4 to 12.12 Å for 60-10-10, an increase of 2.6%, and a corresponding decrease in the crystal density of 3.8%, from 2.072 g cm^{-3} for 60-10-4 to 1.993 g cm^{-3} for 60-10-10. It is remarkable that the very different ranges and distributions of ⊙···⊙ distances lead to mean ⊙···⊙ distances and crystal densities that only vary by a few percent. Equally surprising are the differences in the structures of 60-10-4 and 60-10-6. Their shortest ⊙···⊙ distances differ by 1.02 Å (9.93 Å for 60-10-4 and 10.95 Å for 60-10-6) and their longest ⊙···⊙ distances differ by 1.88 Å (12.59 Å for 60-10-4 and 14.47 Å for 60-10-6), yet their crystal densities differ by less than 1%. This is because 60-10-6 has 14 nearest neighbors instead of only 12. Significantly, the distribution of CF_3 groups in the 10 isomers with solvent-free structures is very different. For example, the CF_3 groups in 60-10-4 are essentially confined to one hemisphere of C_{60} whereas the CF_3 groups in 60-10-6 are distributed over the entire surface of C_{60} (there are four $C(sp^2)_6$ hexagons in 60-10-4 that are only bonded to cage $C(sp^2)$ atoms but none in 60-10-6).

Fig. 3.14 shows the distorted hexagonal close-packed (HCP) arrays of fullerene centroids in the X-ray structures of 60-10-9 and 60-10-10, two structures that have the same crystal density and nearly the same mean ⊙···⊙ distance. However, their HCP arrays are organized very differently. The rigorously planar arrays of nearest neighbor centroids exhibit longer ⊙···⊙ distances in 60-10-9 (12.14−13.31 Å) than in 60-10-10 (10.07−12.12 Å), and the perpendicular interplanar spacings are shorter in 60-10-9 (7.89 and 9.27 Å) than in 60-10-10 (11.18 and 11.31 Å).

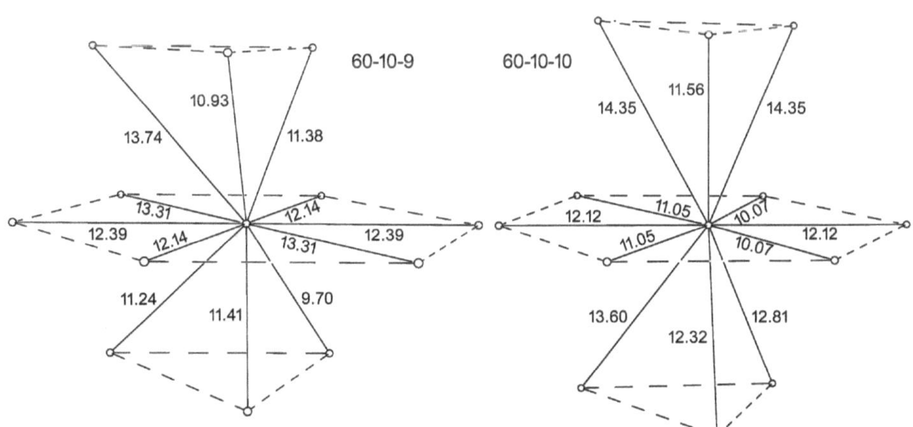

Figure 3.14 The distorted hexagonal close-packed arrays of fullerene centroids (⊙) in the X-ray structures of 60-10-9 and 60-10-10. The ⊙···⊙ distances are rounded off to the nearest 0.01 Å.

A pair of 60-10-9 molecules whose centroids are in parallel planes separated by 7.89 Å exhibit the shortest ⊙···⊙ distance in that structure, 9.703 Å. Those two molecules are shown in Fig. 3.15. They are sufficiently close together that a pair of hexagons, one of each molecule, is π−π stacked with a perpendicular separation of 3.24 Å, shorter than the interplanar separation in graphite. This occurs because the closest CF_3 groups on each molecule do not lie along the ⊙···⊙ vector. Instead, they interdigitate with intermolecular F···F distances no shorter than the sum of van der Waals distances for two F atoms, c. 3 Å. This is the only pair of molecules in the structure of 60-10-9 with a ⊙···⊙ distance shorter than 10.4 Å (recall that this is the longest ⊙···⊙ distance in FHF or PCBM considered to be important for effective electronic coupling). In other words, the hexagons that interact in this way in 60-10-9 are symmetry related.

In contrast, for $C_{60}(CF_3)_{10}$ structures that have two sterically accessible hexagons that can exhibit π−π interactions with nearest neighbors, two short ⊙···⊙ distances are possible. These structures are 60-10-4, 60-10-10, and 60-10-12. By studying the

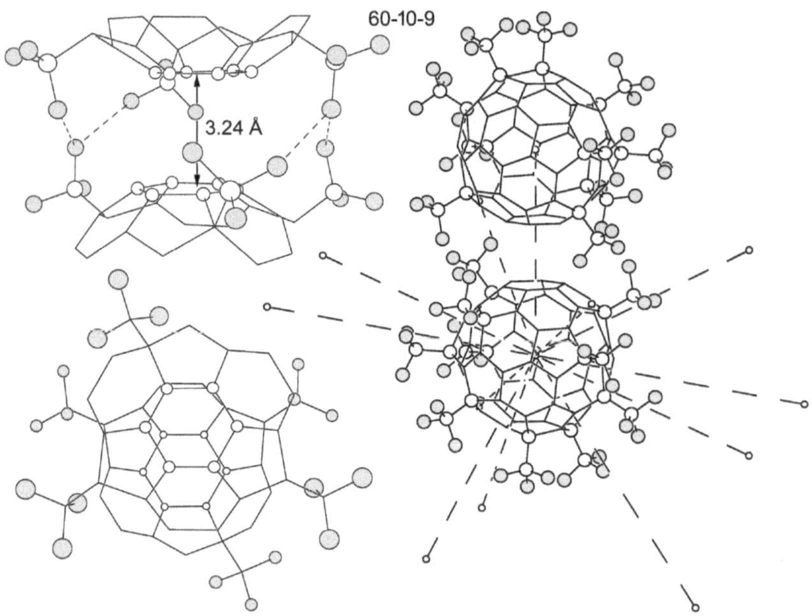

Figure 3.15 Drawings of the X-ray structure of 60-10-9 (all atoms shown as points or spheres of arbitrary size; F atoms highlighted in yellow (light gray in print versions)). The drawing on the right shows the two molecules of 60-10-9 with the shortest fullerene centroid···centroid distance (9.71 Å; centroids shown as *small spheres*). The *dashed lines* indicate the vectors connecting the 12 centroids that surround the bottom molecule in the drawing on the right. The drawings on the left are two orientations of the portions of the two molecules of 60-10-9 the centroids of which are separated by 9.71 Å. The distance between the least-squares planes of the rigorously parallel fullerene hexagons highlighted with their $C(sp^2)$ atoms shown as hollow spheres is 3.24 Å. The dashed lines indicate intermolecular F···F van der Waals contacts of c. 3 Å.

connections between the molecular and electronic structures and the packing patterns of $C_{60}(CF_3)_{10}$ isomers, it may now be possible to design fullerene derivatives with *fewer than* 10 CF_3 groups that are strong electron acceptors *and* have many short ⊙···⊙ distances necessary for good three-dimensional electron mobility in the solid state.

Acknowledgments

Authors are thankful for partial financial support from National Science Foundation (grants CHE-1362302, CHE-1012468).

References

[1] X.-B. Wang, C.-F. Ding, L.-S. Wang, High resolution photoelectron spectroscopy of C_{60}^-, J. Chem. Phys. 110 (1999) 8217–8220.
[2] X.B. Wang, H.K. Woo, X. Huang, M.M. Kappes, L.S. Wang, Direct experimental probe of the on-site Coulomb repulsion in the doubly charged fullerene anion C_{70}^{2-}, Phys. Rev. Lett. 96 (2006) 143002/1–143002/4.
[3] X.B. Wang, H.K. Woo, J. Yang, M.M. Kappes, L.S. Wang, Photoelectron spectroscopy of singly and doubly charged higher fullerenes at low temperatures: C_{76}^-, C_{78}^-, C_{84}^-, and C_{76}^{2-}, C_{78}^{2-}, C_{84}^{2-}, J. Phys. Chem. C 111 (2007) 17684–17689.
[4] Q. Xie, E. Perez-Cordero, L. Echegoyen, Electrochemical detection of C_{60}^{6-} and C_{70}^{6-}: enhanced stability of fullerides in solution, J. Am. Chem. Soc. 114 (1992) 3978–3980.
[5] L. Echegoyen, L.E. Echegoyen, Electrochemistry of fullerenes and their derivatives, Acc. Chem. Res. 31 (1998) 593–601.
[6] E. Espíldora, J.L. Delgado, N. Martín, Donor-acceptor hybrids for organic electronics, Israel J. Chem. 54 (2014) 429–439.
[7] D. Jariwala, V.K. Sangwan, L.J. Lauhon, T.J. Marks, M.C. Hersam, Carbon nanomaterials for electronics, optoelectronics, photovoltaics, and sensing, Chem. Soc. Rev. 42 (2013) 2824–2860.
[8] J.R. Pinzón, A. Villalta-Cerdas, L. Echegoyen, Fullerenes, carbon nanotubes, and graphene for molecular electronics, in: R.M. Metzger (Ed.), Unimolecular and Supramolecular Electronics I. Chemistry and Physics Meet at Metal-molecule Interfaces, Springer, Heidelberg, 2012.
[9] H.W. Kroto, J.R. Heath, S.C. O'Brien, R.F. Curl, R.E. Smalley, C60: buckminsterfullerene, Nature 318 (1985) 162–163.
[10] W. Krätschmer, L.D. Lamb, K. Fostiropoulos, D.R. Huffman, Solid C_{60} – a new form of carbon, Nature 347 (1990) 354–358.
[11] H. Selig, K. Kniaz, G.B.M. Vaughan, J.E. Fischer, A.B. Smith, Fluorinated fullerenes: synthesis and characterization, Macromol. Symp. 82 (1994) 89–97.
[12] A. Tuinman, P. Mukherjee, J.L. Adcock, R.L. Hettich, R.N. Compton, Characterization and stability of highly fluorinated fullerenes, J. Phys. Chem. 96 (1992) 7584–7589.
[13] A.A. Gakh, A.A. Tuinman, J.L. Adcock, R.N. Compton, Highly fluorinated fullerenes as oxidizers and fluorinating agents, Tetrahedron Lett. 34 (1993) 7167–7170.
[14] O.V. Boltalina, S.H. Strauss, Fluorinated fullerenes, in: J.A. Schwarz, C. Contescu, K. Putyera (Eds.), Dekker Encyclopedia of Nanoscience and Nanotechnology, Marcel Dekker, New York, 2004.

[15] C. Jin, R.L. Hettich, R.N. Compton, A. Tuinman, A. Derecskei-Kovacs, D.S. Marynick, B.I. Dunlap, Attachment of two electrons to $C_{60}F_{48}$. Coulomb barriers in doubly charged anions, Phys. Rev. Lett. 73 (1994) 2821–2824.
[16] F. Zhou, G.J. Van Berkel, B.T. Donovan, Electron-transfer reactions of $C_{60}F_{48}$, J. Am. Chem. Soc. 116 (1994) 5485–5486.
[17] Y. Matsuo, T. Nakajima, Electrochemical properties of fluorinated fullerene C_{60}, Electrochim. Acta 41 (1996) 15–19.
[18] A.A. Popov, J. Tarabek, I.E. Kareev, S.F. Lebedkin, S.H. Strauss, O.V. Boltalina, L. Dunsch, Poly(trifluoromethyl)fullerene radical anions. An ESR/Vis-NIR spectroelectrochemical study of $C_{60}F_{2,4}$ and $C_{60}(CF_3)_{2,10}$, J. Phys. Chem. A 109 (2005) 9709–9711.
[19] D. Paolucci, F. Paolucci, M. Marcaccio, M. Carano, R. Taylor, Electrochemistry of perfluorinated fullerenes: the case of three isomers of $C_{60}F_{36}$, Chem. Phys. Lett. 400 (2004) 389–393.
[20] X.-B. Wang, C. Chi, M. Zhou, I.V. Kuvychko, K. Seppelt, A.A. Popov, S.H. Strauss, O.V. Boltalina, L.-S. Wang, Photoelectron spectroscopy of $C_{60}F_n^-$ and $C_{60}F_m^{2-}$ ($n = 17, 33,$ 35, 43, 45, 47; $m = 34$, 46) in the gas phase and the generation and characterization of C_1-$C_{60}F_{47}^-$ and D_2-$C_{60}F_{44}$ in solution, J. Phys. Chem. A 114 (2010) 1756–1765.
[21] R. Taylor, A.G. Avent, T.J. Dennis, J.P. Hare, H.W. Kroto, D.R.M. Walton, J.H. Holloway, E.G. Hope, G.J. Langley, No lubricants from fluorinated C_{60}, Nature 355 (1992) 27–28.
[22] P.J. Fagan, P.J. Krusic, C.N. McEwen, J. Lazar, D.H. Parker, N. Herron, E. Wasserman, Production of perfluoroalkylated nanospheres from buckminsterfullerene, Science 262 (1993) 404–407.
[23] O.V. Boltalina, I.V. Kuvychko, N.B. Shustova, S.H. Strauss, Perfluoroalkylation of fullerenes, in: F. Desouza, K.M. Kadish (Eds.), Handbook of Carbon Nanomaterials, Syntheses and Supramolecular Systems, World Scientific Singapore, vol. 1, 2011.
[24] O.V. Boltalina, A.A. Popov, I.V. Kuvychko, N.B. Shustova, S.H. Strauss, Perfluoroalkylfullerenes, Chem. Rev. 115 (2015) 1051–1105.
[25] S.I. Troyanov, A.A. Goryunkov, E.I. Dorozhkin, D.V. Ignat'eva, N.B. Tamm, S.M. Avdoshenko, I.N. Ioffe, V.Y. Markov, L.N. Sidorov, K. Scheural, E. Kemnitz, Higher trifluoromethylated derivatives of C_{60}, $C_{60}(CF_3)_{16}$ and $C_{60}(CF_3)_{18}$. Synthesis, structure, and theoretical study, J. Fluorine Chem. 128 (2007) 545–551.
[26] A.A. Popov, I.E. Kareev, N.B. Shustova, E.B. Stukalin, S.F. Lebedkin, K. Seppelt, S.H. Strauss, O.V. Boltalina, L. Dunsch, Electrochemical, spectroscopic, and DFT study of $C_{60}(CF_3)_n$ Frontier orbitals ($n = 2-18$): the link between double bonds in pentagons and reduction potentials, J. Am. Chem. Soc. 129 (2007) 11551–11568.
[27] A.A. Goryunkov, I.V. Kuvychko, I.N. Ioffe, D.L. Dick, L.N. Sidorov, S.H. Strauss, O.V. Boltalina, Isolation of $C_{60}(CF_3)_n$ ($n = 2$, 4, 6, 8, 10) with high compositional purity, J. Fluorine Chem. 124 (2003) 61–64.
[28] I.E. Kareev, I.V. Kuvychko, S.F. Lebedkin, S.M. Miller, O.P. Anderson, K. Seppelt, S.H. Strauss, O.V. Boltalina, Synthesis, structure, and ^{19}F NMR spectra of 1,3,7,10,14,17,23,28,31,40-$C_{60}(CF_3)_{10}$, J. Am. Chem. Soc. 127 (2005) 8362–8375.
[29] N.B. Shustova, I.V. Kuvychko, R.D. Bolskar, K. Seppelt, S.H. Strauss, A.A. Popov, O.V. Boltalina, Trifluoromethyl derivatives of insoluble small-HOMO-LUMO-gap hollow higher fullerenes. Nmr and DFT structure elucidation of C_2-(C_{74}-D_{3h})$(CF_3)_{12}$, C_s-(C_{76}-T_d)$(CF_3)_{12}$, C_2-(C_{78}-D_{3h}(5))$(CF_3)_{12}$, C_s-(C_{80}-C_{2v}(5))$(CF_3)_{12}$, and C_2-(C_{82}-C_2(5))$(CF_3)_{12}$, J. Am. Chem. Soc. 128 (2006) 15793–15798.
[30] I.E. Kareev, A.A. Popov, I.V. Kuvychko, N.B. Shustova, S.F. Lebedkin, V.P. Bubnov, O.P. Anderson, K. Seppelt, S.H. Strauss, O.V. Boltalina, Synthesis and X-ray or NMR/

DFT structure elucidation of twenty-one new trifluoromethyl derivatives of soluble cage isomers of C_{76}, C_{78}, C_{84}, and C_{90}, J. Am. Chem. Soc. 130 (2008) 13471–13489.

[31] A.A. Popov, I.E. Kareev, N.B. Shustova, S.F. Lebedkin, S.H. Strauss, O.V. Boltalina, L. Dunsch, Synthesis, spectroscopic and electrochemical characterization, and DFT study of seventeen $C_{70}(CF_3)_n$ derivatives ($n = 2, 4, 6, 8, 10, 12$), Chem. Eur. J. 14 (2008) 107–121.

[32] N.B. Shustova, D.V. Peryshkov, I.V. Kuvychko, Y.-S. Chen, M.A. Mackey, C.E. Coumbe, D.T. Heaps, B.S. Confait, T. Heine, J.P. Phillips, S. Stevenson, L. Dunsch, A.A. Popov, S.H. Strauss, O.V. Boltalina, Poly(perfluoroalkylation) of metallic nitride fullerenes reveals addition-pattern guidelines: synthesis and characterization of a family of $Sc_3N@C_{80}(CF_3)_n$ (n = 2-16) and their radical anions, J. Am. Chem. Soc. 133 (2011) 2672–2690.

[33] N.B. Shustova, I.V. Kuvychko, A.A. Popov, M. von Delius, L. Dunsch, O.P. Anderson, A. Hirsch, S.H. Strauss, O.V. Boltalina, Nitrogen directs multiple radical additions to the 9,9′-Bi-1-aza(C_{60}-I_h)[5,6]fullerene: X-ray structure of 6,9,12,15,18-$C_{59}N(CF_3)_5$, Angew. Chem. Int. Ed. 50 (2011) 5537–5540.

[34] E.I. Dorozhkin, A.A. Goryunkov, I.N. Ioffe, S.M. Avdoshenko, V.Y. Markov, N.B. Tamm, D.V. Ignat'eva, L.N. Sidorov, S.I. Troyanov, Synthesis, structure, and theoretical study of lower trifluoromethyl derivatives of [60]Fullerene, Eur. J. Org. Chem. (2007) 5082–5094.

[35] J.B. Whitaker, I.V. Kuvychko, N.B. Shustova, Y.-S. Chen, S.H. Strauss, O.V. Boltalina, An elusive fulvene 1,7,11,24-$C_{60}(CF_3)_4$ and its unusual reactivity, Chem. Commun. 50 (2014) 1205–1208.

[36] A.D. Darwish, A.G. Avent, A.K. Abdul-Sada, R. Taylor, [60]- and [70]Fullerenes are trifluoromethylated across 5:6-bonds, Chem. Commun. (2003) 1374–1375.

[37] A.D. Darwish, A.K. Abdul-Sada, A.G. Avent, N. Martsinovich, J.M. Street, R. Taylor, Novel addition in trifluoromethylation of [70]fullerene, J. Fluorine Chem. 125 (2004) 1383–1391.

[38] M.S. Meier, M. Poplawska, A.L. Compton, J.P. Shaw, J.P. Selegue, T.F. Guarr, Preparation and isolation of three isomeric C_{70} isoxazolines: strong deshielding in the polar region of C_{70}, J. Am. Chem. Soc. 116 (1994) 7044–7048.

[39] L.E. Echegoyen, F.D. Djojo, A. Hirsch, L. Echegoyen, Electrochemical Studies of Seven Regioisomers of Tris[di(ethoxycarbonyl)methano][60]fullerene: Umpolung of Regiochemistry in Retro-Bingel Reactions and Walk-on-the-Sphere Rearrangements, J. Org. Chem. 65 (2000) 4994–5000.

[40] B.W. Larson, Synthesis and Characterization of Fluorine-containing C_{60} Derivatives and Their Charge Carrier Photophysics in Organic Photovoltaics, (Ph.D. dissertation) Colorado State University, 2013.

[41] B.W. Larson, A.A. Popov, S.H. Strauss, O.V. Boltalina, Synthesis and characterization of Twenty-one new isomers of $C_{60}(CF_3)_{10}$, J. Fluorine Chem. (2016) (submitted for publication).

[42] E.V. Bukovsky, B.W. Larson, T.T. Clikeman, Y.-S. Chen, A.A. Popov, O.V. Boltalina, S.H. Strauss, Structures and structure-related electronic properties of $C_{60}(CF_3)_{10}$ isomers, J. Fluorine Chem. 185 (2016) 103–117.

[43] I.E. Kareev, S.F. Lebedkin, S.M. Miller, O.P. Anderson, S.H. Strauss, O.V. Boltalina, 1,6,11,16,18,26,36,44,48,58-Decakis(trifluoromethyl)-1,6,11,16,18,26,36,44,48,58-decahydro(C_{60}-I_h)[5,6]fullerene benzene 2.5-solvate, Acta Crystallogr. E62 (2006) o1498–o1500.

[44] I.E. Kareev, S.F. Lebedkin, A.A. Popov, S.M. Miller, O.P. Anderson, S.H. Strauss, O.V. Boltalina, 1,6,12,15,18,23,25,41,45,57-Decakis(trifluoromethyl)-1,6,12,15,18,23–

25,41,45,57-decahydro(C_{60}-I_h)[5,6]fullerene, Acta Crystallogr. E62 (2006) o1501–o1503.
[45] N.B. Shustova, D.V. Peryshkov, A.A. Popov, O.V. Boltalina, S.H. Strauss, 1,6,11,18,24,27,33,51,54,60-Decakis(trifluoromethyl)-1,6,11,18,24,27,33,51,54,60-decahydro-(C_{60}-I_h)[5,6]fullerene, Acta Crystallogr. E63 (2007) o3129.
[46] L.K. San, E.V. Bukovsky, B.W. Larson, J.B. Whitaker, S.H.M. Deng, N. Kopidakis, G. Rumbles, A.A. Popov, Y.-S. Chen, X.-B. Wang, O.V. Boltalina, S.H. Strauss, A faux hawk fullerene with PCBM-like properties, Chem. Sci. 6 (2015) 1801–1815.
[47] G. Paterno, A.J. Warren, J. Spencer, G. Evans, V.G. Sakai, J. Blumberger, F. Cacialli, Micro-focused X-ray diffraction characterization of high-quality [6,6]-phenyl-C_{61}-butyric acid methyl ester single crystals without solvent impurities, J. Mater. Chem. C 1 (2013) 5619–5623.
[48] M.M. Olmstead, A. de Bettencourt-Dias, H.M. Lee, D. Pham, A.L. Balch, Interactions of metalloporphyrins as donors with the electron acceptors C60, tetracyanoquinomethane (TCNQ) and trinitrofluorenylidenemalonitrile, Dalton Trans. (2003) 3227–3232.
[49] N. Matsuzawa, D.A. Dixon, T. Fukunaga, Semiempirical calculations of dihydrogenated buckminster fullerenes, $C_{60}H_2$, J. Phys. Chem. 96 (1992) 7594–7604.
[50] N. Matsuzawa, T. Fukunaga, D.A. Dixon, Electronic structure of 1,2- and 1,4-$C_{60}X_{2n}$ derivatives with $n = 1$, 2, 4, 6, 8, 10, 12, 18, 24, and 30, J. Phys. Chem. 96 (1992) 10747–10756.
[51] P.W. Fowler, D.E. Manolopoulous, An Atlas of Fullerenes, Dover, Mineola, NY, 2006.
[52] I.E. Kareev, I.V. Kuvychko, A.A. Popov, S.F. Lebedkin, S.M. Miller, O.P. Anderson, S.H. Strauss, O.V. Boltalina, 1,4,10,19,25,41,49,60,66,69-$C_{70}(CF_3)_{10}$. A high-temperature synthesis reveals the surprisingly stable C_1-$C_{70}X_{10}$ isomer with a *para*^{7}-*meta-para* ribbon of nine $C_6(CF_3)_2$ edge-sharing hexagons, Angew. Chem. Int. Ed. 44 (2005) 7984–7987.
[53] N.B. Shustova, B.S. Newell, S.M. Miller, O.P. Anderson, R.D. Bolskar, K. Seppelt, A.A. Popov, O.V. Boltalina, S.H. Strauss, Discovering and verifying elusive fullerene cage isomers: structures of C_2-p^{11}-(C_{74}-D_{3h})$(CF_3)_{12}$ and C_2-p^{11}-(C_{78}-D_{3h}(5))$(CF_3)_{12}$, Angew. Chem. Int. Ed. 46 (2007) 4111–4114.
[54] I.E. Kareev, I.V. Kuvychko, N.B. Shustova, S.F. Lebedkin, V.P. Bubnov, O.P. Anderson, A.A. Popov, S.H. Strauss, O.V. Boltalina, C_1-(C_{84}-C_2(11))$(CF_3)_{12}$: trifluoromethylation yields structural proof of a minor C_{84} cage and reveals a principle of higher fullerene reactivity, Angew. Chem. Int. Ed. 47 (2008) 6204–6207.
[55] I.E. Kareev, G. Santiso-Quinones, I.V. Kuvychko, I.N. Ioffe, I.V. Goldt, S.F. Lebedkin, K. Seppelt, S.H. Strauss, O.V. Boltalina, Synthesis of 1,7- and 1,9-$C_{60}F(CF_3)$: sterically hindered CF_3 rotation and through-space J_{FF} coupling in 1,9-$C_{60}F(CF_3)$ and in C_1- and C_s-$C_{60}F_{17}(CF_3$, J. Am. Chem. Soc. 127 (2005) 11497–11504.
[56] N.B. Shustova, I.V. Kuvychko, D.V. Peryshkov, J.B. Whitaker, B.W. Larson, Y.-S. Chen, L. Dunsch, K. Seppelt, A.A. Popov, S.H. Strauss, O.V. Boltalina, Chem. Commun. 47 (2011) 875.
[57] N.B. Shustova, I.V. Kuvychko, O.V. Boltalina, S.H. Strauss, Acta Crystallogr. E63 (2007) o4575.
[58] H.W. Kroto, The stability of the fullerenes C_{24}, C_{28}, C_{32}, C_{36}, C_{50}, C_{60}, and C_{70},, Nature 329 (1987) 529–531.
[59] R.C. Haddon, Chemistry of the fullerenes: the manifestation of strain in a class of continuous aromatic molecules,, Science 261 (1993) 1545–1550.
[60] A. Hirsch, M. Brettreich, Fullerenes – Chemistry and Reactions, Wiley-VCH, Weinheim, 2005.

[61] O.V. Boltalina, A.A. Popov, S.H. Strauss, Single-crystal X-ray structures of fullerenes and their derivatives, in: H. Dodziuk (Ed.), Strained Hydrocarbons. Beyond the Van't Hoff and Le Bel Hypothesis, Wiley-VCH, Weinheim, 2011.
[62] A.F. Wells, Structural Inorganic Chemistry, fifth ed., Clarendon Press, Oxford, 1984.
[63] S. Liu, Y.-J. Lu, M. Kappes, J.A. Ibers, The structure of the C_{60} Molecule: X-Ray crystal structure determination of a twin at 110 K, Science 254 (1991) 408–410.
[64] F. Cozzi, W.H. Powell, C. Thilgen, Numbering of fullerenes, Pure Appl. Chem. 77 (2005) 843–923.
[65] W.H. Powell, F. Cozzi, G.P. Moss, C. Thilgen, R.J.R. Hwu, A. Yerin, Nomenclature for the C_{60}-I_h and C_{70}-$D_{5h(6)}$ fullerenes (IUPAC recommendations 2002), Pure Appl. Chem. 74 (2002) 629–695.

Trifluoromethylated Corannulene Derivatives: Thermodynamic Stability and Electron-Accepting Properties

4

A.A. Popov
Leibniz Institute for Solid State and Materials Research, Dresden, Germany

Chapter Outline

4.1 Introduction

Since the early steps of its development, the field of molecular organic semiconductors has a pronounced misbalance toward *p*-type materials, despite many efforts dedicated to development of *n*-type organic semiconductors which might be comparable in stability and device performance with their *p*-type counterparts [1−5]. Functionalization with electron-withdrawing groups, such as fluorine or perfluoralkyl radicals, is one of the common methods to increase an electron affinity and improve *n*-type properties of organic semiconductors [1,6−11].

Typically, perfluoroalkylated polycyclic aromatic hydrocarbons (PAHs) are synthesized either by a multistep bottom-up approach starting from smaller organic precursors bearing R_F groups [12,13], or by metal-catalyzed substitution reactions in solution with PAH derivatives bearing reactive substituents (such as chlorine or bromine) [14]. Positions of R_F groups in PAH-R_F derivatives synthesized by these approaches are usually determined by their position in precursors (smaller organic compounds) and by the mechanism of the bottom-up synthesis, rather than thermodynamic stability of the product. Recently, a new method of (poly)perfluoroalkylation was

New Fluorinated Carbons: Fundamentals and Applications. http://dx.doi.org/10.1016/B978-0-12-803479-8.00004-8

developed based on the gas-phase reaction of PAH with the R_F radicals produced by the thermal decomposition of R_FI [15]. The method allows perfluoroalkylation of virtually any PAH and introduces several R_F groups at the perimeter of the PAH core by substitution of hydrogen [15–23]. For instance, trifluoromethylation of corannulene (CORA hereafter) produces a series of CORA$(CF_3)_x$ compounds with x in the range from 3 to 7 [23,24]. The most prominent compound in the series is CORA$(CF_3)_5$ with fivefold symmetric arrangement of the groups [23], and analogous fivefold symmetric derivative is also obtained for perfluorobenzyl ($CF_2C_6F_5$) [18]. In the meantime, the bottom-up approach afforded the synthesis of trifluoromethylated corannulenes with one to three CF_3 [25,26], thus complementing the results of high-temperature trifluoromethylation. Formation of a single product in gas-phase radical trifluoromethylation is not typical. Usually, a range of compositions is obtained, each presented by several isomers. As perfluoroalkylation of PAHs proceeds at high temperatures, the product distribution (eg, location of R_F groups at the perimeter of the PAH core) is close to the thermodynamic control [20,23]. Thus, calculations of relative energies of the isomers with different location of R_F groups allow prediction of the possible products. The factors affecting stability of the isomers can be also revealed by computational studies.

Perfluoroalkylation can dramatically increase the electron-accepting properties of PAHs. For instance, for CORA, functionalization with different perfluoroalkyl groups ($R_F = CF_3$, $C_6F_5CF_2$) increases the electron affinity by c.0.3 eV per R_F group and induces similarly high positive shifts of electrochemical reduction potentials [15,18,20–24]. Similar effects were also observed for other PAHs (the electron affinity increment per CF_3 group depends on the size of the PAH core). Computational studies have been very efficient in the prediction of the electron-accepting properties of perfluoralkylated PAHs [8,17,18,22,23].

In this chapter, we will demonstrate how computational studies are able to provide an in-depth description of the structural and electronic properties of PAH-R_F derivatives. We will use a series of trifluoromethylated corannulene derivatives as a representative example to illustrate how computational studies can reveal the main aspect of thermodynamic stability of different isomers, reliably predict electron-accepting properties, both for gas-phase and solution, and finally explain the variation of the electron-accepting properties in dependence on the way how CF_3 groups are distributed at the perimeter (rim) of the corannulene molecule.

4.2 Thermodynamic Stability of CORA$(CF_3)_x$ Derivatives

The corannulene molecule has 10 positions available for substitution, and since the π-system is not changing when H is substituted by CF_3, thermodynamic stability of different isomers is determined solely by the distribution of the groups on the rim and their mutual interactions. Interestingly, whenever the structures synthesized by high-temperature radical reactions can be compared to the results of the lower-temperature wet-chemistry approaches, very different addition patterns can be seen.

In this section, we analyze thermodynamic stability issues of trifluoromethylated corannulene derivatives based on the results of DFT PBE/TZ2P calculations for all possible isomers with two to eight CF_3 groups (note that derivatives with 1, 9, and 10 CF_3 groups have only one possible isomer). For selected isomers, B3LYP-D/def2-TZVP calculations were also performed and the relative energies were found to be similar to those obtained at the less computationally demanding PBE/TZ2P level.

To simplify a discussion of multiple trifluoromethylated corannulene derivatives, we will use a simplified designation similar to that developed earlier for perfluoralkylated fullerenes [11]. The corannulene core is designated as "CORA," whereas the number following "CORA" after a hyphen shows how many CF_3 groups are attached to the corannulene core (ie, CORA-2 is equivalent to $CORA(CF_3)_2$ and both denote $C_{20}H_8(CF_3)_2$ derivative). Additional number is used for experimentally characterized isomers, and the structures are usually numbered in a temporal order of their characterization (ie, CORA-**3-1** is the first isomer of $CORA(CF_3)_3$ described in the literature). Fig. 4.1A also shows numbering system for carbon atoms of the CORA rim, which will be used to distinguish the $CORA(CF_3)_x$ isomers.

Topology of the corannulene molecule offers seven possible isomers of $CORA(CF_3)_2$. Of them five have CF_3 groups in distant positions and are predicted to be isoenergetic

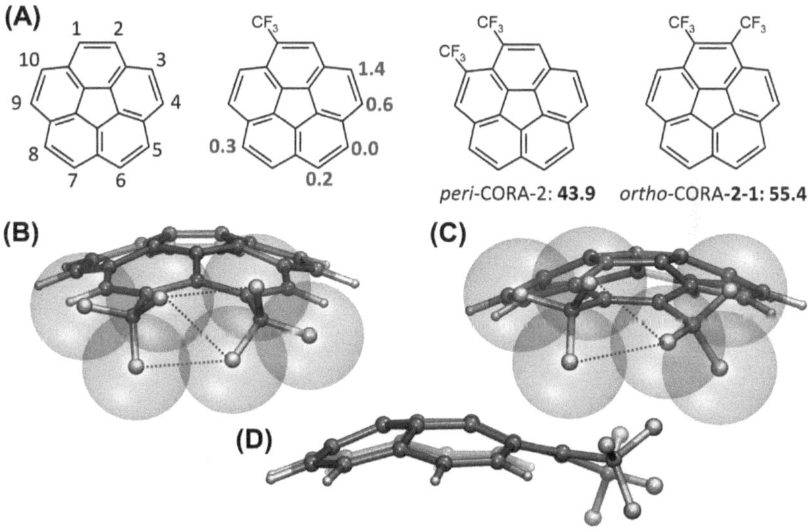

Figure 4.1 (A) Numbering system of the corannulene rim, relative energies of CORA-2 derivatives with distant CF_3 positions (red number (dark gray in print versions)) and of two isomer with close CF_3 contacts (*peri-* and *ortho*-CORA-2); (B and C) DFT-optimized molecular structures of *peri*-CORA-2 (B) and *ortho*-CORA-2 (C) showing van der Waals radii of fluorine atoms (transparent spheres) and the inter-CF_3 F-F contacts at the distances shorter than 3.0 Å (*dotted lines*); (D) molecular structure of *ortho*-CORA-2 emphasizing $C(F_3)-C-C-C(F_3)$ twist. In (B–D), carbon atoms in the CORA backbone and CF_3 groups are *gray* and light green (very dark gray in print versions), respectively, whereas fluorine atoms are yellow (light gray in print versions).

within 2 kJ/mol (Fig. 4.1). The mutual interactions of CF_3 groups in these isomers are negligible. In two other isomers the CF_3 groups are located close to each other, which leads to the dramatic increase of the relative energies. Placing CF_3 groups in 1, 10 positions (ie, interaction through one carbon atom, *peri* hereafter) has a destabilizing effect of 44 kJ/mol, whereas the isomer with two CF_3 groups in adjacent 1, 2 positions (*ortho* hereafter) has relative energy of 55.4 kJ/mol. Interestingly, the latter is the structure of the CORA-**2-1** isomer synthesized in Ref. [25] by a bottom-up approach. Thus, the sole experimentally available $CORA(CF_3)_2$ isomer is the least thermodynamically stable one in the whole series.

Considerable destabilization of *peri-* and *ortho*-isomers can be ascribed to the repulsion of bulky CF_3 groups. Assuming fluorine van der Waals radius of 1.5 Å [27], *peri*-CORA-2 has three F—F contacts shorter than van der Waals distance (2.625, 2.634, and 2.711 Å), whereas *ortho*-CORA-2 has two such contacts (2.515 and 2.639 Å). The overlap of van der Waals radii of the fluorine atom of neighboring groups is visualized in Fig. 4.1B and C. $C(F_3)—C(F_3)$ distances in *peri-* and *ortho*-isomers are 3.482 and 3.123 Å, respectively. Thus, CF_3 groups are closer to each other in the *ortho*-isomer, which agrees with its higher destabilization energy. Furthermore, $CF_3—CF_3$ repulsion in CORA-**2-1** imposes subtle but distinct distortion of the corannulene backbone: the functionalized CC bond on the rim is twisted with respect to the hub CC bond (in pentagon) by 3.5 degree. Whereas C—H bonds in CORA lie in one plane (ie, dihedral angles H—C—C—H are zero), two $C—C(F_3)$ bonds in CORA-**2-1** are twisted by 12.2 degree (Fig. 4.1D).

Bowl depth (defined as the mean perpendicular distance of the rim carbon atoms from the least squares plane of the central pentagon) is also affected by derivatization: the value for DFT-optimized CORA is 0.920 Å, in all CORA-2 derivatives with distant CF_3 groups the depth is reduced to 0.890 Å, whereas in *peri-* and *ortho*-isomers the values are 0.825 and 0.871 Å, respectively.

$CORA(CF_3)_3$ has five almost isoenergetic isomers with distant CF_3 positions (0—2 kJ/mol, Fig. 4.2), including the CORA-**3-2** isomer, obtained by the high-temperature trifluoromethylation of CORA [24]. Next in the energy scale are three isomers with one *peri* moiety and one distant CF_3 group (44—46 kJ/mol), followed by three isomers with one *ortho* moiety and one distant CF_3 group (54—56 kJ/mol), including the CORA-**3-1** isomer synthesized by bottom-up approach [25]. The isomer with CF_3 groups in 1, 2, 3 positions (ie, adjacent *ortho* and *peri* moieties) is the least

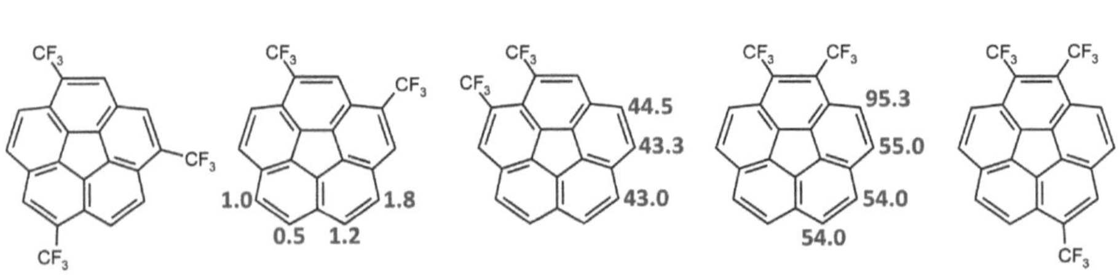

CORA-**3-2**: 0.0 CORA-**3-1**: 54.0

Figure 4.2 Relative energies of all $CORA(CF_3)_3$ isomers. CORA-**3-1** and CORA-**3-2** are characterized experimentally.

stable with the relative energy of 95 kJ/mol. Importantly, the relative energies of the isomers with adjacent CF_3 positions are almost perfectly reproduced by considering energy increments for *peri* and *ortho* moieties from CORA-2 isomers.

The bowl depth in CORA-3 isomer is further reduced to 0.875 Å for isomers with distant positions of CF_3 groups, 0.808−0.811 Å for isomers with one *peri* moiety, 0.835−0.838 Å for isomers with one *ortho* moiety, and 0.825 Å for the 1,2,3-CORA-3 isomer. Distortions induced by CF_3 group in the CORA backbone in *peri*- and *ortho*-isomers of CORA-3 are similar to those described above for CORA-2 isomers, except for the 1,2,3-CORA-3, which has more pronounced changes in the local geometry parameters. In its *ortho*-fragment, the rim CC bond is twisted versus the hub CC bond by 5.5 degree, and the $C(F_3)-C-C-C(F_3)$ dihedral angle is increased to 23.3 degree, whereas in the adjacent *peri*-fragment the $C(F_3)-C-C-C(F_3)$ dihedral angle is increased to 34.7 degree (vs c.2 grad in *peri*-CORA-2).

Multiple isomers of $CORA(CF_3)_4$ span the relative energy range of 150 kJ/mol (Fig. 4.3). Three isoenergetic isomers (0−0.3 kJ/mol), including the CORA-**4-2** isomer synthesized by high-temperature trifluoromethylation [24], have their CF_3 groups at distant positions. They are followed by two groups of isomers (six structures in each group) combining one *peri*- or *ortho*-motif with two distant CF_3 groups, whose relative energies span the ranges of 41−43 and 51−54 kJ/mol, respectively. The energies of the isomers combining two distant *peri* moieties or two distant *ortho* moieties are 83.7 and 104.9 kJ/mol, respectively, whereas the energy of two isoenergetic isomers with one *peri* and one *ortho* moieties at distant positions, including the isomer CORA-**4-1** from the Ref. [25], is 94.4 kJ/mol. Likewise, the isomers with adjacent *peri* and *ortho* moieties (or just 1,2,3-moiety) and a distant CF_3 group span the energy range of 92−94 kJ/mol. The least stable are the two isomers with all CF_3 groups on adjacent carbon atoms (142 kJ/mol for 1,2,3,4-CORA-4 and 147 kJ/mol for 1,2,3,10-CORA-4).

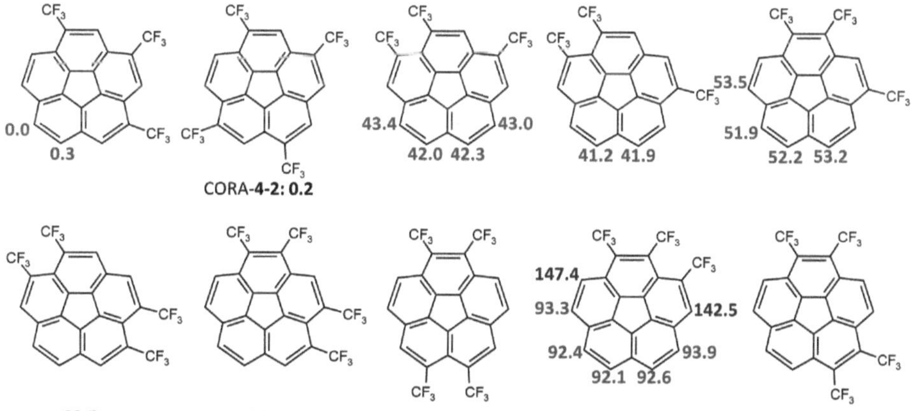

Figure 4.3 Relative energies of a majority of $CORA(CF_3)_4$ isomers. Not shown are 1,2,5,7-CORA-4 and 1,2,5,8-CORA-4 with relative energies of 52.3 and 51.5 kJ/mol, respectively. CORA-**4-1** and CORA-**4-2** are characterized experimentally.

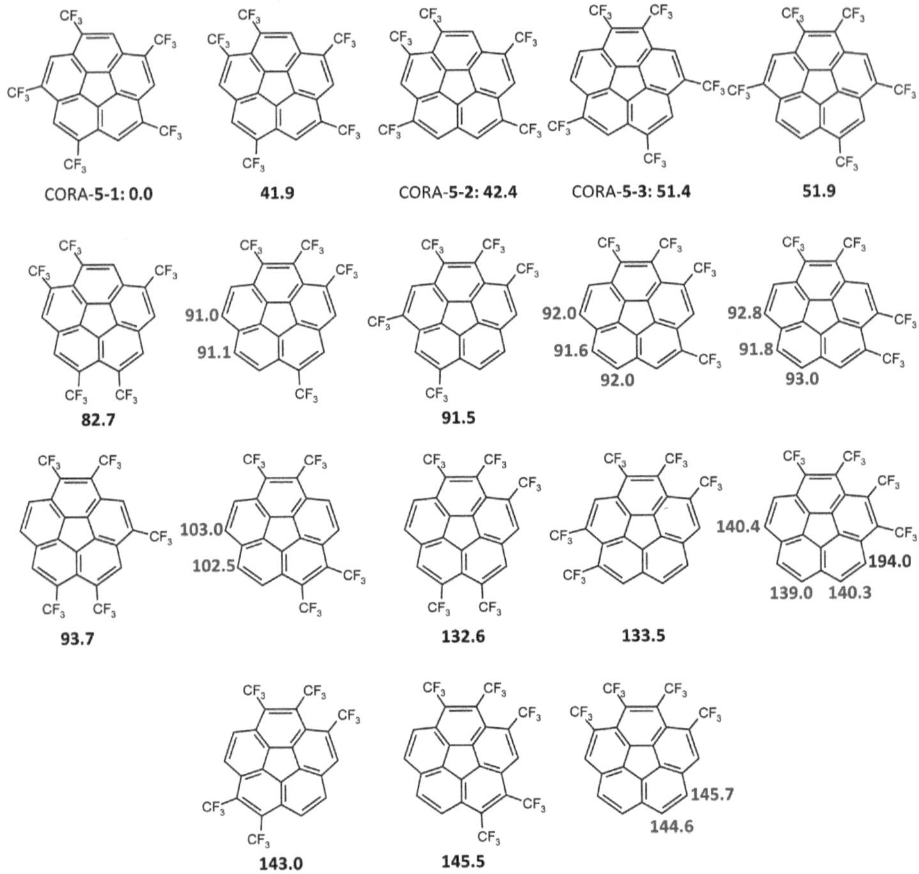

Figure 4.4 Relative energies of all CORA(CF_3)$_5$ isomers. CORA-**5-1,** CORA-**5-2**, and CORA-**5-3** are characterized experimentally.

CORA(CF_3)$_5$ has the largest variety of possible isomers (28, see Fig. 4.4). CORA-**5-1** is the sole isomer of CORA(CF_3)$_5$ with all CF_3 groups in distant positions, and this is the most stable isomer of CORA(CF_3)$_5$. This isomer is produced by high-temperature trifluoromethylation with the highest yield [23]. Two next most stable isomers, including another experimentally characterized isomer CORA-**5-2**, combine one *peri* moiety and three distant CF_3 groups; their relative energies are near 42 kJ/mol. Then follow two isomers with the *ortho* moiety and three distant CF_3 groups with the relative energies of 51.4 and 51.9 kJ/mol; the third experimentally characterized isomer CORA-**5-3** belongs to this group. All other isomers require at least two moieties with closely spaced CF_3 groups and span the energy range of 83–194 kJ/mol (see Fig. 4.4 for a complete list). As might be expected, the least stable is the 1,2,3,4,5-CORA-5 isomer with all CF_3 in adjacent positions.

The bowl depth continues the trend to become smaller with the increase of the number of CF_3 groups at the rim. However, the distribution of CF_3 groups on the rim also has strong effect on the flattening of the bowl. From CORA-1 to CORA-5, the largest depth is observed for the isomers with distant positions of CF_3 groups. When one *peri* or one *ortho* moiety is present, the depth is further decreased, and the flattening effect is more pronounced for the *peri* moiety. For instance, the bowl depth is 0.845 Å in CORA-**5-1**, 0.777 Å in CORA-**5-2**, and 0.805 Å CORA-**5-3**. Continuing the CORA-5 series, the presence of two *peri* and two *ortho* moieties further decreases the bowl depth in an almost incremental fashion to 0.701 and 0.763 Å, respectively. When one *peri* and one *ortho* moiety are present at once, the bowl depth depends on their mutual positions: adjacent position (1,2,3 fragment) results in 0.792 Å, whereas distant position of two moieties give the value of 0.73–0.75 Å. Likewise, for the structures with three to four adjacent CF_3 moieties, the bowl depth varies in the range of 0.72–0.79 Å.

Five is the largest number of groups which can be arranged on the CORA rim at distant positions, and at this stage it is instructive to analyze the variation of relative energies of the whole set of CORA-2 to CORA-5 isomers. We continuously observe the same pattern—the lowest energies are found for isomers with the distant positions of CF_3 groups (and these isomers are virtually isoenergetic), then follow isomers with one *peri* moiety, then isomers with one *ortho* moiety, than two *peri* moieties, etc. Importantly, within each group of isomers, variation of relative energies is very small (few kJ/mol). Furthermore, the isomers with adjacent positions of *ortho* and *peri* moieties, and the isomers when such moieties are located at distant positions, also have very similar energies. All these facts show that relative stability of CORA$(CF_3)_x$ isomers can be rationalized using only two simple parameters—the number of *peri* and *ortho* moieties (N_{peri} and N_{ortho}). Fig. 4.5 plots relative energies of all CORA$(CF_3)_x$ isomers ($x = 2–5$) as a function of N_{peri} and N_{ortho}. The whole set of data can be very well fitted ($R^2 = 0.995$) by an equation of plane with *peri* and *ortho* energy

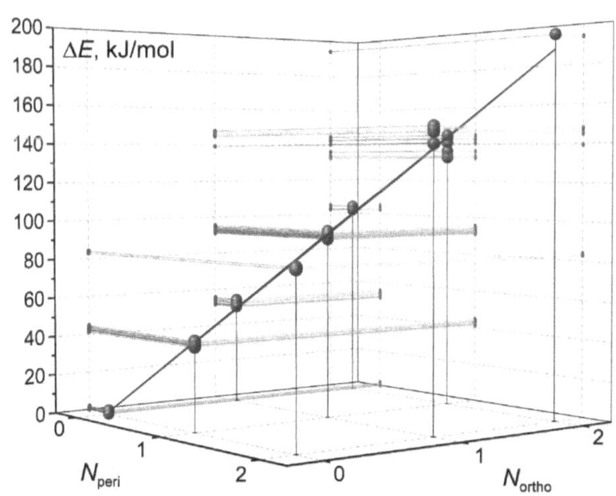

Figure 4.5 Relative energy of CORA$(CF_3)_x$ isomers ($x = 2–5$) as a function of the number of *peri* (N_{peri}) and *ortho* (N_{ortho}) moieties. The plot is shown in the projection parallel to the fitted plane (seen as a purple line (dark gray in print versions) in such projection). Equation of the plane is ΔE(kJ/mol) = 0.8(±0.8) + 42.9(±0.6) N_{peri} + 50.0(±0.6) N_{ortho}. Relative energies are shown as red spheres (gray in print versions), also shown are projections on coordinate planes.

increments of 42.9 ± 0.6 and 50.0 ± 0.6 kJ/mol, respectively. Note that these values are close to but somewhat smaller than the relative energies of *peri*-CORA-2 and *ortho*-CORA-2 isomers (Fig. 4.1).

Starting from six CF_3 groups, it is not possible to arrange them on the rim of the corannulene molecule without at least two fragments with close $CF_3{-}CF_3$ contacts. More precisely, the minimum number of contacts can be realized only with one *peri* and one *ortho* moieties. Thus, three most stable isomers of CORA(CF_3)$_6$ (relative energies all within 3 kJ/mol) have one *ortho* and one *peri* moiety (Fig. 4.5). These structures correspond to the experimentally characterized CORA-**6-1**, CORA-**6-2**, and CORA-**6-3**, all obtained by high-temperature trifluoromethylation [24]. All other isomers have three or more adjacent CF_3 moieties and their relative energies span the range of 41−155 kJ/mol (Fig. 4.6A).

The minimum number of *peri* and *ortho* moieties for seven CORA(CF_3)$_7$ groups is four (2 + 2). Five isomers of CORA(CF_3)$_7$ fulfilling this condition span the relative energy range of 0.0−16.5 kJ/mol. Note that the range of the relative energies of isomers with the same number of close $CF_3{-}CF_3$ contacts is higher than for CORA derivatives with smaller number of groups. This shows that for the already rather "crowded" CORA derivative with seven CF_3 groups at the rim, relative positions of *peri* and *ortho* moieties start to play a role. Three lowest energy isomers are CORA-**7-1**, CORA-**7-2**, and CORA-**7-3** isolated and structurally characterized [24]. The energies of other isomer with five or more close $CF_3{-}CF_3$ contacts fall in the range of 49−113 kJ/mol (Fig. 4.6B).

Among seven isomers of CORA(CF_3)$_8$, five have six close $CF_3{-}CF_3$ contacts (three *peri* and three *ortho* moieties) and relative energies in the range of 0−14 kJ/mol, whereas two others have seven close $CF_3{-}CF_3$ contacts and the relative energies of 61 and 63 kJ/mol. Importantly, the relative energies of CORA-6, CORA-7, and CORA-8 isomers remains roughly an additive function of the number of *peri* and *ortho* moieties in the structure, albeit with larger deviations. The deviations are mainly caused by additional distortions caused by adjacent location of several *peri* and *ortho* moieties.

The inevitable formation of destabilizing $CF_3{-}CF_3$ contacts in CORA(CF_3)$_x$ derivatives with $x > 5$ may have an impact on the possibility to produce such structures in the experiment. Furthermore, with the increase of x beyond five the minimum possible number of such contacts is increasing by two per CF_3 group (2 in CORA(CF_3)$_6$, four in CORA(CF_3)$_7$, six in CORA(CF_3)$_8$, etc.). To evaluate the impact of the topology on thermodynamics of CF_3 substitution, we have computed reaction energies (ΔE_r) for two isodesmic reactions of CF_3/H exchange with trifluoromethylbenzene:

$$1/x\,\mathrm{CORA} + C_6H_5CF_3 = 1/x\,\mathrm{CORA}\,(CF_3)_x + C_6H_6 \tag{4.1}$$

$$\mathrm{CORA}\,(CF_3)_x + C_6H_5CF_3 = \mathrm{CORA}\,(CF_3)_{x+1} + C_6H_6 \tag{4.2}$$

The first reaction describes net reaction energy for the whole CORA(CF_3)$_x$ molecule (normalized to the number of groups), whereas the second one emphasizes incremental changes of the energy with the increase of the number of CF_3 groups. Note that

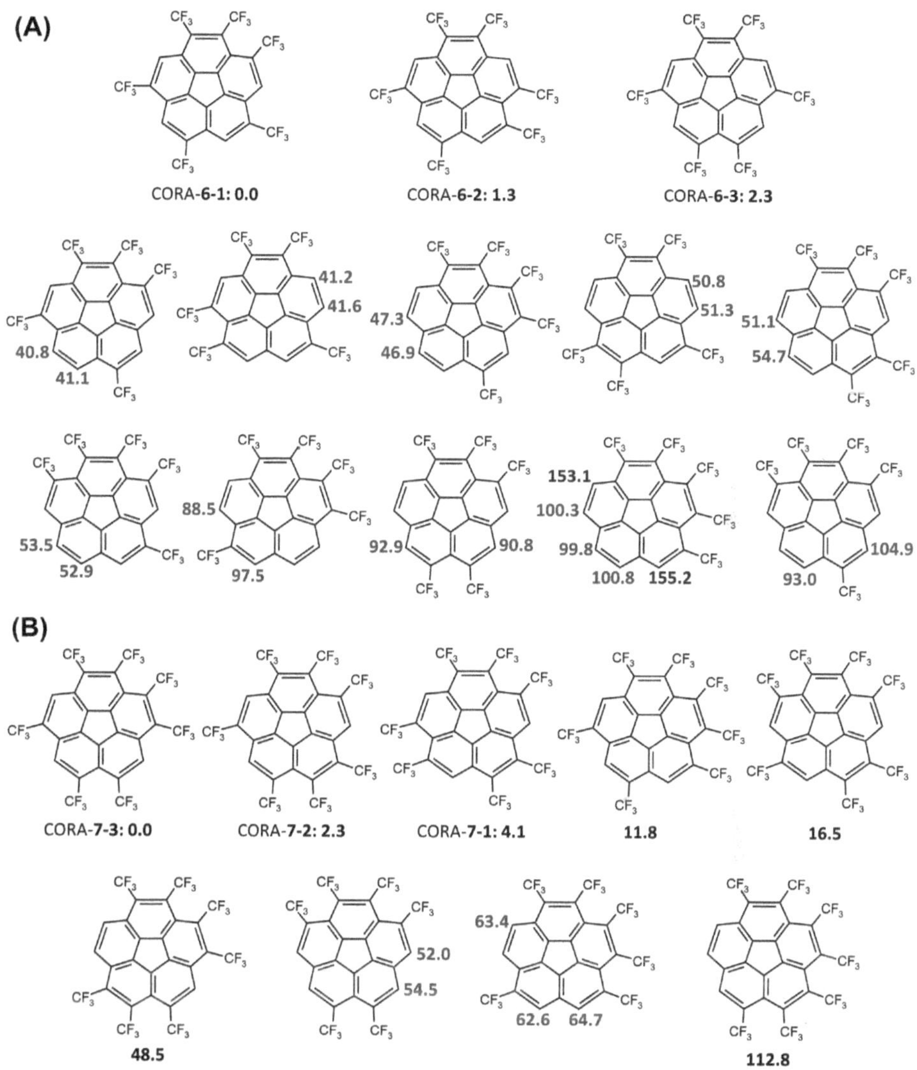

Figure 4.6 Relative energies of all CORA(CF_3)$_6$ (A) and CORA(CF_3)$_7$ (B) isomers. Experimentally characterized are three lowest energy isomers in each series.

the use of benzene as a reference is arbitrary, and the choice of a different reference system will increase or decrease the values uniformly by a constant value.

For x below 5, the increase of the number of added groups results in slight increase of the reaction energy (ie, reaction becomes slightly more endothermic) but remains below 10 kJ/mol (Fig. 4.7A). This shows that CF_3 bonding to the CORA backbone is similar to that in trifluoromethylbenzene and has no dramatic effect on the stability, although the strain is gradually increasing with more and more CF_3 groups being attached to the rim. Above $x = 5$, both reactions show a sharp bend upward of their energies. In other words,

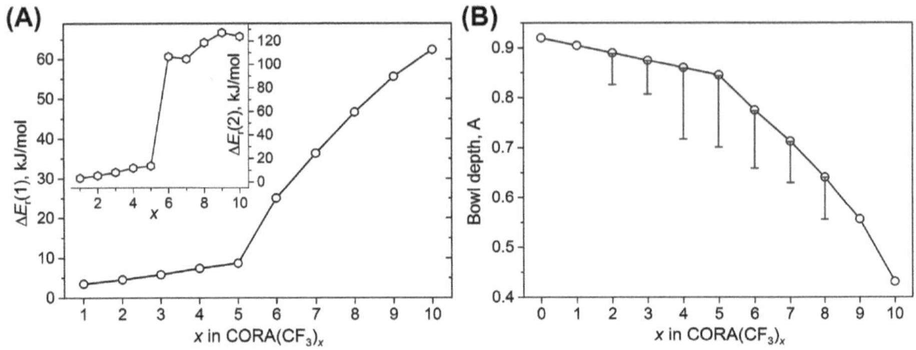

Figure 4.7 (A) Variation of average and incremental (inset) CF_3/H exchange reaction energy ($\Delta E_r(1)$ and $\Delta E_r(2)$, respectively) between CORA$(CF_3)_x$ and $C_6H_5CF_3$ as a function of the number of CF_3 groups; (B) variation of the bowl depth in CORA$(CF_3)_x$ derivatives; open circles show the values for the most stable isomers, vertical bars show the range of values in the whole set of isomers.

each subsequent CF_3/H substitution becomes increasingly more endothermic by 107 and 105 kJ/mol for CORA-5 and CORA-6, respectively (roughly equals the sum of *peri* and *ortho* energy increments). Introduction of further groups requires even large energies (eg, 119 kJ/mol for CORA-7). This dramatic increase of the energy imposes the limit for the maximal number of CF_3 groups, which can be introduced to the corannulene backbone. Experimentally, up to six added groups could be realized in high-temperature reaction of corannulene with CF_3I reaction in appreciable amounts, the yield of CORA-7 isomers is quite low but still sufficient for their structural characterization, whereas possible addition of eight or nine groups is seen only in mass-spectra (ie, only trace amounts of CORA-8 and CORA-9 are formed) and CORA-10 is not detected at all [24].

The bowl depth of the corannulene backbone (defined as the mean perpendicular distance of the rim carbon atoms from the least−squares plane of the central pentagon) is one of the most important geometrical characteristics of corannulene derivatives. Fig. 4.7B shows the change of the maximum bowl depth in the whole CORA$(CF_3)_x$ series; also shown are the ranges of the bowl depth values for different isomers. For the pristine corannulene, DFT predicts the bowl depth of 0.920 Å (compare with the experimentally determined value of 0.875(2) Å [28]). In all CORA-2 derivatives with distant CF_3 groups the depth is reduced to 0.890 Å, whereas in *peri*- and *ortho*-isomers the values are 0.825 and 0.871 Å, respectively. The bowl depth in CORA-3 isomer is further reduced to 0.875 Å for isomers with distant positions of CF_3 groups, 0.808−0.811 Å for isomers with one *peri* moiety, 0.835−0.838 Å for isomers with one *ortho* moiety, and 0.825 Å for the 1,2,3-CORA-3 isomer.

With the further increase of the number of CF_3 groups at the rim, the bowl depth continues the trend to become smaller with strong variation in dependence on the distribution of CF_3 groups on the rim (Fig. 4.7B). Up to CORA-5, the largest depth is observed for the isomers with distant positions of CF_3 groups. When one *peri* or one *ortho* moiety is present, the depth is further decreased, and the flattening effect is more pronounced for the *peri* moiety. For instance, the bowl depth is 0.845 Å in CORA-**5-1** (experimental value is 0.788 Å), 0.777 Å in CORA-**5-2**, and 0.805 Å CORA-**5-3**. Continuing the CORA-5 series, the presence of two *peri* and two *ortho* moieties further decreases the bowl depth in an almost incremental fashion to 0.701 and 0.763 Å, respectively. When one *peri* and one *ortho* moiety are present at once, the bowl depth depends on their mutual positions: adjacent position (1, 2, 3 fragment) results in 0.792 Å, whereas distant position of two moieties give the value of 0.73–0.75 Å. Likewise, for the structures with three to four adjacent CF_3 moieties, the bowl depth varies in the range of 0.72–0.79 Å. Starting with CORA-6 and beyond, the flattening of the CORA bowl with the number of CF_3 groups is more pronounced. The reason is the same as for the reaction energies discussed above: the number of adjacent CF_3 groups, which cause flattening of the CORA backbone, is growing fast with x. For CORA-10, the trend ends in the bowl depth of only 0.431 Å (Figs. 4.7 and 4.8). Note that replacement of CF_3 groups in CORA-10 by less bulky groups, eg, CH_3, results in the increase of the bowl depth to 0.672 Å, which shows that repulsion of bulky groups at the corannulene rim is the main reason for the flattening of the bowl. This conclusion agrees with several experimental observations of flattening the corannulene bowl in derivatives with bulky sterically hindered groups [29–32].

An increase of the strain energy and flattening of the bowl with the number of CF_3 groups parallels stronger distortions caused by adjacent CF_3–CF_3 moieties, especially when such short contacts become inevitable ($x \geq 6$) and *peri*/*ortho* fragments cannot be placed on distant positions from each other. For instance, distortions induced by CF_3 group in the CORA backbone in *peri-* and *ortho*-isomers of CORA-3 are similar to those described above for CORA-2 isomers, except for the 1,2,3-CORA-3, which has more pronounced changes in the local geometry parameters. In its *ortho*-fragment, the rim CC bond is twisted versus the hub CC bond by 5.5 degree, and the $C(F_3)$–C–C–$C(F_3)$ dihedral angle is increased to 23.3 degree, whereas in the adjacent *peri*-fragment the $C(F_3)$–C–C–$C(F_3)$ dihedral angle is increased to 34.7 degree (vs c.2° in *peri*-CORA-2). Starting from CORA-7, multiple adjacent *peri/ortho* moieties are inevitable. The structure of CORA-10 (Fig. 4.8C)

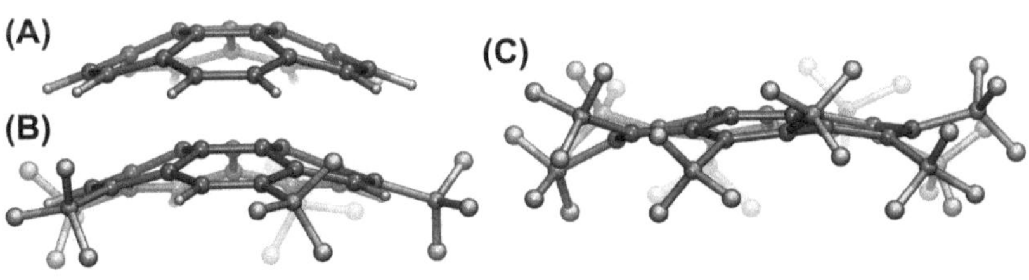

Figure 4.8 DFT-optimized molecular structures of (A) CORA (bowl depth 0.905 Å), (B) CORA-5-1 (bowl depth 0.845 Å), and (C) CORA-10 (bowl depth 0.431 Å). Carbon atoms in the CORA backbone and CF_3 groups are gray (dark gray in print versions) and light green (gray in print versions), respectively, whereas fluorine atoms are yellow (light gray in print versions). Note dramatic flattening of the bowl and strong twist of dihedral angles $C(F_3)$–C–C–$C(F_3)$ (41.0 degree, *ortho* moieties) and $C(F_3)$–C–C–$C(F_3)$ (41.2 degree, *peri* moieties) for CORA-10.

can serve as an example of the situation when all CF_3 groups are in adjacent positions. In this molecule, the rim CC bond is twisted versus the hub CC bond by 6.0 degree, the $C(F_3)-C-C-C(F_3)$ dihedral angles in *ortho* fragments are 41.0 degree, and $C(F_3)-C-C-C(F_3)$ dihedral angles in *peri* fragments are 41.2 degree. Such distortions show that the destabilizing effect of adjacent CF_3-CF_3 moieties is stronger in "crowded" CORA derivatives with large amount of CF_3 groups, which explains an additional increase of the $\Delta E_r(2)$ values for $x = 8-10$ (Fig. 4.7A).

To summarize this section, we conclude that thermodynamic stability of CORA$(CF_3)_x$ derivatives is largely determined by the repulsion of CF_3 groups in adjacent (*ortho* and *peri*) positions, and hence the most stable isomers have the smallest number of CF_3-CF_3 contacts. Energy penalty for the CF_3 groups in *peri* and *ortho* positions can be estimated as c.43 and 50 kJ/mol, respectively. These energy increments allow estimation of the relative energies of the isomers with such substructures with high degree of certainty. High-temperature radical trifluoromethylation always leads to the most thermodynamically stable isomers, whereas bottom-up wet-chemistry approaches usually result in CF_3 less thermodynamically stable isomer with groups in adjacent position. The possibility to have all CF_3 groups in distant positions is exhausted after five groups, and hence addition of each next CF_3 groups requires an additional energy exceeding 100 kJ/mol to overcome the steric repulsion. As a result, isolable amounts of CORA$(CF_3)_x$ are limited to $x = 7$, whereas derivatives with larger numbers of CF_3 groups can hardly be obtained in reasonable yield.

4.3 Electron-Accepting Properties of CORA$(CF_3)_x$ Derivatives and Addition Patterns

Increasing electron-accepting properties and creating an array of PAH derivatives with tunable electron-accepting properties is the main motivation behind multiple perfluoroalkylation of CORA and other PAHs [15–23,25,33–35]. To quantify these electron accepting properties, we will use three closely related parameters—gas phase electron affinity (*EA*), electrochemical reduction potential in solution ($E_{1/2}$), and the energy of the lowest unoccupied molecular orbital (LUMO). The latter is the simplest parameter available from the computation of the neutral molecules (eg, like those described in the previous section) and hence convenient when analysis of many possible structures is needed. Computation of *EA* requires optimization of the molecule in the neutral and anionic state, whereas computation of reduction potential also requires taking into account solvation energies. However, *EA* and $E_{1/2}$ are experimentally available, and hence computations can be directly compared to experiment, whereas LUMO energies are suitable for analysis of correlations between the data but cannot be directly compared to experimental data (not that although experimentally measured quantities are sometimes called LUMO energy, this should be considered a jargon). In this section, we will use *EA* values of a series of CORA$(R)_x$ derivatives computed at the DFT-B3LYP level with the def2-TZVP basis set additionally augmented with diffuse functions for peripheral CF_3 groups and hydrogen atoms. Solvation energies necessary for prediction of redox properties were computed using COSMO polarized continuum model. In the discussion of LUMO, we will also use B3LYP data when discussing limited set of experimentally available structures, and PBE values when discussing the complete set of all CORA$(CF_3)_x$ structures.

In the first report on the synthesis of CORA-**5-1**, which showed a dramatic increase of the *EA* and $E_{1/2}$ of the CF_3-derivative over the pristine corannulene [23], we hypothesized that since substitution of hydrogen by CF_3 on the rim does not affect the π-system, the increase of the electron affinity should be incremental and should not depend on the way how CF_3 groups are distributed on the rim. Fig. 4.9A shows *EA*s computed for a series of arbitrarily chosen CORA$(R)_x$ derivatives with different substituents ($x = 0-10$, $R = CH_3$, F, Cl, CF_3, CN). Similar curves (but damped by 30–40% due to the effect of solvation) were also computed for $E_{1/2}$ values in acetonitrile (not shown). *EA* values in Fig. 4.9A systematically increase with x for $R =$ CN, CF_3, Cl, and F and show a trend of leveling off for large number of added groups. In the range of $x = 1-8$ the data can reasonably be described by a linear function, which gives an increment of 0.30 eV and 0.20 V per CF_3 group for *EA* and $E_{1/2}$, respectively. These increments are surpassed only by CN group (0.39 eV/0.25 V), whereas substitution with chlorine and fluorine atoms gives more modest increase (0.16 eV/0.10 V for Cl and 0.13 eV/0.07 V for F). Note that the counter-intuitively small increase of the electron-accepting properties by fluorine substitution is due to F being a σ-acceptor but a π-donor, which diminishes the electron-withdrawing effect when fluorine is bonded to C-sp^2 atoms forming an extended π-system. In line with experimental findings, CH_3 substitution reduces the electron-accepting properties of CORA [36].

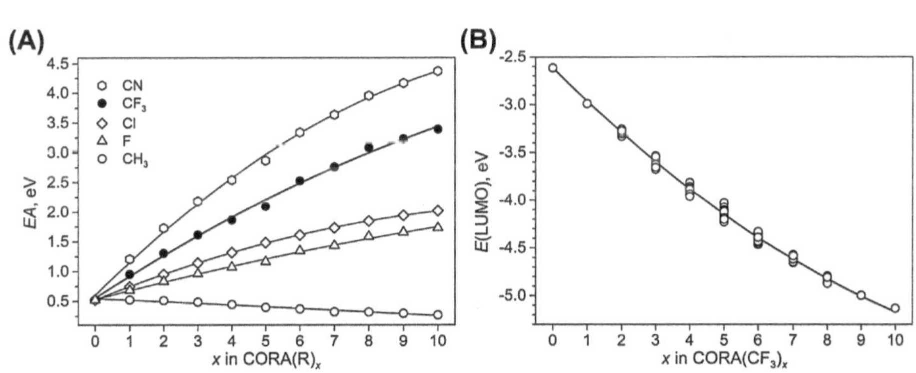

Figure 4.9 (A) Electron affinities (*EA*s) of a series of CORA$(R)_x$ derivatives computed at the B3LYP/def2-TZVP level (*based on the data from Ref. I.V. Kuvychko, S.N. Spisak, Y.-S. Chen, A.A. Popov, M.A. Petrukhina, S.H. Strauss, O.V. Boltalina, A buckybowl with a lot of potential: C_5-$C_{20}H_5(CF_3)_5$, Angew. Chem.-Int. Edit. Engl. 51 (2012) 4939–4942.*). Derivatives were chosen arbitrary with the condition of the uniform distribution of the group on the rim. Lines are quadratic fit to the data. (B) LUMO energies of the complete set of CORA$(CF_3)_x$ isomers ($x = 0-10$) computed at the PBE/TZ2P level. The line is the result of the quadratic fit. *LUMO*, lowest unoccupied molecular orbital.

With more CORA$(CF_3)_x$ derivatives becoming available for the electrochemical and electron-affinity studies, the hypothesis of the independence of the electron-accepting properties on the CF_3 addition motif comes under scrutiny. For instance, Lentz et al. reported that reduction potentials of CORA$(CF_3)_x$ derivatives with CF_3 groups in adjacent positions are noticeably more positive than expected based on the increments discussed above [25]. Isolation of a large series of CORA$(CF_3)_x$ derivatives by high-temperature trifluoromethylation enables a more systematic analysis of the influence of substitution motif on the EA and $E_{1/2}$ [24]. Fig. 4.10A and Table 4.1 compares experimental and computed ΔEA and $\Delta E_{1/2}$ values for all CORA$(CF_3)_x$

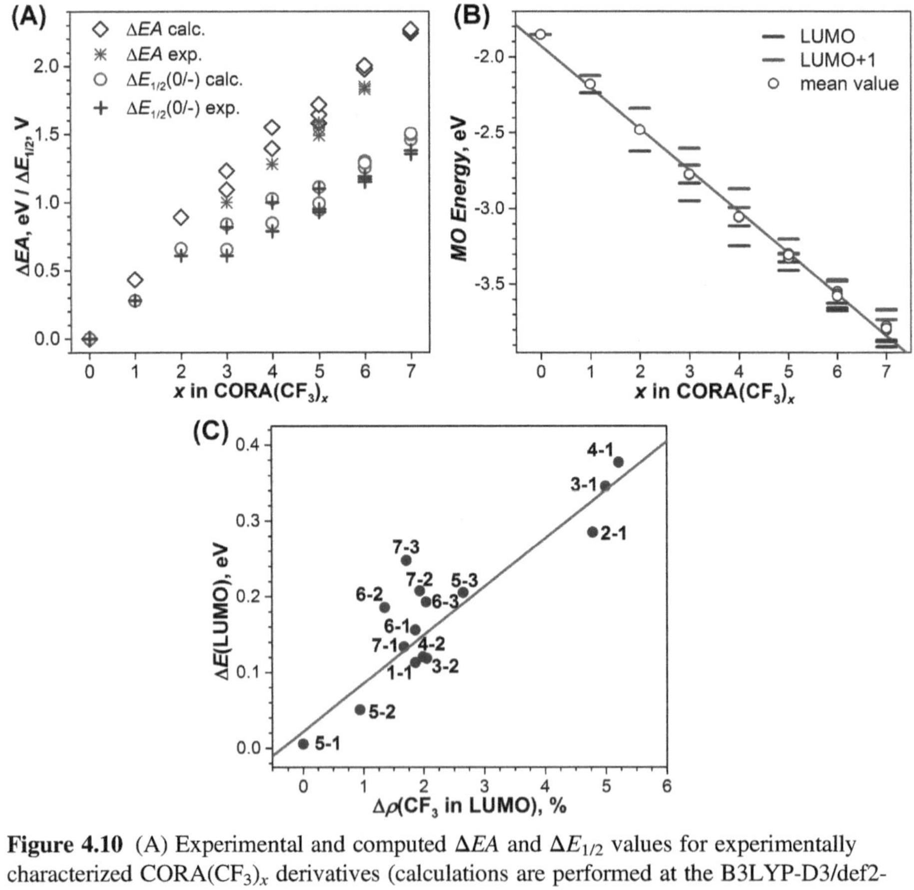

Figure 4.10 (A) Experimental and computed ΔEA and $\Delta E_{1/2}$ values for experimentally characterized CORA$(CF_3)_x$ derivatives (calculations are performed at the B3LYP-D3/def2-TZVP level with additional diffuse function for atoms at the periphery). (B) LUMO and LUMO+1 energies for experimentally characterized CORA$(CF_3)_x$ derivatives, open circles show average values (0.5 E(LUMO) + 0.5 E(LUMO+1)), the line shows linear fit for the average values. (C) Dependence of LUMO/LUMO+1 splitting energy, ΔE(LUMO), on the difference in the CF_3 contribution to LUMO and LUMO+1, $\Delta\rho$(CF_3 in LUMO). The isomers are denoted by two last numbers (ie, "CORA-**5-1**" is denoted as "**5-1**," etc.). *LUMO*, lowest unoccupied molecular orbital.

Table 4.1 Experimental and Calculated Reduction Potentials ($E_{1/2}$) and Electron Affinities (*EA*) in a Series of CORA(CF_3)$_x$ Derivatives, Their LUMO and LUMO+1 Energies, and Net Contributions of CF_3 Groups (ρ) to These Orbitals[a]

	Exp[b]		Calc		LUMO		LUMO+1	
CORA-X	$E_{1/2}$	*EA*	*EA*	$E_{1/2}(0/-)$	*E*	$\rho(CF_3)$	*E*	$\rho(CF_3)$
CORA	0.00	0.73	0.54	0.00	−1.85	0.0	−1.85	0.00
CORA-1	0.28		0.98	0.28	−2.24	1.9	−2.12	0.1
CORA-**2-1**	0.61		1.45	0.66	−2.62	5.2	−2.34	0.4
CORA-**3-1**	0.82		1.79	0.84	−2.95	5.7	−2.60	0.7
CORA-**3-2**	0.61	1.73	1.65	0.66	−2.83	3.6	−2.71	1.6
CORA-**4-1**	1.00		2.11	1.03	−3.25	5.9	−2.87	0.7
CORA-**4-2**	0.79	2.01	1.96	0.85	−3.12	4.4	−2.99	2.4
CORA-**5-1**	0.95	2.22	2.15	0.95	−3.30	3.4	−3.30	3.4
CORA-**5-2**	0.93	2.27	2.21	0.99	−3.35	4.0	−3.30	3.1
CORA-**5-3**	1.10	2.32	2.28	1.11	−3.41	5.6	−3.20	2.9
CORA-**6-1**	1.19	2.58	2.55	1.30	−3.63	5.5	−3.47	3.6
CORA-**6-2**	1.15	2.56	2.54	1.25	−3.66	5.2	−3.47	3.9
CORA-**6-3**	1.17	2.58	2.57	1.29	−3.68	5.7	−3.48	3.6
CORA-**7-1**	1.35		2.76	1.50	−3.87	6.0	−3.73	4.3
CORA-**7-2**	1.38		2.78	1.46	−3.88	6.2	−3.67	4.3
CORA-**7-3**	1.38		2.79	1.50	−3.91	6.4	−3.67	4.6

LUMO, lowest unoccupied molecular orbital.

[a] $E_{1/2}$ values are for acetonitrile and are given in Volts versus $CORA^{0/-}$ redox couple, *EA* and MO energies are in eV, and $\rho(CF_3)$ values are in %. Theoretical values are computed at the B3LYP/def2-TZVP level with additional diffuse functions for CF_3 groups and periphery hydrogen atoms.

[b] Experimental data on CORA-1, CORA-**2-1**, CORA-**3-1**, and CORA-**4-1** are from Ref. [25], the rest of experimental data are from Refs [23,24].

derivatives studied at this moment (note that "Δ" here means that the values are referred to *EA* and $E_{1/2}$ of pristine CORA).

Comparison with experimental data proves high reliability of the computational results (Table 4.1). The largest error is found for the pristine CORA (experimental value of 0.73 eV versus calculated 0.54 eV). The error is decreasing with the increase of the number of CF_3 groups and becomes less than 0.05 eV for the CORA-6 isomers. Computed relative *EA* values (ΔEA) plotted in Fig. 4.10A exhibit the same trend as experimental ones. Systematic overestimation of the ΔEA values by computations is largely caused by the noticeable underestimation of the CORA value. The values listed in Table 4.1 and plotted in Fig. 4.10A show that experimental reduction potentials are also reliably predicted by our computational approach. Calculations slightly overestimate experimental potentials, and the error is increasing with the number of CF_3 groups, from c.0.06 V for two to five CF_3 groups to the maximum of 0.15 V for seven CF_3 groups. Within the sets of isomers, calculations correctly reproduce experimental observation of significant differences in $E_{1/2}$ values for CORA-**3-1** and CORA-**3-2**, or CORA-**4-1** and CORA-**4-2** etc. Importantly, solvation energy is gradually decreasing with the increase of the number of CF_3 groups, and hence relative *EA* values are larger than $E_{1/2}$ values. For instance, the ΔEA values for CORA-7 isomers are near 2.25 eV, whereas the $\Delta E_{1/2}$ values in acetonitrile for the same isomers are near 1.4 V.

Since theory correctly reproduces the difference in reduction potentials of different isomers, the reasons of the substantial variation of *EA* ($E_{1/2}$) values for different isomers can be now addressed. Comparison of the LUMO energies gives the first hint on variation of electron-accepting properties in different isomers. Fig. 4.9B shows LUMO energies of all CORA$(CF_3)_x$ structures as a function of x. As might be expected, gradual stabilization of LUMOs with increase of x is observed, and the whole set of data can be well approximated by a quadratic function. However, it can be also seen that dispersion of the LUMO energies of isomers with the same number of CF_3 groups can be significant. The largest variation (0.198 eV) is found for CORA-5 isomers, whose lowest and highest LUMO values are approaching the LUMO energies of CORA-6 and CORA-4 isomers, respectively. The highest LUMO energy (−4.031 eV) is found for the experimentally available CORA-**5-1** isomer with all CF_3 groups in distant positions, whereas the lowest LUMO energy (−4.229 eV) is predicted for the 1,2,3,6,10-isomer with four adjacent CF_3 groups. From this simple analysis, we can conclude that position of CF_3 groups on the rim indeed plays an important role for the electron-accepting properties. Hence, more detailed analysis of the correlation between the positions of CF_3 groups and LUMO energies is needed.

The high C_{5v} symmetry of the pristine CORA molecule results in the twofold degeneracy of its LUMO (two LUMO components are shown in Fig. 4.11A). Perfluoroalkylation of the CORA reduces the symmetry of the molecule and hence degeneracy of the LUMO is lifted in all CORA$(CF_3)_x$ derivatives except for C_5-symmetric CORA-**5-1**. This results in the gap between the former LUMO components for CORA, which become LUMO and LUMO+1 orbitals in CORA$(CF_3)_x$ derivatives. DFT calculations show that the gap between the LUMO and LUMO+1 (ΔLUMO hereafter) for the experimentally characterized CORA$(CF_3)_x$ structures can reach 0.4 eV (Table 4.1, B3LYP/def2-TZVP level). At the same time, the *mean* energy of the LUMO and LUMO+1 is virtually identical for all isomers and decreases by 0.28 eV per CF_3 group as shown in Fig. 4.10B. In other words, position of CF_3 groups on the rim plays no role as long as the *net influence on both* LUMO components of CORA is considered. But when the twofold degeneracy of the orbital is lifted, only one of the components, with lower energy, becomes the true LUMO of the CORA$(CF_3)_x$ derivative and determines electron-accepting properties of the molecule. The isomers with inhomogeneous distribution of CF_3 groups on the rim may be expected to give a larger LUMO/LUMO+1 splitting, and hence lower LUMO energies.

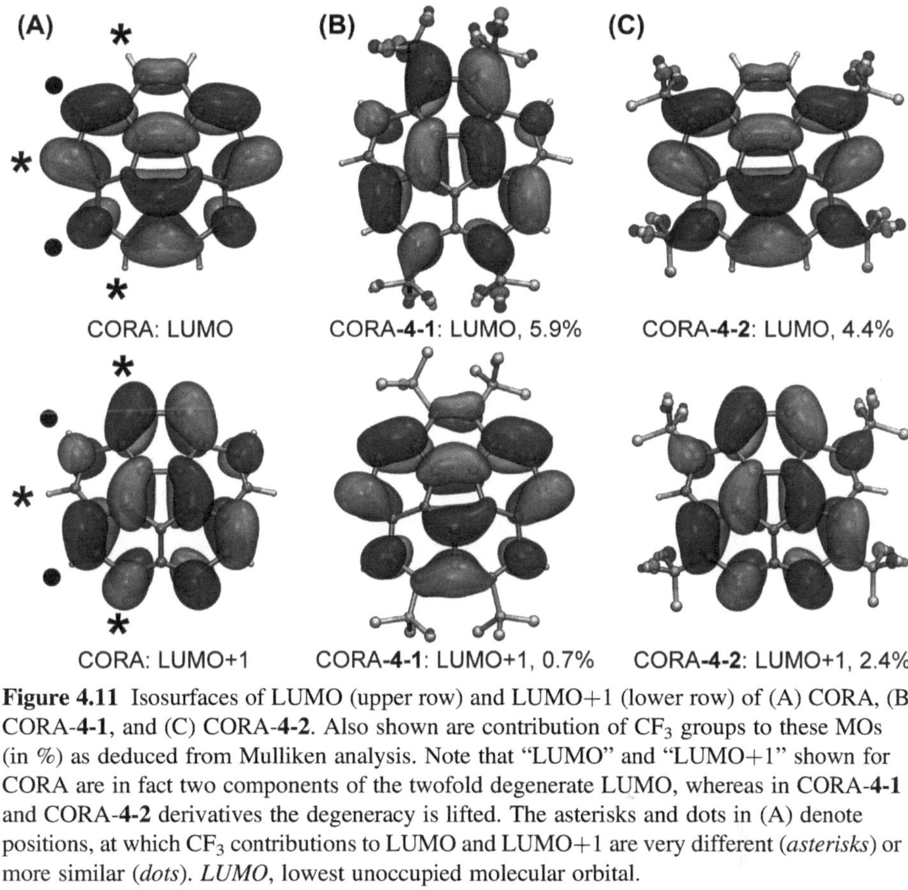

Figure 4.11 Isosurfaces of LUMO (upper row) and LUMO+1 (lower row) of (A) CORA, (B) CORA-**4-1**, and (C) CORA-**4-2**. Also shown are contribution of CF_3 groups to these MOs (in %) as deduced from Mulliken analysis. Note that "LUMO" and "LUMO+1" shown for CORA are in fact two components of the twofold degenerate LUMO, whereas in CORA-**4-1** and CORA-**4-2** derivatives the degeneracy is lifted. The asterisks and dots in (A) denote positions, at which CF_3 contributions to LUMO and LUMO+1 are very different (*asterisks*) or more similar (*dots*). *LUMO*, lowest unoccupied molecular orbital.

LUMO and LUMO+1 of CORA and CORA$(CF_3)_x$ derivatives are essentially π-orbitals and have the same shapes in the whole series (Fig. 4.11). However, Mulliken population analysis shows that 2−11% to these orbitals are delocalized onto CF_3 groups (the CF_3 contributions to LUMOs are systematically increasing with x but not in a linear way). Table 4.1 lists net localization percentage of LUMO and LUMO+1 on CF_3 groups (denoted as $\rho(CF_3)$) in all structurally characterized CORA$(CF_3)_x$ derivatives. It can be immediately noticed that LUMO always has larger $\rho(CF_3)$ contribution than LUMO+1, and that the energy difference between orbitals correlates well with the difference of $\rho(CF_3)$ contributions (Fig. 4.10C). Fig. 4.11 illustrates this situation by showing LUMO and LUMO+1 and corresponding $\rho(CF_3)$ values for CORA-**4-1** and CORA-**4-2**. In the former, $\rho(CF_3)$ for the LUMO, 5.9%, is considerably larger than for the LUMO+1, 0.7%, and the ΔLUMO value is 0.38 eV. In CORA-**4-2**, the CF_3 contributions are more uniformly distributed between LUMO and LUMO+1 with $\rho(CF_3)$ values of 4.4% and 2.4%, respectively, and the ΔLUMO gap is only 0.12 eV. As a result, LUMO of CORA-**4-1** is 0.13 eV lower than the LUMO of CORA-**4-2**, in agreement with more positive reduction potential of the former (Table 4.1). Analogous situation is found for CORA-3 and CORA-5 isomers. In all these structures, *ortho* position of CF_3 groups "attracts" the LUMO and results in inhomogeneous distribution of $\rho(CF_3)$ values between LUMO and LUMO+1 with larger ΔLUMO gap (Fig. 4.11B). This observation can be generalized as follows: each substitution position at the CORA rim can be ascribed a certain increment, denoting how strongly CF_3 group in this position contributes to the LUMO and LUMO+1 (ρ_i^{LUMO} and ρ_i^{LUMO+1}, respectively, where *i* numbers the carbon atom at the rim). For some positions, ρ_i^{LUMO} and ρ_i^{LUMO+1} values are very different, whereas for others are more similar (see Fig. 4.11A; note that the sum of ρ_i^{LUMO} and ρ_i^{LUMO+1} is the same for all *i*). Stabilization of LUMO can be achieved by maximizing the sum of ρ_i^{LUMO} contributions for all positions bearing CF_3 groups. Thus, for the low LUMO energy CF_3 groups should be placed at positions with large $\rho_i^{LUMO} - \rho_i^{LUMO+1}$ difference. One of the ways to achieve this is to place them in *ortho* position, which explains why the isomers with adjacent CF_3 groups usually have lower LUMO energies than the isomers with more uniform distribution of CF_3 groups (CORA-**4-1** vs CORA-**4-2**, Fig. 4.11). However, the presence of the *ortho* moiety is not necessary condition—for instance, the lowest LUMO energy among CORA-2 isomers is achieved in the 1,6-isomer because these equivalent positions have the largest $\rho_i^{LUMO} - \rho_i^{LUMO+1}$ differences.

4.4 Conclusions

In this chapter, we analyzed the factors affecting thermodynamics stability and electron-accepting properties of trifluoromethylated corannulene derivatives. Exhaustive calculations of the whole set of $CORA(CF_3)_x$ ($x = 0-10$) isomers showed that CF_3 groups in adjacent (*ortho* and *peri*) positions are destabilizing, and hence the most stable isomers either have no such fragments at all ($x \leq 5$) or have the minimum number of them ($x = 6-8$). Destabilizing effects of CF_3-CF_3 groups in *peri* and *ortho* moieties are transferable between the isomers and are estimated to be 43 and 50 kJ/mol. Beyond $CORA(CF_3)_5$, the inevitable CF_3-CF_3 contacts increase the energy cost by more than 100 kJ/mol per CF_3 group, which make trifluoromethylation beyond $CORA(CF_3)_7$ hardly possible.

Increase of the number of CF_3 groups systematically enhances the electron-accepting properties of $CORA(CF_3)_x$ derivatives with the average increment for electron affinities of 0.30 eV per CF_3 group. However, detailed analysis of the whole set of isomers shows that position of CF_3 groups on the corannulene rim plays a role, especially for derivatives with $x \leq 5$. The twofold degenerate LUMO of CORA is split in its CF_3 derivatives. The mean LUMO/LUMO+1 energy is invariant for the given number of CF_3 groups, but the magnitude of the splitting depends on the way how CF_3 groups are distributed on the rim. The isomers with nonuniform distribution of CF_3 groups (especially those CF_3 groups in *ortho* positions) have substantially different CF_3 contribution to LUMO and LUMO+1, hence the gap between these orbitals is large, and the LUMO energy is low. On the contrary, the isomers with uniform distribution of CF_3 groups on the rim have smaller LUMO/LUMO+1 gap and hence higher LUMO energies. Thus, the most thermodynamically stable isomers are usually those with the lowest electron affinities.

Although in this work we analyzed only corannulene derivatives, we believe that these results can be generalized to the other types of PAHs as the destabilizing role of CF_3 groups in adjacent positions and substantial contribution of CF_3 groups to LUMOs are expected for all kinds of $PAH(CF_3)_x$ derivatives.

Acknowledgments

The author acknowledges Olga Boltalina, Steve Strauss, and their students and coworkers from Colorado State University (Fort Collins, USA) for fruitful collaboration on perfluorlkylated PAHs and for providing the experimental data on a series of $CORA(CF_3)_x$ derivatives prior to publication. European Research Council is acknowledged for financial support. Computations were performed using computational facilities provided by Technical University of Dresden and on the Cascade cluster in EMSL, a DOE Office of Science User Facility sponsored by the Office of Biological and Environmental Research and located at Pacific Northwest National Laboratory.

References

[1] J.E. Anthony, A. Facchetti, M. Heeney, S.R. Marder, X. Zhan, n-Type organic semiconductors in organic electronics, Adv. Mater. 22 (2010) 3876–3892.
[2] X. Gao, Y. Hu, Development of n-type organic semiconductors for thin film transistors: a viewpoint of molecular design, J. Mater. Chem. C 2 (2014) 3099–3117.
[3] C. Wang, H. Dong, W. Hu, Y. Liu, D. Zhu, Semiconducting π-conjugated systems in field-effect transistors: a material odyssey of organic electronics, Chem. Rev. 112 (2012) 2208–2267.
[4] C.R. Newman, C.D. Frisbie, D.A. da Silva Filho, J.-L. Brédas, P.C. Ewbank, K.R. Mann, Introduction to organic thin film transistors and design of n-channel organic semiconductors, Chem. Mat. 16 (2004) 4436–4451.
[5] M. Mas-Torrent, C. Rovira, Novel small molecules for organic field-effect transistors: towards processability and high performance, Chem. Soc. Rev. 37 (2008) 827–838.
[6] M.L. Tang, Z. Bao, Halogenated materials as organic semiconductors, Chem. Mat. 23 (2011) 446–455.
[7] F. Babudri, G.M. Farinola, F. Naso, R. Ragni, Fluorinated organic materials for electronic and optoelectronic applications: the role of the fluorine atom, Chem. Commun. (2007) 1003–1022.
[8] H. Sun, A. Putta, M. Billion, Arene trifluoromethylation: an effective strategy to obtain air-stable n-type organic semiconductors with tunable optoelectronic and electron transfer properties, J. Phys. Chem. A 116 (2012) 8015–8022.
[9] M.-H. Yoon, A. Facchetti, C.E. Stern, T.J. Marks, Fluorocarbon-modified organic semiconductors: molecular architecture, electronic, and crystal structure tuning of arene- versus fluoroarene-thiophene oligomer thin-film properties, J. Am. Chem. Soc. 128 (2006) 5792–5801.

[10] H. Usta, A. Facchetti, T.J. Marks, n-Channel semiconductor materials design for organic complementary circuits, Acc. Chem. Res. 44 (2011) 501–510.

[11] O.V. Boltalina, A.A. Popov, I.V. Kuvychko, N.B. Shustova, S.H. Strauss, Perfluoroalkylfullerenes, Chem. Rev. 115 (2015) 1051–1105.

[12] M. Schlosser, CF_3-Bearing aromatic and heterocyclic building blocks, Angew. Chem.-Int. Edit. Engl. 45 (2006) 5432–5446.

[13] S. Geib, S.C. Martens, U. Zschieschang, F. Lombeck, H. Wadepohl, H. Klauk, L.H. Gade, 1,3,6,8-Tetraazapyrenes: synthesis, solid-state structures, and properties as redox-active materials, J. Org. Chem. 77 (2012) 6107–6116.

[14] O.A. Tomashenko, V.V. Grushin, Aromatic trifluoromethylation with metal complexes, Chem. Rev. 111 (2011) 4475–4521.

[15] I.V. Kuvychko, K.P. Castro, S.H.M. Deng, X.-B. Wang, S.H. Strauss, O.V. Boltalina, Taming hot CF_3 radicals: incrementally tuned families of polyarene electron acceptors for air-stable molecular optoelectronics, Angew. Chem.-Int. Edit. Engl. 52 (2013) 4871–4874.

[16] K.P. Castro, T.T. Clikeman, N.J. DeWeerd, E.V. Bukovsky, K.C. Rippy, I.V. Kuvychko, G.-L. Hou, Y.-S. Chen, X.-B. Wang, S.H. Strauss, O.V. Boltalina, Incremental tuning up of fluorous phenazine acceptors, Chem.-Eur. J. 22 (2016) 3930–3936.

[17] K.C. Rippy, E.V. Bukovsky, T.T. Clikeman, Y.-S. Chen, G.-L. Hou, X.-B. Wang, A.A. Popov, O.V. Boltalina, S.H. Strauss, Copper causes regiospecific formation of C_4F_8-containing six-membered rings and their defluorination/aromatization to C_4F_4-containing rings in triphenylene/1,4-$C_4F_8I_2$ reactions, Chem.-Eur. J. 22 (2016) 874–877.

[18] L.K. San, T.T. Clikeman, C. Dubceac, A.A. Popov, Y.-S. Chen, M.A. Petrukhina, S.H. Strauss, O.V. Boltalina, Corannulene molecular rotor with flexible perfluorobenzyl blades: synthesis, structure and properties, Chem.-Eur. J. 21 (2015) 9488–9492.

[19] T.T. Clikeman, E.V. Bukovsky, X.-B. Wang, Y.-S. Chen, G. Rumbles, S.H. Strauss, O.V. Boltalina, Core perylene diimide designs via direct bay- and ortho-(poly)trifluoromethylation: synthesis, isolation, x-ray structures, optical and electronic properties, Eur. J. Org. Chem. 2015 (2015) 6641–6654.

[20] L.K. San, E.V. Bukovsky, I.V. Kuvychko, A.A. Popov, S.H. Strauss, O.V. Boltalina, Single-step gas-phase polyperfluoroalkylation of naphthalene leads to thermodynamic products, Chem.-Eur. J. 20 (2014) 4373–4379.

[21] T.T. Clikeman, E.V. Bukovsky, I.V. Kuvychko, L.K. San, S.H.M. Deng, X.-B. Wang, Y.-S. Chen, S.H. Strauss, O.V. Boltalina, Poly(trifluoromethyl)azulenes: structures and acceptor properties, Chem. Commun. 50 (2014) 6263–6266.

[22] I.V. Kuvychko, C. Dubceac, S.H.M. Deng, X.-B. Wang, A.A. Granovsky, A.A. Popov, M.A. Petrukhina, S.H. Strauss, O.V. Boltalina, $C_{20}H_4(C_4F_8)_3$: a fluorine-containing annulated corannulene that is a better electron acceptor than C_{60}, Angew. Chem.-Int. Edit. Engl. 52 (2013) 7505–7508.

[23] I.V. Kuvychko, S.N. Spisak, Y.-S. Chen, A.A. Popov, M.A. Petrukhina, S.H. Strauss, O.V. Boltalina, A buckybowl with a lot of potential: C_5-$C_{20}H_5(CF_3)_5$, Angew. Chem.-Int. Edit. Engl. 51 (2012) 4939–4942.

[24] I.V. Kuvychko, T. Clikeman, C. Dubceac, S.H.M. Deng, X.-B. Wang, A.A. Popov, M.A. Petrukhina, S.H. Strauss, O.V. Boltalina, Understanding Polyarene Trifluoromethylation With Hot CF_3 Radicals: A Corannulene Primer, 2016 (Unpublished Data).

[25] B.M. Schmidt, B. Topolinski, M. Yamada, S. Higashibayashi, M. Shionoya, H. Sakurai, D. Lentz, Fluorinated and trifluoromethylated corannulenes, Chem.-Eur. J. 19 (2013) 13872–13880.

[26] B.M. Schmidt, S. Seki, B. Topolinski, K. Ohkubo, S. Fukuzumi, H. Sakurai, D. Lentz, Electronic properties of trifluoromethylated corannulenes, Angew. Chem.-Int. Edit. Engl. 51 (2012) 11385−11388.
[27] S.S. Batsanov, Van der Waals Radii of Elements, Inorg. Mater. 37 (2001) 871−885.
[28] M.A. Petrukhina, K.W. Andreini, J. Mack, L.T. Scott, X-ray quality geometries of geodesic polyarenes from theoretical calculations: what levels of theory are reliable? J. Org. Chem. 70 (2005) 5713−5716.
[29] Y. Sevryugina, A.Y. Rogachev, E.A. Jackson, L.T. Scott, M.A. Petrukhina, Ray and density functional theory structural study of 1,3,5,7,9-penta-tert-butylcorannulene, $C_{40}H_{50}$, J. Org. Chem. 71 (2006) 6615−6618.
[30] S. Samdal, L. Hedberg, K. Hedberg, A.D. Richardson, M. Bancu, L.T. Scott, Gas-phase molecular structure of decachlorocorannulene, $C_{20}Cl_{10}$. An electron-diffraction study augmented by ab initio, DFT, and normal coordinate calculations, J. Phys. Chem. A 107 (2003) 411−417.
[31] Q. Zhang, K. Kawasumi, Y. Segawa, K. Itami, L.T. Scott, Palladium-catalyzed C−H activation taken to the limit. flattening an aromatic bowl by total arylation, J. Am. Chem. Soc. 134 (2012) 15664−15667.
[32] K.K. Baldridge, K.I. Hardcastle, T.J. Seiders, J.S. Siegel, Synthesis, structure and properties of decakis(phenylthio)corannulene, Org. Biomol. Chem. 8 (2010) 53−55.
[33] B.M. Schmidt, B. Topolinski, S. Higashibayashi, T. Kojima, M. Kawano, D. Lentz, H. Sakurai, The synthesis of hexafluorosumanene and its congeners, Chem.-Eur. J. 19 (2013) 3282−3286.
[34] B.M. Schmidt, S. Seki, B. Topolinski, K. Ohkubo, S. Fukuzumi, H. Sakurai, D. Lentz, Electronic properties of trifluoromethylated corannulenes, Angew. Chem.-Int. Edit. Engl. 51 (2012) 11385−11388.
[35] B.M. Schmidt, D. Lentz, Syntheses and properties of buckybowls bearing electron-withdrawing groups, Chem. Lett. 43 (2014) 171−177.
[36] T.J. Seiders, E.L. Elliott, G.H. Grube, J.S. Siegel, Synthesis of corannulene and alkyl derivatives of corannulene, J. Am. Chem. Soc. 121 (1999) 7804−7813.

Fluorination–Defluorination and Fluorine Storage Properties of Single-Wall Carbon Nanotubes and Carbon Nanohorns

5

Y. Hattori, H. Touhara
Shinshu University, Ueda, Japan

Chapter Outline

5.1 Introduction

Carbon nanotubes (CNTs) have continuously attracted significant interests due to their high potential use that they would offer improvements in many fields of human life, ie, energy, environment, electronics, transportation, medicine, etc. As has been well known, fluorination is one of the most effective chemical methods to modify and control the physical and chemical properties of carbon materials [1], which have been applied without exception to CNTs. Indeed, this interest has been prompted by the first sidewall functionalization of single-wall carbon nanotubes (SWNTs) by direct fluorination [2] which yielded the "unbundling" of the SWNT bundles and "soluble" fluoronanotubes.

New Fluorinated Carbons: Fundamentals and Applications. http://dx.doi.org/10.1016/B978-0-12-803479-8.00005-X

Since then, the development in fluorine chemistry of the SWNTs has been remarkable, particularly in the research on the functionalization with regard to improved chemical reactivity and dispersion in solvents, cutting tubes, metallic/semiconducting tube ratio, changes in average tube diameter, reinforcing fibers and polymer composites, etc. However, one of the most significant advances in the fluorine chemistry of SWNTs in recent years is finding the fluorine storage properties of the SWNTs [3] and of single-wall carbon nanohorns (SWNHs) [4–6]. As another important feature of the recent fluorine chemistry of SWNTs, it can be mentioned that understanding the versatile interaction of fluorine with curved hybrid networks of sp^2-like carbon atoms has been becoming much clearer, which is closely related to the fluorine storage properties of SWNTs and SWNHs. It is noted that the diverse interaction is concerned with the fluorine location, ie, exo- and/or endo-addition of the tubes.

In light of the recent developments in the studies of the fluorine chemistry of the SWNTs and SWNHs, this chapter focuses on the thermal fluorination–defluorination of the SWNTs and SWNHs, and their fluorine storage properties.

5.2 Fluorination–Defluorination and Fluorine Storage Properties of Single-Wall Carbon Nanotubes

The graphite fluoride, $(CF)_n$ with covalent C–F bonds, is formed under the harsh condition of the 600°C fluorination with elemental fluorine whereas fluorofullerenes $C_{60}F_x$ are formed under relatively mild conditions at RT ~ 300°C. These differences in the fluorination behavior are mainly attributable to the difference in sp hybridization of carbon atoms. The deviation from the planar sp^2 hybridization of the spherically curved one in C_{60} is significant enough to cause its electronic properties and chemical reactivity to differ from those of graphite. Because of the lower temperature fluorination of solid C_{60}, forming a face-centered-cubic lattice with the interstitial sites, possibility of using C_{60} as a fluorine storage material was considered in early studies. It has, however, been shown that chemisorbed fluorine in the solid C_{60} is unable to remove due to the formation of the essentially C–F covalent bonds, although the bonds being weakened. SWNTs, discovered after the fullerenes as the second family of nanocarbons, of which carbon frameworks also consist of the curved sp^2 hybridization but the curvature being cylindrical. Accordingly, the reaction of the SWNTs with elemental fluorine also proceeds relatively easily from RT to 300°C, whereas the single-wall tubular structure does not survive at the higher temperature fluorination than around 300°C. The stoichiometry of the fluorinated SWNTs (F-SWNTs) depends on the reaction time, morphological characteristics such as tube diameter, bundle or aggregate, pore structure and adsorptive properties, etc. Most studies of fluorination–defluorination of SWNTs have been performed for the purpose of sidewall functionalization which has induced modified morphological, chemical, and electronic properties [2,7,8]. The nature of C–F bonding, and the thermal recovery and/or decomposition mechanisms of F-SWNTs have also been investigated through the thermal defluorination behaviors.

The thermal recovery behavior of F-SWNTs was investigated in detail by UV–vis-near-IR absorption spectroscopy and Raman spectroscopy [7,9]. The samples used were F-SWNTs with stoichiometry $CF_{0.43-0.50}$ and $CF_{0.43}$ obtained by the 250°C fluorination of SWNTs (prepared by a laser vaporization method, ~1.3 nm in diameter). According to the X-ray photoelectron spectroscopy (XPS), most of the fluorine was covalently bonded to a single carbon atom, the stoichiometry being close to the presumed saturation one. After heat annealing the sample $CF_{0.43-0.50}$ in inert gas atmospheres between 100°C and 350°C for 1 h, spectroscopic features of the pristine SWNTs were observed [7]. The XPS, Raman, and resistance measurements of annealed samples of $CF_{0.43}$ indicated that most of the fluorine was lost at 200–300°C, and was found to be virtually complete at 400°C [9]. In these studies, the gaseous products evolved from the samples were not analyzed. On the other hand, the high temperature thermal defluorination up to 1000°C of the partially fluorinated CF_x ($x \leq 0.2$, prepared by a 50°C fluorination of the Hipco tube with ~1.0 nm diameter) in an argon atmosphere resulted in cutting the nanotubes to a range of short length (average <50 nm) [10]. The thermogravimetric analysis (TGA) and in situ infrared (IR) spectra showed that during the pyrolysis of the F-SWNTs, fluorine was completely driven off the nanotube structure, and the fluorine has taken off a certain amount of carbon from the nanotube sidewall. However, ~55% of the mass survived as pristine nanotubes, even up to 1000°C in the heating process. Interestingly, the IR spectra also showed that the thermal defluorination process occurred in two steps: first, the evolution of CO_2 and COF_2 between ~300°C and ~560°C, and second the evolution of CF_4 at temperatures higher than 560°C. Later, this defluorination behavior was confirmed by the work with an in situ scanning tunneling microscopy (STM) imaging for F-SWNTs with a stoichiometry $CF_{0.5}$ (prepared by a 150°C fluorination of MER-SWNTs with 1.2–1.5 nm diameter) and for the sample after defluorination with thermal treatment at 400°C [8]. The work showed that the desorption of fluorine through annealing can recover the original structure to a certain extent and cut the SWNTs in the radial direction. The STM images of the same F-SWNTs showed different diameter values depending on the sidewall regions which indicate that the fluorine chemisorption is not homogeneous as shown by the banding structure. The boundary of two different domains of chemisorbed and bare regions in the F-SWNTs would induce the cutting in the radial direction.

To gain further insight into the thermal defluorination behavior of F-SWNTs, analysis of the gaseous species evolved during the thermal treatment of F-SWNTs has been required. The thermal decomposition of F-SWNTs with a stoichiometry $CF_{0.5}$ (prepared from the Hipco tube by a 50°C fluorination) was investigated by freezing the gaseous products formed at temperatures between 50°C and 500°C under high vacuum in an argon matrix at 10–20 K and analyzing the trapped species by IR spectroscopy [11]. The major products of the decomposition were COF_2 below 300°C and CF_4 above 300°C. Mass spectrometric investigation of gases evolved from the F-SWNTs with a stoichiometry $CF_{0.22}$ during heating up to 500°C has recently been reported [12]. The sample used was obtained by the 220°C fluorination of the SWNTs (prepared by an electric arc method, ~1.5 nm diameter). The gases containing fluorine appeared in the range from 100°C to 150°C. In the range of 150–300°C, the mixture of disintegration products containing COF_2, and in the range of 300–400°C mostly CF_4 and C_4F_6

appeared. Interestingly, the two studies showed the same major products released from the F-SWNTs samples at the same temperature range, and also showed no evolution of molecular fluorine. In addition to an etching of the tubes observed at temperatures as low as 100–150°C, the CF_4 evolution as a major gas-phase species indicates remarkable damages of the tubes during defluorination.

An important and interesting paper has recently been published, which reports the reversible fluorine storage into SWNTs [3]. Until then, reversible fluorination–defluorination of the SWNTs has never been reported. The SWNTs used were provided by Helix and grown from a chemical vapor deposition (CVD) method. After the standard process of purification by the ultrasonication in a concentrated hydrochloric acid, TEM observations showed open-ended tubes with an average diameter of 2 nm, forming bundles of size of ~40 nm. The fluorine adsorbed SWNTs with the stoichiometry CF_x ($x = 0.12$–0.42) were prepared by the reaction with 1 atm elemental fluorine at the lower temperature range −191–25°C. The weight losses of the TGA curves suggested that the F-SWNTs are almost totally defluorinated upon heating under an inert atmosphere by 300°C. The change in ^{19}F NMR spectra of F-SWNTs by the heating confirmed the removal of fluorine. It should be noted that out-gassing of F-SWNTs under primary vacuum for 48 h also resulted in a nearly total removal of adsorbed fluorine. After thermal defluorination, the resulting sample can be refluorinated similarly with the pristine tubes. Such reversible fluorination–defluorination is presumed to arise from endo-fluorination at internal surfaces of the tube, and the interaction between the fluorine and carbon atoms seems to tend toward physisorption almost without damage of the tube [13] as will be discussed in the next section. Another mechanism has been also proposed for the endo-fluorination via the fluorine condensation into the tube at the lower temperatures of −191°C and −178°C. These fluorination–defluorination behaviors of the SWNTs would be promising for fluorine storage, which will be more certainly by mass spectrometric and IR-UV analysis of the gaseous species evolved from the F-SWNTs upon heating. For further insight into the reversible fluorination, comparative studies with another SWNTs would be interesting to examine whether the reversible fluorination originates from the inherent properties of the SWNTs used or from the operating conditions for fluorination–defluorination.

5.3 Fluorine Storage Properties of Carbon Nanohorns

5.3.1 *Structural Features and Fundamental Aspects of Fluorine Adsorption on Single-Wall Carbon Nanohorns*

5.3.1.1 *Structural Features of Single-Wall Carbon Nanohorns*

SWNHs are the latest member of the SWNT family, whereas they have a variety of features that are quite different from the SWNTs. The SWNHs used in fluorine adsorption–desorption studies were prepared by the high-power CO_2 laser ablation of graphite under atmospheric Ar pressure without catalysts, and the purity of the SWNH samples was determined by a TGA to be 90–92 wt.% without noncarbon

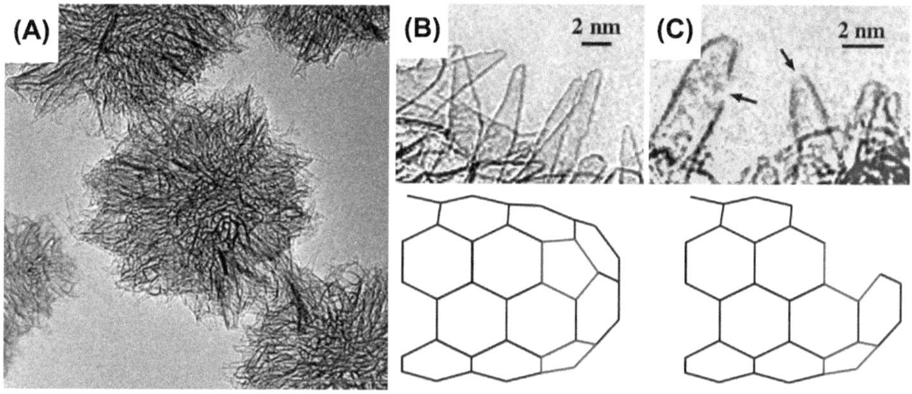

Figure 5.1 TEM images of (A) aggregates of as-grown single-wall carbon nanohorns (SWNHs), (B) individual SWNHs, and (C) hole-opened SWNHs.
Courtesy of Professor Bandow at Meijo University.

materials. The detailed procedure for the preparation is described elsewhere [14]. SWNHs have a characteristic structure composed of a spherical aggregate (80–100 nm diameter) of many single-graphene tubules (~2 nm diameter, 40–50 nm length) whose one end is closed by a horn or a cone with an average cone opening angle at ~20° [15]. The number of carbon atoms per SWNH is therefore in the range of ~10,000 ($C_{\sim 10,000}$) [16]. Individual SWNHs are not synthesized as isolated units, but aggregate to make a "dahlia flower-like" particle (SWNH aggregate). The horn- or cone-like shape is due to the five five-membered carbon rings situated at the horn-shaped end. The aggregates of the carbon nanohorns and individual nanohorns are clearly shown in TEM images of Fig. 5.1A and B, respectively. Hole-opened SWNHs (h-SWNHs) shown in the TEM image of Fig. 5.1C was obtained by oxidation in a 30% aqueous solution of H_2O_2 at 100°C [17]. It is worth noting that the TEM images in Fig. 5.1B and C clearly show the bending sites or kinks which are topological defects associated with the presence of carbon pentagons [18]. Thus SWNHs have various types of adsorption sites which are external and internal surfaces of the nanohorns, and interstitial sites as shown in Fig. 5.2. Based on the nitrogen adsorption analyses at 77 K shown in Fig. 5.3A and TEM observations, the structural properties of the as-grown and h-SWNHs are summarized in Table 5.1. The pore structural parameters in Table 5.1 indicate that the SWNHs have a bimodal porosity consisting of micropores and mesopores, being rich in mesopores. It should be noted that the Brunauer–Emmett–Teller (BET)-specific surface area S_{BET} of the as-grown SWNHs is significantly higher than the value ~200 m^2/g of the typical end-closed SWNTs (~2 nm diameter) prepared by a laser ablation method.

5.3.1.2 Adsorption Sites, Surface Areas, and Fluorine Adsorption Stoichiometry of Single-Wall Carbon Nanohorns

Fig. 5.3B shows the TGA curves of the fluorine adsorption on SWNHs and h-SWNHs at 25°C and 200°C under a fluorine flow of 0.2 cm^3/min. When the

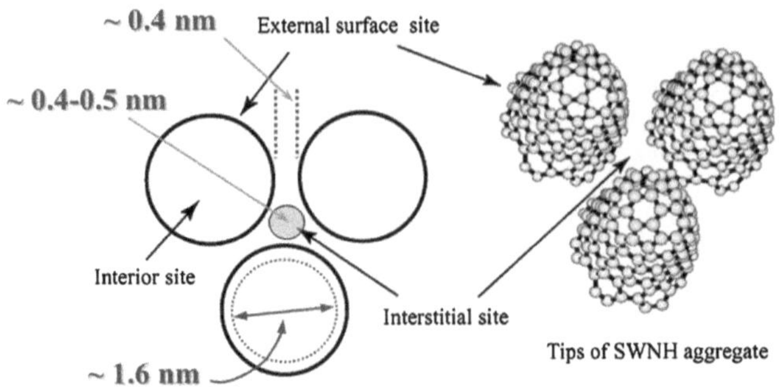

Figure 5.2 Gas adsorption sites of SWNHs. *SWNHs*, single-wall carbon nanohorns. Courtesy of professor Bandow at Meijo University.

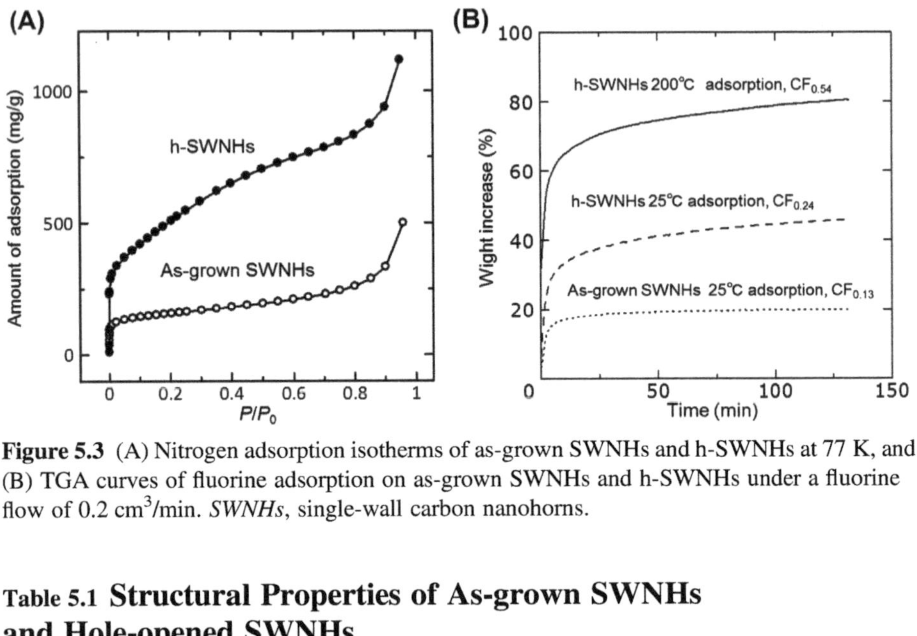

Figure 5.3 (A) Nitrogen adsorption isotherms of as-grown SWNHs and h-SWNHs at 77 K, and (B) TGA curves of fluorine adsorption on as-grown SWNHs and h-SWNHs under a fluorine flow of 0.2 cm^3/min. *SWNHs*, single-wall carbon nanohorns.

Table 5.1 Structural Properties of As-grown SWNHs and Hole-opened SWNHs

Samples	S_{BET} (m^2/g)	V_{micro} (cm^3/g)	V_{meso} (cm^3/g)	Particle Size (nm)	Tube Diameter (nm)	Tube Length (nm)	I_D/I_G
As-grown SWNHs	427	0.18	0.62	80–100	2–3 nm	30–50	1.88
Hole-opened SWNHs	1430	0.46	0.46	80–100	2–3 nm	30–50	2.38

SWNHs, single-wall carbon nanohorns.

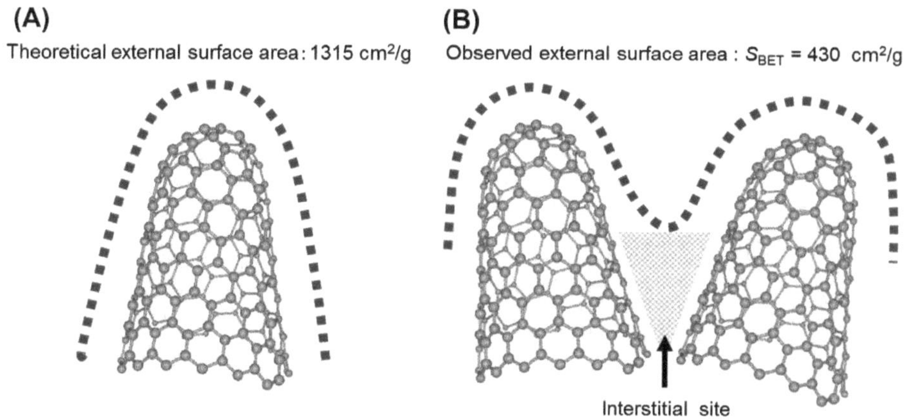

Figure 5.4 Schematic illustration of external surface sites available for N_2 and F_2 adsorption of an isolated SWNH (A) and aggregate of SWNHs (B). *SWNHs*, single-wall carbon nanohorns. Nanohorn model courtesy of Professor Bandow at Meijo University.

fluorine adsorption behavior of the SWNHs is compared with the nitrogen adsorption behavior, very interestingly, both are very similar. That is to say, both the adsorption amounts of fluorine and nitrogen on the SWNHs significantly increased upon hole-opening indicating that internal surfaces are available for adsorption sites. The stoichiometry of the fluorine adsorption at 25°C of the SWNHs $CF_{0.17}$ increases to $CF_{0.24}$ by hole-opening. The adsorption increases up to $CF_{0.54}$ at the higher temperature of 200°C.

For understanding the fluorine adsorption on the SWNHs, we initially investigated the nitrogen adsorption on the SWNHs because the LJ potential parameters used to describe the adsorption behavior of a fluorine molecule (0.36 nm) is almost comparable to those of a nitrogen molecule (0.39 nm) [19]. Fig. 5.4A and B conceptually illustrate the external surface sites available for nitrogen adsorption of an isolated carbon nanohorn, and of aggregates of the nanohorns, respectively. The external specific surface area of an isolated carbon nanohorn of 1315 m^2/g can be considered to be equal in principle to half the surface area of a graphene sheet, 2630 m^2/g. The S_{BET} value of the aggregates given in Table 5.1 corresponds to 33% of the external surface area of an isolated nanohorn. This indicates that nitrogen molecules are adsorbed on the external surfaces of the aggregated SWNHs, not adsorbed at the interstitial sites as illustrated in Fig. 5.4B. Fig. 5.5A and B illustrates the external and internal surface sites available for nitrogen adsorption of an isolated hole-opened nanohorn, and of aggregates of hole-opened nanohorns, respectively. Both the external and internal surface areas of an isolated hole-opened nanohorn equals, in principal, 1315 m^2/g, and then the total surface area of an isolated hole-opened nanohorn equals that of a graphene sheet of 2630 m^2/g. Upon hole-opening, the S_{BET} of h-SWNHs increases by more than three times than for the as–grown SWNHs because the inner surfaces of the h-SWNHs are

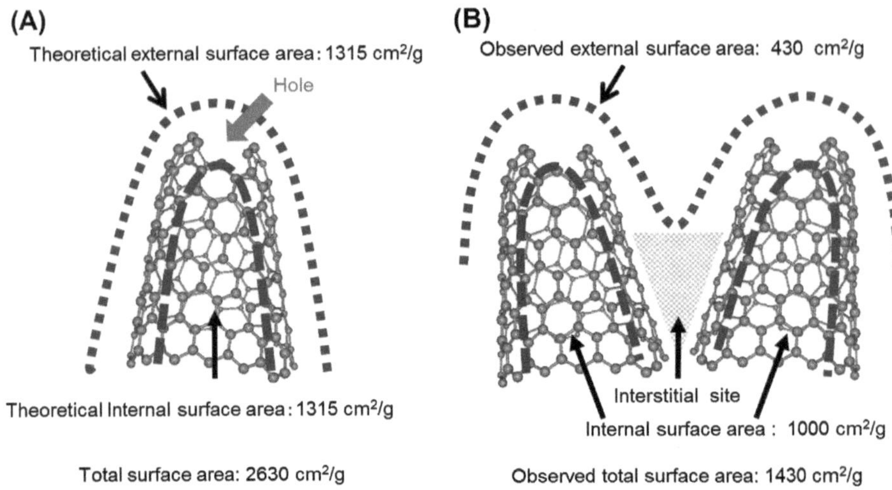

Figure 5.5 Schematic illustration of external and internal surface sites of an isolated h-SWNH (A) and external and internal surface sites available for N_2 and F_2 adsorption of the aggregate of h-SWNHs (B). h-SWNH, hole-opened single-wall carbon nanohorns.

available for nitrogen adsorption through the opened gates (nanowindows) as illustrated in Fig. 5.5B. The difference between the S_{BET} of the h-SWNHs and that of the as-grown SWNHs is c.1000 m^2/g, which is the increment in the S_{BET} by hole-opening, corresponding to 76% (1000/1315) of the theoretical internal surface area. The S_{BET} value of the h-SWNHs corresponds to 83% (1430/(1315 + 430)) of the sum of the external and internal surface areas shown in Fig. 5.5B.

Based on the above consideration of the adsorption sites, the fluorine adsorption stoichiometry of the as-grown SWNHs and h-SWNHs is possible to estimate. In the estimation, the fluorine adsorption behavior at temperatures lower than 200°C is reasonably assumed to be similar to the nitrogen adsorption. However, it is absolutely necessary to take into account that there is a limiting F/C stoichiometry of 0.5 for which the outside wall–fluorinated SWNTs can still maintain their tube-like structure [2]. According to the above analysis for the nitrogen adsorption on the as-grown SWNHs, and because of the comparable molecular size with nitrogen, fluorine can access only 33% of the external surface of the as-grown SWNHs, thus the F/C stoichiometry of the as-grown SWNHs is 0.17 (0.5×0.33). The estimated stoichiometry is almost comparable to the experimentally observed stoichiometry of $CF_{0.13}$ at a 25°C adsorption. In a similar manner, the estimated stoichiometry of the h-SWNHs is obtained as $CF_{0.41}$ (0.5×0.83). The experimental stoichiometry of $CF_{0.24}$ at a 25°C adsorption is smaller than the expected one because the difference in the adsorption behavior between the chemical and physical adsorptions would become remarkable at the outer convex and the inner concave surfaces composed of carbon sp^2-hybridization as will be discussed later.

5.3.2 Fluorine Adsorption and Desorption Properties of Single-Wall Carbon Nanohorns

Fluorination of the SWNHs was first performed with the aim of forming more functionalized nanohorns that have characteristic structural and adsorptive properties [20]. The fluorination was carried out by the direct reaction of the as-grown SWNHs with elemental fluorine in a static reactor using a nickel vessel, and the gravimetric stoichiometry of the fluorinated SWNHs (F-SWNHs) obtained at 30°C, 100°C, and 200°C fluorinations were 0.25, 0.26, and 0.45, respectively. Fig. 5.6 shows the TEM images of the pristine and fluorinated SWNHs. The image of the pristine sample showed that it consists of spherical particles with about an 80 nm diameter. The TEM observations indicated that the nanohorn and dahlia-like structures remain after the fluorination. In addition, no other carbon forms were found in these photographs of the F-SWNHs indicating that no destruction of the carbon skeleton of the SWNHs did occurred up to a 200°C fluorination.

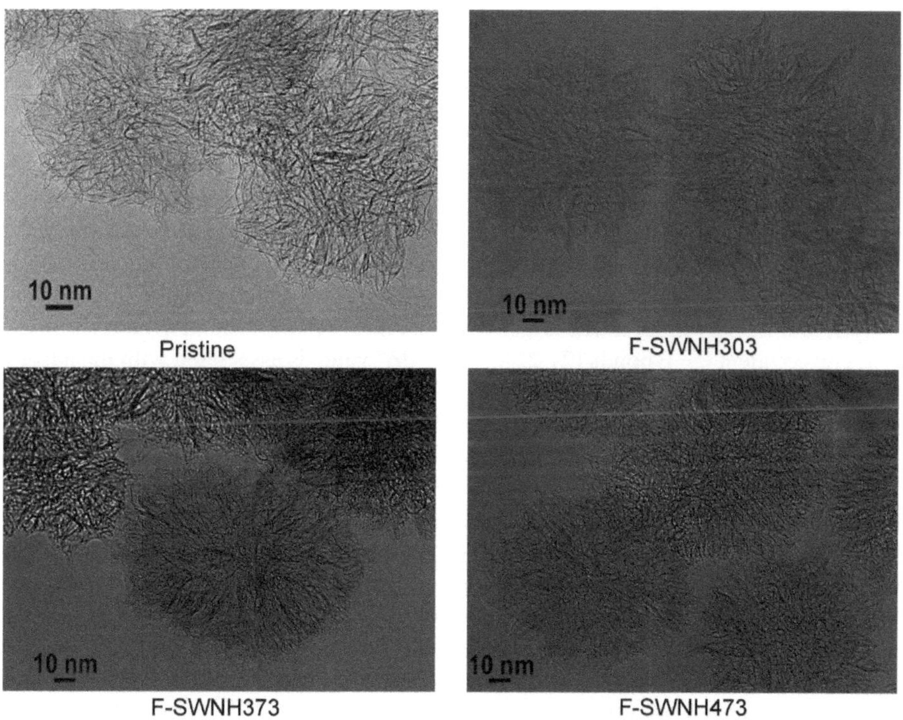

Figure 5.6 High-resolution TEM images of pristine and fluorinated SWNHs at 30 (F-SWNH303), 100 (F-SWNH373), and 200°C (F-SWNH473). *SWNHs*, single-wall carbon nanohorns.
Reproduced with permission from Hattori et al. [20]. Copyright 2004 American Chemical Society.

Table 5.2 Pore Structural Parameters of Pristine and Fluorinated As-grown SWNHs at 30°C, 100°C, and 200°C

Samples	S_{BET} (m²/g)	V_{micro} (cm³/g)	V_{total} (cm³/g)	βE° (kJ/mol)
Pristine	399	0.136	0.473	6.22
F-SWNH 30	192	0.058	0.324	5.37
F-SWNH 100	71	0.033	0.261	3.09
F-SWNH 200	280	0.078	0.316	3.82

β, affinity coefficient (0.33); *E*°, characteristics energy; *SWNHs*, single-wall carbon nanohorns.

The change in the adsorptive properties by the fluorination is very interesting, and the pore structural parameters according to the nitrogen adsorption isotherms analyses at 77 K are summarized in Table 5.2, in which the βE° values were calculated by the Dubinin Radushkevich (DR) method. The total amount of N_2 adsorption on the SWNHs fluorinated at 30°C and 100°C decreased in comparison to the pristine SWNHs, due to the narrowing of the pore width and the weight increase by the fluorination. On the other hand, the amount of N_2 adsorption on the SWNHs fluorinated at 200°C was greater than those of the pristine and other fluorinated samples, which is induced by the significant increase in the S_{BET}. This result implies that nanowindows were formed in the sidewalls of the SWNHs and that the N_2 molecules can be adsorbed on the internal surfaces of the SWNHs through the opened windows. These results are important in connection with the functionalization of the SWNHs; the inner surfaces become available for adsorption by the SWNHs without any change in the nanohorn structure and the dahlia-like morphology, which leads to the ability of fluorine storage in the SWNHs. The βE° of F-SWNH 30 is reasonably smaller than that of pristine SWNH, and interestingly, much larger than that of the fluorinated activated carbon fibers with covalent C−F bonds [21,22]. As the βE° value is associated with the adsorption energy, the change in the βE° value indicates a change in the surface free energy by fluorination. This result suggests that the surface of the interstitial pore is not completely covered with covalent C−F bonds.

When compared to the SWNTs, their higher surface area and higher amounts of fluorine adsorption at a lower temperature have led to a study of the fluorine storage properties of the SWNHs, and the fluorine adsorption−desorption behavior has been investigated by a thermogravimetric analysis-differential thermal analysis (TGA-DTA) apparatus specially designed for use in a fluorine atmosphere. Fig. 5.7A and B shows the fluorine adsorption−TGA curves of the as-grown SWNHs and h-SWNHs, respectively, under a fluorine flow of 0.2 cm³/min at 25°C, 100°C, and 200°C, respectively. Both the as-grown SWNHs and h-SWNHs rapidly adsorb fluorine to form fluorine adsorbed as-grown SWNHs (F-as-grown SWNHs) and fluorine adsorbed h-SWNHs (F-h-SWNHs). The adsorbed stoichiometry is $CF_{0.13}$ (25°C), $CF_{0.14}$ (100°C), and $CF_{0.50}$ (200°C) for the F-as-grown SWNHs, and $CF_{0.29}$ (25°C), $CF_{0.35}$ (100°C), $CF_{0.54}$ (200°C) for F-h-SWNHs. At a fluorination temperature lower than 100°C, the adsorbed amounts of fluorine of the as-grown SWNHs are lower than

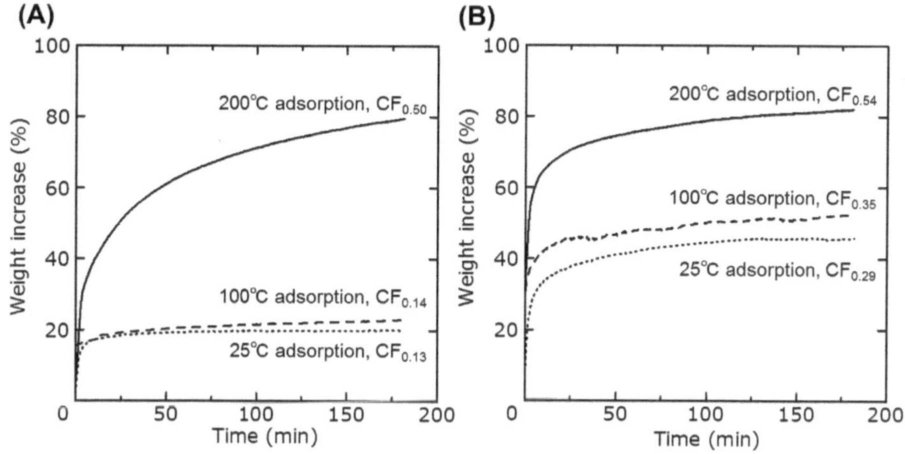

Figure 5.7 TGA curves of fluorine adsorption of as-grown SWNHs (A) and h-SWNHs (B) under a fluorine flow of 0.2 cm^3/min. *SWNHs*, single-wall carbon nanohorns.

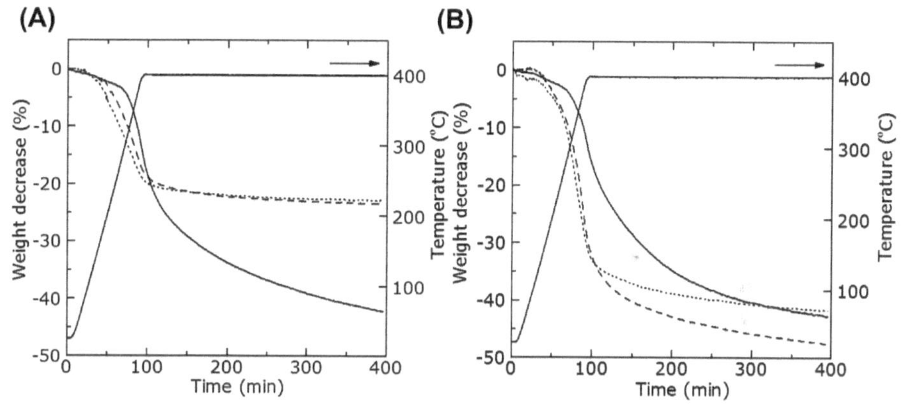

Figure 5.8 TGA curves of fluorine desorption of fluorine adsorbed as-grown SWNHs (A) and fluorine adsorbed hole-opened SWNHs (B) under a nitrogen flow of 300 cm^3/min at the heating rate 5°C/min. The samples are the same as those shown in Fig. 5.7. *SWNHs*, single-wall carbon nanohorns.

those of the h-SWNHs, whereas the stoichiometries of both the as-grown and h-SWNHs are comparable at the 200°C adsorption because the internal surfaces of the as-grown SWNHs become available for adsorption due to the hole-opening by the 200°C fluorination. Fig. 5.8 shows the fluorine desorption–TGA curves which were obtained, followed by the fluorine adsorption–TGA presented in Fig. 5.7. The TGA was performed under a nitrogen flow of 300 cm^3/min at the heating rate of 5°C/min, the temperature being increased from RT to 400°C. The fluorine-adsorbed samples at temperatures below 100°C lose weight which corresponds to 100% of the adsorbed fluorine until the temperature reaches 400°C. The samples that adsorbed fluorine at 200°C lost approximately 45% of their weights by a 300 min hold after raising the temperature to 400°C. This loss corresponds to 100% of the adsorbed fluorine. Such a weight loss at relatively lower temperatures suggests the fluorine release from the samples, and these phenomena have been investigated in detail by TGA–DTA–MS studies. A typical result of the TGA–DTA–MS is shown in Fig. 5.9 for the case of $CF_{0.33}$ obtained by direct fluorination of the h-SWNHs with elemental fluorine at 50°C in a static reactor. Interestingly, the TGA curve reveals about a 35% weight loss of the sample which corresponds to 100% of the adsorbed fluorine by a 350-min hold at 400°C. More remarkable is that the trace of the ion current of the released fluorine ($m/z = 19$) is almost the same as that of all the released gases, which suggests the high purity of the desorbed fluorine. The high purity of the released fluorine is confirmed by the MS spectrum represented in Fig. 5.10; no

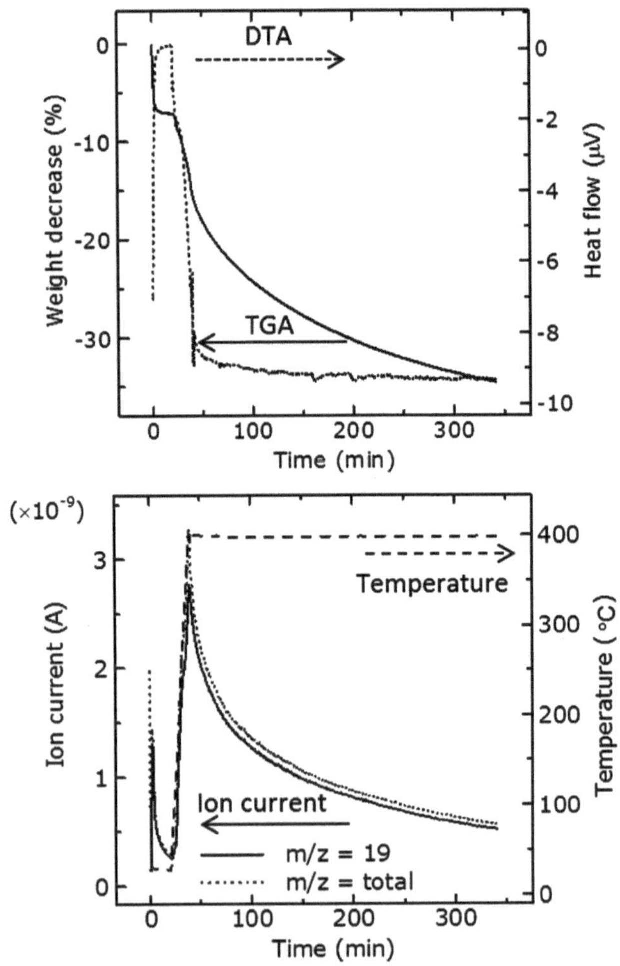

Figure 5.9 TGA-DTA-MS traces of fluorine adsorbed hole-opened -SWNHs $CF_{0.33}$ under a He flow of 300 cm^3/min at the heating rate of 5°C/min. The MS data was obtained by electron impact ionization at 70 eV. *SWNHs*, single-wall carbon nanohorns.

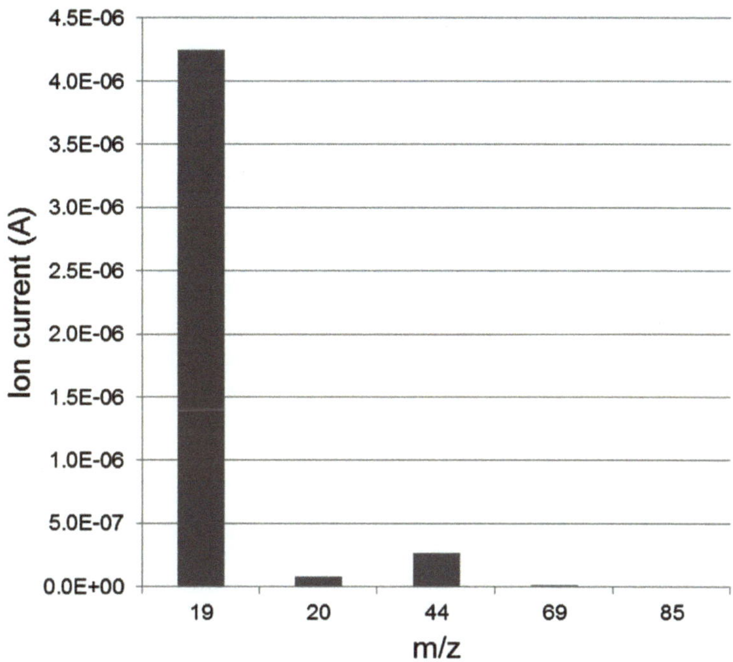

Figure 5.10 MS spectrum of gaseous species released from $CF_{0.33}$. $m/z = 19$ for F, 20 for HF^+, 44 for CO_2^+, 69 for CF_3^+.

fluorocarbons are observed and the purity based on the ion current of the generated gaseous species containing fluorine is approximately 99%. These results indicate that neither etching nor destruction of the carbon framework of the SWNHs occur during the fluorine desorption process suggesting the possible repeated use of the SWNHs. TEM images of the defluorinated F-h-SWNHs are shown in Fig. 5.11 for the case of $CF_{0.49}$ prepared by a 200°C fluorination with defluorination being carried out at 400°C for 2 h under an argon flow of 100 cm^3/min. These images indicate that defluorination does not damage the SWNHs and that individual horn-shaped sheaths of single-walled graphene sheets and the characteristic structure composed of an aggregate of many nanohorns are well preserved after the complete defluorinaion. The high purity of more than 99% of the released fluorine was also confirmed by UV (quantitative analysis of fluorine on the basis of adsorption at 283 nm) and IR (quantitative analysis of fluorocarbons) spectroscopy for the gases generated from the F-h-SWNH CF_{036} up to a 200°C heating under dynamic-pumping [23].

5.3.3 Mechanism of Fluorine Desorption and Nature of C−F Bonding

As part of the investigation to clarify the process and mechanism of fluorine release from F-as-grown SWNHs and F-h-SWNHs, the fluorine desorption behavior was

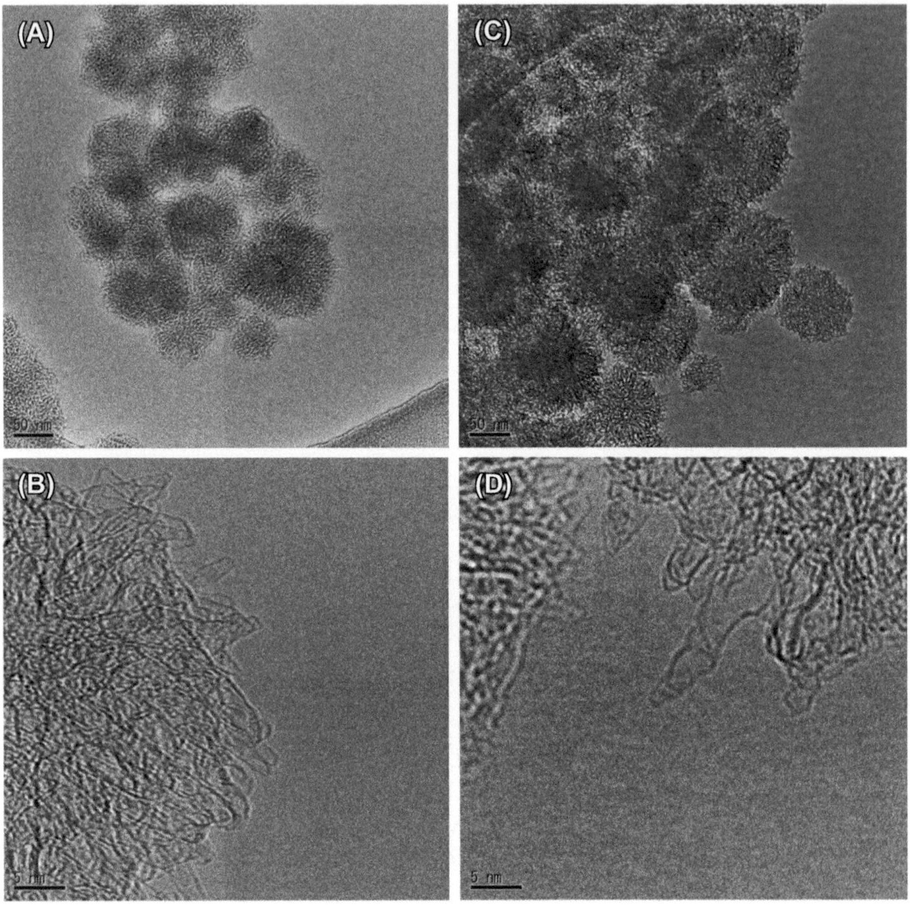

Figure 5.11 TEM images of pristine (A) and (B), and completely defluorinated fluorine adsorbed hole-opened SWNHs ($CF_{0.49}$ prepared by 200°C fluorination) under argon flow at 400°C for 2 h (C) and (D). *SWNHs*, single-wall carbon nanohorns.

analyzed in detail by TGA-DTA under a nitrogen flow of 100 cm^3/min, and graphite fluoride $(C_2F)_n$ was used as the reference material for the DTA. Fig. 5.12 shows the DTA and TGA results of $CF_{0.10}$ (F-as-grown SWNH) and $CF_{0.17}$ (F-h-SWNH) prepared by direct fluorination at 25°C. The most important observation is the endothermic defluorination accompanied by a weight loss in the entire temperature range from RT to 200°C. The weight loss equivalent to the mass of stoichiometric fluorine contained in each sample is also attained by heating to 200°C. To our knowledge, this is the first observation of the endothermic defluorination which indicates the endothermic C–F bond dissociation at the surfaces of the F-SWNHs to atomic fluorine. The dissociated atomic fluorine recombines to form molecular fluorine, and desorbs as fluorine molecule. Interestingly, the normalized DTA profiles of the samples are different from each other indicating the difference in the C–F binding energy; the F-as-grown SWNH sample

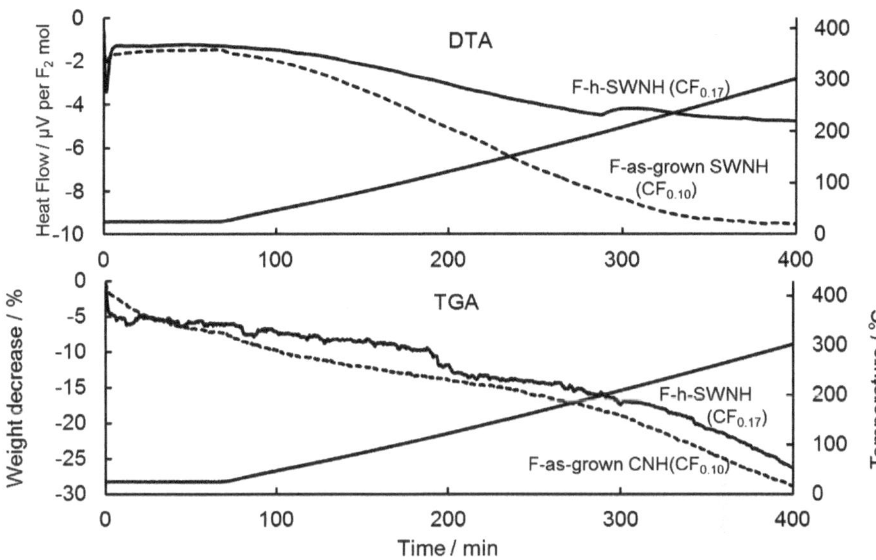

Figure 5.12 TGA-DTA curves of fluorine adsorbed as-grown SWNH $CF_{0.10}$ and fluorine adsorbed hole-opened SWNH $CF_{0.17}$ under a N_2 flow of 100 cm^3/min. The values of the heat flow of the DTA are normalized with regard to per mole of fluorine. *SWNH*, single-wall carbon nanohorn.

shows more pronounced endothermic defluorination than the F-h-SWNH samples. This result suggests a weaker C–F bond of the F-h-SWNH sample than that of the F-as-grown SWNH one. Fig. 5.13 shows expanded views of the TGA and DTA curves before heating in Fig. 5.12. The endothermic weight loss under a nitrogen flow without heating is clearly observed suggesting the very weak interaction of adsorbed fluorine with the SWNHs as will be discussed later. The thermal defluorination of the graphite fluorides, $(CF)_n$ and $(C_2F)_n$, occurs at 500–600°C accompanied by the evolution of fluorocarbons, which is a very strong exothermic reaction [24]. As for F-SWNTs, $CF_{0.5}$, obtained by the 50°C fluorination [11], and $CF_{0.22}$ by the 220°C fluorination also release fluorocarbon as major gaseous species during the thermolysis in the temperature range of 50–500°C [12]. Taking into consideration these thermal behaviors of the fluorinated carbon materials, the endothermic fluorine release of the F-SWNHs without fluorocarbon impurities and etching of carbon framework at lower temperature suggests the weaker interaction between the fluorine and carbon atoms.

Based on a series of solid-state ^{19}F NMR studies of the C–F bonding in fluorine–carbon materials including CNTs and fullerene C_{60}, it has been revealed that the ^{19}F NMR chemical shift δ_F is very sensitive to the interaction between the fluorine and carbon atoms, and the chemical shift becomes higher as the C–F covalence becomes weaker [3,13,25]. Fig. 5.14 shows the MAS ^{19}F NMR spectra of the same samples used in the TG-DTA study depicted in Fig. 5.12 [5], the 50% defluorinated $CF_{0.10}$ (F-as-grown SWNH) and a comparison to the material graphite fluoride $(CF)_n$. No band ascribed to molecular fluorine is observed at $\delta_F = +423$ ppm/$CFCl_3$, and due to

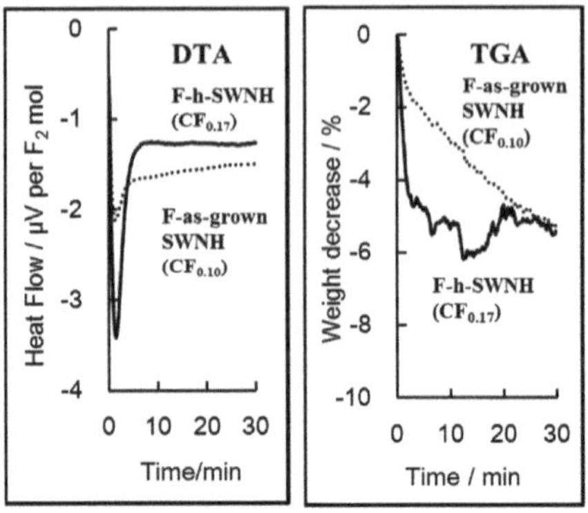

Figure 5.13 Expanded views of the TGA and DTA curves under a N_2 -flow of 100 cm^3/min, before the heating in Fig. 5.12. *SWNH*, single-wall carbon nanohorn; *F-as-grown SWNH*, fluorine adsorbed as-grown SWNH; *F-h-SWNH*, fluorine adsorbed hole-opened SWNH.

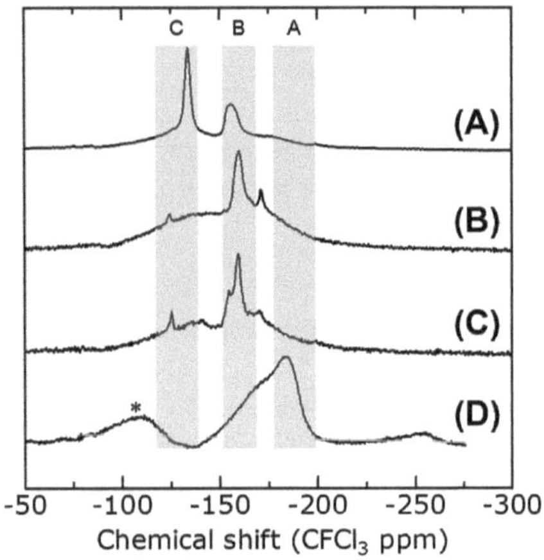

Figure 5.14 ^{19}F MAS NMR spectra of fluorine adsorbed as-grown SWNH $CF_{0.10}$ (A), fluorine adsorbed hole-opened SWNH $CF_{0.17}$ (B), 50% defluorinated $CF_{0.10}$ (C), and comparison material graphite fluoride $(CF)_n$ (D), at the spinning speed of 25–26 kHz. *, spinning side band. *SWNH*, single-wall carbon nanohorn.

Reproduced with permission from Touhara et al. [5]. Copyright 2012 The Society of Fluorine Chemistry, Japan.

the highly versatile C—F bonding nature in the fluorinated carbon materials, three bands, A, B, and C, are observed in the spectra. The line centered at $\delta_F = -183.7$ ppm in band A observed for $(CF)_n$ is well known to be assigned to the covalent C—F bond with sp^3 carbon hybridization [1,3,25]. The spectra of the two samples, $CF_{0.10}$ (F-as-grown SWNH) and $CF_{0.17}$ (F-h-SWNH), show no line in band A, whereas lines appear at the lower chemical shifts of $-153.9 \sim -158.6$ ppm in band B and $-133.6 \sim -122.4$ ppm in band C. Consequently, the spectra indicate that two weaker bonding modes in the fluorine adsorption are simultaneously present in the F-SWNHs. The δ_F values of band B are appreciably higher than those of the fluorinated SWNTs CF_x ($\delta_F = -163 \sim -180$ ppm for $x = 0.32-0.61$) but a little lower than those of $C_{60}F_x$ ($\delta_F = -151$ ppm for $x = 45 \sim 48$) [25]. These results mean that the C–F bond strength in the F-SWNHs is weaker than that in the F-SWNTs whose C–F bond has been understood as a weakened covalent bond with a longer bond length compared to the covalent C–F bond of the graphite fluoride. The weakened covalent bond would be defined as the C($sp^{2+\delta}$)-F ($0 < \delta < 1$) bond with a bond order less than one and the C–F covalent as C(sp^3)-F with a bond order of unity [1]. The weakened bond, the so-called "semi-ionic" C–F bonding in the stage-1 fluorine-graphite intercalation compounds of C_xF ($x = 2.47-3.61$) has been described as a substantially covalent bond due to the hyperconjugation involving C–C bonds on the planar carbon sheets and C–F bonds [26]. This interpretation is not necessarily applicable to the F-SWNHs because of their highly distorted carbon framework with cylindrical and conical curvatures. The more weakened C–F bond in the F-SWNHs than the F-SWNTs will be discussed in detail in the next paragraph. More noticeable is the higher δ_F values of band C, which is shifted toward the lower field by 20–30 ppm from band B, and by 50–60 ppm from the line of the C–F covalent bond. Such lower δ_F values have so far been reported for only two cases; one is $CF_{0.4}$ ($\delta_F = -120$ ppm) prepared by the room temperature fluorination of the open-ended CVD-SWNT with an average diameter of 2 nm [13], and the other is $CF_{0.18-0.42}$ ($\delta_F = -128$ ppm) prepared by fluorination at the lower temperature range 25 ~ −191°C of the purified Helix-tubes [3]. These significantly high chemical shifts have been assigned to the unusual weak bonding mode of fluorine to the SWNTs [3,13]. Importantly, the fluorinated SWNTs were IR-silent. The exact nature of the C–F interaction and fluorine location have not yet been clarified, but presumed to be ascribed to a metastable interaction like physisorption at the internal surface of the tube [3,13]. The endo-fluorine addition was also proposed based on TGA and IR analyses of the F-SWNTs [27,28] and predicted by theoretical studies [29,30]. What are the most interesting results in the ^{19}F NMR spectra of the F-SWNHs are that band C is observed in both the as-grown and hole-opened samples, indicating that fluorine adsorption with the metastable interaction does occur at the outer surfaces of the SWNHs. Very interestingly, these fluorine adsorbed as-grown and hole-opened SWNHs showed no band in the IR region except the band at 1080–1120 cm^{-1} which corresponds to the weakened covalent bond. It is worthy to recall here that, in the case of the SWNTs, band C is reported only for open-ended tubes. The intensity of band C significantly decreases after the 50% defluorination (Fig. 5.14C).

5.3.4 Distinguishing Features of Single-Wall Carbon Nanohorns Associated With the Fluorine Storage Properties

The fluorine storage properties of the SWNHs are attributed to their structural and electronic properties which are considerably different from those of the SWNTs, and are characterized by the following:

1. SWNHs possess two different bending modes of the carbon frameworks in the individual nanohorns, one being cylindrical curvature and the other a conical one. The cylindrical curvature of the SWNHs is common to the SWNTs, whereas the conical one is SWNHs

inherent to the bending mode (Fig. 5.1), and as a consequence, the SWNHs become thinner toward the tip accompanied by an increase in the curvature. The curvature allows the sp^2 σ-bonds at the carbon atom to deviate from planarity; the hybridization of the 2s and 2p orbitals is not pure sp^2 and, hence, the π-orbitals are no longer of pure p-orbital character, but include a finite 2s component. Therefore, SWNTs and SWNHs have an intermediate hybridization between sp^2 and sp^3, that is $sp^{2+\delta}$, δ being $0 < \delta < 1$. Such a rehybridization is characterized by the pyramidalization angle θ_p ($\theta_{\sigma\pi} - 90°$) which becomes larger from 0° for the planar sp^2 carbon orbital to 19.5° for the pure sp^3. The θ_p of the SWNTs and SWNHs with a cylindrical curvature is inversely proportional to their diameters [29], the upper limit θ_p being considered to be equal to 11.6° of C_{60} with the spherical curvature. The rehybridization results from the curvature induces strain in the carbon framework [30], which leads to the increase in the total energy of the SWNTs and SWNHs, and consequently, their carbon atoms are in an energetically higher state [31]. Indeed, we have found room temperature fluorine chemisorption by the reaction of high purity open-end and closed-end SWNTs (prepared by a laser ablation method, tube diameter 1.4–1.5 nm) with elemental fluorine, the stoichiometry being $CF_{0.23}$ and $F_{0.28}$, respectively [32]. As already mentioned, the lower temperature fluorine adsorption at 25 ~ −191°C was also recently reported for the open-ended SWNTs with the diameter of 2 nm. In the case of the SWNHs, in addition to the cylindrical curvature of the tube with ~2 nm diameter, each nanohorn includes the conical curvature with a gradient where the tube diameter becomes smaller toward the tube tip. In other words, the horn-shaped carbon framework consists of consecutive sets of differently pyramidalized carbon atoms. Such a peculiar form of the nanohorns induces further stress in the carbon framework, and as a consequence, the reactivity of the SWNHs would be considerably higher than the SWNTs. It may be noted that $E_s \propto D^{-2}$, where E_s is the strain energy per carbon atom, and D is the tube diameter, which is computationally shown [30].

2. The carbon framework of the SWNHs includes five carbon pentagons which produce the conical shape situated at the conical tip. The carbon pentagons at the bending sites or kinks (Fig. 5.1) are topological defects with the strain energy and are highly reactive compared with the other carbon-hexagon sites [33].

3. As-grown SWNHs are active in the electron spin resonance (ESR), their spectra revealing localized spins (spin density $N_s = 6 \times 10^{18}$ spins/g) which seems to be in the vicinity of the pentagons [34,35]. It should be noted here that purified SWNTs are inactive in the ESR [35]. We have investigated changes in the ESR spectra of the SWNHs by fluorine adsorption in the temperature range of 0–200°C, and Fig. 5.15 show the stoichiometry F/C dependence of the localized spin density N_s and line width ΔH_{pp} [36]. Up to the stoichiometry of $CF_{0.27}$ for the as-grown SWNHs, N_s monotonically decreases to 4.7×10^{16} spins/g as the fluorine adsorption proceeds, and then slightly increases to 9.1×10^{16} spins/g with the increasing fluorine concentration up to $CF_{0.43}$. The significant decrease in N_s of up to c.99% of that of pristine by fluorine adsorption indicates the fairly mild reaction of fluorine and the SWNHs without dangling bond formation. Hole-opened SWNHs are also active in the ESR, the N_s value of 1.8×10^{18} spins/g being comparable to that of the as-grown ones, while changes in the spectra due to the fluorine adsorption are appreciably different. In particular, the reduction of N_s upon fluorine adsorption up to $CF_{0.35}$ remains at c.90% which would be attributable to the fluorine adsorption at the internal surfaces with a very weak Van der Waals type interaction.

The fluorine storage properties of SWNHs are provided by the following two fluorine adsorption modes. One is the chemisorption with a weakened covalent

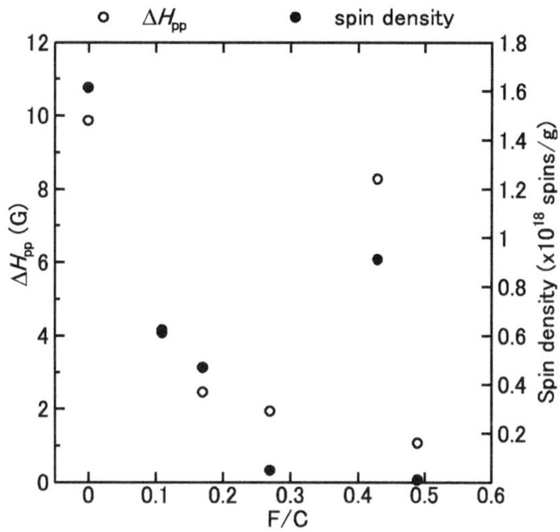

Figure 5.15 ESR line width and spin density of fluorine adsorbed as-grown SWNH (prepared by fluorination at 0–200°C) as a function of F/C. *SWNH*, single-wall carbon nanohorn.

type bond defined as the C(sp$^{2+\delta}$)-F bond with a bond order less than one. This kind of fluorine adsorption occurs at the external surfaces with a cylindrical curvature of the SWNHs, and the topological defects with localized spins. The other is the physisorption with a metastable Vav der Waals like interaction of the fluorine atom and carbon atoms which was also predicted by the theoretical study [37]. This adsorption mode is presumed to occur at the external surfaces of the horn-shaped carbon framework with the curvature having a gradient. These two adsorption modes are common to both the as-grown SWNHs and h-SWNHs. As for the h-SWNHs, physisorption with the Van der Waals interaction possibly arises from a metastable endo-location of the fluorine addends, which was discussed in the earlier studies for the case of the open-ended SWNTs [3,13]. However, the concave inner surfaces at the conical shape of the h-SWNHs would be strained more highly than that of the cylindrical curvature of the SWNTs. This is possibly more favorable for the endo-adsorption with a metastable Vav der Waals like interaction of the fluorine and carbon atoms.

As a concluding remark, adsorbed fluorine on the SWNHs by these interaction modes can be removed under an inert gas flow and/or dynamic pumping up to a 400°C heating, which suggests that the SWNHs have a high potential use as fluorine storage materials.

Acknowledgment

This work was partly supported by NEDO (New Energy and Industrial Technology Department Organization) Project Code P0523.

References

[1] H. Touhara, F. Okino, Carbon 38 (2000) 241–267.
[2] E.T. Mickelson, C.B. Huffman, A.G. Rinzler, R.E. Smalley, R.H. Hauge, J.L. Margrave, Chem. Phys. Lett. 296 (1998) 188–194.
[3] F. Chamssedine, K. Guerin, M. Dubois, E. Disa, E. Petit, Z. El Fawl, A. Hamwi, J. Fluorine Chem. 132 (2011) 1072–1078.
[4] H. Touhara, K. Ymamoto, S. Yanagisawa, Y. Hattori, F. Okino, T. Yamaguchi, S. Iijima, S. Bando, T. Isoga, Y. Etou, M. Itano, T. Wada, T. Satou, in: Abstracts of Papers of Am. Chem. Soc., vol. 242, August 2011. 36-FLUO, Denver.
[5] H. Touhara, S. Yanagisawa, Y. Hattori, K. Gotoh, K. Takeda, Y. Etou, T. Isogai, M. Itano, Y. Yamaguchi, S. Iijima, in: 20th Int. Nat. Sym. Fluorine Chem. Abstracts, 2012, p. 179. Kyoto, Japan.
[6] Y. Hattori, S. Yanagisawa, Y. Nishi, Y. Etou, T. Isogai, M. Itano, Y. Yamaguchi, S. Iijima, H. Touhara, ibid, p.180.
[7] W. Zhao, C. Song, B. Zheng, J. Liu, T. Viswanathan, J. Phys. Chem. B 106 (2002) 293–296.
[8] J. Kang, D. Dharmpal, O.V. Kuznetsov, V.N. Khabashesku, K.F. Kelly, Chem. Phys. Lett. 534 (2012) 43–47.
[9] P.E. Pehrsson, W. Zhao, J.W. Baldwin, C.H. Song, J. Liu, S. Kooi, B. Zheng, J. Phys. Chem. B 107 (2003) 5690.
[10] Z. Gu, R.H. Hauge, R.E. Smally, J.L. Margrave, Nano Lett. 2 (2002) 1009–1013.
[11] H.F. Bettinger, H. Peng, J. Phys. Chem. B 109 (2005) 23218–23224.
[12] Y.M. Shulga, V.M. Martynenko, A.V. Krestinin, A.P. Kharitonov, G.I. Davidova, E.I. Knerelman, V.I. Krastev, D.V. Schur, Int. J. Hydrogen Energy 36 (2011) 1349–1354.
[13] D. Claves, H. Li, M. Dubois, Y. Ksari, Carbon 47 (2009) 2557–2562.
[14] S. Iijima, M. Yudasaka, R. Yamada, S. Bandow, K. Suenaga, F. Kokai, K. Takahashi, Chem. Phys. Lett. 309 (1999) 165–170.
[15] M. Zhang, M. Yudasaka, K. Ajima, J. Miyawaki, S. Iijima, ACS Nano 1 (2007) 265–272.
[16] S. Bandow, A.M. Rao, G.U. Sumanasekera, P.C. Eklund, F. Kokai, K. Takahashi, M. Yudasaka, S. Iijima, J. Appl. Phys. A 71 (2000) 561–564.
[17] S. Utumi, J. Miyawaki, H. Tanaka, Y. Hattori, T. Itoi, N. Ichikuni, H. Kanoh, M. Yudasaka, S. Iijima, K. Kaneko, J. Phys. Chem. B 109 (2005) 14319–14324.
[18] K. Urita, S. Seki, S. Utumi, D. Noguchi, H. Kanoh, H. Tanaka, Y. Hattori, Y. Ochiai, N. Aoki, M. Yudasaka, S. Iijima, K. Kaneko, Nano. Lett. 6 (2006) 1325–1328.
[19] F. Cuadros, I. Cohadina, W. Ahamada, Mol. Eng. 6 (1996) 319–325.
[20] Y. Hattori, H. Kanoh, F. Okino, H. Touhara, D. Kasuya, M. Yudasaka, S. Iijima, K. Kaneko, J. Phys. Chem. B 108 (2004) 9614–9618.
[21] G. Li, K. Kaneko, S. Ozeki, F. kino, R. Ishikawa, M. Kanda, H. Touhara, Langmuir 11 (1995) 716–717.
[22] G. Li, K. Kaneko, F. Okino, R. Ishikawa, M. Kanda, H. Touhara, J. Colloid Interface Sci. 172 (1995) 539–540.
[23] T. Isogai, Y. Etou, M. Itano, Unpublished Results.
[24] N. Watanabe, T. Nakajima, H. Touhara, Graphite Fluorides, Elsevier, 1988.
[25] W. Zhang, M. Dubois, K. Guerin, P. Bonnet, H. Kharbache, F. Masin, A.P. Kharitonov, A. Hamwi, Phys. Chem. Chem. Phys. 12 (2010) 1388–1398.
[26] Y. Sato, K. Itoh, R. Hagiwara, T. Fukunaga, Y. Ito, Carbon 42 (2004) 3243–3249.
[27] F. Chamssedine, D. Claves, Chem. Phys. Lett. 443 (2007) 102–106.

[28] D. Claves, J. Rossignol, Chem. Phys. Lett. 468 (2009) 231–233.
[29] Z. Chen, W. Thiel, A. Hirsch, Chem. Phys. Chem. 1 (2003) 93–97.
[30] H.F. Bettinger, K.N. Kudin, G.E. Scuseria, J. Am. Chem. Soc. 123 (2001) 12849–12856.
[31] K. Kanamitsu, S. Saito, J. Phys. Soc. Jpn. 71 (2002) 483–486.
[32] S. Kawasaki, K. Komatsu, F. Okino, H. Touhara, H. Kataura, Phys. Chem. Chem. Phys. 6 (2004) 1769–1772.
[33] H.F. Bettinger, J. Phys. Chem. 109 (2005) 6922–6924.
[34] S. Garaji, L. Thien-Nga, R. Gaal, L. Forro, K. Takahashi, F. Kokai, M. Yudasaka, S. Iijima, Phys. Rev. B 62 (2001) 17115–17119.
[35] S. Bandow, F. Kokai, K. Takahashi, S. Iijima, Appl. Phys. A 73 (2001) 281–285.
[36] Y. Hattori, Y. Nishii, H. Touhara, Unpublished Results.
[37] N.G. Levedev, I.V. Zaporotskova, L.A. Chernozatonskii, Int. J. Quantum Chem. 96 (2004) 142–148.

Synthesis and Characterization of Fluorinated Carbon Fibers and Nanotubes

6

Y.-S. Lee
Chungnam National University, Daejon, Korea

Chapter Outline

6.1 Introduction

Many studies are performed with various surface modifications to improve the properties of carbon materials [1,2]. Possible surface modifications of carbon materials include anodization [3–5], plasma treatment [6–8], and solution- [9] and gas-phase oxidation [10].

In the past 5 years, the surface fluorination of carbon materials has attracted considerable interest because the fluorination significantly changes the chemical and electrical properties of the surface of carbon materials.

The diversity of carbon materials must also be considered to understand the wide range of intercalation between C and F. The most important feature of a C–F system is that there are two types of bonding natures: ionic and covalent. Experimental data demonstrate that graphite intercalated with F (CF_x, $x < 0.2$), establishing C–F ionic bonds, possesses extremely high electrical conductivity (close to that observed for

New Fluorinated Carbons: Fundamentals and Applications. http://dx.doi.org/10.1016/B978-0-12-803479-8.00006-1

metals, eg, 2×10^5 S/cm) [11,12]. For high F concentrations, the transport properties of CF_x compounds decrease due to the formation of covalent bonds between the C and F atoms. At the same time, the conductivity of the material decreases and other physicochemical properties also change. Therefore, the nature of the C—F bond, as well as the physical and chemical properties, must be considered to broaden the application of fluorinated carbon materials.

Many previous studies have investigated the synthesis of fluorinated carbon materials and the changes in their electrical properties. The present chapter reviews the recent results on the synthesis and electrical behaviors of fluorinated carbon nanotubes (CNTs) and fibers as well as their sensing properties based on electrical properties.

6.2 Synthesis of Fluorinated Carbon Materials

6.2.1 Direct Fluorination

Direct fluorination is one of the modification methods performed on the surfaces of various materials. The fluorination reaction can use a variety of gases, including F species and diluted mixtures of fluorine and nitrogen, argon, or helium are used to control the introduction of functional groups onto the carbon surface. The fluorine and fluorine-containing gases have strong reactivity and can easily react with the carbon surface at room temperature without a catalyst [13]. Initially, direct fluorination was mainly used to modify the surfaces of polymers [14], but it has also been applied to the surface treatment of various carbon materials [15—18].

The chemical composition and structure of fluorinated carbon materials are dependent on the structure and characteristics of the starting material and on the fluorination conditions, such as temperature, reaction time, pressure, and composition of the gaseous reactants [19,20].

6.2.1.1 Direct Fluorination of Carbon Nanotubes

Fluorinated CNTs have been produced by direct fluorination with gaseous F_2 or vaporized BrF_3. The fluorination of CNTs with gaseous F_2 is performed at elevated temperatures [21], and the fluorination of CNTs with BrF_3 vapors is performed at room temperature [22].

The surface compositions (F/C) of fluorinated single-walled carbon nanotubes (SWCNTs) determined by XPS spectra are shown in Fig. 6.1. The F/C value increases with increasing temperature, approaching 0.5 at 250°C [23] and finally reaching 0.65 at 300°C. Since the CF_x composition with $x > 0.5$ is prohibited in sidewall fluorination because of electrostatic repulsion between fluorine atoms, we assume that this result indicates a new phase formation (Fig. 6.2) [23].

Consequently, the amount of fluorine introduced increases with increasing reaction temperature, and the fluorine atoms may be covalently attached to the sidewalls of the SWCNTs. If some part of the SWCNTs is destroyed by fluorination at high temperature, the insides of the nanotubes might be able to be fluorinated. However, this

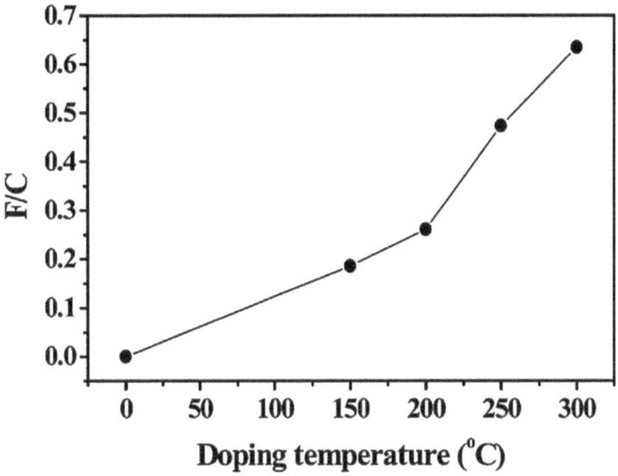

Figure 6.1 The surface compositions (F/C) of fluorinated single-walled carbon nanotubes determined by X-ray photoelectron spectroscopy spectra as a function of fluorination temperature [23].

hypothesis cannot be confirmed because it is not easy to verify whether fluorine is doped on the inside of the SWCNTs or not. At temperatures below 200°C, fluorinated SWCNTs show an ionic bonding character (semi-ionic), which implies sidewall functionalization. At high temperatures, the binding energy shifts to the higher binding value, revealing a covalent bonding character of $(CF)_n$, which implies that the resistivity increases by enhancing the sp^3 bonding and disintegration of the nanotube walls [24].

The fluorination temperature of multi-walled carbon nanotubes (MWCNTs) is significantly increased compared to that of SWCNTs because of the presence of graphitic multilayers surrounding the nanotube core. This temperature depends on the number of MWCNT shells and is generally close to 400°C [11,25]. In the fluorination of MWCNTs, fluorine is intercalated into the outer part of the CNTs, where

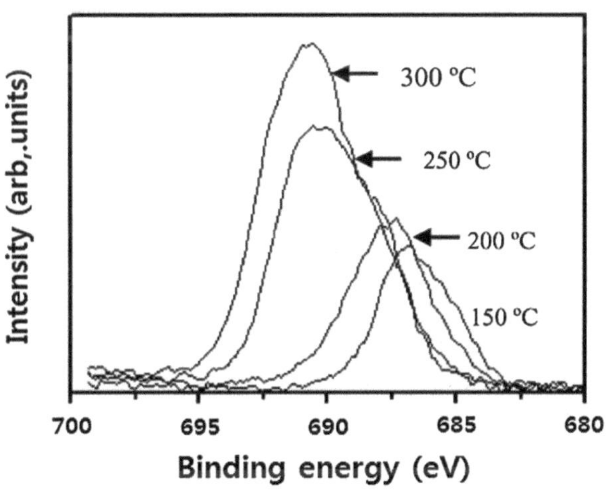

Figure 6.2 X-ray photoelectron spectroscopy F1s spectra of the fluorinated single-walled carbon nanotubes at various fluorination temperatures [23].

graphene layers are planar and the repeat distance is approximately 0.60 nm. In contrast, the inner part of the CNTs is not intercalated. However, the graphitic structure is not maintained in the fluorinated CNTs at 430°C [26].

The stoichiometry of fluorinated CNTs and the direct fluorination conditions are shown in Table 6.1. According to studies on annealing fluorinated carbon materials [27], the largest fluorine loss is known to occur between 200°C and 300°C, and defluorination is virtually complete at 400°C. Therefore, the largest x value of CF_x appears at a fluorination reaction temperature of approximately 300°C. Chamssedine

Table 6.1 Stoichiometry of Fluorinated Carbon Nanotube According to Direct Fluorination Condition

CNT	Reaction Gas	Reaction Time	Reaction Temperature (°C)	CF_x	Analytic Method	References
SWCNT	F_2 and He	5 h	150	$CF_{0.114}$	TGA-DTA	[21]
			250	$CF_{0.521}$		
			325	$CF_{0.495}$		
			400	$CF_{0.565}$		
SWCNT	F_2 and He	19 h	200	$CF_{0.256}$	XPS	[29]
		9 h	220	$CF_{0.435}$		
		5 h	260	$CF_{0.526}$		
SWCNT	F_2 and He	–	50	$CF_{0.2}$		[30]
SWCNT	F_2	3 h	280	$CF_{0.27}$	TGA	[28]
			25	$CF_{0.42}$		
			−100	$CF_{0.28}$		
			−178	$CF_{0.23}$		
			−191	$CF_{0.18}$		
SWCNT	F_2 and N_2	10 h	250	$CF_{0.43}$	XPS	[27]
MWCNT	F_2	2 h	Room temperature	$CF_{0.22}$	–	[25]
MWCNT	F_2 and N_2	10 min	Room temperature	$CF_{0.053}$	XPS	[31]
			100	$CF_{0.08}$		
			200	$CF_{0.152}$		
			400	$CF_{0.02}$		

CNT, carbon nanotubes; *MWCNT*, multi-walled carbon nanotubes; *SWCNT*, single-walled carbon nanotubes; *TGA*, thermogravimetric analysis; *TGA-DTA*, thermogravimetric analysis-differential thermal analysis; *XPS*, X-ray photoelectron spectroscopy.

et al. [28] reported that at a fluorination temperature below 25°C, the x value of the CF_x of CNTs can rise to the maximum of 0.5 without damaging the tubes (I_D/I_G is quasi-equal to that for the original SWCNTs).

6.2.1.2 Direct Fluorination of Carbon Fibers

In the case of carbon fibers, direct fluorination has been used to improve their surface wettability. However, oxyfluorination using fluorine−oxygen mixtures has proven to be an attractive method for increasing the hydrophilicity of carbon fiber surfaces [32−35]. The stoichiometry of fluorinated carbon fibers and the direct fluorination conditions are shown in Table 6.2. The degree of fluorination depends on the nature of the fibers and the influence of various physical properties, such as the layer fluorinating method used.

A higher content of intercalated fluorine can be obtained using either a better-graphitized structure or a more extensive fluorine treatment at higher temperatures and pressures [36]. The reaction of fluorine and graphite between c.100°C and 300°C is kinetically hindered by the formation of covalent C−F bonds at the outer surface, which prevents further migration of fluorine into the interplanar space and thus further fluorination [37].

Fischer et al. reported that polyacrylonitrile (PAN) or mesophase pitch-based carbon fibers were prepared in a nitrogen−fluorine gas mixture stream at 406°C and normal pressure. The partial pressure of fluorine was held constant at approximately 80 torr. CF_x compositions with x ranging from 0.25 to 1.1 were obtained by varying the reaction time from 3 to 24 h [38]. According to Ho et al. [39], the two reactions are the desired reaction leading to the formation of solid fluorinated carbon and a combustion reaction leading to the formation of gaseous fluorides, such as CF_4 and C_2F_6, ie, the removal of carbon from the fibers.

The carbon fibers could be fluorinated with an F_2-HF mixture or ClF_3 [40,41]. The surface of the PAN-based carbon fibers treated with an F_2-HF mixture shows the highest degree of fluorination at approximately 60% and a C/F ratio in the range of 1.5−1.8. According to the results of Bismarck et al., the amount of fluorine on the surface after fluorination with only F_2 gas was highly similar to that for fluorination with ClF_3. This result indicates that fluorination with ClF_3 can penetrate deeper portions of the fiber, not just the surface [41].

Park et al. reported the fluorination and oxyfluorination of PAN-based carbon fibers [42,43]. The carbon fibers treated with F_2/O_2 mixtures showed a higher fluorine content value of approximately 60% and a C/F ratio of 1.5%. Lee et al. [18] reported that both the fluorine/carbon and oxygen/carbon ratios increased as the oxyfluorination time increased at a total pressure of 5 kPa. However, the fluorine/carbon ratios decreased as the total pressure increased. This result indicates that fluorine and oxygen are easily introduced into the sample even at a low pressure in a short time period at room temperature.

6.2.2 Plasma Fluorination

Plasma treatment in a gas, including fluorine, can easily change the graphite surface wettability, producing significantly hydrophobic surfaces of fluorinated graphite,

Table 6.2 Stoichiometry of Fluorinated Carbon Fiber According to Direct Fluorination Condition

Carbon Fiber	Reaction Gas	Reaction Time	Reaction Temperature (°C)	CF_x	Analytic Method	References
PAN and mesophase pitch-based carbon fiber	F_2 and N_2	3–24 h	406	$CF_{0.25-1.1}$	Gravimetrically	[38]
PAN-based carbon fiber	F_2	First step: 1 h	First step: 25	$CF_{0.1}$	XPS	[39]
		Second step: 2 h	Second step: 100			
	F_2–HF (9:1)	24 h	200	$CF_{0.16}$		
	F_2–N_2 (1:9)		380	$CF_{1.82}$		
				$CF_{1.54}$		
				$CF_{1.75}$		
				$CF_{1.6}$		
PAN-based carbon fiber	F_2–HF	1 h	Room temperature	$CF_{0.013}$	XPS	[41]
		5 h		$CF_{0.043}$		
		9 days		$CF_{0.1}$		
	ClF_3	60 h	25	$CF_{0.55}$		
PAN-based carbon fiber	F_2 and O_2 (1:1) 5 kPa	30 min	25	$CF_{0.275}$	XPS	[18]
		60 min	25	$CF_{0.316}$		
	F_2 and O_2 (1:1) 80 kPa	30 min	25	$CF_{0.2}$		

PAN, polyacrylonitrile; *XPS*, X-ray photoelectron spectroscopy.

$(CF)_n$, $(C_2F)_n$ [44,45]. The interaction of plasmas generated by fluorinated monomers, such as CF_4, C_2F_6, and SF_6, with polymers gives rise to the fluorination of the treated surface.

There are two types of plasma fluorination: low-pressure plasma treatment and atmospheric plasma fluorination. Low-pressure plasma fluorination has been used industrially for the surface modification and functionalization of various types of surfaces [46–50]. A disadvantage of low-pressure plasma treatment is that it requires a well-designed system that includes vacuum pumps as well as sealed and contained handling equipment. Therefore, low-pressure plasma treatments are often only used in batch processes given that the size of the reactor chamber is limited due to capital and operational costs [51]. Thus, some researchers have developed an atmospheric plasma fluorination procedure [51,52], which allows for the introduction of small amounts of fluorine-containing functionalities onto the surfaces of carbon fibers. Atmospheric plasma instruments generate and sustain a stable plasma at atmospheric pressure in such a way that permits continuous surface treatment [53,54].

6.2.2.1 Plasma Fluorination of Carbon Nanotubes

In general, the degree of fluorination appears to increase with plasma duration and voltage. The stoichiometry of plasma-fluorinated CNTs and the plasma fluorination conditions are shown in Table 6.3. The semi-ionic and covalent C–F bonds were introduced in the CF_4 plasma-exposed CNTs. The presence of covalently bonded C–F increases as the CF_4 gas flow rate increases during CF_4 plasma exposure and/or as the exposure time increases [55].

However, in the results by Plank et al., the plasma-treated SWCNT for 60 s shows a clear decrease in F/C relative to the plasma-treated SWCNT for 40 s [57]. The plasma functionalization process is complex, and the mechanism for fluorine attachment to nanotubes has been studied [55,60]. The low F/C results at plasma-treated SWCNT for 60 s may be attributed to the nonlinearity of the plasma process, an effect also observed by others using CF_4 plasma to functionalize SWCNTs [57]. As discussed previously, the plasma process exposes the SWCNTs to ion bombardment [55,60]. The flux of ions reaching the SWCNT sample surface should not vary considerably over time after equilibrating during the first few seconds of bias voltage application. Therefore, functionalization via fluorination requires a trade-off between the creation of defects and the production of sufficient concentrations of radicals capable of bonding to sites on the SWCNTs [57].

In the case of CF_4 plasma treatment of MWCNTs, the atomic concentration of fluorine can reach up to 18.7% when fluorine is plasma treated for 90 s. After 90 s treatment, the fluorine atomic percentage tends to saturate [58].

Hong et al. reported that super-hydrophobic CNT powders are formed by NF_3 glow-discharge plasma at low pressure [56]. Furthermore, SF_6 could be used as the plasma source to introduce fluorine functional groups onto the CNT surface [55,61]. The SWCNTs treated by SF_6 plasma were more fluorinated compared to those treated by CF_4 plasma [59].

Table 6.3 Stoichiometry of Fluorinated Carbon Nanotube According to Plasma Fluorination Condition

CNT	Method	Plasma Source	Reaction Time (s)	Reaction Temperature (°C)	CF_x	Analytic Method	References
SWCNT	Low-pressure plasma	CF_4	–	Room temperature	$CF_{0.14}$	XPS	[56]
SWCNT	Low-pressure plasma	CF_4	10	Room temperature	$CF_{0.30}$	XPS	[57]
			20		$CF_{0.51}$		
			40		$CF_{0.63}$		
			60		$CF_{0.55}$		
MWCNT	Low-pressure plasma	CF_4	30	Room temperature	$CF_{0.07}$	XPS	[58]
			60		$CF_{0.16}$		
			90		$CF_{0.23}$		
			120		$CF_{0.24}$		
SWCNT	Low-pressure plasma	CF_4 or SF_6	100	Room temperature	–	–	[59]

CNT, carbon nanotubes; *MWCNT*, multi-walled carbon nanotubes; *SWCNT*, single-walled carbon nanotubes; *XPS*, X-ray photoelectron spectroscopy.

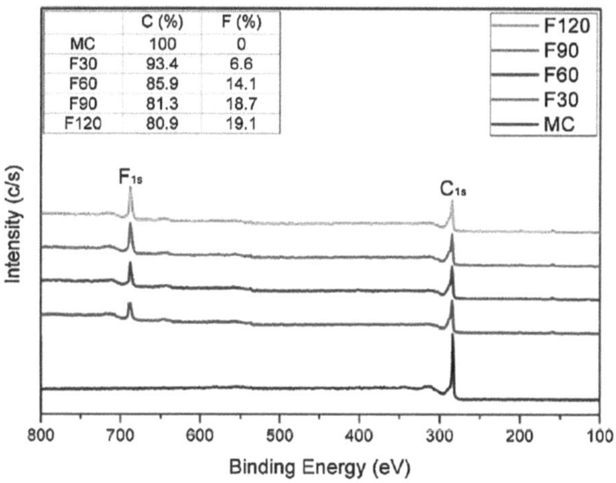

Figure 6.3 X-ray photoelectron spectroscopy survey spectra of multi-walled carbon nanotubes treated by fluorine plasma for 0, 30, 60, 90, and 120 s [58].

6.2.2.2 Plasma Fluorination of Carbon Fibers

Atmospheric plasma treatment has been used to continuously modify carbon fibers. SF_6 is an excellent fluorinating agent for low-pressure plasma fluorination because of its high fluorine content [62]. It was impossible to create a stable atmospheric plasma jet using only SF_6 (or any other heavy fluorine-containing gas species); therefore, it was necessary to dilute the fluorine-containing gas. Suitable carrier gases needed to be identified are nitrogen, hydrogen, helium, and argon [63], which are already investigated. In atmospheric plasma, the plasma jet was generated by ionizing feed gas mixtures of SF_6 or chlorodifluoromethane ($CHClF_2$) diluted by inert gases [51,64,65]. The surface fluorine content of continuous atmospheric plasma fluorination—treated carbon fibers was found to depend on the residual time of the fibers in the plasma environment, as shown in Fig. 6.3. Furthermore, continuous atmospheric plasma fluorination was more effective when exposing the fibers repeatedly to the plasma jet, ie, a higher degree of surface fluorination could be achieved [52]. The stoichiometry of plasma-fluorinated CNTs and the plasma fluorination conditions are shown in Table 6.4.

6.3 Electrical Characteristics of Fluorinated Carbon Materials

6.3.1 Electrical Characteristics of Fluorinated Carbon Fibers

In fluorinated carbon fibers, the electrical conductivity changes according to C—F bonding, not just the degree of fluorination. Conductivity changes were induced to change the π-electron system of the graphitic portions of the carbon fibers. Such changes in the π-electron system may be attributed to intercalation of fluorine. The formation of graphite fluoride resulted in a significant decrease in the conductivity [66].

Table 6.4 Stoichiometry of Fluorinated Carbon Fiber According to Plasma Fluorination Condition

Carbon Fiber	Method	Plasma Source	Reaction Time (min)	Reaction Temperature (°C)	F (at. %)	Analytic Method	References
PAN-based carbon fiber	Atmospheric plasma	$N_2 + CHClF_2$	1–8	Room temperature	1.1–1.5	XPS	[64]
PAN-based carbon fiber	Atmospheric plasma	$N_2 + CHClF_2$	0.8–4	Room temperature	2.9–4.9	XPS	[52]

PAN, polyacrylonitrile; *XPS*, X-ray photoelectron spectroscopy.

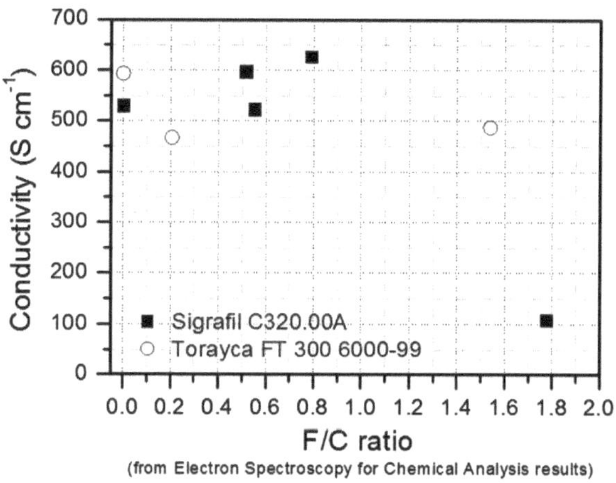

Figure 6.4 The conductivity changes of fluorinated PAN-based carbon fibers with various fluorine contents. *ESCA*, Electron spectroscopy for chemical analysis; *PAN*, polyacrylonitrile.

Bismarck et al. [41] reported conductivity changes in fluorinated carbon fibers with various fluorination methods. As shown in Fig. 6.4, slight fluorination increases the conductivity, whereas higher degrees of fluorination are associated with a lower conductivity. Strong fluorination conditions (using ClF or HF-F_2) lead to fluorination of not only the surface but also the bulk phase of the carbon fibers. In this case, the sp^2 state of the carbon layers generating the π-electron system is changed to an sp^3 state (fluorocarbon-like graphite fluoride), and the conductivity decreases considerably. Moreover, even if the same fluorination is performed, maintenance of the sp^2 structure can vary with respect to the type of carbon fibers, and changes in the conductivity can also be observed.

Ho et al. [67] reported that changes in the conductivity of fluorinated carbon fibers took on different aspects depending on the type of carbon fibers: PAN-based standard modulus T300 or high-tensile strength directly fluorinated T700. The conductivity of fluorinated T300 fibers decreased by up to 8%, whereas that of fluorinated T700 fibers increased, as shown in Fig. 6.5. The increase in electrical conductivity observed for the

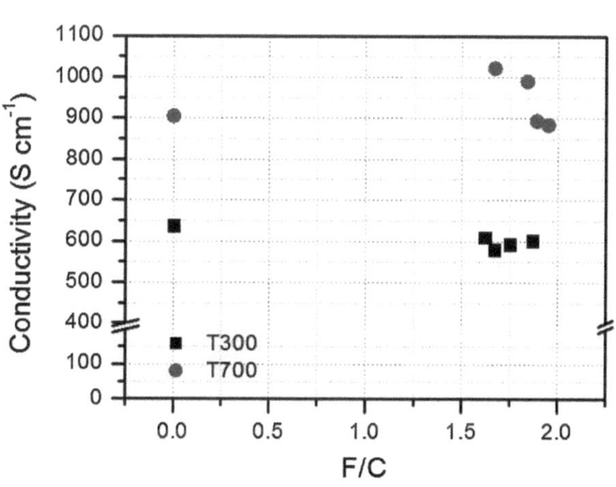

Figure 6.5 The conductivity of fluorinated PAN-based carbon fibers took on a different aspect according to kind of carbon fibers. *PAN*, polyacrylonitrile.

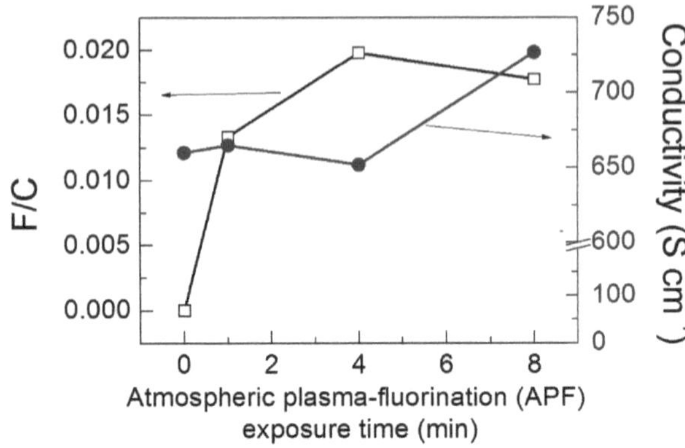

Figure 6.6 The F/C ratio and conductivity of fluorinated polyacrylonitrile (PAN)-based carbon fibers by atmospheric plasma fluorination.

fluorinated T700 fibers could be attributed to the intercalation of fluoride species into the carbon fibers [68]. The decrease in the conductivity of fluorinated T300 fibers indicates the formation of covalent C—F bonds, which corresponds with the results obtained by Tressaud et al. [69,70].

In the case of fluorinated PAN-based carbon fibers, which were treated by atmospheric plasma fluorination using ionized $CHClF_2$ diluted by N_2, atmospheric plasma fluorination only marginally affected the conductivity of the carbon fibers. A slight increase in the mean conductivity can be observed after exposing the fibers to atmospheric plasma fluorination for 8 min [51]. Fig. 6.6 shows that atmospheric plasma fluorination only marginally affected the conductivity, in which a slight increase of the mean conductivity can be observed after exposing the fibers to atmospheric plasma fluorination for 8 min. These results can be explained by an increased conductivity because the carbon fibers are more graphitic due to plasma fluorination. These minor changes are also likely attributed to the changes occurring at the carbon fiber surface, namely, the original desized and mildly oxidized fibers over sputtered fibers leading to a more graphitic structure. During the first minute of surface modification, the energized and ionized gas began to sputter the carbon fiber surface, which caused significant damage to the fiber surface. However, as the plasma treatment time increased, more ordered graphite structures of carbon fiber were exposed, consequently decreasing the d-mode, as shown in Fig. 6.7 [51].

Lee et al. [18] reported changes in the conductivity of oxyfluorinated carbon fibers; the conductivity of oxyfluorinated carbon fibers was 1.8 times higher than that of pristine fibers at room temperature, as shown in Fig. 6.8. This increased conductivity likely occurs because oxyfluorination only changes the surface region of the carbon fibers. Fluorine species could be intercalated into the graphene layers of the carbon fibers at room temperature; a high-quality graphite could have a high electrical conductivity, eg, close to that of metal [71].

Therefore, fluorination can be used to modify carbon fibers for biosensor materials to improve their electrical properties. Lee et al. [72,73] reported that PAN-based activated carbon fibers (CNT embedded) and pitch-based activated carbon fibers used as biosensor electrode materials were modified by oxyfluorination. In Fig. 6.9A and B,

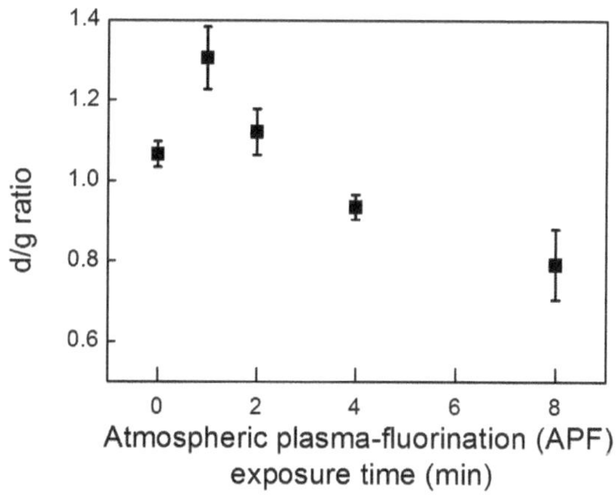

Figure 6.7 Ratio of degenerate deformation mode (d) and crystallite mode (g) of desized carbon fibers and atmospheric plasma-fluorinated carbon fibers as function of treatment time [51].

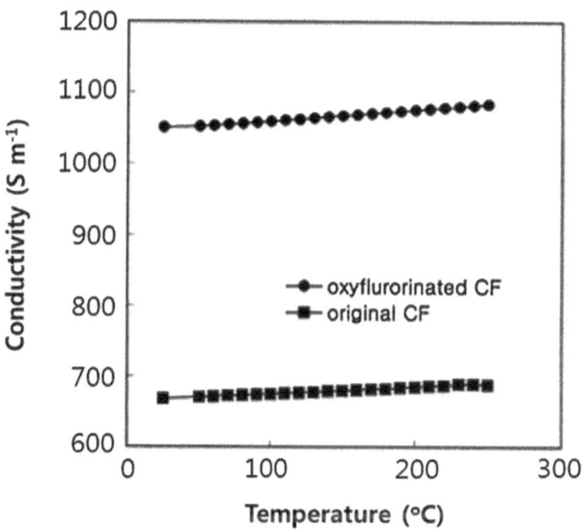

Figure 6.8 Temperature dependence of the electrical conductivity for pristine and oxyfluorinated carbon fibers [18].

the intercept values of the calibration plots, which are related to the electron transfer rate of the carbon fibers, increased because of the effects of oxyfluorination.

6.3.2 Electrical Characteristics of Fluorinated Single-Walled Carbon Nanotubes

Lee et al. [23] reported changes in the electrical properties of SWCNTs according to fluorination temperatures. As shown in Fig. 6.10, the resistivity increased with increasing reaction temperature, ie, the F-doping content on the SWCNT surface. The electrical properties of SWCNTs were altered by fluorination to decrease the conductivity of the individual SWCNTs.

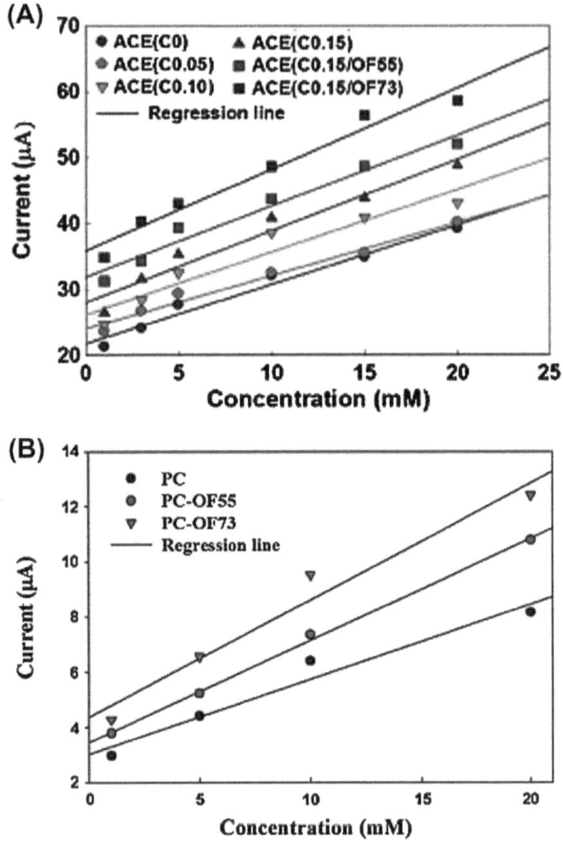

Figure 6.9 Calibration plots from the amperometric response of oxyfluorinated polyacrylonitrile (PAN)-based carbon fibers (A) and pitch-based carbon fibers (B) [72,73].

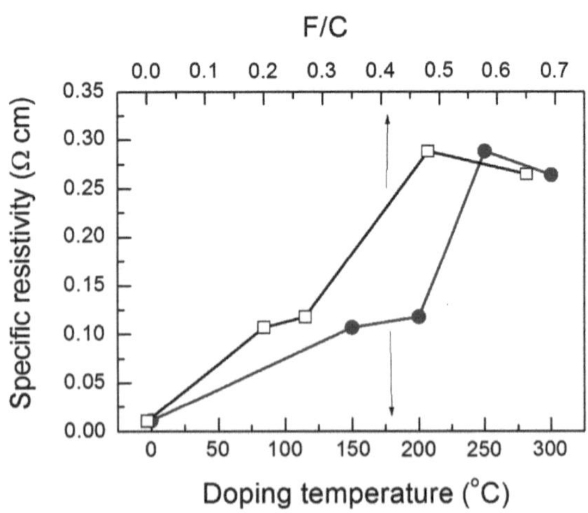

Figure 6.10 The resistivity of fluorinated single-walled carbon nanotubes at room temperature [23].

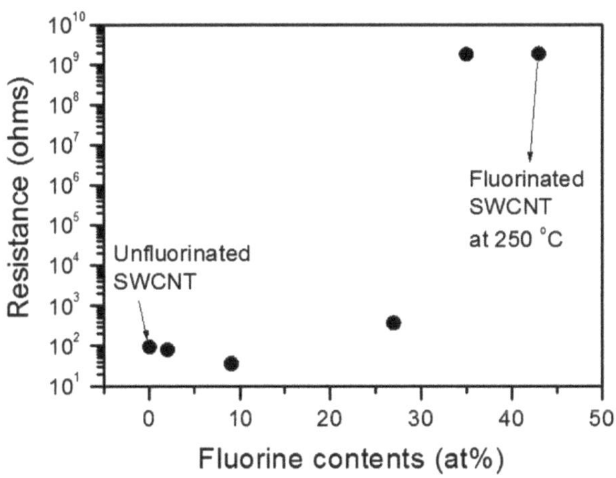

Figure 6.11 The resistance of fluorinated single-walled carbon nanotubes (SWCNTs) with fluorine contents.

Following the research of Pehrsson et al., Fig. 6.11 shows that the fluorination increased the resistance of the SWCNT films by $\sim 10^7$ [27]. The fluorination decreased the electrical conductivity of the SWCNT by disrupting its π conjugation. The conductivity of the fluorinated SWCNTs depends on the fluorinated carbon atoms, and hence the π bonds on the nonfluorinated carbon atoms, form continuous chains along the tube axis [74]. Most studies predict that fluorinated SWCNTs are semiconducting or insulating, except when a continuous π-bonded chain runs along the SWCNT axis, rendering them metallic.

Dong et al. reported the NO_2 and NH_3 gas-sensing properties of fluorinated SWCNTs treated by plasma using CF_4 or SF_6 [59]. NH_3 is an electron donor, whereas NO_2 is an electron-withdrawing molecule. The NO_2 gas-sensing results (Fig. 6.12A) showed that the resistance of the pristine and Ar- or O_2-treated SWCNTs decreased upon exposure to NO_2, corresponding to a typical p-type semiconducting behavior. However, the fluorinated SWCNTs showed a reverse change in the resistance, different from the reaction of the pristine SWCNT with NO_2 gas. The C–F bond is highly polarized, forming a significant electrostatic attraction between C^+ and F^- rather than the typical covalent bond [75]. Accordingly, the fluorinated SWCNTs were more hole doped, and the resistance of the fluorinated SWCNTs increased. However, these changes occurred only when NO_2 gases were chemisorbed on the fluorinated SWCNTs. When NO_2 is chemisorbed on the fluorinated SWCNT, the strong bonding creates an empty level above the valence band maximum [76]. Therefore the band structure of the fluorinated SWCNTs is altered, decreasing the conductance of this p-type semiconductor.

In response to NH_3 exposure (Fig. 6.12B), the fluorinated SWCNTs showed significant enhancements of sensing response [59]. This significant increase in response is attributed to the fact that the adsorption energies and charge transfers were caused by the C–F bonds on the fluorinated SWCNTs. As mentioned above, the C–F bond was highly polarized, and therefore the resistance change of the fluorinated SWCNTs in NH_3 gases was further increased compared to that of the pristine SWCNT paste.

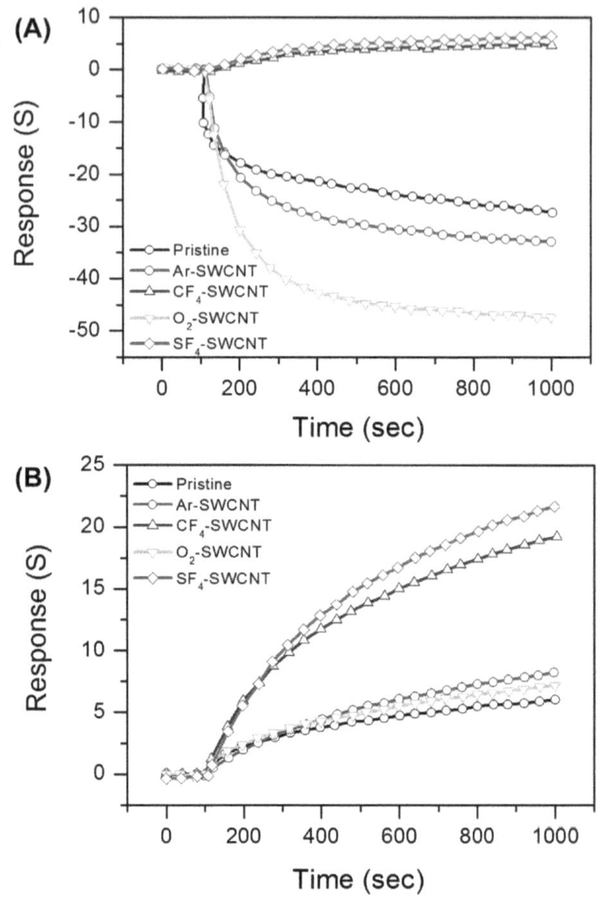

Figure 6.12 Electrical resistance change of the plasma-treated single-walled carbon nanotubes (SWCNTs) in response to NO_2 (A) and NH_3 (B) at room temperature [59].

6.3.3 Electrical Characteristics of Fluorinated Multi-Walled Carbon Nanotubes

Yue et al. [26] reported the electrochemical performance of fluorinated MWCNTs with different F/C ratios synthesized at different fluorination temperatures. The conductivities of the fluorinated MWCNTs were measured by impedance as shown in Fig. 6.13. When the fluorination temperature was increased to 400°C, the conductivity of the fluorinated MWCNTs decreased sharply to approximately 1.8×10^{-8} S/cm.

Lee et al. also reported the electrochemical properties of fluorinated MWCNTs, in which the resistance of MWCNTs decreased only when the MWCNTs were fluorinated below 200°C as shown in Fig. 6.14 [31].

The fluorination of MWCNTs could be proposed to occur as follows: fluorine atoms intercalate step-by-step into the carbon layer with the π-type electrons, forming a covalent C—F bond and resulting in the carbon hybridization from sp^2 to sp^3 [26].

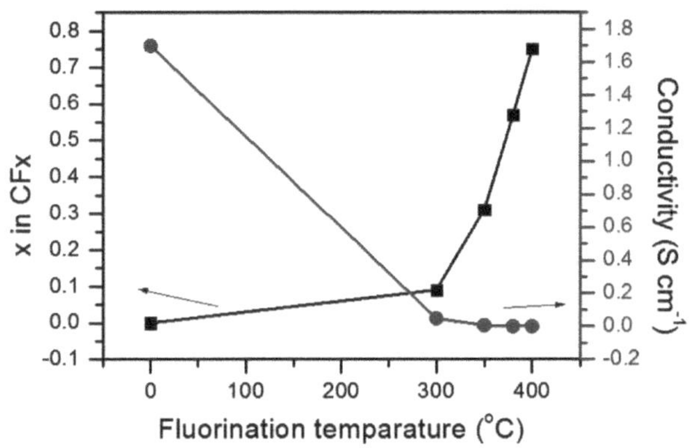

Figure 6.13 The conductivity of fluorinated multi-walled carbon nanotubes with different fluorine contents synthesized at different fluorination temperatures.

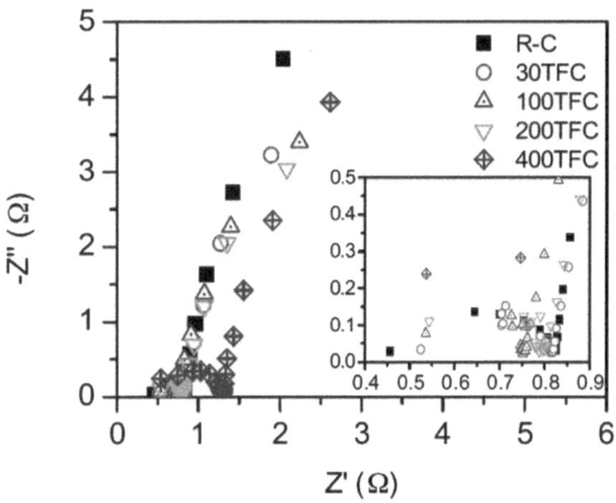

Figure 6.14 Nyquist impedance plots of the fluorinated multi-walled carbon nanotube-based electrodes [31].

These chemical changes of MWCNTs increased with increasing fluorination temperature, and the conductivity of the fluorinated MWCNTs also decreased. However, the nonfluorinated regions in the core should act as an easy electron transport path within the fluorinated CNTs, which could improve the electrical conductivity of the fluorinated MWCNTs [26].

Lee et al. [77] found that the NO gas-sensing properties of fluorinated MWCNTs were as shown in Fig. 6.15. The NO gas-sensing properties were shown to depend on the fluorination temperature in a way that could be divided into three regions separated at 400°C and 1000°C. In fluorinated MWCNTs at 100–400°C, the fluorinated MWCNTs showed the opposite trend in electrical resistance change compared to traditional p-type MWCNTs. In fluorinated MWCNTs at 600°C, the fluorinated MWCNTs indicated traditional p-type gas-sensing behavior. In fluorinated MWCNTs at 1000°C, a reoriented carbon structure was observed because of the high temperature. Furthermore, the NO-sensing properties of fluorinated MWCNTs at 1000°C showed a similar phenomenon as pristine MWCNTs. The inversion of electrical properties in fluorinated

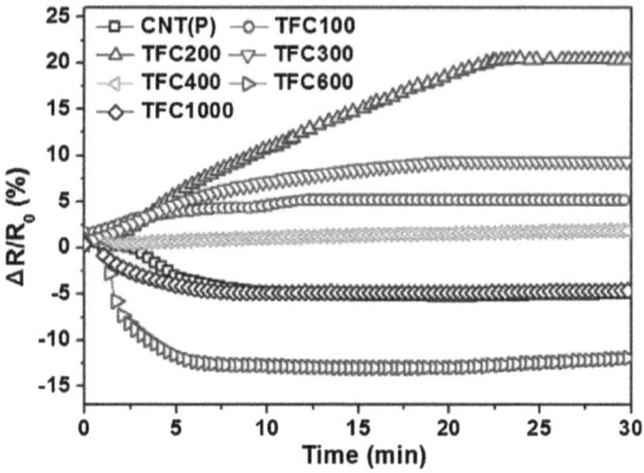

Figure 6.15 Electrical resistance change of the fluorinated multi-walled carbon nanotubes in response to NO [77].

MWCNTs occurred because fluorinated MWCNTs have a lower Fermi level compared to pristine MWCNTs, as noted above. The electrons in the valance band move to the lowest unoccupied molecular orbital (LUMO) of fluorinated MWCNTs because the high electronegativity of fluorine results in the generation of more holes in the valance band. When NO gas is adsorbed on fluorinated MWCNTs, the electrons in the highest occupied molecular orbital of NO molecules transfer to the LUMO of fluorinated CNTs, resulting in an increase in the Fermi level [77].

References

[1] K.S. Yang, D.J. Lee, S.K. Ryu, Y. Korai, Y.J. Kim, I. Mochida, Korean J. Chem. Eng. 16 (1999) 518–524.
[2] B.K. Sea, S.Y. Choo, T.J. Lee, S. Morooka, S.K. Song, Korean J. Chem. Eng. 12 (1995) 416–420.
[3] A. Fukunaga, S. Ueda, M. Nagumo, Carbon 37 (1999) 1081–1085.
[4] M. Delamar, G. Desarmot, O. Fagebaume, R. Hitmi, J. Pinson, J.M. Savent, Carbon 35 (1997) 801–807.
[5] M.R. Alexander, F.R. Jones, Carbon 33 (1995) 569–580.
[6] M.A. Montes-Moran, A. Martiez-Alonso, J.M.D. Tascon, R.J. Young, Compos. Pt. A-Appl. Sci. Manuf. 32 (2001) 361–371.
[7] K.F. Köster, P. Schwartz, Composites Sci. Tech. 60 (2000) 2005–2010.
[8] G. Bogoeva-Gaceva, E. MÄder, L. Haüssler, A. Dekanski, Compos. Pt. A-Appl. Sci. Manuf. 28 (1997) 445–452.
[9] Z. Wu, C.U. Pittman, S.D. Gardner, Carbon 33 (1995) 597–605.
[10] W.H. Lee, J.G. Lee, P.J. Peucroft, Appl. Surf. Sci. 171 (2001) 136–142.
[11] T. Nakajima, S. Kasamatsu, Y. Matsuo, Eur. J. Solid State Inorg. Chem. 33 (1996) 831–840.
[12] T. Nakajima, N. Watanabe, I. Kameda, M. Endo, Carbon 24 (1986) 343–361.
[13] M.J. Jung, J.W. Kim, J.S. Im, S.J. Park, Y.S. Lee, J. Ind. Eng. Chem. 15 (2009) 410–414.
[14] A.P. Kharitonov, R. Taege, G. Ferrier, V.V. Teplyakov, D.A. Syrtsova, G.H. Koops, J. Fluor. Chem. 126 (2005) 251–263.

[15] H. Touhara, F. Okino, Carbon 38 (2000) 241−267.
[16] Y.S. Lee, J. Fluor. Chem. 128 (2007) 392−403.
[17] A. Tressaud, E. Durand, C. Labrugère, J. Fluor. Chem. 125 (2004) 1639−1648.
[18] Y.S. Lee, B.K. Lee, Carbon 40 (2002) 2461−4268.
[19] T. Nakajima, N. Watanabe, Graphite Fluorides and Carbon-fluorine Compounds, CRC Press, Florida, 1991.
[20] N. Watanabe, T. Nakajima, H. Touhara, Graphite Fluorides, Elsevier, Amsterdam, 1988.
[21] E.T. Mickelson, C.B. Huffman, A.G. Rinzlera, R.E. Smalley, R.H. Hauge, J.L. Margrave, Chem. Phys. Lett. 296 (1998) 188−194.
[22] A.V. Okotrub, N.F. Yudanov, A.L. Chuvilin, I.P. Asanov, YuV. Shubin, L.G. Bulusheva, A.V. Gusel'nikov, I.S. Fyodorov, Chem. Phys. Lett. 322 (2000) 231−236.
[23] Y.S. Lee, T.H. Cho, B.K. Lee, J.S. Rho, K.H. An, Y.H. Lee, J. Fluor. Chem. 120 (2003) 99−104.
[24] K.H. An, J.G. Heo, K.G. Jeon, D.J. Bae, C. Jo, C.W. Yang, C.Y. Park, Y.H. Lee, Y.S. Lee, Y.S. Chung, Appl. Phys. Lett. 80 (2002) 4235−4237.
[25] A. Hamwi, H. Alvergnat, S. Bonnamy, F. Béguin, Carbon 35 (1997) 723−728.
[26] H.J. Yue, W. Zhang, H.D. Liu, Z. Liu, G.M. Zhong, Y. Yang, Nanotechnology 24 (2013) 424003.
[27] P.E. Pehrsson, W. Zhao, J.W. Baldwin, C. Song, J. Liu, S. Kooi, B. Zheng, J. Phys. Chem. B 107 (2003) 5690−5695.
[28] F. Chamssedine, K. Guérin, M. Dubois, E. Disa, E. Petit, Z. El Fawal, A. Hamwi, J. Fluor. Chem. 132 (2011) 1072−1078.
[29] P.R. Marcoux, J. Schreiber, P. Batail, S. Lefrant, J. Renouard, G. Jacob, D. Albertini, J.Y. Mevellec, Phys. Chem. Chem. Phys. 4 (2002) 2278−2285.
[30] Z. Gu, H. Peng, R.H. Hauge, R.E. Smalley, J.L. Margrave, Nano Lett. 2 (2002) 1009−1013.
[31] M.J. Jung, E. Jeong, Y.S. Lee, Appl. Surf. Sci. 347 (2015) 250−257.
[32] Y.S. Lee, B.K. Lee, J.S. Rho, Korean J. Chem. Eng. 20 (2003) 151−156.
[33] A.P. Kharitonv, J. Fluor. Chem. 104 (2000) 97−107.
[34] F.J. Toit du, R.D. Sanderson, J. Fluor. Chem. 98 (1999) 107−114.
[35] H. Schonherr, Z. Hruska, G.J. Vancso, Macromolecules 31 (1998) 3679−3685.
[36] A. Tressaud, B. Chevalier, L. Piraux, M. Cassart, J. Fluor. Chem. 72 (1995) 165−170.
[37] T. Nakajima, Synthesis, structure, and physicochemical properties of fluorine graphite intercalation compounds, in: T. Nakajima (Ed.), Fluorine−carbon and Fluoride−carbon Materials: Chemistry, Physics, and Applications, Marcel Dekker, Inc., New York, 1995, pp. 1−32.
[38] L. Fischer, U. Siemann, W. Ruland, Colloid Polym. Sci. 261 (1983) 744−749.
[39] K.K.C. Ho, G. Kalinka, M.Q. Tran, N.V. Polyakova, A. Bismarck, Compos. Sci. Technol. 67 (2007) 2699−2706.
[40] R.B. Mathur, V. Gupta, O.P. Bahl, A. Tressaud, S. Flandrois, Synth. Met. 114 (2000) 197−200.
[41] A. Bismarck, R. Tahhan, J. Springer, A. Schulz, T.M. Klapötke, H. Zeil, W. Michaeli, J. Fluor. Chem. 84 (1997) 127−134.
[42] S.J. Park, M.K. Seo, Y.S. Lee, Carbon 41 (2003) 723−730.
[43] M.K. Seo, S.J. Park, J. Colloid Interface Sci. 330 (2009) 237−242.
[44] N. Watanabe, Y. Ashida, T. Nakajima, Bull. Chem. Soc. Jpn. 55 (1982) 3197−3199.
[45] T. Nakajima, M. Kawaguchi, N. Watanabe, Carbon 20 (1982) 287−291.
[46] Y. Khairallah, F. Arefi, J. Amouroux, Thin Solid Films 241 (1994) 295−300.
[47] H. Loh, R.E. Cohen, R.F. Baddour, J. Mater. Sci. 22 (1987) 2937−2947.

[48] H. Kasai, M. Kogoma, T. Moriwaki, S. Okazaki, J. Phys. D-Appl. Phys. 19 (1986) 225–228.
[49] H. Höcker, Pure Appl. Chem. 3 (2002) 423–428.
[50] E. Selli, G. Mazzone, C. Oliva, F. Martini, C. Riccardi, R. Barni, B. Marcandalli, M.R. Massafra, J. Mater. Chem. 11 (2001) 1985–1991.
[51] K.K.C. Ho, A.F. Lee, A. Bismarck, Carbon 45 (2007) 775–784.
[52] K.K.C. Ho, A.F. Lee, S. Lamoriniere, A. Bismarck, Compos. Pt. A-Appl. Sci. Manuf. 39 (2008) 364–373.
[53] M.J. Shenton, M.C. Lowell-Hoare, G.C. Stevens, J. Phys. D-Appl. Phys. 34 (2001) 2761–2768.
[54] S. Kanazawa, M. Kogoma, T. Moriwaki, S. Okazaki, J. Phys. D-Appl. Phys. 21 (1988) 838–840.
[55] N.O.V. Plank, L. Jiang, R. Cheung, Appl. Phys. Lett. 83 (2003) 2426–2428.
[56] Y.C. Hong, D.H. Shin, S.C. Cho, H.S. Uhm, Chem. Phys. Lett. 427 (2006) 390–393.
[57] N.O.V. Plank, G.A. Forrest, R. Cheung, A.J. Alexander, J. Phys. Chem. B 109 (2005) 22096–22101.
[58] C.K. Liu, J.M. Wu, H.C. Shih, Sens. Actuator B-Chem. 150 (2010) 641–648.
[59] K.Y. Dong, D.J. Ham, B.H. Kang, K. Lee, J. Choi, J.W. Lee, H.H. Choi, B.K. Ju, Talanta 89 (2012) 33–37.
[60] N.O.V. Plank, R. Cheung, Microelectron. Eng. 578 (2004) 73–74.
[61] Z. Hou, B. Cai, H. Liu, D. Xu, Carbon 46 (2008) 405–413.
[62] S. Borisov, V.S. Khotimsky, A.I. Rebrov, S.V. Rykov, D.I. Slovetsky, Y.M. Pashunin, J. Membr. Sci. 125 (1997) 319–329.
[63] M. Nakahara, Y. Sanada, Carbon 33 (1995) 735–736.
[64] K.K.C. Ho, S. Lamoriniere, G. Kalinka, E. Schulz, A. Bismarck, J. Colloid Interface Sci. 313 (2007) 476–484.
[65] K.K.C. Ho, S.R. Shamsuddin, M. Laffan, A. Bismarck, Compos. Pt. A-Appl. Sci. Manuf. 42 (2011) 453–461.
[66] K. Tsutsumi, S. Ishida, K. Shibita, Colloid Polym. Sci. 268 (1990) 31–37.
[67] K.K.C. Ho, G. Beamson, G. Shia, N.V. Polyakova, A. Bismarck, J. Fluor. Chem. 128 (2007) 1359–1368.
[68] M. Zayat, D. Davidov, H. Selig, Carbon 32 (1994) 485–491.
[69] A. Tressaud, V. Gupta, L. Piraux, L. Lozano, E. Marquestaut, S. Flandrois, A. Marchand, O.P. Bahl, Carbon 32 (1994) 1485–1492.
[70] A. Tressaud, C. Guimon, V. Gupta, F. Moguet, Mater. Sci. Eng. B 30 (1995) 61–68.
[71] T. Nakajima, Fluorine–carbon and Fluoride–carbon Materials, Marcel Dekker, New York, 1995, pp. 25–29.
[72] J.S. Im, J. Yun, J.G. Kim, T.S. Bae, Y.S. Lee, Appl. Surf. Sci. 258 (2012) 2219–2225.
[73] J.G. Kim, J.S. Im, T.S. Bae, J.H. Kim, Y.S. Lee, J. Ind. Eng. Chem. 19 (2013) 94–98.
[74] K.N. Kudin, H.F. Bettinger, G.E. Scuseria, Phys. Rev. B 63 (2001) 045413.
[75] D. O'Hagan, Chem. Soc. Rev. 37 (2008) 308–319.
[76] L. Bai, Z. Zhou, Carbon 45 (2007) 2105–2110.
[77] J.S. Im, S.C. Kang, B.C. Bai, T.S. Bae, S.J. In, E. Jeong, S.H. Lee, Y.S. Lee, Carbon 49 (2011) 2235–2244.

Perfluoroalkylated PAH n-Type Semiconductors: Theory and Experiment

7

H. Sun, J.H. Kramer
University of South Dakota, Vermillion, SD, United States

Chapter Outline

7.1 Introduction

Incorporating fluorine onto organic compounds can drastically alter their physical and chemical properties due to fluorine's high electronegativity and small atomic radius [55,69,90]. Fluorine's high electronegativity leads to both strong C—F bonds and corresponding C—C bonds within the backbone where the C—F bonds are located

New Fluorinated Carbons: Fundamentals and Applications. http://dx.doi.org/10.1016/B978-0-12-803479-8.00007-3

[35,55,83–85]. These strong C–F and corresponding C–C bonds provide higher stability for organofluorine molecules compared to corresponding hydrocarbon compounds. Fluorinated organic materials offer many advantages over nonfluorinated ones including high thermal and photochemical stability, high tolerance to extreme environmental conditions, and potential high efficiency in optoelectronic devices [4,49,52,56,65,75,87,97,101]. Fluorination can have profound tuning effects upon the electronic and optical properties of organic compounds; thus fluorine is found in many state-of-the-art materials. Partially fluorinated liquid crystals are used in active matrix liquid crystal displays; optically transparent fluorinated polymers enable the manufacturing of increasingly integrated microelectronic devices by 16 nm photolithography; and fluorinated membranes serve as proton transport elements in current generation fuel cells [68]. Fluorinated electrolyte additives are used in lithium batteries to improve their long-term stability and cell voltage [8,29,96].

Beyond the current applications of fluorinated materials, recent research work in organofluorine chemistry has gone into the fields of organic semiconductors [42], energy conversion and storage [4,36,76], and ^{18}F-labeled radiotracers for positron emission tomography scan [23], as well as fluorinated pharmaceuticals for ^{19}F magnetic resonance imaging applications [31]. Chart 7.1 shows examples of fluorinated compounds with organic semiconductor applications [20,21,32–34,46,55,61,62,90,91,99]. Significant changes in electronic properties upon fluorination make fluorinated materials a metaphorical goldmine [9,18,40, 46,60,83–86]. In this chapter, we will focus on the recent development in perfluoroalkylated polyaromatic hydrocarbons (PAHs) molecules and discuss how perfluoroalkylation affects the physicochemical properties of the PAHs, as well as the resulting challenges caused by perfluoroalkylation, by analyzing recent development in both theoretical and experimental works.

Differentiating between direct fluorine substitution of PAHs (ie, fluorine on an sp^2 hybridized carbon (eg, compounds **1** [71], **2** [79], **3** [62], and **7** [75] in Chart 7.1)) and

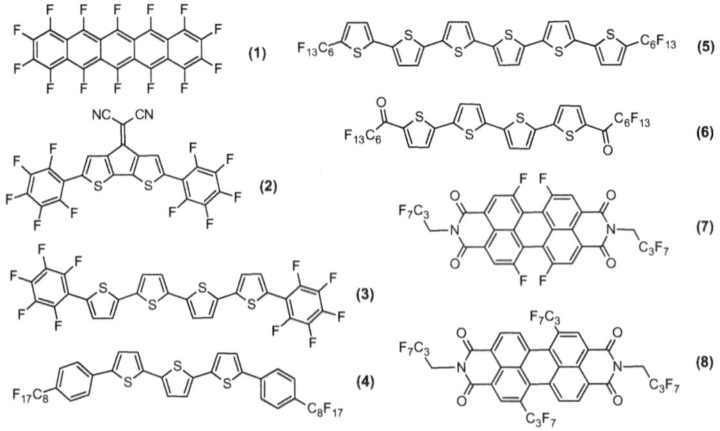

Chart 7.1 Fluorinated organic semiconductor compounds.

perfluoroalkylation of PAHs (ie, fluorine is on an sp^3 hybridized carbon chain (eg, compounds **4** [20], **5** [21], **6** [98], **7** [75], and **8** [14] in Chart 7.1)) is essential for understanding the fundamental physicochemical properties and potential applications of these two types of aromatic compounds.

Fluorine possesses a size similar to that of hydrogen but with the highest electronegativity, allowing for a fluorinated molecule to acquire profound electronic effects while avoiding unnecessary steric hindrance [69]. Though fluorine's electronegativity is 4.0, direct fluorination on aromatic rings (sp^2 carbons) only introduces a moderate electron-withdrawing effect (σ_p value for fluorine is 0.06) due to the coexisting σ-accepting and π-donating effects of fluorine atom. In contrast, C_{sp^3}—F containing functional groups have a much greater electron-withdrawing ability (Table 7.1). For example, a $-CF_2CF_3$ substitute has a σ_p value of 0.52 and a $-CF_2OCF_3$ substituent has a σ_p value of 0.81 [27]. Just for comparison, the CN group generally utilized in n-type organic semiconductor materials has a σ_p value of 0.66. Furthermore, the perfluoroalkyl and perfluoropolyether substituents possess superhydrophobicity, oleophobicity, and fluorophilicity, but other common electron-withdrawing substitutes (eg, CN, NO_2, and COOH) do not. C_{sp^3}—F functional groups are much more chemically resistant compared to C_{sp^2}—F functional groups, which are susceptible to nucleophilic aromatic substitution (S_NAr) when strong nucleophiles are present [80—82,88,93].

Table 7.1 Hammett Substituent Constants of Perfluoroalkyl and Related Substituents[a]

Substituents	σ_p	σ_m
F	0.06	0.34
CH_2F	0.11	0.12
CHF_2	0.32	0.29
CF_3	0.54	0.43
C_2F_5	0.52	0.47
n-C_3F_7	0.48	0.44
n-C_4F_9	0.52	0.47
CF_2OCF_3	0.81	0.81
OCF_3	0.35	0.38
SCF_3	0.50	0.40
$SOCF_3$	0.69	0.63
SF5	0.68	0.61
C_6F_5	0.27	0.26
CH_2CF_3	0.09	0.12

[a]Ref. [27].

In addition to its profound electron-withdrawing ability, the greater steric effect of the perfluoroalkyl chain, compared to fluorine itself, results in a stronger ability to modulate the molecular assembly in solid state [2,6,69,83–85]. Fluorine's high electronegativity results in the low polarizability of C_{sp^3}—F and corresponding C—C bonds in perfluoroalkyl chains, resulting in superior superhydrophobicity, fluorophilicity, and oleophobicity which are important factors for many applications ranging from building air-stable n-type organic semiconductor devices [91] to antifouling coating applications [30,104]. Excellent fluorophilicity and oleophobicity of perfluoroalkyl groups also allow for orthogonal solution processability of these perfluoralkylated organic semiconductor materials [100].

We should also notice that there are a large number of research articles that utilize perfluoroalkyl chains with methylene ($-CH_2-$) spacer(s) between the aromatic backbone and the perfluoroalkyl chain where superhydrophobicity, fluorophilicity, and oleophobicity of these partial perfluoroalkyl groups are emphasized more than their electronic effects [11,24,35,63,78,102,103]. The usage of methylene spacer(s) in perfluoroalkyl chains changes their electronic properties in comparison to the pure, non-spaced perfluoroalkyl chains. The increased number of $-CH_2-$ spacer places the perfluoroalkyl chain further away from the aromatic core, weakening the inductive effects of the electron-withdrawing perfluoroalkyl chain (Table 7.1) [27]. Though including this portion of the perfluoroalkylated PAHs would provide a more comprehensive picture, the large quantity of literature reports in this area and limited chapter space precludes an in-depth review and discussion of perfluoroalkylated PAHs in which combined electron deficiency, superhydrophobicity, and oleophobicity are generally more important for their applications than just superhydrophobicity and oleophobicity alone.

With the strong electron-withdrawing effect, perfluoroalkyl substituents can change the corresponding PAHs from p-type to n-type organic semiconductors or ambipolar, and, in many cases, also become air-stable n-type organic semiconductors [91]. In this chapter, we will focus on an in-depth review and discussion of perfluoroalkylated PAHs involving (1) theoretical consideration of stereoelectronic effects of perfluoroalkylated PAHs; (2) synthesis, characterization, and crystal engineering of perfluoroalkylated PAHs; and (3) physicochemical properties of perfluoroalkylated PAHs.

7.2 Stereoelectronic Consideration of Perfluoroalkylated Polyaromatic Hydrocarbons

As we discussed earlier, perfluoroalkyl substituents, including trifluoromethyl groups, possess greater inductive effect than direct fluorine substitution on aromatic compounds. Due to the tetrahedron structure of the sp^3 hybridized carbon (C_{sp^3}), fluorine on the C_{sp^3} barely has any π-donating effect to aromatic rings, which results in profound electron-withdrawing effect to the aromatic rings. It is beneficial to understand the electronic and steric effects of the perfluoroalkyl substituents to gain solid understanding of the fundamental physicochemistry properties of perfluoroalkylated PAHs. Here we will briefly discuss how perfluoroalkylation affects the basic thermodynamic properties first, then further discuss how perfluoroalkylation affects the electron transfer (ET) kinetics. Given the complexity of the ET in solid-state organic semiconductors, the first principle prediction of the ET properties is still a significant challenge if it is not impossible at all. Up to date, there is only very limited theoretical work that has been done indirectly to access how perfluoroalkylation affects the organic semiconductor charge transfer properties.

7.2.1 *The Electronic Effect of Perfluoroalkyl Substituents*

Early discussion on electronic effects of perfluoroalkyl substituents could correlate back to the Hammett linear free-energy relationships (LFER) [27], where the electronic effect of the trifluoromethyl group was initially studied. With more perfluoroalkylated PAHs being available as a result of advancements in fluorine chemistry, the electronic effects of perfluoroalkyl substituents are quantified through the LFER study compared to initial qualitative assessment. Table 7.1 summarizes the electronic effect of some perfluoroalkyl substituents and a few other common functional groups with similar electronic effect as a comparison.

Theoretical calculations show both the highest occupied molecular orbital (HOMO) and lowest unoccupied molecular orbital (LUMO) energies are lowered after perfluoroalkylation of PAHs [83–85], and multiple perfluoroalkyl substitution generally gives cumulative electron-withdrawing effect on PAHs. However, such summed electron-withdrawing effect does not follow the linear correlation with the total number of perfluoroalkyl substituents on the same aromatic ring, particularly, in the case when most of the hydrogen atoms were substituted by perfluoroalkyl groups (Fig. 7.1A and B).

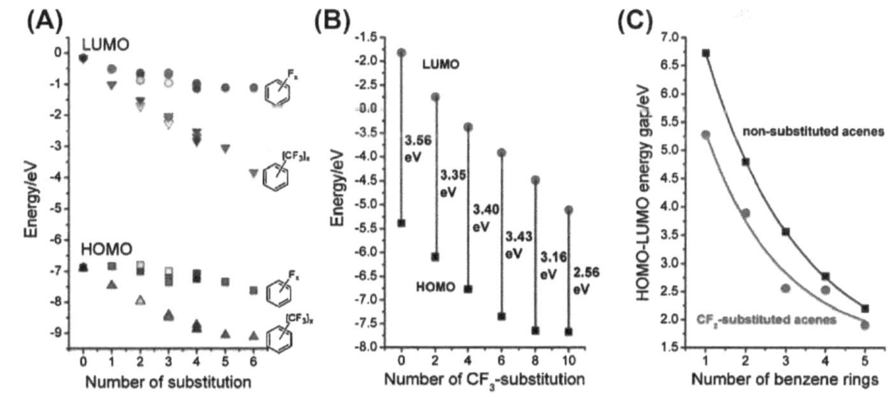

Figure 7.1 Calculated HOMO and LUMO energies and energy gaps for fluorinated and trifluoromethylated benzenes (A), trifluoromethylated anthracenes (corresponding structures are shown in Chart 7.2) (B), and pertrifluoromethylated acenes with the *x* axis as the total number of benzene rings in acenes (C). HOMO–LUMO energies of isomers of bis-, tri-, and tetra-substituted benzenes are also shown in panel (A). *HOMO*, highest occupied molecular orbital; *LUMO*, lowest unoccupied molecular orbital.

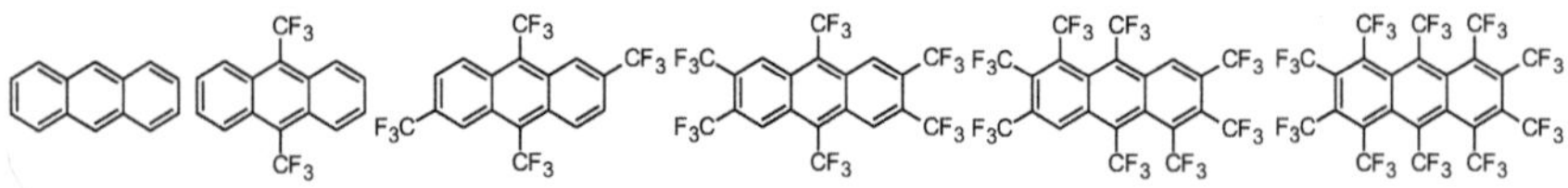

Chart 7.2 Trifluoromethylated anthracene derivatives used for computing highest occupied molecular orbital, lowest unoccupied molecular orbital, ionization potential, electron affinity, reorganization, and bandgap energies.

This is perhaps due to the nonplanarity nature of PAHs with multiple perfluoroalkyl substituents caused by the steric hindrance of the perfluoroalkyl groups [83–85].

Another result of multiple perfluoroalkylation of PAHs is that the bandgap of PAHs which is generally estimated by the HOMO and LUMO energy difference decreases as the number of perfluoroalkyl groups increases for the same PAH molecule (Fig. 7.1B). According to the definition of the bandgap, we should alert readers that high accuracy bandgap calculation should be done by calculating the density of states (DOS) of the solid-state materials by using at least one unit cell with periodic boundary conditions if the solid-state structure is known for the materials [94]. However, the DOS calculation normally is expensive, and it requires a large amount of computational resources [57]. Fortunately, the rough estimate value from the difference between calculated HOMO and LUMO energies for a single molecule in the gas phase (vacuum) normally gives us a good idea about the bandgap energy in the solid state, though it might be coincidental [83–85]. This estimation method has been used and accepted by most experimental chemists working on organic semiconductor materials [15]. The bandgap of perfluoroalkylated PAHs decreases in a similar trend of that of corresponding nonsubstituted PAHs as the π–conjugation increases (Fig. 7.1C). In addition to the changes of HOMO and LUMO energies, the gas-phase adiabatic ionization potential (IP/eV) and electron affinity (EA/eV) increases as a result of perfluoroalkylation when compared to nonsubstituted PAHs. These results demonstrate the profound ability of perfluoroalkyl groups in tuning these thermodynamic parameters over a very large range. In contrast, direct fluorination of PAHs shows fine-tuning effect over these thermodynamic parameters as shown in Fig. 7.1A. Thus a combination of perfluoroalkylation and fluorination of PAHs could, in principle, serve the purpose of optimization of the basic thermodynamic properties of organic semiconductor materials through computational chemistry–assisted molecular design.

The profound electron-withdrawing effect of perfluoroalkyl substituents is also reflected from the electrochemical results as we will discuss later in this chapter. Kuvychko *et al.* demonstrated that calculated EA and the experimental first reduction potential of a trifluoromethylated corannulene matches well [41]. This result provides confidence to both theoretical and experimental chemists when comparing gas-phase calculation results and redox potential measurement.

7.2.2 Changes in Reorganization Energy Associated With Electron Transfer Upon Perfluoroalkyation

While the HOMO and LUMO energies, IPs and EAs, and bandgaps represent some of the basic thermodynamic information, the reorganization energy associated with ET is

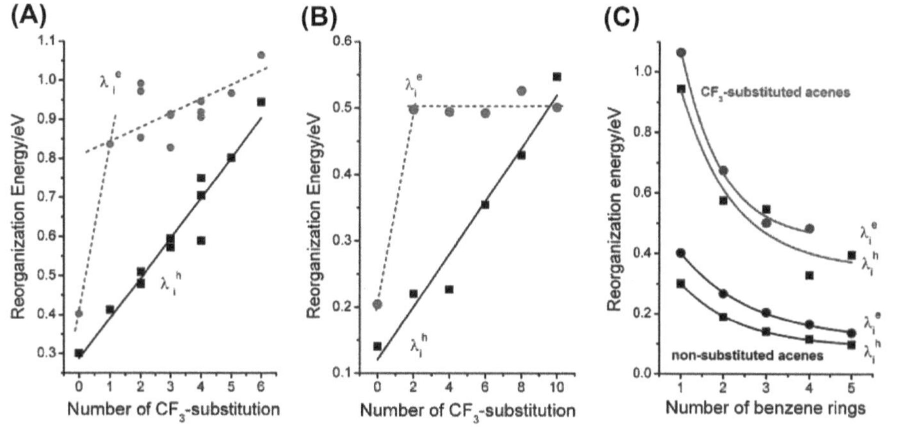

Figure 7.2 Adiabatic λ_i of all trifluoromethylated benzenes (A), trifluoromethylated anthracenes (structures are shown in Chart 7.2) (B), and pertrifluoromethylated acenes and nonsubstituted acenes (from benzene to pentacene) (C). The *solid lines* in panels (A) and (B) are the linear fitting results for λ_i^{h} with $R = 0.95$, and the short dashed lines are employed to represent the λ_i^{e} changing trend in panels (A) and (B).
Reprinted with permission from H. Sun, A. Putta, M. Billion, J. Phys. Chem. A 116 (2012) 8015−8022. Copyright (2012) American Chemical Society.

another important parameter to perfluoroalkylated PAHs for organic semiconductor applications. We first demonstrated how perfluoroalkylation affects the adiabatic inner-sphere reorganization energy associated with the ET of PAHs [83−85].

Both λ_i^{e} (reorganization energy for electron transport) and λ_i^{h} (reorganization energy for hole transport) increase as a result of increasing number of CF_3 substitution on the same aromatic ring (Fig. 7.2A and B). λ_i^{h} increases linearly as increasing number of trifluoromethyl substitution on the same aromatic core, however, λ_i^{e} levels out after the first trifluoromethyl substitution. This result indicates that one can go ahead with adding more trifluoromethyl groups after paying this one-time unavoidable penalty to create more electron-deficient PAHs for air-stable n-type semiconductors.

7.2.3 Changes in Intermolecular Interactions Upon Perfluoroalkylation

Because of the substantial electronic effects and steric hindrance, perfluoroalkyl groups not only affect the electronic properties of the PAHs, but also affect the intermolecular interactions through both their electronic and steric effects [86]. Our initial computational work on trifluoromethylated benzene and pyridine derivatives clearly shows that introducing strong electron-withdrawing CF_3 groups onto PAHs increases the intermolecular quadruple−quadruple interactions which lead to stronger $\pi-\pi$ stacking interactions in the solid-state materials.

We utilized unconstrained dimer optimizations with second order Møller−Plesset perturbation theory (MP2) and dispersion-corrected density functional theory (DFT-

Table 7.2 Calculated Dimerization Energies (E_{int}/kcal/mol) and $\pi-\pi$ Stacking Distances of Various Dimers Between Benzene Derivatives at MP2/aug-cc-pvDZ//MP2/6−31g(d) Level of Theory[a]

$\pi-\pi$ Stacked Dimers	E_{int} (kcal/mol)	$\pi-\pi$ Distance/Å
Benzene	−2.07	4.281
Monofluorobenzene	−4.20	3.758
Trifluoromethylbenzene	−6.74	3.646
1,3-Bistrifluoromethylbenzene	−8.05	3.665

[a]Ref. [86].

D) methods to probe the $\pi-\pi$ interaction energies of a series of dimers of benzene, pyridine, and bipyridine derivatives that serve as computational models for investigating how $\pi-\pi$ interaction energies were affected by methylation, fluorination, and trifluoromethylation. Enhancement of the intermolecular $\pi-\pi$ interaction energies, reduction of $\pi-\pi$ distances, and complementary cofacial orientations were found with the trifluoromethylated dimers compared to fluorinated and methylated dimers. These computational studies illustrate the strong potential of trifluoromethylation and, possibly, perfluoroalkylation of acenes and heteroacenes to enhance ET properties through better $\pi-\pi$ stacked structures, and making perfluoroakylated PAHs viable candidates for n-type organic semiconductor materials (Table 7.2).

7.3 Perfluoroalkylated Polyaromatic Hydrocarbons: Synthesis, Characterization, and Crystal Engineering

Developing new synthetic methodologies to improve the yield and selectivity of perfluoroalkylation reactions of PAHs has been a growing point of interest in organofluorine chemistry. Various perfluoroalkylating reagents offer synthetic and material chemists an impressive assortment of methods to synthesize new perfluoroalkylated compounds [1,55,58−60,89]. For example, perfluoroalkyl chains can be introduced to substrates through metal-mediated reaction pathways. Transition metal (Cu, Zn, Fe, etc.) complexes can be utilized to generate perfluoroalkyl anions that react with the substrate through addition−elimination reactions [38,44,45,55,67,74] or the metals can act as a reductive reagent to generate a perfluoroalkylating reagent *in situ* [25,59]. Additionally, metal-mediated perfluoroalkylation reactions may undergo a radical reaction pathway in which perfluoroalkyl radicals generated from the corresponding perfluoroalkyl halides are the active species.

In the meantime, metal-free perfluoroalkylation pathways were investigated to promote cleaner reactions. These pathways not only can give regioselective perfluoroalkylated PAHs [53], but also allow the usage of aqueous media, photocatalytic

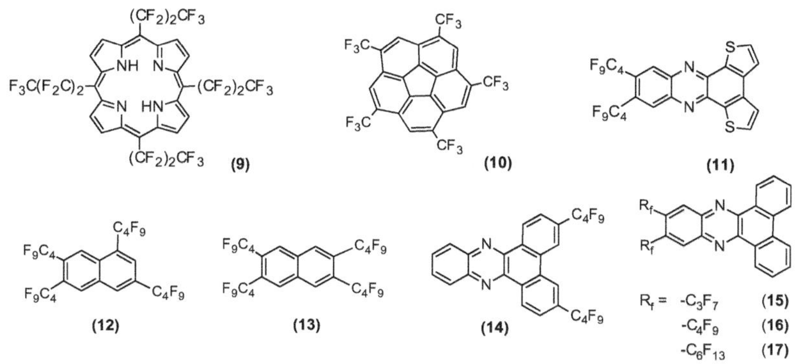

Chart 7.3 Examples of perfluoroalkylated polyaromatic hydrocarbons.

reactions, and simplified purification procedures [7,13,66,77]. Some special reagents can generate highly stable and valuable perfluoroalkyl and trifluoromethyl cations for synthetic purposes [50,89,95].

With the development of various synthetic strategies, many perfluoroalkylated PAHs were prepared for diverse applications, including from early work on perfluoroalkylated porphyrins for fluorous phase catalysis [16,39,47] to recent work on perfluoroalkylated perylene bisimide for n-type organic semiconductor applications [9,28,66,72]. Some examples of perfluoroalkylated PAHs are shown in Chart 7.3.

7.3.1 Synthesis and Purification: Early Stage Versus Late-Stage Perfluoroalkylation

Perfluoroalkylated porphyrin (**9**) has attracted considerable attention during the establishment of fluorous chemistry in the early 1990s due to significant interest in their unique electronic and catalytic properties that work in fluorous solvents [17]. β-Perfluoroalkylated porphyrins and phthalocyanine were also prepared since then by several groups. In addition to these macrocycle compounds, relatively smaller perfluoroalkylated PAHs were later prepared for the purpose of exploring air-stable organic semiconductor materials [91]. Recently, Strauss/Boltalina's group utilized high temperature thermal-induced radical reactions to perfluoroalkylate PAHs in a one-pot reaction directly from the nonsubstituted PAHs followed by high-performance liquid chromatography separation [72]. This high-temperature radical reaction strategy provides a quick pathway to access a large library of perfluoroalkylated PAHs for initial investigations of functionalized PAHs.

Many perfluoroalkylated PAHs were made from their corresponding brominated or iodinated PAHs with great control over regioselectivity by metal-mediated perfluoroalkylation. Recently, our group has synthesized a series of perfluoroalkylated PAHs and polyheteroaromatics from corresponding brominated precursors for applications in organic semiconductor materials and crystal engineering [2,6,83–85]. Due to significant solubility problems for both brominated starting materials and

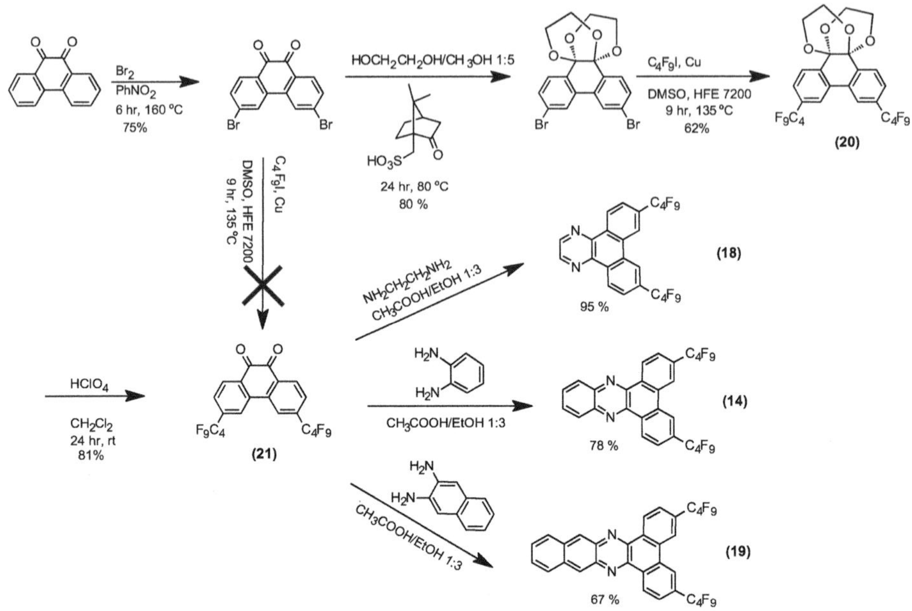

Scheme 7.1 Example of early stage introduction of perfluoroalkyl groups onto polyaromatic hydrocarbons.

perfluoroalkylated final products with large conjugation, late-stage introduction of perfluoroalkyl groups, particularly the long perfluoroalkyl chains, is problematic for both reaction yield and product purification [2]. Early stage introduction of perfluoroalkyl groups onto a key precursor (compound **21**) provides a practical solution for large-scale production of perfluoroalkylated PAHs, provided a proper pathway for cyclization is available for the final PAHs and polyheteroaromatics products [6] (Scheme 7.1).

It is interesting to observe that these perfluoroalkylated PAHs form lamellar $\pi-\pi$ stacked solid-state structures with the steering power from the perfluoroalkyl chains, which aggregate together similar to what is observed in the liquid phase of fluorous solvents and hydrocarbon solvents [2,83–85] (via infra).

7.3.2 Characterization: NMR, MS, and Single Crystal X-Ray Diffraction

Characterization of perfluoroalkylated PAHs is typically done by traditional characterization techniques and has used in organic chemistry including ^{1}H and ^{19}F nuclear magnetic resonance (NMR) spectroscopy, mass spectrometry (MS), elemental analysis, infrared (IR) spectroscopy, and X-ray single crystal diffraction. Though ^{13}C NMR is very useful for hydrocarbon-based compounds, the use of ^{13}C NMR to characterize perfluoroalkylated PAHs is almost impractical due to following reasons: (1) the relaxation time for the quaternary carbons in the perfluoroalkyl chains is too long, (2) without ^{19}F decoupling, the ^{13}C signals for the perfluoroalkyl chain is

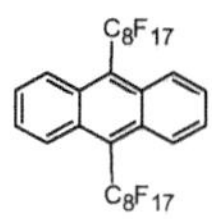

(22)

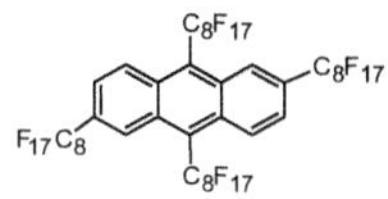

(23)

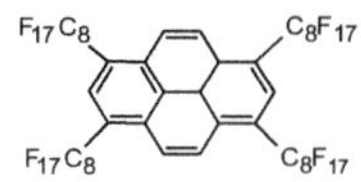

(24)

Chart 7.4 Polyperfluoroalkylated polyaromatic hydrocarbons with high fluorescence yield and photostability.

nonresolvable, very weak, and availability of NMR probes that are capable of ^{19}F-decoupled ^{13}C NMR experiments is limited, (3) solubility of the perfluoroalkylated PAHs is typically low, particularly for those large size PAHs in which the solubility could be below 1 mg/mL in hydrocarbon-based NMR solvents. Because of the solubility problem, sometimes a mixture of deuterated solvents and fluorinated solvents is useful as NMR solvents to characterize the final product [83−85].

For MS, due to their high electron-deficient structure caused by strong electron-withdrawing perfluoroalkyl groups, one may need to use the negative ion detection for mass spectroscopic measurement in order to observe the molecular ion. This is particularly important for highly perfluoroalkylated PAHs, for example, compounds **12** and **13** in Chart 7.3 and compound **23** in Chart 7.4. Other than this precaution, the general MS characterization technique and instrumentation should be effective for perfluoroalkylated PAHs. In addition to MS measurement, elemental analysis is also used to identify molecular composition and the purity of the compounds.

Single crystal X-ray diffraction experiments provide the molecular structure and solid-state packing which are important for the material function in solid state. Since the development of perfluoroalklyated PAHs, many crystal structures were obtained; we will discuss in detail the molecular structures, and more importantly, the molecular organization in the solid state in the following section.

7.3.3 Crystal Engineering: How Perfluoroalkyl Substitution Affects the Solid-State Structures of the Polyaromatic Hydrocarbons

One of the important aspects in materials science is focused on controlling the solid-state structure (crystalline structure in most cases) to generate materials with specific functions, commonly referred as "materials by design" [10,15]. As we discussed earlier, PAHs offer a promising future in organic electronics where the large π-conjugation offers increased charge mobility, an essential parameter for organic electronic devices. Formation of a specific solid-state structure is critical for the performance of organic semiconductor devices. For example, organic field-effect transistors (OFETs) require lamellar π−π stacked structures for improving charge mobility, which is one of the most important parameters for OFETs [26,37,43,54,62,71,73,75]. Recent work by our group shows that perfluoroalkyl chains offer an efficient solution to this predicament, when substituted onto PAHs, the perfluoroalkyl chain can act as a steering group to control the molecular orientation in the solid state (Scheme 7.2) [2].

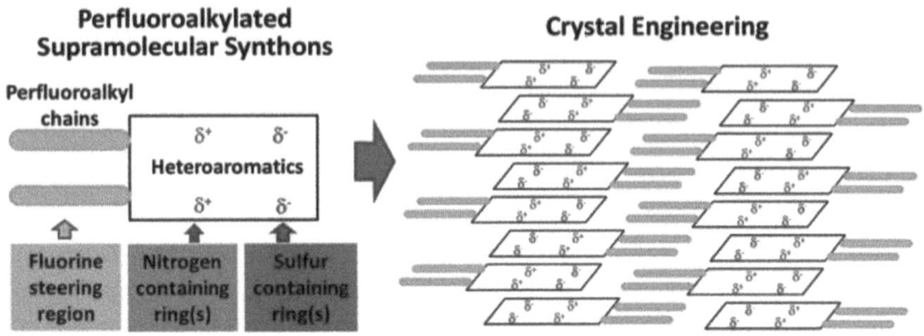

Scheme 7.2 Perfluoroalkyl steering for the assembly of $\pi-\pi$ stacked organic molecules [2]. Reprinted with permission from A. Putta, J.D. Mottishaw, Z. Wang, H. Sun, Cryst. Growth Des. 14 (2014) 350–356. Copyright (2014) American Chemical Society.

This concept was proved by many of the new perfluoroalkylated PAHs (Chart 7.3) synthesized in our lab over the past few years [2,6,83–85] and recent works from Strauss/Boltalina's group [10,41,72]. Introducing perfluoroalkyl chains onto one side of the PAH molecule offers one step forward to achieve the "materials by design" goal in organic semiconductor materials; though many areas of work need to be further investigated before the strategy can be further used in practical applications. The perfluoroalkyl chains help in steering the molecular orientation in the solid state through multiple F ••• F interactions among perfluoroalkyl chains from different molecules, yet suppress the C–H ••• π interactions through their steric hindrance and relatively weaker C–F ••• H–C interactions compared to C–F ••• F–C interactions [2,5,64,69,83–85]. The arrangement of these organic molecules allows for the $\pi-\pi$ stacking on partial positive and partial negative regions to enhance the charge mobility of the organic semiconductor. Further introducing soft polarizable sulfur atoms into the PAH ring assists in shortening the interplanar distance down to 3.247 Å (Fig. 7.3).

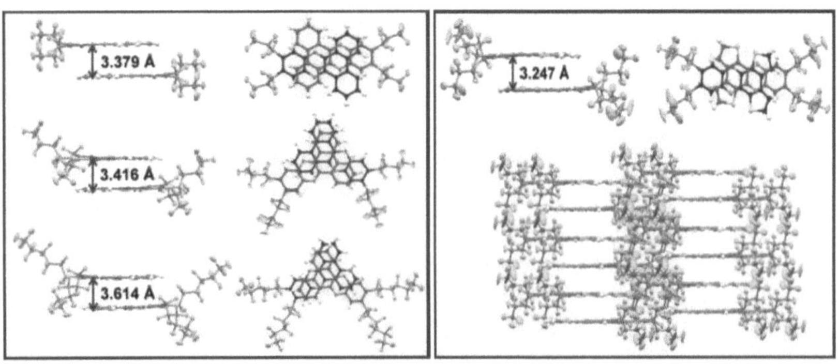

Figure 7.3 Examples of crystal packing for compounds **15**, **16**, and **17** (left) and **11** (right) shown in Chart 7.3.
Reprinted with permission from A. Putta, J.D. Mottishaw, Z. Wang, H. Sun, Cryst. Growth Des. 14 (2014) 350–356. Copyright (2014) American Chemical Society.

The F—F interactions between perfluoroalkyl chains are quite important for the formation of lamellar structures. All perfluoroalkylated PAH crystalline structures of our group have been solved and reported in the literature showing the perfluoroalkyl chains aggregate together to form a "fluorine" phase [2,5,6,10,41,72,83—85]. These interactions, though weak, could act as "glue" to bond together perfluoroalkylated PAHs. One analog of this phenomenon is what has been observed in the liquid phase where fluorocarbon solvents are immiscible with hydrocarbon solvents.

The packing motif of perfluoroalkylated large conjugated aromatic and heteroaromatic systems, as previously discussed, is typically the lamellar stacked structure. The advantage of this structure is that it not only hinders or prevents T-shaped and herringbone packing, it has π—π stacking while shifted or displaced so the stacking is not directly on top of one another. This reduces the steric hindrance of the perfluoroalkyl chains, preserves the stability effects of the perfluoroalkyl chains, and maintains the π—π stacked interactions to potentially increase the charge mobility [48,83—85].

7.4 Physicochemical Properties of Perfluoroalkylated Polyaromatic Hydrocarbons

Earlier, we discussed how the electronegativity of fluorine can result in unique properties in the corresponding PAHs. These properties, such as superhydrophobicity and fluorophilicity, are the results of incorporating fluorine in addition to its electron-withdrawing ability. Electrochemical analysis of perfluoroalkylated PAHs clearly demonstrates the electron-withdrawing ability of the perfluoroalkyl substituents. These results demonstrate greater induction effects of perfluoroalkyl groups; yet, with almost no resonance effect compared to other commonly used electron-withdrawing groups, such as —F, —Cl, —CN, $-NO_2$, —COOR, where both induction and resonance effects coexist.

7.4.1 Changes in Electrochemical Properties Upon Perfluoroalkylation

Early work done by Therien and DiMagno et al. shows the first reduction potential of meso-perfluoropropylporphyrin (**9**) shifts more than 670 mV to positive compared to that of tetraphenylporphyrin [16]. It is also interesting to observe positive shifts in the redox potential for other perfluoroalkylated PAHs. Strauss/Boltalina's group reported a 950 mV positive shift in its first reduction potential of compound **10** when compared to nonsubstituted one [41]. Furthermore, the degree of redox potential shifts seems to not only depend on the number and substitution position of the perfluoroalkyl chains, but also depends on the π-conjugation size of the PAHs involved. This is perhaps not a surprise as the HOMO and LUMO energy levels will change as a result of increasing π-conjugation.

Our recent work systematically studied the predicted redox potential changes as a result of perfluoroalkylation of anthracene (Fig. 7.4). The first reduction potentials of

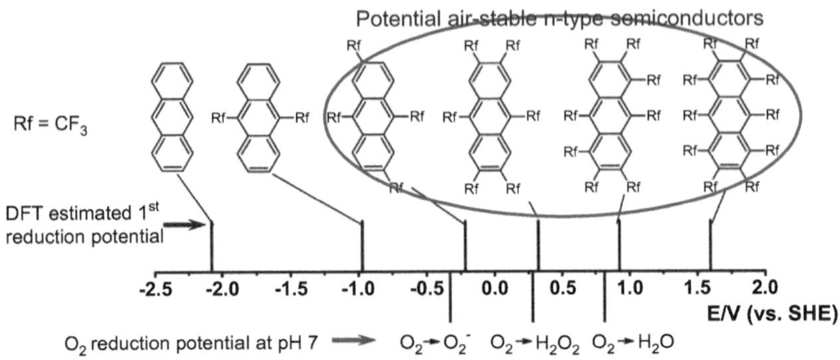

Figure 7.4 Estimated first reduction potential of trifluoromethylated anthracenes from DFT calculated EA based upon the semiempirical correlation between EA and redox potential for polyaromatics, assuming the solvation free energy is −1.99 eV [70] in organic solvents. *DFT*, discrete Fourier transform; *EA*, electron affinity.
Reprinted with permission from H. Sun, A. Putta, M. Billion, J. Phys. Chem. A 116 (2012) 8015−8022. Copyright (2012) American Chemical Society.

perfluoroalkylated anthracenes could shift over 3 V compared to that of nonsubstituted anthracene. The last two compounds on the right side of this figure are hypothetically proposed, and likely will be very difficult to prepare, if not impossible at all. Tetra- and hexa-perfluoroalkylated anthracenes are potentially air-stable n-type organic semiconductor materials due to the fact that perfluoroalkyl groups can potentially block the proton source, water, from the moisture in the air [83−85].

7.4.2 Changes in Photophysical and Photochemical Properties Upon Perfluoroalkylation

In addition to the dramatic changes in electronic properties of the PAHs upon perfluoroalkylation, their photophysical and photochemical properties are also changed. Early work by DiMagno and coworkers studied both electrochemical and photophysical properties of the perfluoroalkylated porphyrin (**9**) and its corresponding metalloporphyrins [16,17]. These studies clearly demonstrated the influence of perfluoroalkyl substitution on redox potential and photophysical properties of the perfluoroalkylated porphyrins. Our recent work showed that the fluorescence quantum yield of anthracene and pyrene significantly increases after perfluorooctylation (Compounds 22−24, Chart 7.4) [83−85].

Furthermore, we observed that photochemical stability of these perfluoroalkylated PAHs increases significantly upon perfluoroalkylation. This increase in photostability for perfluoroalkylated PAHs is a monumental step in the applications of organic electronics, particularly for organic solar cells in which stability is paramount for real world application. Increases in photochemical stability of the aromatic core are believed to be through the inhibition of the endoperoxide formation [83−85]. The use of fluorinated solvents further improves the photochemical stability of the PAHs. These results point out the possibility of use fluorinated media, for example, fluorinated polymers, to stabilize the organic solar cells.

7.4.3 n-Type Semiconductor Performance of Perfluoroalkylated Polyaromatic Hydrocarbons

The performance of n-type organic semiconductors is generally characterized by measuring electron mobility in OFETs under either vacuum or in air depending on the materials stability. So far, the most used and successful n-type organic semiconductor materials among perfluoroalkylated PAHs are perylenediimide- and naphthalenediimide-based small molecules and corresponding polymeric derivatives [91]. Perfluoroalkylated thiophene and thiazole–thiophene oligomers are another type of promising n-type organic semiconductors which showed high charge mobility and sometime air-stability [22,44,91,92].

In a recent review on n-type organic semiconductor materials, Marks and coworkers clearly summarized the recent progresses with pinpoint of both successes and challenges in this field [92] . Resummarizing detailed charge mobilities, on/off ratio, threshold voltage, and potential applications in different n-type organic semiconductor devices seems redundant since readers can easily refer to those original and review articles to access the device properties [3,12,19,51]. At current development stage of n-type organic semiconductor materials, it is still too early to sketch a clear picture of the correlation between molecular structures and device performance. Out of the coexisting opportunities and challenges in today's n-type organic semiconductor fields, we would argue that one of the key solutions for these challenges is to fundamentally understand the weak noncovalent intermolecular interactions that govern the solid-state structures of the bulk organic semiconductor materials and the structures of the interface between different solid-state materials (eg, semiconductors and metal electrodes) and apply this understanding to the systematic design of the high performance air-stable n-type organic semiconductor devices.

7.5 Summary and Perspective

As a result of the development of new perfluoroalkylation methods and reagents over the past 30 plus years, synthesis, characterization, and semiconductor applications of perfluoralkylated PAHs have gradually evolved. The strong electron-withdrawing and low polarizable nature of the perfluoroalkyl substituents lead to the unique properties of the perfluoroalkylated PAHs including formation of highly electron-deficient π-conjugated systems, significantly positive shifts of the redox potentials, IPs and EAs, changes in the bandgaps, and increasing both chemical and photochemical stabilities. The superhydrophobicity and oleophobicity of the perfluoralkylated PAHs further provides the basis for solution processability. The unique ability of perfluoroalkyl groups to control crystalline structures of the PAH cores is a quickly growing point that is expected to benefit the organic materials sector through materials by design. Some of the bigger challenges in this area are still to be solved: (1) fundamentally understanding on how F ••• F interactions affect the solid-state structure; (2) access the ET kinetics in these perfluoroalkylated PAHs at solid state; and (3) how the interface between perfluoroalkylated PAHs and metal electrodes affect the organic semiconductor device performance which is still untouched.

Acknowledgment

Authors are thankful for the financial support from the NSF CAREER program (grant number CHE1355677).

References

[1] K. Aikawa, Y. Nakamura, Y. Yokota, W. Toya, K. Mikami, Stable but reactive perfluoroalkyl zinc reagents: application in ligand-free copper-catalyzed perfluoroalkylation of aryl iodides, Chem. Eur. J. 21 (2015) 96−100.
[2] A. Putta, J.D. Mottishaw, Z. Wang, H. Sun, Rational design of lamellar pi-pi stacked organic crystalline materials with short interplanar distance, Cryst. Growth Des. 14 (2014) 350−356.
[3] J.E. Anthony, Functionalized acenes and heteroacenes for organic electronics, Chem. Rev. 106 (2006) 5028−5048.
[4] F. Babudri, G.M. Farinola, F. Naso, R. Ragni, Fluorinated organic materials for electronic and optoelectronic applications: the role of the fluorine atom, Chem. Commun. (2007) 1003−1022.
[5] R.J. Baker, P.E. Colavita, D.M. Murphy, J.A. Platts, J.D. Wallis, Fluorine-fluorine interactions in the solid state: an experimental and theoretical study, J. Phys. Chem. A 116 (2012) 1435−1444.
[6] M.O. BaniKhaled, J.D. Mottishaw, H. Sun, Steering power of perfluoroalkyl substituents in crystal engineering: tuning the pi-pi distance while maintaining the lamellar packing motif for aromatics with various sizes of pi-conjugation, Cryst. Growth Des. 15 (2015) 2235−2242.
[7] B. Lantano, S. B.-V, M.R. Torviso, S.M. Garcia, A. Tinnirello, Al. Postigo, Direct C-H perfluoroalkylation of (di)benzo(hetero)arenes in aqueous media, J. Fluorine Chem. 161 (2014) 149−155.
[8] M. Bolloli, J. Kalhoff, F. Alloin, D. Bresser, M.L.P. Le, B. Langlois, S. Passerini, J.-Y. Sanchez, Fluorinated carbamates as suitable solvents for LiTFSI-based lithium-ion electrolytes: physicochemical properties and electrochemical characterization, J. Phys. Chem. C 119 (2015) 22404−22414.
[9] O.V. Boltalina, A.A. Popov, I.V. Kuvychko, N.B. Shustova, S.H. Strauss, Perfluoroalkylfullerenes, Chem. Rev. 115 (2015) 1051−1105.
[10] T.T. Clikeman, E.V. Bukovsky, I.V. Kuvychko, L.K. San, S.H.M. Deng, X.-B. Wang, Yu-S. Chen, S.H. Strauss, O.V. Boltalina, Poly(trifluoromethyl)azulenes: structures and acceptor properties, Chem. Commun. 50 (2014) 6263−6266.
[11] B. Cornils, Fluorous biphase systems: the new phase-separation and immobilization technique, Angew. Chem. Int. Ed. Engl. 36 (1997) 2057−2059.
[12] V. Coropceanu, J. Cornil, D.A. Da Silva Filho, Y. Olivier, R. Silbey, J.-L. Bredas, Charge transport in organic semiconductors, Chem. Rev. 107 (2007) 926−952.
[13] L. Cui, Y. Matusaki, N. Tada, T. Miura, B. Uno, A. Itoh, Metal-free direct C-H perfluoroalkylation of arenes and heteroarenes using a photoredox organocatalyst, Adv. Synth. Catal. 355 (2013) 1615−4150.
[14] E. Di Donato, R.P. Fornari, S. Di Motta, Y. Li, Z. Wang, n-Type charge transport and mobility of fluorinated perylene bisimide semiconductors, J. Phys. Chem. B 114 (2010) 5327−5334.

[15] Y. Diao, K.M. Lenn, W.-Y. Lee, M.A. Blood-Forsythe, J. Xu, Y. Mao, Y. Kim, J.A. Reinspach, S. Park, A. Aspuru-Guzik, G. Xue, P. Clancy, Z. Bao, C.B. Mannsfeld Stefan, Understanding polymorphism in organic semiconductor thin films through nanoconfinement, J. Am. Chem. Soc. 136 (2014) 17046–17057.
[16] S.G. DiMagno, R.A. Williams, M.J. Therien, Facile synthesis of meso-Tetrakis (perfluoroalkyl)porphyrins: spectroscopic properties and x-ray crystal structure of highly electron-deficient 5,10,15,20-Tetrakis(heptafluoropropyl)porphyrin, J. Org. Chem. 59 (1994) 6943–6948.
[17] S.G. DiMagno, P.H. Dussault, J.A. Schultz, Fluorous biphasic singlet oxygenation with a perfluoroalkylated photosensitizer, J. Am. Chem. Soc. 118 (1996) 5312–5313.
[18] W. Dmowski, 1,2-Dibromotetrafluoroethan (Freon 114B2) as a building block for fluorine compounds, J. Fluorine Chem. 142 (2012) 6–13.
[19] L. Dou, Y. Liu, Z. Hong, G. Li, Y. Yang, Low-bandgap near-IR conjugated polymers/molecules for organic electronics, Chem. Rev. 115 (2015) 12633–12665.
[20] A. Facchetti, J. Letizia, M.-H. Yoon, M. Mushrush, H.E. Katz, T.J. Marks, Synthesis and characterization of diperfluorooctyl-substituted phenylene-thiophene oligomers as n-type semiconductors. Molecular structure-film microstructure-mobility relationships, organic field-effect transistors, and transistor nonvolatile memory elements, Chem. Mater. 16 (2004) 4715–4727.
[21] A. Facchetti, M.-H. Yoon, L. Charlotte, G.R. Hutchison, A. Mark, T.J. Marks, Building blocks for N-Type molecular and polymeric electronics. Perfluoroalkyl- versus alkyl-functionalized oligothiophenes (nTs; $n = 2-6$). Systematic synthesis, spectroscopy, electrochemistry, and Solid-State organization, J. Am. Chem. Soc. 126 (2004) 13480–13501.
[22] A. Facchetti, M. Mushrush, M.-H. Yoon, G.R. Hutchison, M.A. Ratner, T.J. Marks, Building blocks for n-type molecular and polymeric electronics. Perfluoroalkyl- versus alkyl-functionalized oligothiophenes (nT; $n = 2-6$). Systematics of thin film microstructure, semiconductor performance, and modeling of majority charge injection in field-effect transistors, J. Am. Chem. Soc. 126 (2004) 13859–13874.
[23] M.L. Frost, J.E. Compston, D. Goldsmith, A.E. Moore, G.M. Blake, M. Siddique, L. Skingle, I. Fogelman, 18F-fluoride positron emission tomography measurements of regional bone formation in hemodialysis patients with suspected adynamic bone disease, Calcified Tissue Int. 93 (2013) 436–447.
[24] J.A. Gladysz, C. Emnet, Fluorous solvents and related media, in: J.A. Gladysz, D.P. Curran, I.T. Horvath (Eds.), Handbook of Fluorous Chemistry, Wiley-VCH Verlag GmbH & Co. KGaA, Weinheim, FRG, 2004, pp. 11–23.
[25] A. Lishchynskyi, V.V. Grushin, Cupration of C2F5H: isolation, structure, and synthetic applications of [K(DMF)2][(t-BuO)Cu(C2F5)]. Highly efficient pentafluoroethylation of unactived aryl bromides, J. Am. Chem. Soc. 135 (2013) 12584–12587.
[26] M. Gsaenger, J.H. Oh, M. Koenemann, H.W. Hoeffken, A.-M. Krause, Z. Bao, F. Wuerthner, A crystal-engineered hydrogen-bonded octachloroperylene diimide with a twisted core: an n-channel organic semiconductor, Angew. Chem. Int. Ed. 49 (2010) 740–743. S740/741-S740/710.
[27] C. Hansch, A. Leo, R.W. Taft, A survey of hammett substituent constants and resonance and field parameters, Chem. Rev. 91 (1991) 165–195.
[28] L. He, K. Natte, J. Rabeah, C. Taeschler, H. Neumann, A. Bruckner, M. Beller, Heterogeneous platinum-catalyzed C-H perfluoroalkylation of arenes and heteroarenes, Angew. Chem. Int. Ed. 54 (2015) 4320–4324.
[29] M. He, L. Hu, Z. Xue, C. Cheung Su, P. Redfern, L.A. Curtiss, B. Polzin, A.von Cresce, K. Xu, Z. Zhang, Fluorinated electrolytes for 5-V Li-Ion chemistry: probing voltage

stability of electrolytes with electrochemical floating test, J. Electrochem. Soc. 162 (2015) A1725–A1729.
[30] K.G. Huang, M. Julie, Stability of nonfouling electroless nickel-polytetrafluoroethylene coatins after exposure to commercial dairy equipment sanitizers, J. Dairy Sci. 98 (2015) 5983–5994.
[31] Y. Ji, H. Waiczies, L. Winter, P. Neumanova, D. Hofmann, J. Rieger, R. Mekle, S. Waiczies, T. Niendorf, Eight-channel transceiver RF coil array tailored for 1H/19F MR of the human knee and fluorinated drugs at 7.0 T, NMR Biomed. 28 (2015) 726–737.
[32] B.A. Jones, M.J. Ahrens, M.-H. Yoon, A. Facchetti, T.J. Marks, M.R. Wasielewski, Electron transport: high-mobility air-stable n-type semiconductors with processing versatility: dicyanoperylene-3,4:9,10-bis(dicarboximides), Angew. Chem. Int. Ed. 43 (2004) 6363–6366.
[33] B.A. Jones, A. Facchetti, M.R. Wasielewski, T.J. Marks, Tuning orbital energetics in arylene diimide semiconductors. Materials design for ambient stability of n-type charge transport, J. Am. Chem. Soc. 129 (2007) 15259–15278.
[34] J. Xu, X. Liu, Y. Fu, Recent advance in transition-metal-mediated trifluoromethylation for the construction of $C(sp^3)$-CF_3 bonds, Tetrahedron Lett. 55 (2014) 585–594.
[35] J.A. Gladysz, M. Jurisch, Structural, physical, and chemical properties of fluorous compounds, Top Curr. Chem. 308 (2012) 1–24.
[36] T. Kamata, H. Sasabe, Y. Watanabe, D. Yokoyama, H. Katagiri, J. Kido, A series of fluorinated phenylpyridine-based electron-transporters for blue phosphorescent OLEDs, J. Mater. Chem. C 4 (2015) 1104–1110.
[37] P.P. Kapadia, L.R. Ditzler, J. Baltrusaitis, D.C. Swenson, A.V. Tivanski, F.C. Pigge, Semiconducting organic assemblies prepared from tetraphenylethylene tetracarboxylic acid and bis(pyridine)s via charge-assisted hydrogen bonding, J. Am. Chem. Soc. 133 (2011) 8490–8493.
[38] H. Kato, K. Hirano, D. Kurauchi, N. Toriumi, M. Uchiyama, Dialkylzinc-mediated cross-coupling reactions of perfluoroalkyl and perfluoroaryl halides with aryl halides, Chem. Eur. J. 21 (2015) 3895–3900.
[39] V.N. Korotchenko, K. Senerin, M.R. Gagne, The aerobic oxidation of alcohols with a ruthenium porphyrin catalyst in organic and fluorinated solvents, Org. Biomol. Chem. 6 (2008) 1961–1965.
[40] I.V. Kuvychko, S.H. Strauss, O.V. Boltalina, Solution-phase perfluoroalkylation of C60 leads to efficient and selective synthesis of bis-perfluoroalkylated fullerenes, J. Fluorine Chem. 143 (2012) 103–108.
[41] I.V. Kuvychko, S.N. Spisak, Y.-S. Chen, A.A. Popov, M.A. Petrukhina, S.H. Strauss, O.V. Boltalina, A buckybowl with a lot of potential: C5-C20H5(CF3)5, Angew. Chem. Int. Ed. 51 (2012) 4939–4942.
[42] Y. Le, Y. Umemoto, M. Nitani, Y. Aso, Perfluoroalkyl-annelated conjugated systems toward n-type organic semiconductors, Pure Appl. Chem. 80 (2009) 589–597.
[43] J.A. Letizia, M.R. Salata, C.M. Tribout, A. Facchetti, M.A. Ratner, T.J. Marks, n-Channel polymers by design: optimizing the interplay of solubilizing substituents, crystal packing, and field-effect transistor characteristics in polymeric bithiophene-imide semiconductors, J. Am. Chem. Soc. 130 (2008) 9679–9694.
[44] Y. Li, L. Tan, Z. Wang, H. Qian, Y. Shi, W. Hu, Air-stable n-type semiconductor: core-perfluoroalkylated perylene bisimides, Org. Lett. 10 (2008) 529–532.
[45] Y. Li, C. Li, W. Yue, W. Jiang, R. Kopecek, J. Qu, Z. Wang, Direct functionalization of polycyclic aromatics via radical perfluoroalkylation, Org. Lett. 12 (2010) 2374–2377.

[46] T. Liang, C.N. Neumann, T. Ritter, Introduction of fluorine and fluorine-containing functional groups, Angew. Chem. Int. Ed. 52 (2013) 8214–8264.
[47] C. Liu, D.-M. Shen, Q.-Y. Chen, Fluorous biphasic catalytic oxidation of alkenes and aldehydes with air and 2-Methylpropanal in the presence of (B–Perfluoroalkylated tetraphenylporphyrin)cobalt complexes, Eur. J. Org. Chem. (2006) 2703–2706.
[48] L. Liu, G. Yang, Y. Duan, Y. Geng, Y. Wu, Z. Su, The relationship between intermolecular interactions and charge transport properties of trifluoromethylated polycyclic aromatic hydrocarbons, Org. Electron. 15 (2014) 1896–1905.
[49] M. Losurdo, M.M. Giangregorio, P. Capezzuto, G. Bruno, F. Babudri, A. Cardone, C. Martinelli, G.M. Farinola, F. Naso, M. Buechel, Impact of fluorinated vinylene units on supramolecular organization and optical properties of poly(p-phenylenedifluorovinylene) thin films as a class of blue band gap conjugated polymers, Polymer 49 (2008) 4133–4140.
[50] Y. Mace, E. Magnier, The new age of electrophilic perfluoroalkylation reactions, Eur. J. Org. Chem. (2012) 2479–2494.
[51] L. Martiradonna, Organic semiconductors fast crystal patterning, Nat. Mater. 14 (2015) 1186.
[52] Y. Matsubara, A. Kimura, Y. Yamaguchi, Z.-I. Yoshida, Meso-disubstituted anthracenes with fluorine-containing groups: synthesis, light-emitting characteristics, and photostability, Org. Lett. 10 (2008) 5541–5544.
[53] M. Matsugi, M. Hasegawa, S. Hasebe, S. Takai, R. Suyama, Y. Wakita, K. Kudo, H. Imamura, T. Hayashi, S. Haga, Direct perfluoroalkylation of non-activated C-H bonds of phenols, Tetrahedron Lett. 49 (2008) 4189–4191.
[54] H. Meng, F. Sun, M.B. Goldfinger, F. Gao, D.J. Londono, W.J. Marshal, G.S. Blackman, K.D. Dobbs, D.E. Keys, 2,6-Bis[2-(4-pentylphenyl)vinyl]anthracene: a stable and high charge mobility organic semiconductor with densely packed crystal structure, J. Am. Chem. Soc. 128 (2006) 9304–9305.
[55] D.Y. Mikhaylov, Y.H. Budnikova, Fluoroalkylation of organic compounds, Russ. Chem. Rev. 82 (2013) 835–864.
[56] B. Milian Medina, D. Beljonne, H.-J. Egelhaaf, J. Gierschner, Effect of fluorination on the electronic structure and optical excitations of pi-conjugated molecules, J. Chem. Phys. 126 (2007) 111101.
[57] S. Mohr, L.E. Ratcliff, L. Genovese, D. Caliste, P. Boulanger, S. Goedecker, T. Deutsch, Accurate and efficient linear scaling DFT calculations with universal applicability, PCCP 17 (2015) 31360–31370.
[58] K. Monir, A.K. Bagdi, M. Ghosh, A. Hajra, Regioselective oxidative trifluoromethylation of imidazoheterocycles via C(sp^2)-H bond functionalization, J. Org. Chem. 80 (2015) 1332–1337.
[59] H. Morimoto, T. Tsubogo, N.D. Litvinas, J.F. Hartwig, A broadly applicable copper reagent for trifluoromethylations and perfluoroalkylations of aryl iodides and bromides, Angew. Chem. Int. Ed. 50 (2011) 3793–3798.
[60] P.M. Murphy, C.S. Baldwin, R.C. Buck, Syntheses utilizing n-perfluoroalkyl iodides [R_fI, C_nF_{2n+1}-I] 2000-2010, J. Fluorine Chem. 138 (2012) 3–23.
[61] M. Mushrush, A. Facchetti, M. Lefenfeld, H.E. Katz, T.J. Marks, Easily processable phenylene-thiophene-based organic field-effect transistors and solution-fabricated nonvolatile transistor memory elements, J. Am. Chem. Soc. 125 (2003) 9414–9423.
[62] M.-H. Yoon, A. Facchetti, C.E. Stern, T.J. Marks, Fluorocarbon-Modified organic semiconductors: molecular architecture, electronic, and crystal structure tuning of arene- versus fluoroarene-thiophene oligomer thin-film properties, J. Am. Chem. Soc. 128 (2006) 5792–5801.

[63] K. Olofsson, S.-Y. Kim, M. Larhed, D.P. Curran, A. Hallberg, High-speed, highly fluorous organic reactions, J. Org. Chem. 64 (1999) 4539–4541.

[64] T. Ono, Y. Hayakawa, Fluorine-based crystal engineering, ACS Symp. Ser. 911 (2005) 498–513 (Fluorine-Containing Synthons).

[65] V. Percec, M. Glodde, T.K. Bera, Y. Miura, I. Shiyanovskaya, K.D. Singer, V.S.K. Balagurusamy, P.A. Heiney, I. Schnell, A. Rapp, H.W. Spiess, S.D. Hudson, H. Duan, Self-organization of supramolecular helical dendrimers into complex electronic materials, Nature 419 (2002) 384–387.

[66] A. Postigo, Aromatic radical perfluoroalkylation reactions, Can. J. Chem. 90 (2012) 493–497.

[67] Q. Qi, Q. Shen, L. Lu, Copper-mediated aerobic fluoroalkylation of arylboronic acids with fluoroalkyl iodides at room temperature, J. Am. Chem. Soc. 134 (2012) 6548–6551.

[68] L. Qiao, D. Agoumba, A. Waterfeld, J.S. Thrasher, Improvement of polymer electrolyte membranes for polymer electrolyte membrane and direct methanol fuel cell applications, Proc.–Electrochemical Soc. 2000–22 (2001) 92–102 (Power Sources for the New Millennium).

[69] K. Reichenbacher, H.I. Süssa, J. Hulliger, Fluorine in crystal engineering-"the little atom that could", Chem. Soc. Rev. 34 (2005) 22–30.

[70] R.S. Ruoff, K.M. Kadish, P. Boulas, E.C.M. Chen, Relationship between the electron affinities and half-wave reduction potentials of fullerenes, aromatic hydrocarbons, and metal complexes, J. Phys. Chem. 99 (1995) 8843–8850.

[71] Y. Sakamoto, T. Suzuki, M. Kobayashi, Y. Gao, Y. Fukai, Y. Inoue, F. Sato, S. Tokito, Perfluoropentacene: high-performance p-n junctions and complementary circuits with pentacene, J. Am. Chem. Soc. 126 (2004) 8138–8140.

[72] L.K. San, E.V. Bukovsky, I.V. Kuvychko, A.A. Popov, S.H. Strauss, O.V. Boltalina, Single-step gas-phase polyperfluoroalkylation of naphthalene leads to thermodynamic products, Chem. Eur. J. 20 (2014) 4373–4379.

[73] R.S. Sanchez-Carrera, P. Paramonov, G.M. Day, V. Coropceanu, J.-L. Bredas, Interaction of charge carriers with lattice vibrations in oligoacene crystals from naphthalene to pentacene, J. Am. Chem. Soc. 132 (2010) 14437–14446.

[74] Y. Ye, M.S. Sanford, Merging visible-light photocatalysis and transition-metal catalysis in the copper-catalyzed trifluoromethylation of boronic acids with CF3I, J. Am. Chem. Soc. 134 (2012) 9034–9037.

[75] R. Schmidt, J.H. Oh, Ya-S. Sun, M. Deppisch, A.-M. Krause, K. Radacki, H. Braunschweig, M. Koenemann, P. Erk, Z. Bao, F. Wuerthner, High-performance air-stable n-channel organic thin film transistors based on halogenated perylene bisimide semiconductors, J. Am. Chem. Soc. 131 (2009) 6215–6228.

[76] Y. Shiroata, Organic materials for electronic and optoelectronic devices, J. Mater. Chem. 10 (2000) 1–25.

[77] M. Slodowicz, S. Barata-Vallejo, A. Vazquez, N.S. Nudelman, A. Postigo, Light-induced iodoperfluoroalkylation reactions of carbon-carbon multiple bonds in water, J. Fluorine Chem. 135 (2012) 137–143.

[78] N. Spetseris, S. Hadida, D.P. Curran, T.Y. Meyer, Organic/fluorous phase extraction: a new tool for the isolation of organometallic complexes, Organometallics 17 (1998) 1458–1459.

[79] J. Sreedhar Reddy, T. Kale, G. Balaji, A. Chandrasekaran, S. Thayumanavan, Cyclopentadithiophene-based organic semiconductors: effect of fluorinated substituents on electrochemical and charge transport properties, J. Phys. Chem. Lett. 2 (2011) 648–654.

[80] H. Sun, S.G. DiMagno, Anhydrous tetrabutylammonium fluoride, J. Am. Chem. Soc. 127 (2005) 2050–2051.
[81] H. Sun, S.G. DiMagno, Room-temperature nucleophilic aromatic fluorination: experimental and theoretical studies, Angew. Chem. Int. Ed. 45 (2006) 2720–2725.
[82] H. Sun, S.G. DiMagno, Fluoride relay: a new concept for the rapid preparation of anhydrous nucleophilic fluoride salts from KF, Chem. Commun. 5 (2007) 528–529.
[83] H. Sun, A. Putta, M. Billion, Arene trifluoromethylation: an effective strategy to obtain air-stable n-type organic semiconductors with tunable optoelectronic and electron transfer properties, J. Phys. Chem. A 116 (2012) 8015–8022.
[84] H. Sun, A. Putta, J.P. Kloster, U.K. Tottempudi, Unexpected photostability improvement of aromatics in polyfluorinated solvents, Chem. Commun. 48 (2012) 12085–12087.
[85] H. Sun, U.K. Tottempudi, J.D. Mottishaw, P.N. Basa, A. Putta, A.G. Sykes, Strengthening pi-pi interaction while suppressing Csp2-H–pi (T-Shaped) interactions via perfluoroalkylation: a crystallographic and computational study that supports the beneficial formation of 1-D pi-pi stacked aromatic materials, Cryst. Growth Des. 12 (2012) 5655–5662.
[86] J.D. Mottishaw, H. Sun, Effects of aromatic trifluoromethylation, fluorination, and methylation on intermolecular pi-pi interactions, J. Phys. Chem. A 117 (2013) 7970–7979.
[87] A. Synytska, D. Appelhans, Z.G. Wang, F. Simon, F. Lehmann, M. Stamm, K. Grundke, Perfluoroalkyl end-functionalized oligoesters: correlation between wettability and end-group segregation, Macromolecules 40 (2007) 297–305.
[88] T. Taldone, P.D. Patel, H.J. Patel, G. Chiosis, About the reaction of aryl fluorides with sodium sulfide: investigation into the selectivity of substitution of fluorobenzonitriles to yield mercaptobenzonitriles via SnAr displacement of fluorine, Tetrahedron Lett. 53 (2012) 2548–2551.
[89] T. Umemoto, Electrophilic perfluoroalkylating agents, Chem. Rev. 96 (1996) 1757–1777.
[90] T.A. Unzner, T. Magauer, Carbon-fluorine bond activation for the synthesis of functionalized molecules, Tetrahedron Lett. 56 (2015) 877–883.
[91] H. Usta, A. Facchetti, T.J. Marks, n-Channel semiconductor materials design for organic complementary circuits, Acc. Chem. Res. 44 (2011) 501–510.
[92] H. Usta, W.C. Sheets, M. Denti, G. Generali, R. Capelli, S. Lu, X. Yu, M. Muccini, A. Facchetti, Perfluoroalkyl-functionalized thiazole-thiophene oligomers as N-channel semiconductors in organic field-effect and light-emitting transistors, Chem. Mater. 26 (2014) 6542–6556.
[93] V.M. Vlasov, Fluoride ion as a nucleophile and a leaving group in aromatic nucleophilic substitution reactions, J. Fluorine Chem. 61 (1993) 193–216.
[94] V. Vlcek, H.R. Eisenberg, G. Steinle-Neumann, L. Kronik, R. Baer, Deviations from piecewise linearity in the solid-state limit with approximate density functionals, J. Chem. Phys. 142 (2015), 034107/034101-034107/034110.
[95] M.S. Wiehn, E.V. Vinogradova, A. Togni, Electrophilic trifluoromethylation of arenes and N-heteroarenes using hypervalent iodine reagents, J. Fluorine Chem. 131 (2010) 951–957.
[96] J. Xia, M. Nie, J.C. Burns, A. Xiao, W.M. Lamanna, J.R. Dahn, Fluorinated electrolyte for 4.5 V Li(Ni0.4Mn0.4Co0.2)O2/graphite Li-ion cells, J. Power Sources 307 (2016) 340–350.
[97] S. Yanagida, Y. Hasegawa, K. Murakoshi, Y. Wada, N. Nakashima, T. Yamanaka, Strategies for enhancing photoluminescence of Nd^{3+} in liquid media, Coord. Chem. Rev. 171 (1998) 461–480.

[98] M.-H. Yoon, S.A. Dibenedetto, A. Facchetti, J. Tobin, Organic thin-film transistors based on carbonyl-functionalized quaterthiophenes: high mobility N-Channel semiconductors and ambipolar transport, J. Am. Chem. Soc. 127 (2005) 1348–1349.
[99] J. Youn, S. Vegiraju, J.D. Emery, B.J. Leever, S. Kewalramani, S.J. Lou, S. Zhang, K. Prabakaran, Y. Ezhumalai, C. Kim, P.-Y. Huang, C. Stern, W.-C. Chang, M.J. Bedzyk, L.X. Chen, M.-C. Chen, A. Facchetti, T.J. Marks, Diperfluorophenyl Fused thiophene semiconductors for n-type organic thin Film transistors (OTFTs), Adv. Electron. Mater. 1 (2015), 1500098/1500091-1500098/1500016.
[100] Z. Yuan, J. Li, Y. Xiao, Z. Li, X. Qian, Core-perfluoroalkylated perylene diimides and naphthalene diimides: versatile synthesis, solubility, electrochemistry, and optical properties, J. Org. Chem. 75 (2010) 3007–3016.
[101] D. Zeng, J. Chen, Z. Chen, W. Zhu, J. He, F. Yu, H. Huang, H. Wu, C. Liu, S. Ren, J. Du, J. Sun, E. Xu, A. Cao, Q. Fang, Efficient improvement of fluorescence quantum yield of fluoreneethynylene-based polymers by introducing a perfluoroalkylbenzene unit to the polymers, Macromol. Rapid Commun. 28 (2007) 772–779.
[102] Q. Zhang, Z. Luo, D.P. Curran, Separation of "light fluorous" reagents and catalysts by fluorous solid-phase extraction: synthesis and study of a family of triarylphosphines bearing linear and branched fluorous tags, J. Org. Chem. 65 (2000) 8866–8873.
[103] W. Zhang, Fluorous technologies for solution-phase high-throughput organic synthesis, Tetrahedron 59 (2003) 4475–4489.
[104] X. Zhao, Y. Su, Y. Liu, R. Zhang, Z. Jiang, Multiple antifouling capacities of hybrid membranes derived from multifunctional titania nanoparticles, J. Membr. Sci. 495 (2015) 226–234.

Electronic Structure of Fluorinated Graphene

8

L.G. Bulusheva, A.V. Okotrub
Nikolaev Institute of Inorganic Chemistry, Siberian Branch of Russian Academy of Sciences, Novosibirsk, Russia

Chapter Outline

8.1 Introduction

Graphene, an ultrathin sheet consisting of a single element, is an ideal object for an investigation of distinctive features of the reactivity of a two-dimensional crystal. Moreover, attachment of any chemical species to the graphene is essential for opening the energy gap, which is normally zero [1]. A control of the gap is a decisive factor for making graphene-based transistors and optoelectronic devices.

A planar honeycomb lattice of graphene, where every carbon atom strongly interacts with three neighbors through σ– and π–electrons, is quite inert to a chemical action. Experimentally, spending of π–electron from each graphene atom for covalent bonding was achieved only in the case of fluorination and hydrogenation [2,3]. Oxidation is a more complex process, which produces a set of oxygen-containing groups inhomogeneously distributed on the basal graphene plane and edges [4].

New Fluorinated Carbons: Fundamentals and Applications. http://dx.doi.org/10.1016/B978-0-12-803479-8.00008-5

Among the chemical elements, fluorine possesses the largest electronegativity and owing to this, it readily attaches to carbon forming a polar covalent bond. Each such act removes an electron from delocalized π−electron cloud thus modifying the initially semimetal state of graphene.

For the precise tuning of electronic structure and properties of graphene, a certain amount of fluorine atoms should be placed in a specific location of the plane. Currently, engineering of the properties of graphene with a functionalization is a rapidly developing field [5] because, particularly, the experimental efforts are supported by extensive quantum-chemical calculations. As the calculations are more abundant and often give a clearer and more representative picture, we will focus primarily on these results. Throughout the chapter, the calculations will be compared with the available experimental data. We begin with a description of modern concepts of the structure and chemical bonding in graphite fluorides, which can serve as a source of graphene fluorides, highlight the character of the graphene in a fluorination, and then discuss every specific feature of the fluorinated graphene in details.

8.2 Brief Guide to Graphite Fluorides

Graphite interacts with elemental fluorine at elevated temperatures and this reaction has been reported for the first time in 1934 [6]. The progress in synthesis and structural characterization of graphite fluorides achieved during the next 50 years has been described in the book by N. Watanabe et al. [7]. Fluorine bonding to each carbon atom of a graphene plane requires a temperature of ~600°C. The produced compound has an approximate composition of $(CF)_n$ and is named poly(carbon monofluoride) or graphite monofluoride. A decrease of the temperature to a value not lower than 350°C causes bonding of the fluorine atom with every second carbon atom, which yields poly(dicarbon monofluoride) $(C_2F)_n$. Due to a difficulty to grow single crystals for X-ray diffraction (XRD) study, the exact structures of these compounds are still unknown.

An analysis of XRD patterns for powders indicated that fluorine atoms are located at both sides of the graphene sheet in $(CF)_n$ [8]. A fully fluorinated carbon hexagon may take chair- or boat-like form and quantum-chemical calculations showed that the former conformation is more stable thermodynamically [9]. To convert the planar carbon hexagon into the chair conformation, fluorine atoms should be

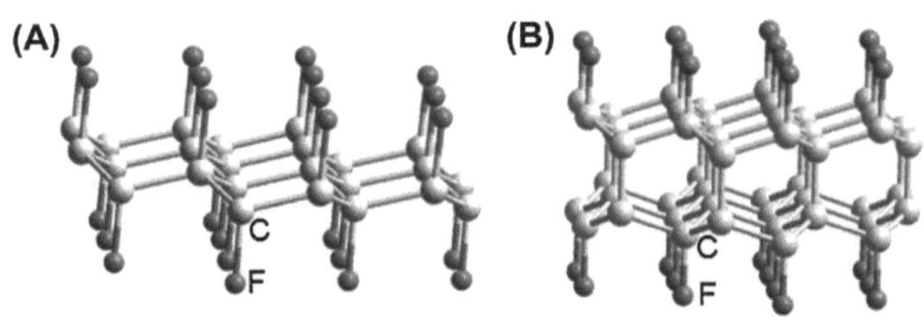

Figure 8.1 Structure of the layers of poly(carbon monofluoride) $(CF)_n$ (A) and poly(dicarbon monofluoride) $(C_2F)_n$ (B).

consistently attached above and below the graphene sheet (Fig. 8.1A). Experimental conditions and, particularly, disorder in the initial graphite may interrupt such fluorine alternation resulting in the formation of the boat-like fluorinated hexagons. The structure of $(C_2F)_n$ is presented by surface-fluorinated graphene bilayers, where the neighboring bare carbon atoms from different layers are covalently bonded (Fig. 8.1B). The sp^3-hybridization of all carbon atoms in $(CF)_n$ and $(C_2F)_n$ makes these graphite fluorides insulating. The high energy of the C—F bond and layered structure of the compounds caused their applications as electrode materials in primary lithium batteries and solid lubricants [10].

Graphite can be fluorinated at a substantially lower temperature by adding HF to F_2 or using metallic or nonmetallic fluorides [11]. The fluorine content in such products is commonly less than 50% and it can be varied depending on synthesis conditions [12]. Since the interlayer space usually contains the species existed in the reaction mixture or formed as result of the synthesis, such kinds of the fluorinated graphite products are in fact the fluorinated graphite intercalation compounds (FGICs). They are interesting because various molecules can substitute the accommodating species [13] that is attractive for storage of strong oxidants and for making novel products in such two-dimensional reactors.

Fluorine distributions in the layers of FGIC have been proposed by analyzing and modeling the spectroscopic data. Three types of graphite fluorides, high temperature–produced $(CF)_n$ and $(C_2F)_n$ and room temperature–produced $(C_{2.5}F)_n$, were compared using the methods of near-edge X-ray absorption fine structure (NEXAFS) and nuclear magnetic resonance (NMR) spectroscopy [14]. NEXAFS provides an information about the partial density of unoccupied electronic states, while the line shift in the NMR spectrum depends on the bond strength in the compound. Both these methods are sensitive to the local atomic arrangement and chemical bonding. NEXAFS spectra measured near the CK-edges of $(CF)_n$ and $(C_2F)_n$ showed the same set of resonances (Fig. 8.2A). The inequality in the relative intensities of the resonance at ~289.5 eV

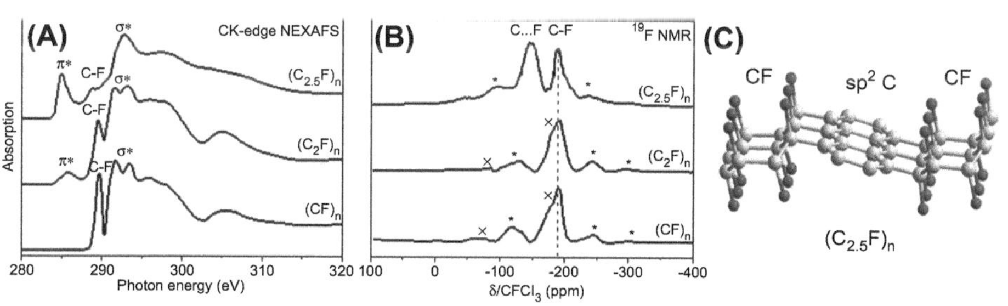

Figure 8.2 CK-edge NEXAFS spectra (A) and ^{19}F NMR spectra (B) measured for graphite fluorides produced at high temperature, $(CF)_n$ and $(C_2F)_n$, and at room temperature, $(C_{2.5}F)_n$. Proposed structure of the layer of $(C_{2.5}F)_n$ (C). *NEXAFS*, near-edge X-ray absorption fine structure.

Adapted with permission from Y. Ahmad, M. Dubois, K. Guérin, A. Hamwi, Z. Fawal, A.P. Kharitonov, A.V. Generalov, A.Yu. Klyushin, K.A. Simonov, N.A. Vinogradov, I.A. Zhdanov, A.B. Preobrajenski, A.S. Vinogradov, NMR and NEXAFS study of various graphite fluorides, J. Phys. Chem. C 117 (2013) 13564–13572. Copyright 2013 American Chemical Society.

was related to different amount of C—F covalent bonds in the compared graphite fluorides. Actually, fluorine is linked to every carbon atom in $(CF)_n$ and every second carbon atom in $(C_2F)_n$. In contrast to the spectra of these samples, the CK-edge spectrum of $(C_{2.5}F)_n$ showed an intense resonance at ~285 eV characteristic of π*—electron interactions. On the other hand, the high-energy resonances repeated those in the spectra of $(CF)_n$ and $(C_2F)_n$. The ^{19}F magic-angle spinning NMR spectra of $(CF)_n$ and $(C_2F)_n$ had an intense line at −190 ppm thus evidencing the covalent bonding between C and F in both graphite fluorides (Fig. 8.2B). The spectrum of the room temperature—produced $(C_{2.5}F)_n$ exhibited an additional line at −147 ppm, which was attributed to weakening of the C—F bonds due to a hyperconjugation with the π—system of nonfluorinated carbon regions [15]. Based on these results, the structure of $C_{2.5}F$ layers was presented as coexistence of planar aromatic hexagons and buckled CF regions (Fig. 8.2C).

Further clarification of C—F bonding in room-temperature fluorinated graphite has been done using the angle-resolved NEXAFS in a combination with quantum-chemical calculations [16]. The composition of the matrix was ~$C_{2.3}F$ as evaluated by X-ray photoelectron spectroscopy (XPS). The strong angular dependence of the intensity of the low-energy peaks located at 686.7 and 688.8 eV in the FK-edge spectrum (Fig. 8.3A) ambiguously pointed to the nearly perpendicular orientation of the C—F bonds to the basal plane of layers. Given that the fluorinated layers keep the π—electron regions, this orientation is possible only in the case of subsequent attachment of fluorine atoms to both sides of the graphene sheet. The model, which best fitted the spectroscopy data, consisted of zigzag chains from fluorinated carbon atoms alternated with the chains of bare sp^2-hybridized carbon atoms (Fig. 8.3B). The

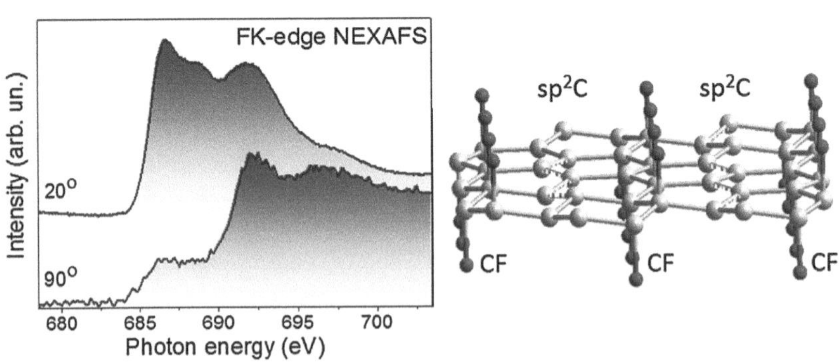

Figure 8.3 FK-edge NEXAFS spectra of room temperature—fluorinated graphite $(C_{2.3}F)_n$ measured at an incidence angle of radiation of 20 and 90 degree (A). Proposed structure of the layer of graphite fluoride with an alternation of zigzag sp^2-hybridized carbon chains and CF chains (B). *NEXAFS*, near-edge X-ray absorption fine structure.
Adapted with permission from A.V. Okotrub, N.F. Yudanov, I.P. Asanov, D.V. Vyalikh, L.G. Bulusheva, Anisotropy of chemical bonding in semifluorinated graphite C_2F revealed with angle-resolved X-ray absorption spectroscopy, ACS Nano 7 (2013) 65—74. Copyright 2013 American Chemical Society.

similar fluorine arrangement has been proposed for FGICs analyzed by means of neutron diffraction [15]. The concurrent adsorption of fluorine atoms at multiple sites may cause the seeding of the CF chains in a random distribution across the graphene sheets. The chain length comprises a few fluorine atoms as this was defined from the XPS analysis and confirmed by the atomic force microscopy (AFM) image of a fresh sample surface [17].

Just as the formation of the covalent C–F bonds in the high-temperature $(CF)_n$ and $(C_2F)_n$ is currently a common concept, the nature of the bonding in the layers of graphite fluorides synthesized at lower temperatures is strongly debated. This is due to the substantial differences in the XPS, IR, and NMR spectroscopy data for both types of graphite fluorides. The XPS F1s spectrum is peaked at ~689.2 eV for $(CF)_n$ and it is downshifted to ~686 eV for the room temperature–fluorinated graphites [18,19]. Such a shift was attributed to the formation of "semi-covalent" or "semi-ionic" C–F bonds [19] because the F1s spectrum of ionic compound LiF is at ~684.5 eV. Moreover, additional lines appear in the ^{19}F NMR spectra (Fig. 8.2B) and in the IR spectra of the FGICs as compared to the spectra of high temperature–produced $(CF)_n$ and $(C_2F)_n$ [20]. All these spectroscopic facts were assigned to a strong pulling of the electron density from carbon to fluorine, so that the latter atom becomes nearly ionic. However, the calculations of F1s binding energies, vibrational frequencies and NMR shifts for the fluorinated graphite models with different fluorine patterns and fluorine loadings have shown that the experimentally observed shifts of the lines can find another explanation.

A downshifting of the F1s level in the XPS spectra of the FGIC may be attributed to low density of a graphene network coverage by fluorine. Actually, the calculated energy value gradually decreased from 691 to 685.7 eV with a reduction of the length of the CF chain from the infinite one to that comprising a fluorine pair and isolated F atom [21]. Wherein, the C–F distance in the models relaxed within density functional theory (DFT) increased from 1.36 to 1.49 Å that is characteristics of covalent bonding. Moreover, as it was correctly pointed out by Claves [22], an increase in the charge separation between atoms in the "semi-ionic" C–F bond should cause opposite energetic shifts of the F1s and C1s levels, while as the XPS measurements showed they shift in one direction.

A value of the C–F bond stretching vibration in a three-dimensional $(CF)_n$ model was calculated within DFT to be 1165 cm^{-1}, that is in a reasonable agreement with the experimental value of 1215 cm^{-1} [23]. This value was reduced to 935 cm^{-1} for a $(C_3F)_n$ model where every carbon hexagon had two fluorine atoms in ortho-positions. The optimized length of the C–F bonds in this model was ~1.43 Å. It is obvious, that the frequency of the C–F bond vibration should be influenced by local surrounding of the corresponding CF group in the fluorinated layer. The experimental IR spectrum of FGIC with a matrix composition of $C_{2.5}F$ showed a band with four components (Fig. 8.4A) [17]. They were tentatively assigned to the vibrations of the CF group having three CF neighbors (CF3 at 1230 cm^{-1}), two CF neighbors and one sp^2-hybridized carbon atom (CF2 at 1132 cm^{-1}), one CF neighbor and two sp^2-hybridized carbon atoms (CF1 at 1095 cm^{-1}), and three sp^2-hybridized

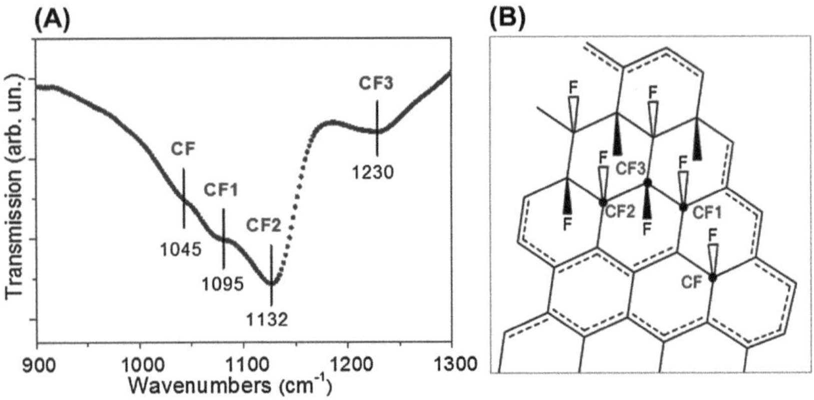

Figure 8.4 IR-spectrum of room temperature—fluorinated graphite $(C_{2.5}F)_n$ showing four components (A), which correspond to stretching vibrations of C—F bonds with different local surroundings (B). CF3, CF2, CF1, and CF assign the bonds, which have three, two, one, and zero CF neighbors.
Reprinted from I.P. Asanov, L.G. Bulusheva, M. Dubois, N.F. Yudanov, A.V. Alexeev, T.L. Makarova, A.V. Okotrub, Graphene nanochains and nanoislands in the layers of room temperature—fluorinated graphite, Carbon 59 (2013) 518—529. Copyright 2013, with permission from Elsevier.

carbon atoms (CF at 1045 cm^{-1}). The corresponding surroundings are schematically shown in Fig. 8.4B.

Panich, in his critical review, has noted that involving of semi-ionic and semi-covalent C—F bonds is not necessary for explanation of ^{13}C and ^{19}F chemical shifts in the NMR spectra of FGICs [24]. An assignment of ^{19}F chemical shifts in room temperature—fluorinated graphites was made based on DFT calculations of the ^{19}F absolute shielding in models with different fluorination motives [25]. It was shown that the shift is caused by variation of the π—electron density formed in the vicinity of the fluorine atom and related to the local concentration of the fluorine distribution. Based on the evolution of the ^{19}F chemical shift of the C—F bond length, Giraudet et al. have proposed to change the denomination "semi-covalent" by "weakened covalent" C—F bonding [26].

8.3 Key Issues Studied for Fluorinated Graphene

Graphite fluorides, where the layers are separated by a distance of 0.6 nm at least, can be considered as sources of fluorinated graphenes. Nevertheless, we note that there are no promising reports on the mechanical extraction of a fluorinated monolayer using, for example, an adhesive tape, which had allowed successful isolating graphene from graphite [27]. The attempts to cleave the graphite monofluoride $(CF)_n$ have yielded multilayered flakes. The monolayers have been obtained from partially fluorinated graphite (with a composition of about C_4F) but their surface areas did not exceed a few squared microns [28]. The reason is a loss of the stiffness of graphene layer with destruction of extended delocalized π−electron system in the result of the fluorine attachment to the basal plane. Moreover, the fluorination procedure may induce many structural defects in graphite fluoride causing the fragility of the crystal. Our experiments showed that fluorinated layers are easily wrinkled under mechanical action that prevents their transferring on a substrate in the form suitable for further investigations. A deposition of fluorinated graphene on a substrate from a colloidal dispersion of graphite fluoride seems to be more productive [29,30]. However, the size of the obtained layers is also small and their surfaces are often etched. In addition, graphite fluorides interact with some common solvents with detachment of fluorine atoms and this is observable as a solution blackening [31,32]. As a result, fluorine content in the layers isolated using liquid-phase exfoliation may differ from that in the parent graphite fluoride. This is difficult to check, because a search of small transparent piece on a substrate for focusing of the X-ray beam on the sample is a challenging task.

In the case of a very gentle exfoliation procedure, the structure of the fluorinated graphene will be determined by starting material, for which not many various configurations are realized in the synthesis as this was overviewed in Section 8.2. Thereby, the direct fluorination of graphene samples opens new possibilities for distribution of fluorine atoms and creation of the configurations, which are impossible to obtain using graphite. Actually, you can attach fluorine atoms at one side of graphene taking a supported sample [33], make a certain fluorine pattern by applying a mask on the graphene [34], or selectively recover the fluorinated surface [35,36].

The issues explored for graphite fluorides are actual for fluorinated graphene and some of them are even more debating. The most often considering aspect is an influence of fluorine content and fluorine pattern on the properties of fluorinated graphene. The question about the nature of C−F bonding is raised again and it is especially acute for one-side low-fluorinated graphenes. Moreover, new heterostructures can be synthesized fluorinating bi- and few-layer graphenes, where only the surface of stack is usually available for the fluorination. We will analyze all these issues in the context of thermodynamical stability of certain configurations, a possible fluorination mechanism, electron interactions between carbon and fluorine, value of band gap energy, and optical transitions.

8.4 Fluorographene

The first paper presented fluorination of a graphene monolayer and was issued by a group headed by the Nobel prize winners, Andre Geim and Kostya Novoselov in 2010 [37]. Graphene was obtained by micromechanical cleavage and fluorinated using XeF_2 at 70°C for a few weeks. Note that fluorination of graphite using XeF_2 occurs only with addition of HF, which works as a catalyst [38]. For the fluorination of a monolayer, it is difficult to find a suitable support because fluorine reacts with almost all solids. In Ref. [37] the graphene was located onto a gold grid covered by a holey polymer film typically used for transmission electron microscopy

(TEM) experiments. At the used fluorination conditions, the support was chemically inert and at the same time, it permitted access of fluorine to both sides of graphene. The fluorinated monolayer had a hexagonal structure and was optically transparent to green laser light that resulted in the absence of Raman scattering peak characteristics for partially fluorinated graphene. The fully fluorinated graphene CF with identical fluorine coverage on both sides was called fluorographene. An attachment of fluorine atoms in a CF system to the same side of the graphene plane is impossible due to the constraints related to the sp^3-bond orientations and strong repulsion between neighboring fluorine atoms.

The absence of band gap in graphene leads to a finite light absorption (~2.3%) below an energy of ~2.5 eV (Fig. 8.5). The energy bands are flattened in the vicinity of the M point of the first Brillouin zone [39] that results in van Hove singularities in the density of states. Electron transitions between these singularities produce a peak located at 4.6 eV in optical absorption spectrum. The fluorographene showed no opacity up to ~3 eV and started absorbing light only in the blue range [37]. This behavior proved that the CF monolayer is a wide band gap semiconductor. Electrical measurements confirmed the highly insulating property of fluorographene.

The published results caused a huge interest to the fluorinated graphene because they demonstrated a higher reactivity of the 2D material in comparison to its 3D counterpart and possibility to open the band gap in graphene with covalent modification. The band gap determined from the optical absorption spectrum well matched to the

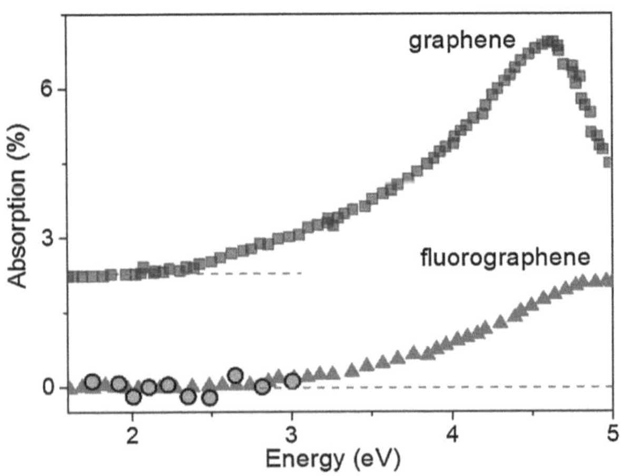

Figure 8.5 Absorption spectra of graphene and fluorographene CF. The *open circles* are the measurements for fluorographene membranes on a TEM grid using filter spectroscopy. The *dashed lines* indicate zero and $\pi\alpha$ opacities. *TEM*, transmission electron microscopy. Reproduced from R.R. Nair, W. Ren, R. Jalil, I. Riaz, V.G. Kravets, L. Britnell. P. Blake, F. Schedin, A.S. Mayorov, S. Yuan, M.I. Katsnelson, H.-M. Cheng, W. Strupinski, L.G. Bulusheva, A.V. Okotrub, I.V. Grigorieva, A.N. Grigorenko, K.S. Novoselov, A.K. Geim, Fluorographene: A two-dimensional counterpart of Teflon, Small 6 (2010) 2877–2884. Copyright 2010, with permission from John Wiley and Sons.

result of the band structure calculation of fully fluorinated graphene performed in the paper published in the same journal issue [29]. The DFT calculations showed that fluorographene, where the fluorinated hexagons are in the chair configuration, is a wide-gap semiconductor with a direct band of ~3.1 eV at the Γ point. These calculations accompanied the investigations of fluorinated monolayers isolated from graphite monofluoride $(CF)_n$ in a liquid. The Raman spectrum of the product had the D and G bands in contrast to the abovementioned results.

In addition to the chair configuration of the fluorographene layer, where the fluorine atoms are sequentially attached to both sides of graphene plane (Fig. 8.1A), and the boat configuration with alternation of pairs from neighboring fluorine atoms above and under the plane (Fig. 8.6A), Leenaerts et al. proposed the models consisted of armchair and zigzag CF chains (Fig. 8.6B and C) [40]. Although the chair configuration was found to be the most stable one, similar to that for $(CF)_n$ [9], the formation energy of the zigzag configuration was only slightly higher. The band gap in this structure is ~3.6 eV that is larger than that predicted for the chair configuration. The DFT calculations generally underestimate the band gap value for a solid due to an uncorrect accounting of electron–electron interactions. This is corrected by using the one-particle Green's function in combination with a dynamically screened Coulomb interaction (GW). The calculations based on the GW approximation are supposed to result in more accurate band gaps for semiconductors and insulators. The GW band gap for the chair configuration of fluorographene is ~7.4 eV that is more than twice larger as compared to the measured value of the optical gap [37]. One of the reasons of the discrepancy between experiment and theory is an exciton nature of optical transitions. On the other hand, experimental samples may contain appreciable amounts of defects, such as missing of fluorine atoms or realization of different fluorine distributions. DFT calculation of a CF system with randomly distributed fluorine atoms detected a reduction of the band gap by ~0.8 eV as compared to the chair configuration [41]. To resolve this issue, the improvement of the crystal quality of fluorographene should be done in the future.

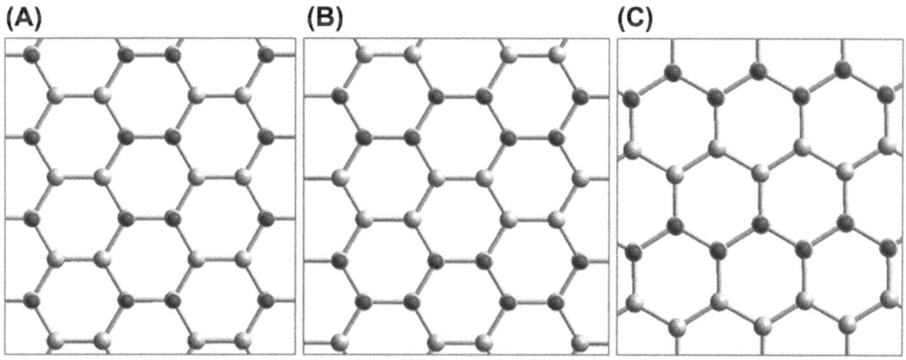

Figure 8.6 Models of boat (A), armchair (B), and zigzag (C) configurations of fluorographene CF. The different colors (shades) represent fluorine adsorbates above and below the graphene plane.

8.5 One-Side Graphene Fluorination

Fluorination of graphene located on a substrate, which does not pass a fluorinating agent, can result in bonding of fluorine to the top surface only. One-side fluorinated systems have been obtained by exposing the chemical vapor deposition (CVD) graphene as grown on a copper foil to XeF_2 gas at 30°C [42] and 250°C [43]. Independent on the fluorination temperature, the composition of products determined by XPS was ~C_4F that assumes realization of the structure with a limiting fluorine coverage (25%). A band gap determined for the C_4F using a conductive probe AFM measurement was ~2.92 eV [43] and this indicated that an appreciable band gap in graphene is open even at a moderate uniform coverage of fluorine. Among others, ordered one-side fluorinated graphene structures with a composition ranged from C_8F to C_2F, the C_4F structure had the largest binding energy per fluorine atom (Fig. 8.7A) [42]. In that structure, every CF group had no fluorinated carbon atoms at second neighbor spacing (Fig. 8.7B). The DFT calculations gave a band gap of 2.93 eV for the one-side C_4F, while the GW correction resulted in a value of 5.99 eV [44].

Two configurations were considered for the half-fluorinated supported graphene (Fig. 8.8A and B) and the GW band gap in the C_2F boat structure was 5.68 eV, while the chair configuration was predicted to be a metal. The predicted band gap value is in good agreement with the value of ~6.2 eV determined for one-side fluorinated epitaxial graphene, formed on the SiC surface [45]. A content of fluorine in the sample was ~43% as determined from the analysis of XPS C1s spectrum measured at a photon excitation of 350 eV. The band gap was obtained as a distance between the peak in the valence band photoemission spectrum and CK-edge NEXAFS spectrum, probing the unoccupied states of a substance.

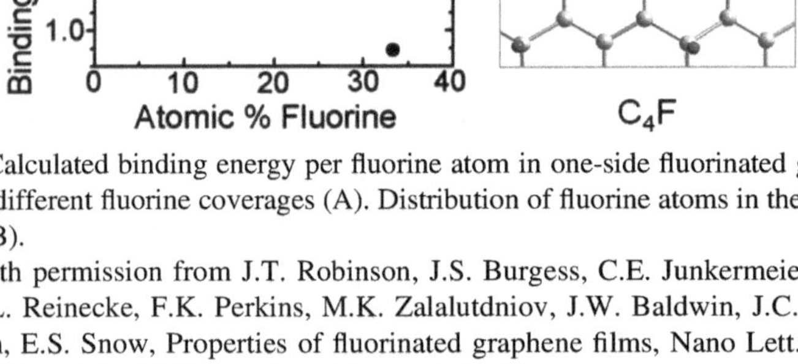

Figure 8.7 Calculated binding energy per fluorine atom in one-side fluorinated graphene models with different fluorine coverages (A). Distribution of fluorine atoms in the most stable C_4F model (B).

Reprinted with permission from J.T. Robinson, J.S. Burgess, C.E. Junkermeier, S.C. Badescu, T.L. Reinecke, F.K. Perkins, M.K. Zalalutdniov, J.W. Baldwin, J.C. Culbertson, P.E. Sheehan, E.S. Snow, Properties of fluorinated graphene films, Nano Lett. 10 (2010) 3001–3005. Copyright 2010 American Chemical Society.

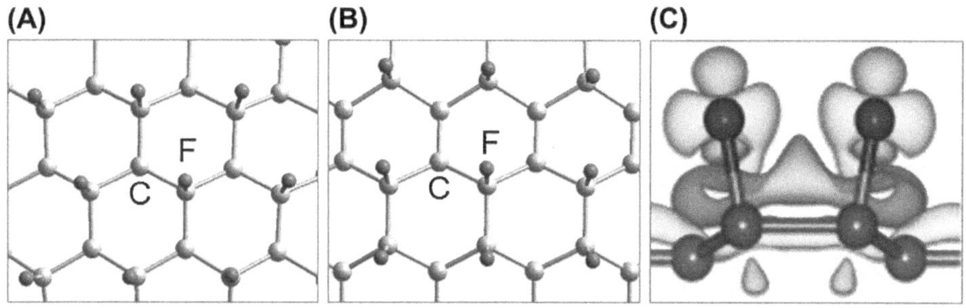

Figure 8.8 Models of chair (A) and boat (B) configurations of one-side fluorinated graphene of C_2F composition. Difference charge density distribution calculated for the C_2F boat configuration (C). Positive (green) and negative (yellow) values represent the electron gain and depletion, respectively.
Reprinted from N. Jiao, C. He, P. Zhou, C.X. Zhang, H.P. Xiao, L.Z. Sun, Surface work function of chemically derive graphene: A first-principles study, Phys. Lett. A 377 (2013) 1760–1765. Copyright 2013, with permission from Elsevier.

The charge density distribution calculated for the C_2F boat structure detected the concentration of electrons at the fluorine-decorated side and electron depletion at the bottom graphene side (Fig. 8.8C) [46]. This should result in different work functions from the fluorine-free and fluorine-decorated graphene sides.

The stability of one-side fluorinated graphene depending on the density of fluorine coverage and position of the fluorine atoms relative to the graphene lattice have been studied in detail in Ref. [47]. The unit cell of graphene consists of two carbon atoms denoted as A and B. Translation of these atoms creates two interpenetrating triangular sublattices of a hexagonal crystal. The fluorine atoms were located in the same sublattice (AA configuration) or bonded with carbon atoms from different sublattices (AB configuration). The stoichiometry of the fluorinated graphene varied from $C_{50}F$ to C_2F. The binding energy per fluorine atom in fluorinated graphene system was calculated as a difference between the total energies of an isolated graphene plus atomic fluorine and C_xF system. The AB configurations with fluorine atoms belonging to different sublattices were found to be more stable for all considered range of the stoichiometries (Fig. 8.9). Starting from C_9F composition, the binding energy of the AB configuration increased almost linearly with the fluorine density indicating that, energetically, fluorine atoms tend to form areas with a high coverage. For low coverage, the difference between AB and AA configurations vanished and their binding energy was close to 2 eV.

The electronic structure of one-side C_4F was studied depending on the fluorine atom position [47]. It was shown that location of fluorine atoms in the same sublattice (AA configuration) produces almost flat bands crossing the Fermi level (Fig. 8.10A) due to localization of the electron density on the bare carbon atoms constituting the B sublattice. The AA system is magnetic with a magnetic moment of $\sim 0.799\ \mu_B$. The more stable AB configuration is semiconducting (Fig. 8.10B) because the $\pi-$systems of aromatic carbon hexagons are isolated from each other by the surrounding CF groups (Fig. 8.10C). The valence band edge of this C_4F system consists of the

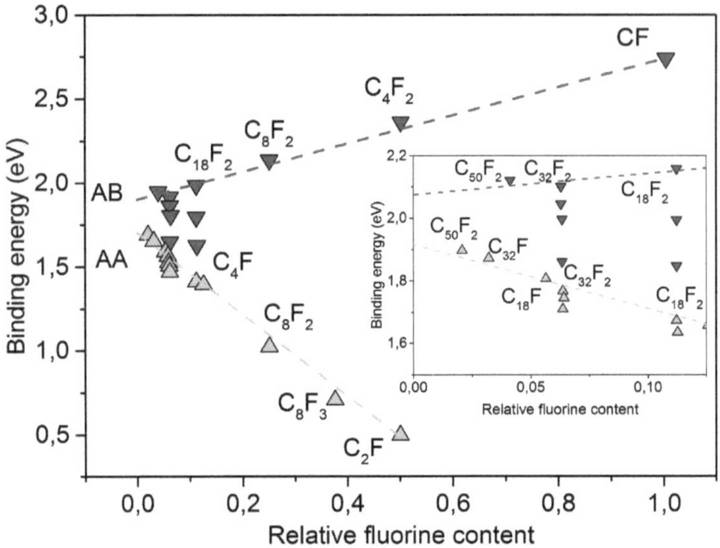

Figure 8.9 Binding energy of a CF_x as a function of the relative fluorine content for the one-side fluorinated graphene models with fluorine atoms located in the same sublattice (AA configuration, triangle up) or in different sublattices (AB configuration, triangle down). In the structures C_4F_2 and CF, the fluorine atoms were attached to both sides of graphene plane. Adapted with permission from H. Santos, L. Henrard, Fluorine adsorption on single and bilayer graphene: Role of sublattice and layer decoupling, J. Phys. Chem. C 118 (2014) 27074–27080. Copyright 2014 American Chemical Society.

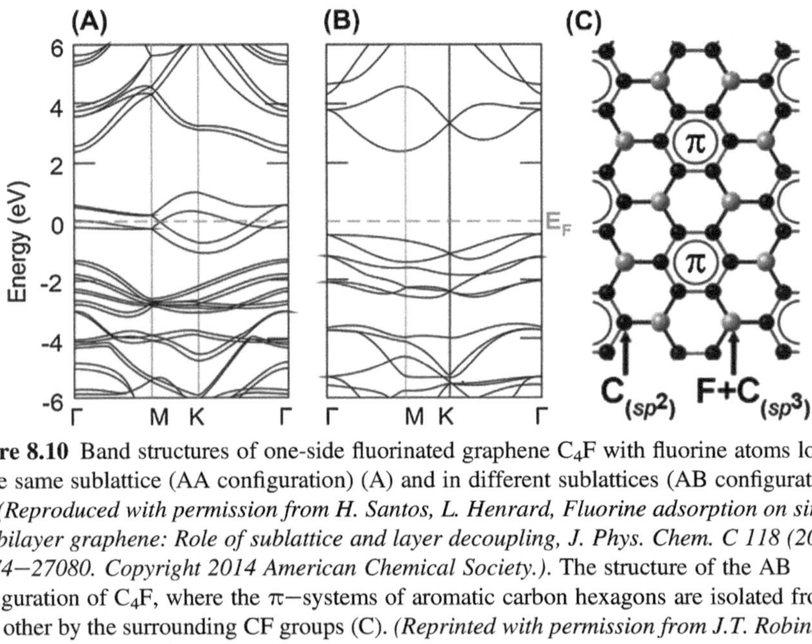

Figure 8.10 Band structures of one-side fluorinated graphene C_4F with fluorine atoms located in the same sublattice (AA configuration) (A) and in different sublattices (AB configuration) (B). *(Reproduced with permission from H. Santos, L. Henrard, Fluorine adsorption on single and bilayer graphene: Role of sublattice and layer decoupling, J. Phys. Chem. C 118 (2014) 27074–27080. Copyright 2014 American Chemical Society.).* The structure of the AB configuration of C_4F, where the π−systems of aromatic carbon hexagons are isolated from each other by the surrounding CF groups (C). *(Reprinted with permission from J.T. Robinson, J.S. Burgess, C.E. Junkermeier, S.C. Badescu, T.L. Reinecke, F.K. Perkins, M.K. Zalalutdniov, J.W. Baldwin, J.C. Culbertson, P.E. Sheehan, E.S. Snow, Properties of fluorinated graphene films, Nano Lett. 10 (2010) 3001–3005. Copyright 2010 American Chemical Society.).*

combination of π–electrons and 2p fluorine orbitals, which are parallel to the graphene plane [48].

The highest stability of the AB C_4F system for one-side graphene coverage was explained based on the local electronic densities calculated for the $C_{50}F$ system (Fig. 8.11A) [47]. The formation of sp^3-bond on a sublattice (say, the A one) induced an increase of the electron density near the Fermi level on the other sublattice (Fig. 8.11B). The larger density localization was observed on the carbon atoms from the sublattice B being next neighbors to the CF group. These atoms should be the most attractive for the attachment of the second electronegative fluorine atom. However, the first neighbors of A–B fluorination are geometrically constrained and fluorine atoms bind to the third-nearest-neighbor positions also having an excess of electrons. This explains the saturated C_4F stoichiometry observed when the supported graphene was fluorinated [42,43]. A moving of one fluorine atom from the stable position to the A site meets a barrier larger than 1.7 eV [47] preventing the formation of the AA C_4F system.

The fluorine-free surface of partially fluorinated graphene can work as an effective sensor for polar molecules, such as, for example, ammonia and water [49,50]. DFT calculations showed that the NH_3 molecule tends to be located over the root of CF group (Fig. 8.12A) and the adsorption energy varies from 0.236 to 0.275 eV depending on the orientation of the molecule relative to the surface of sensor and backside fluorination degree [49]. A character of noncovalent interactions between ammonia and backside fluorinated graphene was revealed using an approach based on the electron density and its derivatives [51]. The gradient isosurface was colored in accordance with the sign of the Laplacian and the strength of interaction (Fig. 8.12B). Large

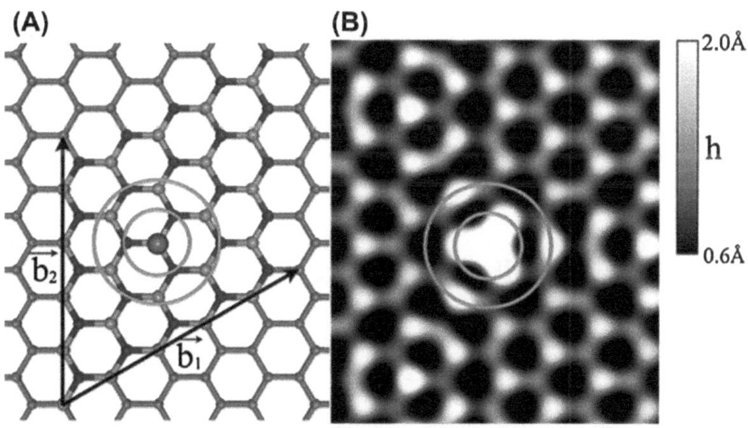

Figure 8.11 The supercell of $C_{50}F$ system (highlighted) with the translation vectors in graphene plane (A). Calculated electron density distribution in $C_{50}F$ (B). The inner and outside circles indicate the first and third neighbors from the F–C bond, respectively.
Reprinted with permission from H. Santos, L. Henrard, Fluorine adsorption on single and bilayer graphene: Role of sublattice and layer decoupling, J. Phys. Chem. C 118 (2014) 27074–27080. Copyright 2014 American Chemical Society.

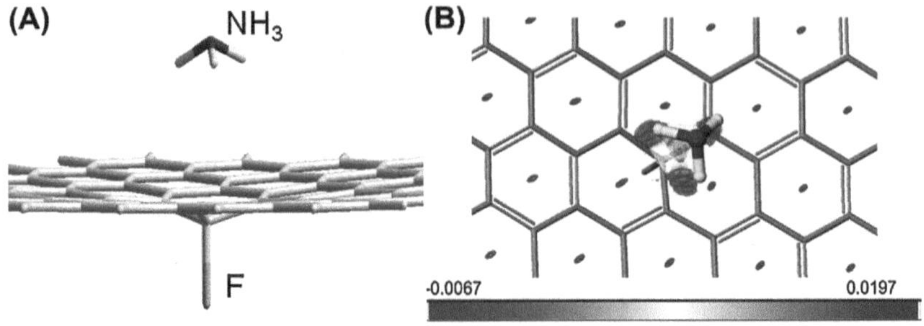

Figure 8.12 Side view of the energetically preferable location of the ammonia molecule relative to the surface of backside-fluorinated graphene (A). Reduced density-gradient-based isosurfaces ($s = 0.35$ a.u.) for van der Waals interactions between ammonia and the fluorinated graphene model (B). Large negative values (in red (light gray in print versions)) correspond to strong attractions and large positive values (in blue (dark gray in print versions)) indicate that the interactions are nonbonding.
Reproduced from M.V. Katkov, V.I. Sysoev, A.V. Gusel'nikov, I.P. Asanov, L.G. Bulusheva, A.V. Okotrub, A backside fluorine-functionalized graphene layer for ammonia detection, Phys. Chem. Chem. Phys. 17 (2015) 444–450. Copyright 2014, with permission from Royal Society of Chemistry.

negative values (in red) correspond to strong attractions and large positive values (in blue) indicate that the interactions are nonbonding. Areas of nonbonding overlapping are located at the center of each hexagonal ring, while a triangular shaped surface between ammonia molecule and fluorinated graphene is deeply red at the edges and weakly red-colored in the center. Thus, there is a slight attraction between the nitrogen atom of NH_3 and carbon atom of CF group despite a large distance between them (~3.5 Å), although the greatest contribution to the adsorption energy comes from the attractions between hydrogen atoms and carbon atoms surrounding the CF group. As these carbon atoms are in the sp^2-hybridization state, adsorption of NH_3 on backside fluorinated graphene occurs via the H–π type interactions, which are dominated by dispersion energies.

8.6 Two-Side Partially Fluorinated Graphene

In mild synthesis conditions, fluorination of both sides of graphene plane can produce a system with fluorination loading less than CF. The two-side fluorination introduces more possibilities of atomic arrangements, thus providing additional freedom to tailor the properties of the fluorinated graphene. DFT calculations revealed that at the same fluorine coverage, the two-side configuration is more stable than the configuration with one-side fluorinated graphene (Fig. 8.13) [52]. Among the considered structures, the fluorographene CF had the largest binding energy per fluorine atoms, while the C_2F and $C_{1.3}F$ coverages were also quite favorable. Note that the formation energies for C_4F with single- and double-side fluorination were similar and the band gap was

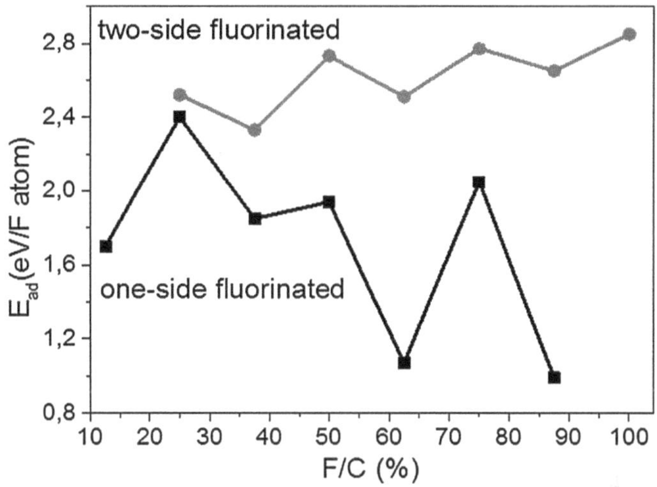

Figure 8.13 The dependence of adsorption energy of fluorine on the F/C ratio in one-side and two-side fluorinated graphene.
Reprinted from Z. Wang, S. Qin, C. Wang, Q. Hui, Fluorine adsorption on the graphene films: From metal to insulator, Comp. Mater. Sci. 97 (2015) 14—19. Copyright 2015, with permission from Elsevier.

only slightly affected by the ratio of the fluorine atoms deposited on different sides of the graphene plane [53]. However, the type of fluorine coverage is crucial for morphology of nonsupporting C_4F. In such nonsupporting systems, the attached fluorine atom produces the out-of-plane forces, and to minimize the surface tension, the initially flat structure is deformed or rolled up into a scroll [54]. Thus, the difficulty in obtaining fluorinated graphene flakes of large-size by exfoliating the partially fluorinated graphite can be explained by a mechanical instability of the extracting layers and their spontaneous folding into a curved structure.

The thermodynamic stability of the partially fluorinated graphene was found to decrease with imbalance of fluorine loading on the graphene sides [55]. The calculations showed that the systems with different number of fluorine atoms located in the A sublattice on one graphene side and in the B sublattice on other side are magnetic and the magnetic moment increases with the imbalance number. The spin splitting comes from the π–electronic states of carbon atoms closest to the edge of the fluorinated region. Such states can induce magnetic ordering around zigzag CF chains [56]. A double-side attachment of fluorine atoms along zigzag direction of the graphene plane breaks the uniformity of the π–system of graphene. As a result, the π–electrons of carbon atoms neighboring with a CF chain align along the chain and their spins interact ferromagnetically (Fig. 8.14). As the carbon atoms at different edges of the zigzag CF chain belong to different graphene sublattices, the spin coupling across the chain is antiferromagnetic.

The half-fluorination of a graphene plane is a saturated limit for the fluorination at room temperature [57]. For the C_2F structures, the double-side fluorination is

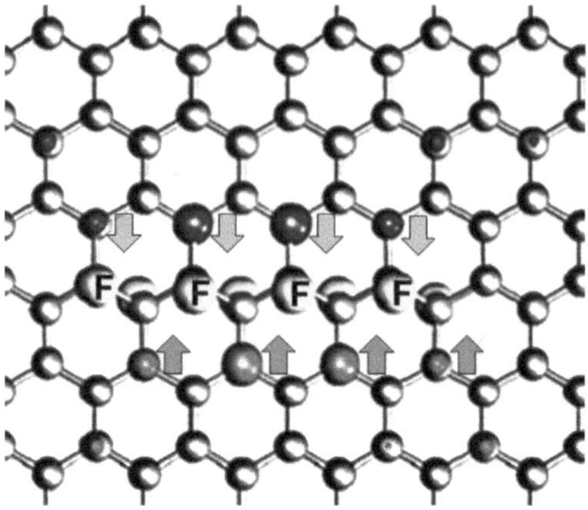

Figure 8.14 A zigzag chain of eight fluorine atoms on graphene with isocontour plots of the spin density obtained from the density functional theory calculations. Regions with opposite spin orientations have different shades and are indicated by the *arrows*.
Adapted with permission from T.L. Makarova, A.L. Shelankov, A.A. Zyrianova, A.I. Veinger, T.V. Tisnek, E. Lähderanta, A.I. Shames, A.V. Okotrub, L.G. Bulusheva, G.N. Chekhova, D.V. Pinakov, I.P. Asanov, Ž. Šljivančanin, Edge state magnetism in zigzag-interfaced graphene via spin susceptibility measurements, Sci. Rep. 5 (2015) 13382. Copyright 2015 Nature Publishing Group.

significantly more preferable than the single-side fluorination [58]. Considerations of the chair, boat, and zigzag distribution of fluorine on the graphene plane revealed the lowest total energy for the last configuration, while imaginary frequencies in a phonon spectrum indicate its instability. This could explain the formation of only short zigzag CF chains with fluorine atoms successively bonded to both sides of the graphene plane, which has been proposed from the analysis of the XPS data [17]. The double-side fluorine ordering as in the chair and boat configurations will produce metallic and semiconducting behavior in C_2F, respectively [58]. The energy bands around the Fermi level in the chair C_2F originated from the π−states of bare carbon atoms.

A comparison of the total energies of the models of double-side C_2F revealed that the chair configuration is second in a stability trend after the structure with alternation of chains from bare carbon atoms and CF chains propagating in the armchair direction [59]. Moreover, the CKα spectrum plotted for this structure agreed with the experimental spectrum better than other seven considered models of C_2F. Based on the analysis of the calculations, electronic interactions in the C_2F monolayer can be schematically represented as in Fig. 8.15. The upper occupied orbitals correspond to the π−system formed by interaction of π–electrons from nonfluorinated carbon atoms. These orbitals extend to ~5 eV in depth of the valence band overlapping with the orbitals being a combination of carbon σ-electrons and 2p fluorine orbitals parallel to the graphene plane. These combinations have antibonding characters respective to the interactions between electrons of carbon and fluorine. Nonbonding 2p orbitals of fluorine occupy a narrow interval (less than 1 eV) that

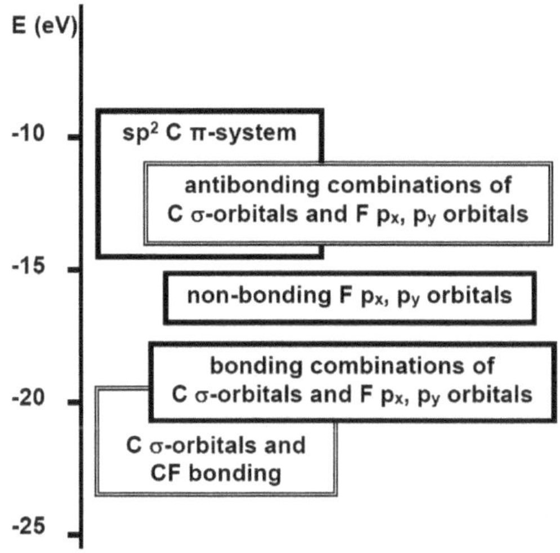

Figure 8.15 Schematic representation of electron interactions in double-side fluorinated C_2F monolayer.

is attributive of lone electron pairs. The next block contains orbitals where tangential 2p fluorine orbitals form bonding with the σ-system of carbon network. The orbitals, providing the σ-bonding between atoms of carbon network and in CF groups, have the lowest energy.

This scheme of the chemical bonding can be applied to other fluorinated graphene systems with taking into account the amount and distribution of fluorine atoms. Thus, to represent the valence band of fluorographene CF, one should remove the upper block corresponding to the π–electron system and the highest occupied orbitals in CF will be presented by antibonding combinations of σ-electrons of carbon and ($p_x + p_y$) orbitals of fluorine [44,48,60]. The carbon π–electron density will dominate in the valence edge levels of partially fluorinated graphene and energy distribution of these levels will be dependent on the fluorine content and fluorine distribution [48,52].

8.7 Fluorinated Bi- and Few-Layer Graphene

When exfoliating fluorinated graphite [61] or fluorinating a stack from several graphene layers [62], the fluorinated bi- and few-layers of graphene can be obtained. A fluorination of monolayer and bilayer graphene in the same conditions allowed revealing a lower reactivity of the bilayer toward CF_4 plasma [62]. However, it is easier to elucidate how different architectures can be formed and which properties should be expected for them using the calculations.

The first theoretical investigation whether it is possible to fluorinate bilayer graphene was done in Ref. [63]. Two graphene layers were stacked in the Bernal configuration, where the A-sublattice of one layer was exactly aligned on the top of the B-sublattice of another layer (Fig. 8.16A). This is the most energetically stable

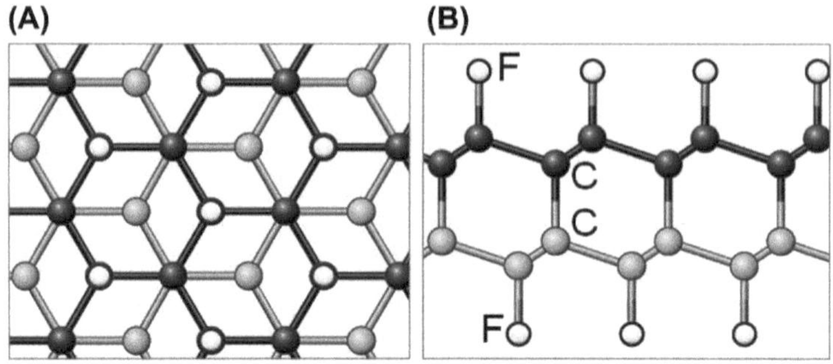

Figure 8.16 Top (A) and side (B) view of bilayer fluorographene C_2F. The carbon atoms of the two layers are given by different colors (shades) for clarity, and the fluorine atoms are shown by open circles with the smallest diameter.
Reprinted with permission from J. Sivek, O. Leenaerts, B. Partoens, F.M. Peeters, First-principles investigation of bilayer fluorographene, J. Phys. Chem. C 116 (2012) 19240–19245. Copyright 2012 American Chemical Society.

stacking of layers, which commonly realized in natural graphite. For a C—C pair from different layers, there are six carbon neighbors. An increase of the number of neighbors involved in the fluorination from one to four resulted in gradual decrease of the interlayer distance. With further increase of the fluorinated neighbors, the distance was changed by a jump of ~1 Å that clearly indicates the formation of an interlayer bonding stronger than the van der Waals one. A change in the hybridization of carbon atoms with the fluorine attachment caused the pushing of neighboring nonfluorinated carbon atoms from different layers with the formation of covalent bonding between them (Fig. 8.16B). All considered configurations were stable against desorption of molecular F_2 from the bilayer surface. However, for the same fluorine content, the configuration in which the fluorine atoms are distributed as equally as possible between the two layers is lower in energy and thus the most stable. The band gap of the fluorinated graphene bilayer was ~3.95 eV as given by DFT that is about ~1 eV larger than the value for fluorographene CF. Furthermore, it was found that fully fluorinated bilayer graphene C_2F is almost as strong as graphene, as its 2D Young's modulus is approximately 300 N/m.

Analysis of charge density in the specified points of the fluorinated bilayer graphene revealed that the valence band maximum is mainly contributed by the $p_x + p_y$ bonding states of C—C atoms in each graphene layer and by those of F—F atoms in each fluorine layers, with no p_z–states involved [64]. This type of the bonding is similar to that at the valence band edge of fluorographene CF [44,48,60] or the character of the orbitals composing the second-energy block in the electronic structure of the half-fluorinated graphene (Fig. 8.15). The conduction band minimum is dominated by the p_z—p_z antibonding states of C—F atoms and interlayer C—C atoms [64]. Based on this observation, it was concluded that the valence band is mainly sensitive to interactions between the x—y planar orbital while the conduction band is only sensitive to the interactions between the electronic orbitals in the z—direction. Hence, the band gap might be modulated by applying biaxial strain. The biaxial compressive strains reduce the band gaps and could transform bilayer C_2F from semiconductor to metal, whereas under the biaxial tensile strains the C_2F should be able to undergo direct-to-indirect band gap transition. Moreover, the calculations show that the fluorinated AB−stacked bilayer is more stable and hence would be more readily formed in comparison to a pair of C_2F monolayers in the boat configuration bonded by the van der Waals forces [65].

A mutual rotation of graphene layers can result in various stacking configurations, which are often produced in the CVD conditions of the graphene growth [66]. A superposition of the two rotated honeycomb lattices is observed as a Moiré pattern in scanning tunneling microscopy images [67]. At certain twist angles, some atoms from the two layers have the same positions along *z*−direction and hence they are at the shortest distance. An external action may induce covalent bonding between these atoms, thus creating diamond-like nanocrystals with the same periodicity as that of the Moiré pattern (Fig. 8.17A). The interlayer C−C bonds can be stabilized by passivating the π−electrons of the neighboring surface atoms by, for example, fluorine. DFT calculations of models of interlayer-bonded twisted graphene bilayer with partially fluorinated surface demonstrated a range of the conductivity behavior from semimetal to semiconducting or insulating depending on the twist angle and the density of interlayer C−C bonds [68]. The valence and conduction bands in bilayer graphene with the twist angle of 30 degree exhibited a linear dispersion around the K point as for pristine and fluorinated interlayer-bonded configuration (Fig. 8.17B). However, significant reduction in the Fermi velocity, reflected as a decrease of the slope of the dispersion bands, was observed in the latter case. Analysis of the density of states showed that these bands are contributed by π−electrons of the carbon atoms

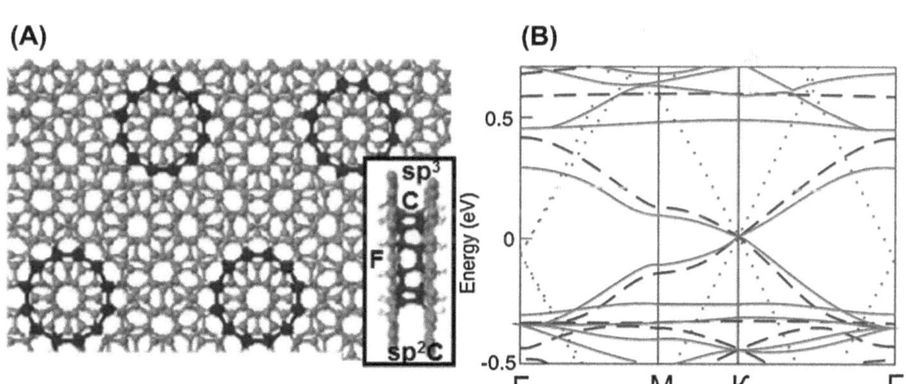

Figure 8.17 Superlattice structure of 30 degree -twisted bilayer graphene with diamond-like nanodomains formed due to covalent bonding between carbon atoms of different layers (A). The insert is a side view of the structure. The sp^3-hybridized carbon atoms are shown by dark balls, the light small balls correspond to fluorine atoms, which are responsible for inducing interlayer C−C bonding. Electronic band structures calculated for fluorinated bilayer (*solid lines*) and pristine 30 degree-twisted bilayer graphene (*dotted lines*) (B).
Adapted with permission from A.R. Muniz, D. Maroudas, Superlattices of fluorinated interlayer-bonded domains in twisted bilayer graphene, J. Phys. Chem. C 117 (2013) 7315−7325. Copyright 2013 American Chemical Society.

in the regions surrounding the interlayer-bonded domain, while the π−electrons of the nonfunctionalized carbon atoms within these domains produce next energy bands, which are almost nondispersive.

The calculations performed in Ref. [69] confirmed that the fluorination of the bilayer graphene surface to the composition of C_4F is not sufficient for the formation of the covalent interlayer bonding. The most stable stacking of the C_4F layers is at where the CF groups from the bottom monolayer are directed to the fluorine-free surface of the top layer (Fig. 8.18A). The binding energy for such C_4F bilayer was only about twice higher than that for the graphene bilayer. The stabilization of this configuration was due to $C^{\delta+}F^{\delta-}...C^{\delta+}F^{\delta-}$ dipole−dipole interactions. The C_4F bilayer had the band gap much smaller than in individual C_4F monolayer and it decreased with the number of layers in the stack (Fig. 8.18B). This effect was attributed to interlayer polarization featured by charge transfer from the upper C_4F layer to the bottom layer and, consequently, the relative shift of the energy levels of the layers resulted in a decrease of the band gap. Actually, computing of the partial charge densities corresponding to the valence band maximum and conduction band minimum indicated their localization on the frontier C_4F layers of the stack (Fig. 8.18A).

8.8 Fluorographene/Graphene Hybrids

The interfacing of areas of graphene and fluorinated graphene creates hybrid structures, which are very attractive since the individual systems are strongly different in

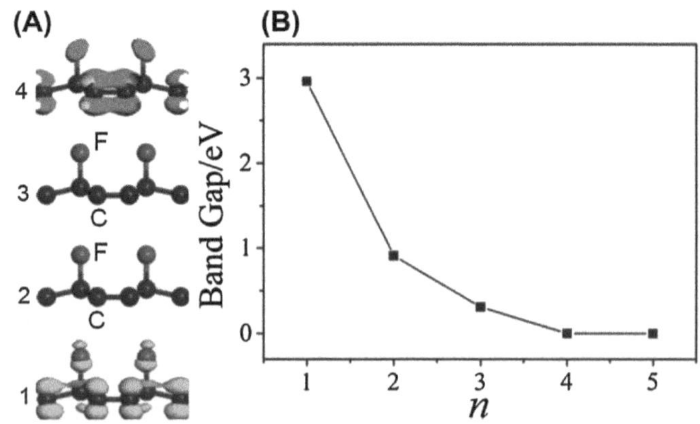

Figure 8.18 Side view of the energetically most stable configuration of C_4F layerstacking (A). The number indicates the layers; the contour plots show localization of the partial charge density from valence band maximum on the top layer and from conductance band minimum on the bottom layer of the stack. Dependence of band gap of $(C_4F)_n$ on the number of layers n plotted by the result of DFT calculations (B).
Reprinted with permission from Y. Li, B.A. Pantoja, Z. Chen, Self-modulated band structure engineering in C_4F nanosheets: First-principles insights, J. Chem. Theory Comput. 10 (2014) 1265−1271. Copyright 2014 American Chemical Society.

conductivity (metal vs insulator) and they both are thermodynamically stable. The latter feature allowed experimentally producing such type of hybrids through the masking of certain graphene areas before their exposure to fluorination [70] or local removal of fluorine atoms by focused ion beam [34]. Note that this is a technically challenging task to obtain the necessary fluorine loading on the nonprotected graphene areas. Actually, it is essential that the fluorinating agent could arrive to both sides of these areas and the mask remained unchanged at the temperature of the fluorination. Particularly, a short exposure of one graphene side to XeF_2 gas was supposed to produce C_4F with isolating aromatic rings (Fig. 8.10C), while no direct evidence for that has been given [70]. Also, a conversion of a fluorinated graphene to the graphene under irradiation [34] is an open question. An increase in the conductivity after the treatment may not be a confirmation of recovery of the intrinsic properties of graphene. An investigation of the irradiated layers by XPS, NEXAFS, Raman and optical reflection spectroscopy methods showed that such layers have poorly atomic ordering and contain sp^3-hybridized carbon states due to, probably, existing of fluorine atoms on the backside of the irradiated graphene [71].

The fluorographene/graphene superlattices can be presented by the alternating stripes [72,73] or embedded small-size graphene areas [73,74]. The considered issues for such superlattices are the atomic arrangement at the interfaces, the width of the strips and the shape of the graphene areas in the CF matrix. Owing to the honeycomb lattice of graphene, the interface in fluorographene/graphene hybrid has the armchair or zigzag structure. The armchair-interfaced hybrids were predicted to be semiconductors with a band gap decreasing with an increase of the width of graphene strip (Fig. 8.19). The band gap behavior was divided into three distinct families $N_{dim} = 3p$, $3p - 1$, and $3p - 2$, where N_{dim} is the number of dimer lines in the graphene strip and p is a positive integer [72,73]. The smaller gap was characteristic for the $3p - 1$ family.

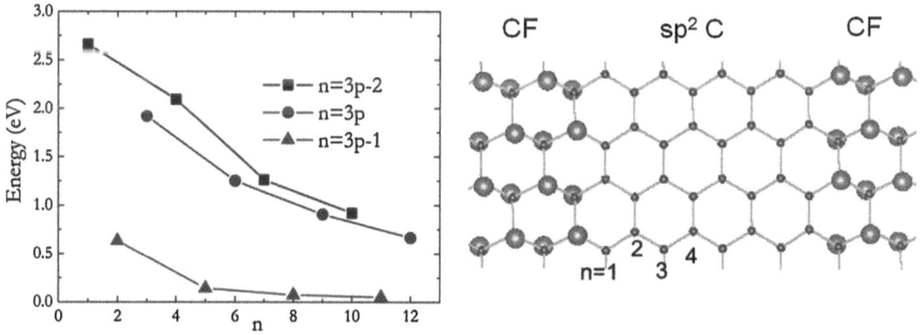

Figure 8.19 Variation in the band gap energy for the $(n,18)$ armchair graphene/fluorographene superlattices as a function of the graphene width n. The small and large balls in superlattice structure show the sp^2-hybridized carbon atoms and fluorine atoms, respectively. Number "*n*" accounts *dimer lines* in the graphene and "*p*" is a positive integer.

Reprinted with permission from H. Shi, H. Pan, Y.-W. Zhang, B.I. Yakobson, Electronic and magnetic properties of graphene/fluorographene superlattices, J. Phys. Chem. C 116 (2012) 18278–18283. Copyright 2012 American Chemical Society.

The zigzag-interfaced fluorographene/graphene hybrids showed magnetic behavior starting from the graphene strip composing three zigzag chains (Fig. 8.20). The spin density was mainly localized on carbon atoms in the graphene strips, especially near the two interfaces. The spins were ferromagnetically coupled at each zigzag interface, because the edge carbon atoms belonged to the same sublattice. The spin orientations at the two interfaces were opposite resulting in the antiferromagnetic ground state in the hybrid. The band gap of hybrids was inversely proportional to the graphene strip width [73]. At a large enough graphene width, the antiferromagnetic and ferromagnetic states of fluorographene/graphene hybrid become nearly degenerated due to weakening of the magnetic interaction between two interfaces.

The formation energy of triangular shaped graphene areas in fluorographene CF was found to decrease with increasing size of the area [74]. It was predicted that hybrids with the embedded armchair-edged graphene are energetically more favorable and they should be semiconductors with size-dependent band gaps. The band structures of hybrids containing zigzag-edged graphenes showed the localized energy levels around the Fermi level due to the imbalance in A− and B−sublattices presented in the embedded graphene area. The magnetic moment scaled linear with the size of the triangular graphene.

The fluorographene/graphene hybrids with hexagonal-shaped graphene areas were semiconductors independently on the edge of the embedded graphene [73]. Configurations with the zigzag edges had lower band gap values (Fig. 8.21). The electronic state of the embedded graphene areas was shown to be very confined due to the insulator character of the CF surrounding.

8.9 Insights Into Fluorination Mechanisms

The gas-phase fluorination for the graphene surface is usually carried out using XeF_2, while it is possible to add HF to F_2 also [75]. XeF_2 likes to bind perpendicular to

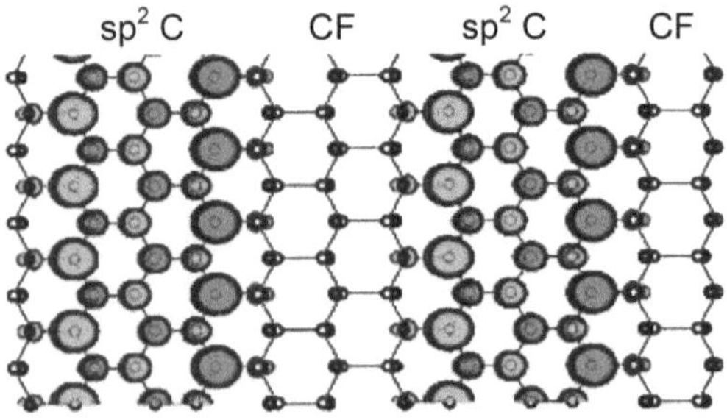

Figure 8.20 Contour plots of the spin density of the (3,6) armchair graphene/fluorographene superlattice (see designation of the model in Fig. 8.19) in the antiferromagnetic state. The different shades correspond to different orientations of the spins.
Reprinted with permission from H. Shi, H. Pan, Y.-W. Zhang, B.I. Yakobson, Electronic and magnetic properties of graphene/fluorographene superlattices, J. Phys. Chem. C 116 (2012) 18278−18283. Copyright 2012 American Chemical Society.

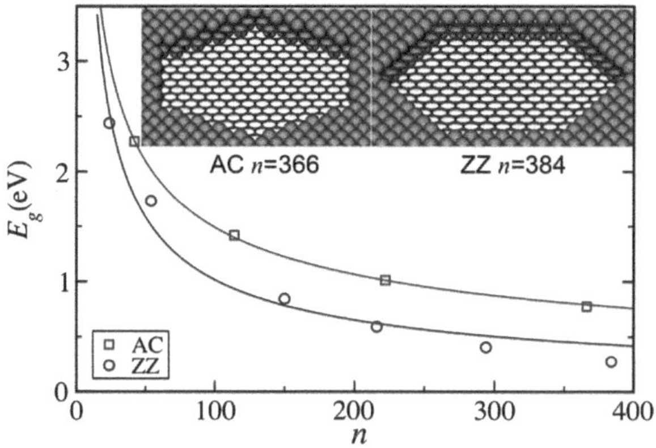

Figure 8.21 Plots showing how the energy of the band gap for fluorographene/graphene hybrids with embedded hexagonal shaped graphene areas having armchair (AC) and zigzag (ZZ) edges decreases with the graphene size *n*. Insert: Configurations of the largest optimized graphene areas. Nano Res., Patterning nanoroads and quantum dots on fluorinated graphene, 4 (2011) 143–152.
Copyright Tsinghua University Press and Springer–Verlag Berlin Heidelberg 2010 with permission of Springer.

graphene, although the configuration does not lead to fluorination [76]. Whereas the structures, where the XeF_2 molecule sits more or less parallel to the graphene plane, act as prereaction complexes for bonding of two fluorine atoms on the surface of graphene with activation barriers of 21.0–22.6 kcal/mol via a single transition state. In postreaction structures, the fluorine atoms are separated by two to four carbon atoms. Subsequent interaction of XeF_2 with these structures results in bonding of two F atoms at (1,4) position that is equivalent to 25% fluorine coverage observed in the case of one-side fluorinated graphene (Fig. 8.7B).

The F_2 molecule alone is also preferably adsorbed on the top of a carbon atom orthogonally to the graphene surface [77]. When the HF molecule presents, a hydrogen bonding is formed between HF and the "upper" fluorine atom of the F_2 molecule, leading to elongation of the F–F bond and thus to the formation of the covalent C–F bond. After that, the obtained FHF^- intermediate interacts with graphene leading to (1,2) or (1,4) addition products. As a result, HF is restored for further catalysis. Both addition possibilities are equally favored as they have approximately the same activation barrier. From a kinetic point of view, the same energy of the transition states producing the different additions indicates that fluorine addition for graphene is nonregioselective. However, the (1,2) addition is thermodynamically more favored by 10 kcal/mol for the final intermediates, and also, the (1,2) addition product is more stable by 10 kcal/mol than the (1,4) product. These calculations thus suggest that HF-catalyzed F_2 addition to graphene will result in a mixture of (1,2) and (1,4) addition patterns.

The fluorine atom prefers to sit atop to a carbon atom, where it shares the single electron with the carbon atom underneath [78]. The binding energy of fluorine in a dilute fluorinated graphene is lower than in fluorographene CF. This makes the fluorine atom mobile along hexagonal paths between different equilibrium positions. An estimated diffusion rate of an isolated neutral fluorine atom is several microns per second. This picture changes for a certain fluorine pattern. The most stable configuration of a pair of fluorine atoms is (1,4) or (para-) position in hexagon [21]. To move one F atom from the stable position to the neighboring A site (path 1 from B to A in Fig. 8.22), the F(B) atom has to overcome a barrier that is larger than 1.7 eV [47]. The diffusion to the first-neighbor position (path 2) is less likely even if the final position is more stable (diffusion barrier 254 meV). Thereafter the fluorine atoms will not be able to approach any closer until they are able to pass through the (1, 3) pair configuration.

8.10 Nature of C—F Bonding

A variation in stability [79] and chemical reactivity [49,80] of the fluorinated graphene samples depending on the fluorine content has raised the issue about the character of C—F bonding. The XPS data also motivated the opening of the discussion. In case of the fluorinated graphitic materials, the F1s spectrum is usually presented by a single line [15,22,81], while its width may be quite big indicating realization of different chemical states of fluorine atoms. The spectra of low-fluorinated graphene materials sometimes show two peaks separated by 1—2 eV [82,83]. By analogy with the early works on the fluorinated graphitic materials, the peak with the higher energy is assigned to the fluorine covalently bonded with carbon, while the low-energy peak is attributed to semi-ionic fluorine atoms. An analysis of the literature shows that two components in the XPS F1s spectrum are also appeared with a heat treatment of fluorinated graphene samples [84,85]. In this case, the components can be attributed

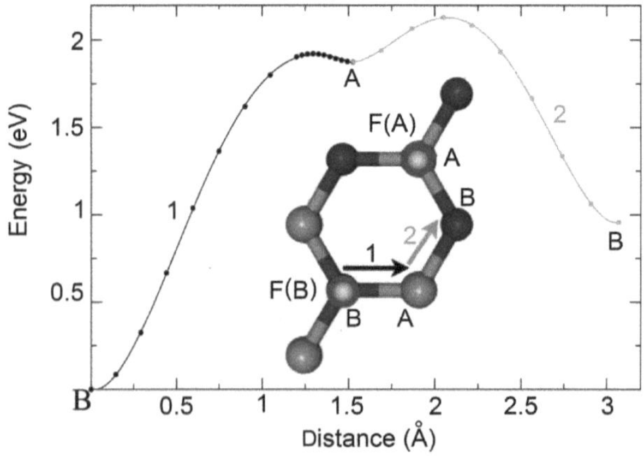

Figure 8.22 Energy barrier in a C_8F_2 system (fluorine atoms are in para-position) when an F(B) atom moves along paths 1 and 2. A and B indicate different sublattices in graphene. Adapted with permission from H. Santos, L. Henrard, Fluorine adsorption on single and bilayer graphene: Role of sublattice and layer decoupling, J. Phys. Chem. C 118 (2014) 27074—27080. Copyright 2014 American Chemical Society.

to fluorine atoms normally attached to the basal graphene plane and those at the defect sites such as, for example, graphene edges and vacancy boundaries. A detectable shift (~0.5 eV) of the main F1s line to lower binding energies was recorded when the fluorine content in the sample was reduced from C_4F to $C_{14}F$ [86].

A calculation of the shifts in F1s level energy was done for various fluorine patterns on graphene sheet using the core-excited pseudopotential technique [87]. The correlation between the level shift calculated respect to an isolated fluorine atom and the C—F bond length is presented in Fig. 8.23. The shorter the bond the higher the shift is. The length of the C—F bond is sensitive to local environment of the CF group and the energy of the F1s level varies about 3.5 eV when the fluorine coverage is changed from very dilute fluorination degree to CF. Since the F1s level energy for isolated fluorine was about midway between those for the fluorine anion in a bilayer

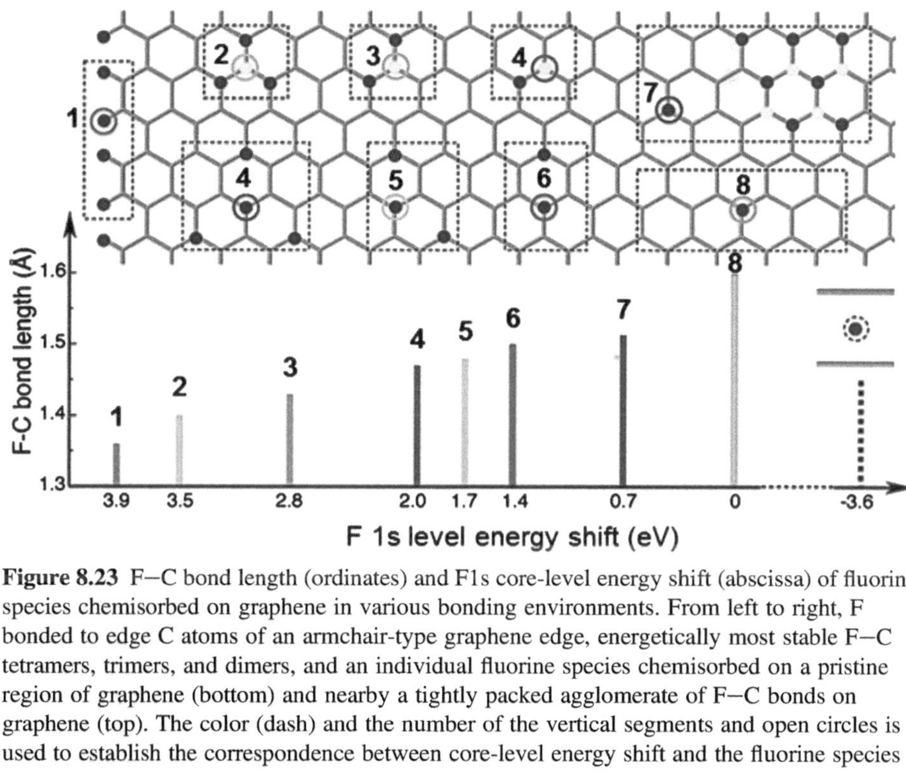

Figure 8.23 F—C bond length (ordinates) and F1s core-level energy shift (abscissa) of fluorine species chemisorbed on graphene in various bonding environments. From left to right, F bonded to edge C atoms of an armchair-type graphene edge, energetically most stable F—C tetramers, trimers, and dimers, and an individual fluorine species chemisorbed on a pristine region of graphene (bottom) and nearby a tightly packed agglomerate of F—C bonds on graphene (top). The color (dash) and the number of the vertical segments and open circles is used to establish the correspondence between core-level energy shift and the fluorine species drawn in the ball-and-stick illustration. The height of the vertical segments is used to show the length of the F—C bonds in the various bonding environments. The F1s core-level binding energies calculated from DFT are referred to that of F in single F—C species (8) chemisorbed on a pristine region of graphene. The core-level energy shift of an F anion in bilayer graphene is also reported on the right (vertical dashed segment).
Reprinted with permission from S. Zhou, S.D. Shera, D.W. Hess, A. Bongiorno, Chemical bonding of partially fluorinated graphene, J. Phys. Chem. C 118 (2014) 26402—26408. Copyright 2014 American Chemical Society.

graphene and the covalent C—F bond at an armchair graphene edge, the authors concluded that this type of the bond has a semi-ionic nature. The C—F bonds with the length larger than 1.5 Å were also attributed to semi-ionic bonds with different interaction strength. They are formed by fluorine atoms at the periphery of closely packed CF areas (the calculated shift of F1s level is 0.7 eV) and fluorine atoms located in the meta-position of hexagon in one-side fluorinated graphene (the calculated shift of F1s level is 1.4 eV). In the regions of C_4F, where fluorine atoms attached in para-position one to another on one side of graphene, the C—F bonding has a character intermediate between semi-ionic and covalent.

In our opinion, the shift values of the F1s level energy, calculated for the fluorinated graphene models (Fig. 8.23), are underestimated. From the calculation, the distance between the F1s peaks in fluorographene CF (model 2 in Fig. 8.23) and double-side fluorinated graphene C_2F should be ~0.7 or ~1.5 eV depending on the fluorine patterns (models 3 and 4 in Fig. 8.23). These distances are markedly smaller than the experimental value of ~2 eV [18]. Thus, when interpreting the XPS data the estimated shifts should be used with caution keeping in mind the possible error of the calculation.

The fluorine atom readily interacts with the sp^2-hybridized carbon atom stabilizing at a distance of ~1.56 Å [55,78,88]. This causes a local distortion of the graphene lattice with the highest lift of the carbon atom bonded to fluorine atom by ~0.4 Å. The F—C—C angle is ~102 degrees and this value is closer to the tetragonal angle of 109.47 degrees in fluorographene CF than to the angle of 90 degrees characteristic of fluorine ion adsorption [89]. Hence, in low-fluorinated graphene, the hybridization of the carbon atom bonded to fluorine has a dominant sp^3 character. This hybridization is more stable since the energy gain due to the C—F bonding overcomes the energy lost due to the graphene lattice distortion.

For a very dilute fluorine content, the estimated charge on the fluorine atom is about $-0.45e$ [88]. The negative charging of fluorine is due to electron depletion on the nearest carbon atoms and three first neighbors in the graphene lattice. The charge redistribution was found to be closely related to the variation in the bond lengths. With elongation of the C—F bond, more charge was pulled out of the graphene backbone [55]. Fig. 8.24 shows projected charge densities for fluorinated graphenes of CF, C_4F, and C_9F composition [80]. The charge density in the middle between carbon and fluorine atoms decreases with a lowering of fluorine content indicating a weakening of the C—F bonding.

Thus, considering the change in the electron density distribution in the fluorinated graphene with the fluorine loading, we can make the following explanation of the decrease of the F1s level energy in the low-fluorinated samples as compared to fluorographene CF. This idea was partially borrowed from Ref. [90]. The fluorographene is characterized by the lowest polarization of the C—F bond because multiple strongly electron-withdrawing fluorine atoms compete for π–electron density of the graphene sheet. In the low fluorine-populated graphene lattice, the fluorine may take the density of the next neighboring bare carbon atoms that causes an increase of the polarization of the C—F bond and consequently, a decrease of the F1s electron binding energy. This picture is in line with the hyperconjugation mechanism involving the electron

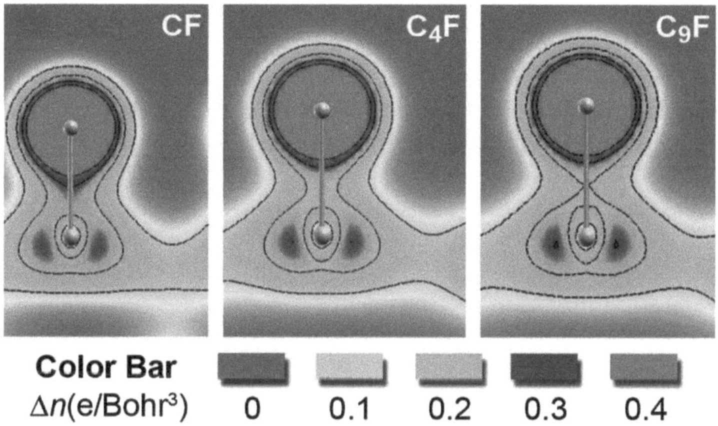

Figure 8.24 Charge density contours of C−F bonds in CF, C_4F, and C_9F.
Reproduced from P. Wang, H. Wang, W. Yang, Anomalous high adsorption energy of H_2O on fluorinated graphenes: a first-principles study, Phys. Chem. Chem. Phys. 16 (2014) 20464−20470. Copyright 2014, with permission from Royal Society of Chemistry.

interactions between C−F bonds and carbon π−bonds from graphene areas [15]. It is evident that the position of the XPS F1s line is not solely dependent on the fluorine content but also on the distribution of fluorine atoms. For example, even at a relatively low-fluorine content the fully fluorinated graphene regions can be formed on the graphene plane and a F1s spectrum of such material should have the energy position similar to the fluorographene CF spectrum [87].

8.11 Optical Properties

A tunable energy gap of the fluorinated graphene makes this material attractive for electro-optical applications. The DFT calculations predict that the most stable structure of the fully fluorinated graphene CF is the chair configuration (Fig. 8.1A) with a direct band gap of 3.1 eV [41]. Although this energy agrees with the experimental value of ~3 eV [37], the coincidence may be accidental because standard local density approximation and generalized gradient approximation functionals commonly used for the solid-state calculations systematically underestimate the gap [91]. The use of the screened hybrid functionals, which include a portion of the exact Hartree−Fock exchange [91], or accounting of the electron−electron interactions through the GW approach [41] increases a band gap of fluorographene up to ~5.1 and 7.5 eV, respectively. The large quasiparticle corrections are due to the reduced dimensionality of fluorographene and enhanced Coulomb interaction. Note that even at the same DFT functional used, the GW band gap value depends on the calculation parameters, such as k−point sampling of Brillouin zone, cut-off energy, thickness of vacuum layer, etc. and a way to account the Green's functions and screened Coulomb potential [92].

The experimental band gap of fluorographene CF was extracted from an optical absorption spectrum and the calculated results should be corrected on the electron–hole interactions. Currently, the photo-excited states are obtained by solving the Bethe–Salpeter equation (BSE) on top of the GW band-structure calculation [93]. In this case, a theoretical absorption spectrum directly corresponds to the imaginary part of the macroscopic dielectric function.

Fig. 8.25 compares the optical absorption spectra calculated within the random-phase approximation (RPA) with inclusion of local field effects, which account for macroscopic inhomogeneity of the charge density in a crystal, GW–RPA and GW–BSE [41]. The RPA spectrum corresponds to the single-particle ground-state calculation, the GW–RPA takes into account the electron–electron (*e–e*) repulsions, the GW–BSE includes the electron–hole (*e–h*) attractions. The first excitonic peak A in the CF spectrum is located at ~5.4 eV and this value is substantially lower than the GW band gap of 7.5 eV (shown by *dotted arrow* in Fig. 8.25). This difference corresponds to an exciton with a binding energy of 2.1 eV. The onset of the optical absorption for the RPA spectrum is ~3.1 eV (shown by *dashed arrow* in Fig. 8.25). Thus, there is a large cancellation of *e–e* and *e–h* effects, which might explain why the experimental absorption value is significantly red-shifted with respect to the GW band gap, whereas the DFT band gap values correspond to the experiments.

The peak A in the GW–BSE spectrum calculated for fluorographene CF (Fig. 8.25) is actually a superposition of two optically active (bright) excitons possessing similar oscillator strength [94]. They are formed by dipole-allowed vertical transitions from the two highest occupied bands to the lowest empty band. The spectrum is dominated by a peak around 9.8 eV that is well below the onset of single-particle transition continuum. The absence of this peak in the GW–RPA and RPA spectra evidences its connection to strong *e–h* coupling and formation of bound excitons. An examination of the excitonic function revealed that the major transitions contributing to this peak originated from two highest occupied bands and four lowest empty bands [92]. Calculations predict that CF should exhibit

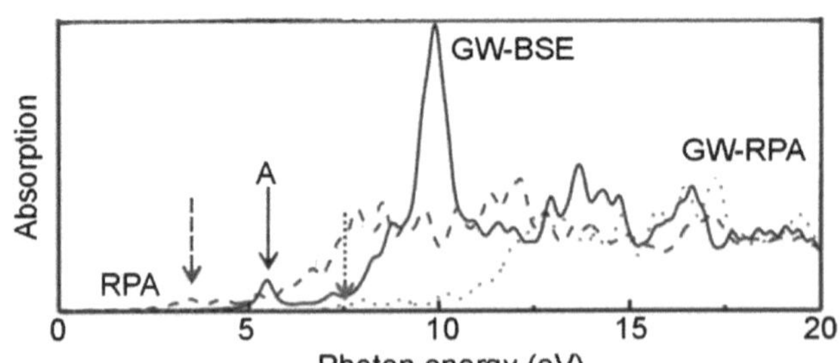

Figure 8.25 Calculated absorption spectra using RPA (*dashed lines*), GW–RPA (*dotted lines*), and GW–BSE (*solid lines*) for flurographene CF. *RPA*, random-phase approximation. Reproduced from D.K. Samarakoon, Z. Chen, C. Nicolas, X.-Q. Wang, Structural and electronic properties of fluorographene, Small 7 (2011) 965–969. Copyright 2011, with permission from John Wiley and Sons.

strong optical anisotropy because completely different absorption spectra were obtained for light polarization perpendicular and parallel to the surface plane [92,94]. For the perpendicular light polarization, fluorographene is almost completely transparent below 10 eV.

The main features of the optical absorption spectra for fluorographene CF, two CF layers with AA stacking, and graphite monofluoride $(CF)_n$ were predicted to be similar [95]. Hence, weak van der Waals interactions between the layers have negligible effect on the optical response. In fluorinated graphene, the strong binding of excitons is due to the weak and specifically nonlocal two-dimensional dielectric screening. Whereas the position of the first exciton peak was independent on the number of layers, the dominant high-energy peak was systematically shifted to higher energies on going from single-layer CF to double layer and then to bulk material. This may be useful for spectroscopic discrimination between samples with different layer numbers.

The GW−BSE spectrum of on-side fluorinated C_4F showed the first optically active exciton at 4.21 eV with a binding energy of 1.31 eV [94]. Exciton wave functions in C_4F were found to be spatially more delocalized than those in CF.

An accounting of excitonic effects in the calculation allowed approaching the calculated band gap for fluorographene to the optical experiment, but the obtained value of ~5.1 eV is still too large. Another explanation of the discrepancy between theory and experiment is a disorder in fluorinated graphenes. To clarify this issue, the optical conductivity spectra were plotted for the models with different types of structural disorder [96]. The fluorinated graphene systems consisting of 2400×2400 carbon atoms were computed using tight-binding (TB) parameters derived by fitting the band electronic structures and local density of states of graphene, partially and fully fluorinated graphene obtained from ab initio calculations carried out in the GW approximation. The fluorine atoms were randomly distributed over the graphene lattice, while two neighbors were attached from the opposite sides of graphene. The in-plane spectra in an interval from 0 to 10 eV are presented in Fig. 8.26 in comparison to graphene spectrum, which exhibits a single resonance originated from electron transitions between van Hove singularities located around the Fermi level (the insert in Fig. 8.26). In fluorographene CF, the optical absorption sets in the electronic band gap of 6.3 eV. The common effect due to the presence of structural disorders such as missing of fluorine atoms or groups of fluorine atoms is the appearance of the intragap states, which positions are defined by the type of disorder. For example, a pronounced peak at ~5 eV corresponds to fluorine vacancies, ie, missing of a fluorine atom from the CF matrix. The presence of graphene regions in CF surrounding produces many peaks between 2–4 eV. Remember, that energies of optical excitations were obtained in the GW approximation, which neglects many body effects like the occurrence of excitons.

Graphene has no photoluminescence (PL) emission due to the absence of the energy gap. The PL spectrum of a fluorographene CF dispersed in acetone showed two peaks at ~3.80 and 3.65 eV [97]. The former peak was attributed to an exciton recombination that gives the bottom limit for the band gap in CF. The lower energy peak was related to excitation of the C−F vibration mode as the result of

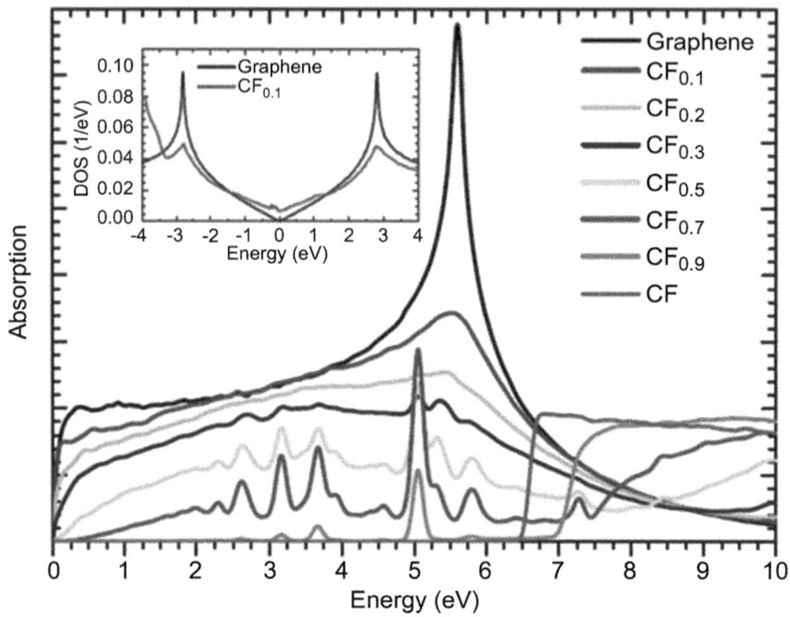

Figure 8.26 In-plane optical conductivity of partially and fully fluorinated graphene with different concentration of randomly distributed paired fluorine atoms. Insert presents the density of states of graphene and $CF_{0.1}$.
Reprinted with permission from S. Yuan, M. Rösner, A. Schulz, T.O. Wehling, M.I. Katsnelson, Phys. Rev. Lett. 114 (2015) 047403. Copyright 2015 by the American Physical Society.

recombination of the *e–h* pair. A PL study of nanocrystalline graphite monofluoride $(CF)_n$ detected six emission modes from 1.72 to 2.79 eV [98]. Since the positions of the lines were before the band-edge emission expected for $(CF)_n$, they were assigned to midgap states produced by defects. Primary candidates for such defects are isolated or clustered nonfluorinated carbon atoms, and unsaturated bonds at nanocrystalline domain boundaries.

A full-color emission from 1.65 to 3.1 eV was observed in the PL spectrum of fluorinated graphene sheets obtained by liquid exfoliation of graphite fluoride with a fluorine content of about 20% [99]. This confirms dependence of the PL property of fluorinated graphene on the fluorination degree and loading. It was proposed that the fluorinated graphene layer always contains sp^2-hybridized nanometer-sized carbon clusters with different shapes randomly surrounded by CF bonds [100]. These clusters can be formed in the fluorination process, after the sample contacts with surrounding environment, during the atomic rearrangement. The DFT calculations of aromatic carbon clusters show that the energy gap varies from 7 to 3 eV when the number of hexagons in cluster increases from 1 to 37. The recombination of the localized *e—h* pairs in the sp^2-clusters with different number of aromatic rings could produce the peaks observed in the PL spectra of the fluorinated graphene samples.

8.12 Conclusions

An energy gain from the formation of covalent C–F bond is sufficient to overcome a barrier needed for local distortion of graphene network when carbon atom changes the hybridization from sp^2 to sp^3. A donation of electron density from graphene to fluorine leads to opening of band gap in fluorinated graphene and this may be a turning point for the development of graphene-based electronics. Fluorinated graphene can be extracted from parent graphite fluoride or obtained by direct fluorination of a graphene specimen. In both cases, the control of crystallinity and composition of the product is crucially important to avoid defects, which can influence conductivity and optical response of the fluorinated graphene.

The theory promises many exciting properties for fluorinated graphene systems. Particularly, it is expected that energy gap in the system is finely tuned with chemisorption of fluorine atoms on certain positions of the graphene lattice. Since the optical gap in fluorographene CF is about 3 eV, a gradual decrease in the fluorine loading will provide the photoluminescence covering the whole visible light spectrum. A coupling of graphene and fluorographene areas in a hybrid structure or imbalance in fluorine attachment to two sides of graphene may cause magnetic ordering in a fluorinated system. Realization of ferromagnetic behavior in the fluorinated graphene will open a door for graphene spintronics.

As opposite to graphite, two-dimensional graphene crystal can be fluorinated from one side only. From a theoretical point of view, this may result in different work functions from the fluorine-free and fluorine-decorated graphene sides and nanoscrolls formation. Experimentally shown that backside-fluorinated graphene layer has a prospect for the elaboration of gas sensors.

The expected properties of fluorinated graphene can be fully realized when a required quantity of fluorine atoms will be placed with necessary density on the graphene plane. This is an actually challenging task, which demands use of nanoscale manipulations.

Acknowledgment

We thank the Russian Science Foundation for partial financial support through a grant #14-13-00813.

References

[1] K.S. Novoselov, Nobel lecture: Graphene: materials in the flatland, Rev. Mod. Phys. 83 (2011) 837–849.

[2] D.K. Samarakkon, Z. Chen, C. Nicolas, X.-Q. Wang, Structural and electronic properties of fluorographene, Small 7 (2011) 965–969.

[3] D.C. Elias, R.R. Nair, T.M.G. Mohiuddin, S.V. Morozov, P. Blake, M.P. Halsall, A.C. Ferrari, D.W. Boukhvalov, M.I. Katsnelson, A.K. Geim, K.S. Novoselov, Control of graphene's properties by reversible hydrogenation: evidence for graphene, Science 323 (2009) 610–613.

[4] W. Gao, L.B. Alemany, L. Ci, P.M. Ajayan, New insights into the structure and reduction of graphite oxide, Nat. Chem. 1 (2009) 403–408.
[5] M.F. Craciun, I. Khrapach, M.D. Barnes, S. Russo, Properties and applications of chemically functionalized graphene, J. Phys. Condens. Matter 25 (2013) 423201.
[6] O. Ruff, O. Bretschneider, F. Evert, Die Reaktionsprodukte der vershiedenen Kohlenstofformen mit Fluor II (Kohlenstoffmonofluorid), Z. Anorg. Allg. Chem. 217 (1934) 1–18.
[7] N. Watanabe, T. Nakajima, H. Touhara, Graphite Fluorides: Studies in Inorganic Chemistry, Elsevier, Amsterdam, 1988, 263 pp.
[8] V.K. Mahajan, R.B. Badachhape, J.L. Margrave, X-ray powder diffraction study of poly(carbon monofluoride), $CF_{1.12}$, Inorg. Nucl. Chem. Lett. 10 (1974) 1103–1109.
[9] J.-C. Charlier, X. Gonze, J.-P. Michenaud, First-principles study of graphite monofluoride $(CF)_n$, Phys. Rev. B 47 (1993) 16162–16168.
[10] P. Kamarchik Jr., J.L. Margrave, Poly(carbon monofluoride): a solid, layered fluorocarbon, Acc. Chem. Res. 11 (1978) 296–300.
[11] A. Hamwi, Fluorine reactivity with graphite and fullerenes. Fluorine derivatives and some practical electrochemical applications, J. Phys. Chem. Solids 57 (1996) 677–688.
[12] T. Nakajima, Synthesis, structure, and physicochemical properties of fluorine–graphite intercalation compounds, in: T. Nakajima (Ed.), Fluorine–Carbon and Fluoride–Carbon Materials, Chemistry, Physics, and Applications A314, Marcel London, New York, 1985, pp. 179–187.
[13] V.N. Mitkin, Types of inorganic fluorocarbon polymer materials and structure-property correlation problems, J. Struct. Chem. 44 (3002) (2003) 82–115.
[14] Y. Ahmad, M. Dubois, K. Guérin, A. Hamwi, Z. Fawal, A.P. Kharitonov, A.V. Generalov, A.Yu Klyushin, K.A. Simonov, N.A. Vinogradov, I.A. Zhdanov, A.B. Preobrajenski, A.S. Vinogradov, NMR and NEXAFS study of various graphite fluorides, J. Phys. Chem. C 117 (2013) 13564–13572.
[15] Y. Sato, K. Itoh, R. Hagiwara, T. Fukunaga, Y. Ito, On the so-called "semi-ionic" C–F bond character in fluorine–GIC, Carbon 42 (2004) 3243–3249.
[16] A.V. Okotrub, N.F. Yudanov, I.P. Asanov, D.V. Vyalikh, L.G. Bulusheva, Anisotropy of chemical bonding in semifluorinated graphite C_2F revealed with angle-resolved X-ray absorption spectroscopy, ACS Nano 7 (2013) 65–74.
[17] I.P. Asanov, L.G. Bulusheva, M. Dubois, N.F. Yudanov, A.V. Alexeev, T.L. Makarova, A.V. Okotrub, Graphene nanochains and nanoislands in the layers of room-temperature fluorinated graphite, Carbon 59 (2013) 518–529.
[18] I.P. Asanov, V.M. Paasonen, L.N. Mazalov, A.S. Nazarov, X-ray photoelectron study of fluorinated graphite intercalation compounds, J. Struct. Chem. 39 (1998) 928–932.
[19] T. Mallouk, N. Bartlett, Reversible intercalation of graphite by fluorine: a new bifluoride, $C_{12}HF_2$, and graphite fluorides, C_xF ($5>x>2$), J. Chem. Soc. Chem. Commun. (1983) 103–105.
[20] A. Hamwi, M. Daoud, J.C. Cousseins, Graphite fluorides prepared at room temperature 1. Synthesis and characterization, Synth. Met. 26 (1988) 89–98.
[21] C.P. Ewels, G. Van Lier, J.-C. Charlier, M.I. Heggie, P.R. Briddon, Pattern formation on carbon nanotube surfaces, Phys. Rev. Lett. 96 (2006) 216103.
[22] D. Claves, Spectroscopic study of fluorinated carbon nanostructures, New J. Chem. 35 (2011) 2477–2482.
[23] H.F. Bettinger, K.N. Kudin, G.E. Scuseria, Structural models of fluorine-graphite intercalation compounds from density functional theory, J. Phys. Chem. A 108 (2004) 3016–3018.

[24] A.M. Panich, Nuclear magnetic resonance study of fluorine-graphite intercalation compounds and graphite fluorides, Synth. Met. 100 (1999) 169–185.
[25] A. Vyalikh, L.G. Bulusheva, G.N. Chekhova, D.V. Pinakov, A.V. Okotrub, U. Scheler, Fluorine patterning in room-temperature fluorinated graphite determined by solid-state NMR and DFT, J. Phys. Chem. C 117 (2013) 7940–7948.
[26] J. Giraudet, M. Dubois, K. Guérin, C. Delabarre, A. Hamwi, F. Masin, Solid-state NMR study of the post-fluorination of $(C_{2.5}F)_n$ fluorine-GIC, J. Phys. Chem. B 111 (2007) 14143–14151.
[27] A.K. Geim, K.S. Novoselov, The rise of graphene, Nat. Mater. 6 (2007) 183–191.
[28] F. Withers, M. Dubois, A.K. Savchenko, Electron properties of fluorinated single-layer graphene transistors, Phys. Rev. B 82 (2010) 073403.
[29] R. Zbořil, F. Karlický, A.B. Bourlinos, T.A. Sterioris, A.K. Stubos, V. Georgakilas, K. Šafářová, D. Jančík, C. Trapalis, M. Otyepka, Graphene fluoride: a stable stoichiometric graphene derivative and its chemical conversion to graphene, Small 6 (2010) 2885–2891.
[30] A.B. Bourlinos, A. Bakandritsos, N. Liaros, S. Couris, K. Safarova, M. Otyepka, R. Zbořil, Water dispersible functionalized graphene fluoride with significant nonlinear optical response, Chem. Phys. Lett. 543 (2012) 101–105.
[31] M. Zhang, Y. Ma, Y. Zhu, J. Che, Y. Xiao, Two-dimensional transparent hydrophobic coating based on liquid-phase exfoliated graphene fluoride, Carbon 63 (2013) 149–156.
[32] X. Wang, W. Wang, Y. Liu, M. Ren, H. Xiao, X. Liu, Controllable defluorination of fluorinated graphene and weakening of C–F bonding under the action of nucleophilic dipolar solvent, Phys. Chem. Chem. Phys. 18 (2016) 3285–3293.
[33] R.J. Kashtiban, M.A. Dyson, R.R. Nair, R. Zan, S.L. Wong, Q. Ramasse, A.K. Geim, U. Bangert, J. Sloan, Atomically resolved imaging of highly ordered alternating fluorinated graphene, Nat. Commun. 5 (2014) 4902.
[34] W.H. Lee, J.W. Suk, H. Chou, J. Lee, Y. Hao, Y. Wu, R. Piner, D. Akinwande, K.S. Kim, R.S. Ruoff, Selective-area fluorination of graphene with fluoropolymer and laser irradiation, Nano Lett. 12 (2012) 2374–2378.
[35] F. Withers, T.H. Bointon, M. Dubois, S. Russo, M.F. Craciun, Nanopatterning of fluorinated graphene by electron beam irradiation, Nano Lett. 11 (2011) 3912–3916.
[36] W.-K. Lee, M. Haydell, J.T. Robinson, A.R. Laracuente, E. Cimpoiasu, W.P. King, P.E. Sheehan, Nanoscale reduction of graphene fluoride via thermochemical nanolithography, ACS Nano 7 (2013) 6219–6224.
[37] R.R. Nair, W. Ren, R. Jalil, I. Riaz, V.G. Kravets, L. Britnell, P. Blake, F. Schedin, A.S. Mayorov, S. Yuan, M.I. Katsnelson, H.-M. Cheng, W. Strupinski, L.G. Bulusheva, A.V. Okotrub, I.V. Grigorieva, A.N. Grigorenko, K.S. Novoselov, A.K. Geim, Fluorographene: a two-dimensional counterpart of Teflon, Small 6 (2010) 2877–2884.
[38] V.G. Makotchenko, A.S. Nazarov, Structural characteristics of dicarbon polyfluoride, J. Struct. Chem. 50 (2010) 1088–1095.
[39] L.G. Bulusheva, O.V. Sedelnikova, A.V. Okotrub, Many-body effects in optical response of graphene-based structures, Int. J. Quant. Chem. 116 (2016) 270–281.
[40] O. Leenaerts, H. Peelaers, A.D. Hernández-Nieves, B. Partoens, F.M. Peeters, First-principles investigation of graphene fluoride and graphane, Phys. Rev. B 82 (2010) 195436.
[41] D.K. Samarakoon, Z. Chen, C. Nicolas, X.-Q. Wang, Structural and electronic properties of fluorographene, Small 7 (2011) 965–969.
[42] J.T. Robinson, J.S. Burgess, C.E. Junkermeier, S.C. Badescu, T.L. Reinecke, F.K. Perkins, M.K. Zalalutdniov, J.W. Baldwin, J.C. Culbertson, P.E. Sheehan, E.S. Snow, Properties of fluorinated graphene films, Nano Lett. 10 (2010) 3001–3005.

[43] S. Kwon, J.-H. Ko, K.-J. Jeon, Y.-H. Kim, J.Y. Park, Enhanced nanoscale friction on fluorinated graphene, Nano Lett. 12 (2012) 6043−6048.
[44] H. Sahin, M. Topsakal, S. Ciraci, Structures of fluorinated graphene and their signatures, Phys. Rev. B 83 (2011) 115432.
[45] A.L. Walter, H. Sahin, K.-J. Jeon, A. Bostwick, S. Horzum, R. Koch, F. Speck, M. Ostler, P. Nagel, M. Merz, S. Schupler, L. Moreschini, Y.J. Chang, T. Seyller, F.M. Peeters, K. Horn, E. Rotenberg, Luminescence, patterned metallic regions, and photon-mediated electronic changes in single-sided fluorinated graphene sheets, ACS Nano 8 (2014) 7801−7808.
[46] N. Jiao, C. He, P. Zhou, C.X. Zhang, H.P. Xiao, L.Z. Sun, Surface work function of chemically derive graphene: a first-principles study, Phys. Lett. A 377 (2013) 1760−1765.
[47] H. Santos, L. Henrard, Fluorine adsorption on single and bilayer graphene: role of sublattice and layer decoupling, J. Phys. Chem. C 118 (2014) 27074−27080.
[48] H.Y. Liu, Z.F. Hou, C.H. Hu, Y. Yang, Z.Z. Zhu, Electronic and magnetic properties of fluorinated graphene with different coverage of fluorine, J. Phys. Chem. C 116 (2012) 18193−18201.
[49] M.V. Katkov, V.I. Sysoev, A.V. Gusel'nikov, I.P. Asanov, L.G. Bulusheva, A.V. Okotrub, A backside fluorine-functionalized graphene layer for ammonia detection, Phys. Chem. Chem. Phys. 17 (2015) 444−450.
[50] H. Guo, Y. Liu, Y. Xu, N. Meng, H. Wang, T. Hasan, X. Wang, J. Luo, B. Yu, Fluorinated graphene and hexagonal boron nitride as ALD seed layer for graphene-based can der Waals heterostructures, Nanotechnol. 25 (2014) 355202.
[51] E.R. Johnson, S. Keinan, P. Mori-Sanchez, J. Contreras-Garsia, A.J. Cohen, W. Yang, Revealing noncovalent interactions, J. Am. Chem. Soc. 132 (2010) 6498−6506.
[52] Z. Wang, S. Qin, C. Wang, Q. Hui, Fluorine adsorption on the graphene films: from metal to insulator, Comp. Mater. Sci. 97 (2015) 14−19.
[53] A.N. Enyashin, A.L. Ivanovskii, Layers and tubes of fluorographene C_4F: stability, structural and electronic properties from DFTB calculations, Chem. Phys. Lett. 576 (2013) 44−48.
[54] H. Zhan, G. Zhang, J.M. Bell, Y. Gu, Graphene with patterned fluorination: morphology modulation and implications, J. Phys. Chem. C 119 (2015) 27562−27568.
[55] Y. Zheng, X. Wan, N. Tang, Q. Feng, F. Liu, Y. Du, Magnetic properties of double-side partially fluorinated graphene from first principles calculations, Carbon 89 (2015) 300−307.
[56] T.L. Makarova, A.L. Shelankov, A.A. Zyrianova, A.I. Veinger, T.V. Tisnek, E. Lähderanta, A.I. Shames, A.V. Okotrub, L.G. Bulusheva, G.N. Chekhova, D.V. Pinakov, I.P. Asanov, Ž. Šljivančanin, Edge state magnetism in zigzag-interfaced graphene via spin susceptibility measurements, Sci. Rep. 5 (2015) 13382.
[57] D.V. Pinakov, V.A. Logvinenko, The relationship between properties of fluorinated graphite intercalates and matrix composition. Intercalates with acetonitrile, J. Therm. Anal. Calometry 86 (2006) 173−178.
[58] W. Yu, S.-P. Gao, Effect of configuration and biaxial strain to electronic structure of half-fluorinated graphene, Surf. Sci. 635 (2015) 78−84.
[59] L.G. Bulusheva, A.V. Okotrub, N.F. Yudanov, Atomic arrangement and electronic structure of graphite fluoride C_2F, Phys. Low-Dim. Struct. 7/8 (2002) 1−14.
[60] L.G. Bulusheva, A.V. Okotrub, Nature of chemical bonding in the fluorinated carbon compounds, Rev. Inorg. Chem. 19 (1999) 79−115.
[61] S.-H. Cheng, K. Zou, H.R. Gutierrez, A. Gupta, N. Shen, P.C. Eklund, J.O. Sofo, J. Zhu, Reversible fluorination of graphene: evidence of a two-dimensional wide bandgap semiconductor, Phys. Rev. B 81 (2010) 205435.

[62] A. Felten, A. Eckmann, J.-J. Pireaux, R. Krupke, C. Casiraghi, Controlled modification of mono- and bilayer graphene in O_2, H_2 and CF_4 plasmas, Nanotechnology 24 (2013) 355705.
[63] J. Sivek, O. Leenaerts, B. Partoens, F.M. Peeters, First-principles investigation of bilayer fluorographene, J. Phys. Chem. C 116 (2012) 19240−19245.
[64] C.H. Hu, Y. Zhang, H.Y. Liu, S.Q. Wu, Y. Yang, Z.Z. Zhu, Tunable band gap in half-fluorinated bilayer graphene under biaxial strains, Comput. Mater. Sci. 65 (2012) 165−169.
[65] C.H. Hu, P. Zhang, H.Y. Liu, S.Q. Wu, Y. Yang, Z.Z. Zhu, Structural stability and electronic and magnetic properties of fluorinated bilayer graphene, J. Phys. Chem. C 117 (2013) 3572−3579.
[66] L. Brown, R. Hovden, P. Huang, M. Wojcik, D.A. Muller, J. Park, Twinning and twisting of tri- and bilayer graphene, Nano Lett. 12 (2012) 1609−1615.
[67] J. Hass, F. Varchon, J.E. Millán-Otoya, M. Sprinkle, N. Sharma, W.A. de Heer, C. Berger, P.N. First, L. Magaud, E.H. Conrad, Why multilayer graphene on 4H-SiC(0001) behaves like a single sheet of graphene, Phys. Rev. Lett. 100 (2008) 125504.
[68] A.R. Muniz, D. Maroudas, Superlattices of fluorinated interlayer-bonded domains in twisted bilayer graphene, J. Phys. Chem. C 117 (2013) 7315−7325.
[69] Y. Li, B.A. Pantoja, Z. Chen, Self-modulated band structure engineering in C_4F nanosheets: first-principles insights, J. Chem. Theory Comput. 10 (2014) 1265−1271.
[70] W.-K. Lee, J.T. Robinson, D. Gunlycke, R.R. Stine, C.R. Tamanaha, W.P. King, P.E. Sheehan, Chemically isolated graphene nanoribbons reversibly formed in flurographene using polymer nanowire masks, Nano Lett. 11 (2011) 5461−5464.
[71] I.P. Asanov, A.V. Okotrub, A.V. Gusel'nikov, I.V. Yushina, D.V. Vyalikh, L.G. Bulusheva, Charge-induced formation of thin conducting layers on fluorinated graphite surface, Carbon 82 (2015) 446−458.
[72] H. Shi, H. Pan, Y.-W. Zhang, B.I. Yakobson, Electronic and magnetic properties of graphene/fluorographene superlattices, J. Phys. Chem. C 116 (2012) 18278−18283.
[73] M.A. Ribas, A.K. Singh, P.B. Sorokin, B.I. Yakobson, Patterning nanoroads and quantum dots on fluorinated graphene, Nano Res. 4 (2011) 143−152.
[74] H. Xie, W.-C. Lu, W. Zhang, P.-H. Qin, C.Z. Wang, K.M. Ho, Electronic and magnetic properties of triangular graphene nanoflakes embedded in fluorographene, Chem. Phys. Lett. 572 (2013) 48−52.
[75] K.F. Kelly, E.T. Mickelson, R.H. Hauge, J.L. Margrave, N.J. Halas, Nanoscale imaging of chemical interactions: fluorine on graphite, PNAS 97 (2000) 10318−10321.
[76] S.-S. Lee, S.-W. Jang, K. Park, E.C. Jang, J.-Y. Kim, D. Neuhauser, S. Lee, A mechanistic study of graphene fluorination, J. Phys. Chem. C 117 (2013) 5407−5415.
[77] S. Osuna, M. Torrent-Sucarrat, M. Solà, P. Geerlings, C.P. Ewels, G. Van Lier, Reaction mechanisms for graphene and carbon nanotube fluorination, J. Phys. Chem. C 114 (2010) 3340−3345.
[78] A. Sadeghi, M. Neek-Amal, G.R. Berdiyorov, F.M. Peeters, Diffusion of fluorine on and between graphene layers, Phys. Rev. B 91 (2015) 014304.
[79] M. Ren, X. Wang, C. Dong, B. Li, Y. Liu, T. Chen, P. Wu, Z. Cheng, X. Liu, Reduction and transformation of fluorinated graphene induced by ultraviolet irradiation, Phys. Chem. Chem. Phys. 17 (2015) 24056−24062.
[80] P. Wang, H. Wang, W. Yang, Anomalous high adsorption energy of H_2O on fluorinated graphenes: a first principles study, Phys. Chem. Chem. Phys. 16 (2014) 20464−20470.

[81] G. Nansé, E. Papirer, P. Fioux, F. Moguet, A. Tressaud, Fluorination of carbon blacks: an X-ray photoelectron spectroscopy study: I. A literature review of XPS studies of fluorinated carbons. XPS investigation of some reference compounds, Carbon 35 (1997) 175–194.
[82] S.D. Sherpa, G. Levitin, D.W. Hess, Effect of the polarity of carbon-fluorine bonds on the work function of plasma-fluorinated epitaxial graphene, Appl. Phys. Lett. 101 (2012) 111602.
[83] B. Wang, J. Wang, J. Zhu, Fluorination of graphene: a spectroscopic and microscopic study, ACS Nano 8 (2014) 1862–1870.
[84] S.D. Costa, J.E. Weis, O. Frank, Z. Bastl, M. Kalbac, Thermal treatment of fluorinated graphene: an in situ Raman spectroscopy study, Carbon 84 (2015) 347–354.
[85] L.G. Bulusheva, V.A. Tur, E.O. Fedorovskaya, I.P. Asanov, D. Pontiroli, M. Riccò, A.V. Okotrub, Structure and supercapacitor performance of graphene materials obtained from brominated and fluorinated graphites, Carbon 78 (2014) 137–146.
[86] K. Hou, P. Gong, J. Wang, Z. Yang, Z. Wang, S. Yang, Structural and tribological characterization of fluorinated graphene with various fluorine content prepared by liquid-phase exfoliation, RSC Adv. 4 (2014) 5643–56551.
[87] S. Zhou, S.D. Shera, D.W. Hess, A. Bongiorno, Chemical bonding of partially fluorinated graphene, J. Phys. Chem. C 118 (2014) 26402–26408.
[88] A. Natan, M.C. Hersam, T. Sedeman, Insights into graphene functionalization by single atom doping, Nanotechnology 24 (2013) 505715.
[89] J.O. Sofo, A.M. Suarez, G. Usaj, P.S. Cornaglia, A.D. Hernández-Nieves, C.A. Balseiro, Electrical control of the chemical bonding of fluorine on graphene, Phys. Rev. B 83 (2011) 081411.
[90] R. Stine, W.-K. Lee, K.E. Whitener Jr., J.T. Robinson, P.E. Sheehan, Chemical stability of graphene fluoride produced by explosure to XeF_2, Nano Lett. 13 (2013) 4311–4316.
[91] F. Karlický, R. Zbořil, M. Otyepka, Band gaps and structural properties of graphene halides and their derivates: a hybrid functional study with localized orbital basis sets, J. Chem. Phys. 137 (2012) 034709.
[92] F. Karlický, M. Otyepka, Band gaps and optical spectra of chlorographene, fluorographene and graphene from G_0W_0, GW_0 and GW calculations on top of PBE and HSE06 orbitals, J. Chem. Theory Comput. 9 (2013) 4155–4164.
[93] S. Albrecht, L. Reining, R. Del Sole, G. Onida, Ab initio calculation of excitonic effects in the optical spectra of semiconductors, Phys. Rev. Lett. 80 (1998) 4510–4513.
[94] W. Wei, T. Jacob, Electronic and optical properties of fluorinated graphene: a many-body perturbation theory study, Phys. Rev. B 87 (2013) 115431.
[95] F. Karlický, M. Otyepka, Band gaps and optical spectra from single- and double-layer fluorographene to graphite fluoride: many-body effects and excitonic states, Ann. Phys. (Berlin) 526 (2014) 408–414.
[96] S. Yuan, M. Rösner, A. Schulz, T.O. Wehling, M.I. Katsnelson, Electronic structures and optical properties of partially and fully fluorinated graphene, Phys. Rev. Lett. 114 (2015) 047403.
[97] K.-J. Jeon, Z. Lee, E. Pollak, L. Moreschini, A. Bostwick, C.-M. Park, R. Mendelsberg, V. Radmilovic, R. Kostecki, T.J. Richardson, E. Rotenberg, Fluorographene: a wide bandgap semiconductor with ultraviolet luminescence, ACS Nano 5 (2011) 1042–1046.
[98] B. Wang, J.R. Sparks, H.R. Gutierrez, F. Okino, Q. Hao, Y. Tang, V.H. Crespi, J.O. Sofo, J. Zhu, Photoluminescence from nanocrystalline graphite monofluoride, Appl. Phys. Lett. 97 (2010) 141915.

[99] Z. Wang, J. Wang, Z. Li, P. Gong, J. Ren, H. Wang, X. Han, S. Yang, Cooperatively exfoliated fluorinated graphene with full-color emission, RSC Adv. 2 (2012) 11681–11686.

[100] P. Gong, J. Wang, W. Sun, D. Wu, Z. Wang, Z. Fan, H. Wang, X. Han, S. Yang, Tunable photoluminescence and spectrum split from fluorinated to hydroxylated graphene, Nanoscale 6 (2014) 3316–3324.

Nature of C—F Bonds in Fluorinated Carbons

9

M. Dubois[1,2], *K. Guérin*[1,2], *A. Hamwi*[1,2], *A. Vinogradov*[3]
[1]Université Blaise Pascal, Aubière, France; [2]CNRS, Aubière, France; [3]St. Petersburg State University, St. Petersburg, Russia

Chapter Outline

9.1 Introduction

Because of the high reactivity of fluorine atoms, their electronegativity (the highest of the periodic table), their small size, on one hand, and of the various allotropic forms of carbon, on the other hand, their reaction results in numerous fluorinated compounds (CF_x) where C—F bonding depends on both the reaction conditions and the nature of the carbonaceous precursor. Both the versatility of C—F bonding and the diversity of fluorinated carbons explain the extensive uses of CF_x (nano)materials. The applications concern the whole fluorine content range, ie, the molar ratio F/C or x in the CF_x composition (Fig. 9.1) [1—8]. x (= F/C) from 0.05 to 1 is useful according to the

New Fluorinated Carbons: Fundamentals and Applications. http://dx.doi.org/10.1016/B978-0-12-803479-8.00009-7

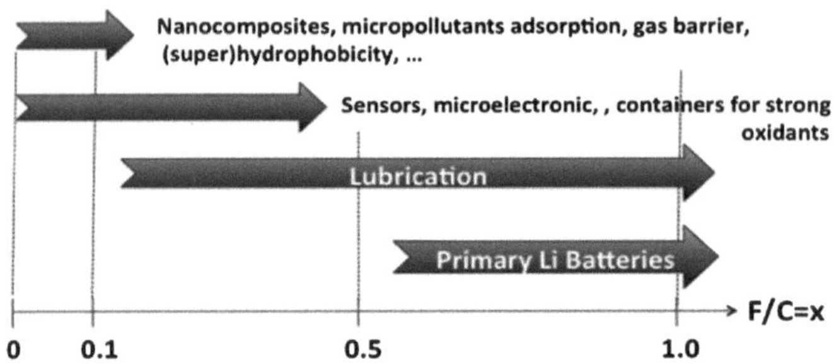

Figure 9.1 Some applications of fluorinated carbonaceous (nano)materials.

application. When x is low ($0.05 < x < 0.2$), fluorination appears as a surface functionalization that enhanced the hydrophobicity, and CF_x may be used as fillers or loads with enhanced interaction with the polymer matrix in (nano)composites, for organic micropollutant remediation, to reach gas barrier properties (especially for polymers), containers for strong oxidants, (super)hydrophobicity, as active or filtering materials in sensors, for tuning the gap of graphene and in microelectronics. Lubrication needs higher x values because C—F bonds are consumed to form the tribofilm and reorganize the carbon lattice. Use of CF_x as electrode material in primary lithium batteries requires also high fluorine content because the C—F bonds are the electroactive species during the discharge. Most of these applications are known for long time but the use of nanomaterials and/or enhancement of the fluorination processes (subfluorination, catalytic fluorination followed by a postfluorination, controlled fluorination with solid fluorinating agent decomposition) results recently in a gain of interest on CF_x and allows significant improvement of the applicative properties. For all the cases, C—F bonding is a key parameter that acts both on the chemical and thermal stability, on the durability of the tribological properties, and on the discharge potentials (the stronger the C—F bond, the lower the potential). The control of the C—F bonding allows adjusting the applicative properties: C—F bonds may be either broken (primary lithium batteries, lubrication) or conserved in the application conditions (hydrophobicity, (nano)composites). The C—F bonding may be tuned according to the fluorination conditions using either molecular fluorine F_2 or catalytic processes ($F_{2(g)}$/$HF_{(g)}$/ $MF_{n(g)}$ $MF_n = IF_5$, BF_3, or ClF_x). The nature of the starting materials also acts on the C—F bonding taking into account a curvature effect of the carbon lattice. The first part of this chapter will be devoted to the fluorination processes. To evaluate and even quantify the strength of the C—F bonds, ie, their covalence, methods must be developed. Because solid state nuclear magnetic resonance (NMR) investigates the electronic shielding effect on nuclei (^{19}F and ^{13}C), ie, the electronic density around these nuclei, that depends on the C—F bonding, NMR appears then as a powerful technique for both analytic purposes taking into account the chemical shifts and investigation of the C—F bond length regarding ^{19}F—^{13}C dipolar coupling. NMR will be discussed in the second part and compared to infrared (IR) spectroscopy, near-edge X-ray absorption fine structure (NEXAFS), and electrochemical data. Finally, three examples will be given to underline the effects of C—F bonding on the applicative properties: in primary lithium battery, to adjust the band gap by defluorination starting with fluorinated graphene and as solid lubricants.

9.2 Fluorination Methods: From Room Temperature to 600°C

9.2.1 Conventional Fluorination With F_2: From C_{60} to Graphite as Precursor

The conventional and most used fluorination route consists in the gas/solid reaction with molecular fluorine F_2. The reaction conditions to reach given fluorine content, ie, the F/C molar ratio or x in CF_x, strongly depends on the nature of the starting materials. Fig. 9.2 gives a schematic view of the resulting F/C ratio according to the fluorination temperature when pure F_2 gas is used. The data, which are all extracted from works in our laboratory, allows a comparison because the fluorinations were conducted with similar gas flux in a dynamic mode and duration (for most of them in the 2–6 h range). Fig. 9.2 highlights three main parameters for the reactivity of F_2 gas with carbonaceous materials: (1) the crystalline order (ie, the graphitization degree), (2) the curvature of the carbon lattice, (3) the specific surface area. Each of these parameters will be discussed in the following part.

9.2.1.1 Graphitization

The higher the graphitization, the higher the fluorination temperature. The reaction of pure F_2 gas with graphite starts at temperatures higher than 300°C. Graphite fluorides with $(C_2F)_n$ and $(CF)_n$ structural types are formed at 350 and 600°C, respectively. Table 9.1 summarizes the fluorination conditions (temperature and duration), the

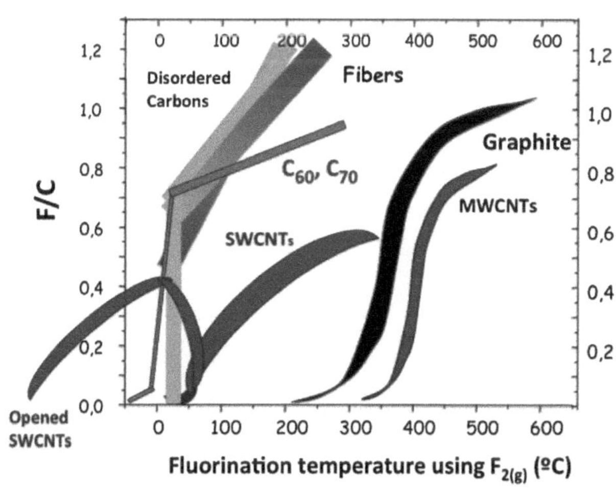

Figure 9.2 Schematic view of reactivity in pure F_2 gas of carbonaceous materials.

Table 9.1 Synthesis Methods of Various Graphite Fluorides, ^{19}F Chemical Shift, and IR Wave Numbers

Starting Materials	Producer of the Raw Material	Synthesis Method	T_{F2} (°C)	F/C	Notation	NMR Chemical Shifts $\delta_{19F}/CFCl_3$ (ppm)	ν_{IR} (cm^{-1})	References
Graphite UF4	Mersen	*F2* >99%	600	1	$(CF)_n$	−190	1215	[10–12]
Graphite UF4	Mersen	*F2* >99%	380	0.6	$(C_2F)_n$	−189	1215	[13]
Homemade $F_2/HF/IF_5$ $(CF_{0.72})(IF_5)_{0.02}$			—	0.72[a]	$CF(IF_5)$	−154	1082	[11,14,15]
			400	0.72	$CF(IF_5)$-400	−189	1140	
			550	0.92	$CF(IF_5)$-550	−189	1215	
$(C_{2.5}F)_n$ Homemade $(CF_{0.32})^a$			—	0.39[a]	$(C_{2.5}F)_n$	−151.3	1082	[16]
			250		$(C_{2.5}F)_n$-250			

[a]From ^{13}C solid echo nuclear magnetic resonance.

starting materials, the fluorine content, and the notations used in this chapter. On the contrary, high fluorine contents can be obtained at low temperature for disordered carbonaceous materials. Because of the high content of both sheet edges and structural defects (vacancies, sheet disruptions), the F/C ratio may reach values higher than one. CF_2 and CF_3 groups are then formed on sheet edges and defects that increase F/C. The case of the fluorination of carbon nanodiscs (CNDs) well evidences the effect of the structural order [9].

Comparison of different precursors (graphite or petroleum cokes) is complicated because various parameters could change, ie, granulometry, specific surface area, surface chemistry, presence of physisorbed water molecules, and so on. In our comparison between as-prepared and graphitized CNDs (graph-CNDs), only the graphitization degree changes whereas the other parameters are still nearly constant, such as the specific surface area, which remains close to 30 $m^2\ g^{-1}$, and the geometries of both cones and discs. The commercially available product for nanocones/discs, provided by N-Tec Corporation, consists of a mixture of 70 wt% of CNDs, 20% of carbon nanocones (CNCs) and 10% of amorphous carbons. Due to the content of nanodiscs, the mixture will be called CNDs. Two series of fluorinated CNCs/discs have been prepared from either as-product CNDs or CNDs posttreated at 2700°C under argon to highlight the different mechanisms of fluorination as a function of the graphitization degree. The fluorination temperatures have been selected to obtain similar fluorine content for the two series, allowing the comparison. The CNCs and CNDs were produced by pyrolysis of heavy oil. The size of the crystalline domains, ie, the coherence length along the c-axis (Lc) determined by X-ray diffraction (XRD), is around 2 nm which means that the as-product exhibits low degree of order. The annealing treatment at 2700°C (the resulting sample is denoted as G-CNDs) improves the crystallinity since Lc reaches the value of 39 nm. The chemical compositions for the two series, ie, the fluorination content, were obtained by weight uptake. It is to note that, in the selected temperature domains (280—340°C for CNDs and 450—500°C for G-CNDs), the exfoliation was not observed. Regarding the fluorination level, CNDs are more reactive in comparison with G-CNDs. In other words, to obtain similar F/C ratio, G-CNDs need a higher fluorination temperature than CNDs. For instance, F/C ratio of 0.78 is reached at 340 and 500°C for CNDs and G-CNDs, respectively. Starting from graphitized samples, the order can be partly maintained upon fluorination, although this process obligatory results in a progressive disorganization of the carbonaceous matrix due to the accommodation of the fluorine atoms and the change of carbon hybridization from sp^2 to sp^3. The more remarkable benefits of the use of graphitized sample as starting material are, on one hand, that cracks and irregularities of both cones and discs are avoided contrary to as-prepared CNDs after fluorination and, on the other hand, structural defects such as CF_2 and CF_3 and dangling bonds are significantly reduced. Because of the fluorine atom accommodation, the fluorination process induces the swelling of the nanoparticles which leads to cracks propagation in the case of as-produced CNDs. Fluorinated samples are strongly rougher than untreated ones in both cases as evidenced by AFM. This is due to the conversion of carbon hybridization from sp^2 to sp^3 and/or the separation of crystallite units due to the decomposition of the amorphous parts in F_2 gas in the case of raw CNDs. The main difference is the possible digging of the surface through the decomposition of amorphous parts (such disordered carbons exhibit a higher reactivity with fluorine gas; this could result in a weak decomposition into CF_4 or C_2F_6 volatile species when the fluorination temperature is high). With the same fluorination level, the aspects of their surfaces are very different for fluorinated CNDs and G-CNDs.

9.2.1.2 Curvature Effect

Curvature of the carbon lattice and the resulting $\sigma-\pi$ rehybridization act on chemical reactivity. Curvature results in strain in the carbon lattice as revealed by the pyramidalization and misalignment of the π-orbitals [17–19]; the pyramidalization angles θ_p in single-walled carbon nanotubes (SWCNTs), which quantitatively characterizes the $\sigma-\pi$ rehybridization, is ranged in between 0 degree for a pure sp^2 and 19.47 degrees for a pure sp^3 carbon atom whereas is equal to 11.6° for all the carbon atoms in fullerene. This parameter θ_p, which is inversely proportional to the diameter, has been used to evaluate the chemical reactivity of the tubes. Nanotubes are expected to be more reactive than a flat graphene sheet and the smaller the diameter, the higher the reactivity. Experimental data confirm this fact because to obtain similar fluorine content (F/C $\approx$ 0.4, see Table 9.2 for experimental details), the temperature must be increased for HiPCO SWCNTs (200°C to reach $CF_{0.37}$), double walled carbon nanotubes (DWCNTs) (300°C, $CF_{0.42}$), multi-walled carbon nanotubes (MWCNTs) (380°C, $CF_{0.43}$), and carbon nanofibers (CNFs) (420°C, $CF_{0.39}$).

Table 9.2 Nanocarbon Synthesis Methods and Fluorination Conditions Using F_2 Gas

	Notation	Synthesis Method	Purity	Average Diameter[b] ϕ (nm)	Duration (h)	T (°C)	F/C
SWCNTs	Helix	CVD[a] Co catalyzed	90%	$\approx$1.3	3	300	0.61
SWCNTs	HiPCO	CVD Co	>95%	1.1	3	200	0.37
SWCNTs	EA	Electro-arc Ni–Y catalyzed[c]	>95%	1.5	5	190	0.32
DWCNTs	Helix	CVD	90%	$\approx$4	3	300	0.42
MWCNTs	Helix	CVD	95%	10–100	3	380	0.43
CNFs	MER	CVD	>90%	140 ± 30	16	420	0.39

SWCNTs, single-walled carbon nanotubes; *DWCNTs*, double walled carbon nanotubes; *MWCNTs*, multi-walled carbon nanotubes; *CNFs*, carbon nanofibers.
[a]Chemical vapor deposition.
[b]Producer data.
[c]See synthesis details in Ref. 45.

The exposure of open-ended SWCNTs under 1 atm of fluorine gas was carried out at low temperature in the −191/25°C range. In such conditions, unusual weak C−F bonds were achieved as revealed by both TGA and ^{19}F NMR [20]. The interaction between fluorine and carbon atoms seems to tend toward physisorption whereas in the case of SWCNTs fluorinated at high temperature, chemisorption takes place. Fluorine atoms can be then removed from the host structure either by moderate heating or by outgassing in primary vacuum. Raman spectroscopy underlines that the carbon lattice is slightly changed upon fluorination; whereas after a conventional fluorination with F_2 gas at higher temperature (280°C), the disorder resulting from the presence of fluorine atoms is increased, a treatment at a very low fluorination temperature near the fluorine condensation temperature appears as beneficial to the ordering of SWCNTs.

9.2.1.3 Porous Materials

Only a few studies related to the fluorination of porous carbonaceous materials were reported. Activated carbons or carbon nanofibers (CNFs) were fluorinated and subsequently used for energy and CH_4 storage [21,22]. Various amounts of fluorine were introduced into ordered mesoporous carbonaceous materials by tuning the fluorination temperature (25−300°C) [23,24] to improve the performance of primary Li^+ ion batteries. The fluorination of an activated carbon and in particular the influence of its surface chemistry on the obtained fluorinated carbonaceous materials as well as their interaction with water during water intrusion and adsorption/desorption measurements were investigated [25,26]. These works highlighted that fluorination induces a decrease of textural properties (surface area and pore volume), sometimes even under mild fluorination conditions. High fluorine content can be reached even at room temperature ($F/C = 0.9$). The fluorination takes place on the entire pore surface, and the formed C−F bonds require to be stabilized, to some extent, by pre- or posttreatments. The covalent grafting of fluorine atoms and decomposition of the carbonaceous lattice always compete with each other. The exposure time and the relative pressure of F_2 during the fluorination must be optimized to avoid drastic decomposition to CF_4 and C_2F_6 gases. Such a decomposition results in the destructuration of the carbonaceous lattice. Different C−F bonds are created irrespective of the fluorination route, the differences being related to various neighbors, more or less fluorinated or oxygenated, and the curvature of the carbonaceous lattice.

9.2.2 Catalytic Fluorination

More recently, graphite fluorides were prepared using a synthesis at room temperature by treatment with gaseous mixture of F_2, HF, and volatile fluorides (BF_3, IF_5, ClF_x) (Table 9.1).

Fig. 9.3 shows the XRD patterns of room temperature $CF(IF_5)$, $CF(BF_3)$, $CF(ClF_x)$, and $(C_{2.5}F)_n$ and conventional high temperature graphite fluorides $(C_2F)_n$ and $(CF)_n$. Four main reflections can be observed. The first ones at 2θ around 10 and 14 degrees could be indexed in a hexagonal system either as the 001 reflection [27] or as a 002 reflection as suggested elsewhere [28]. This reflection corresponds to an average

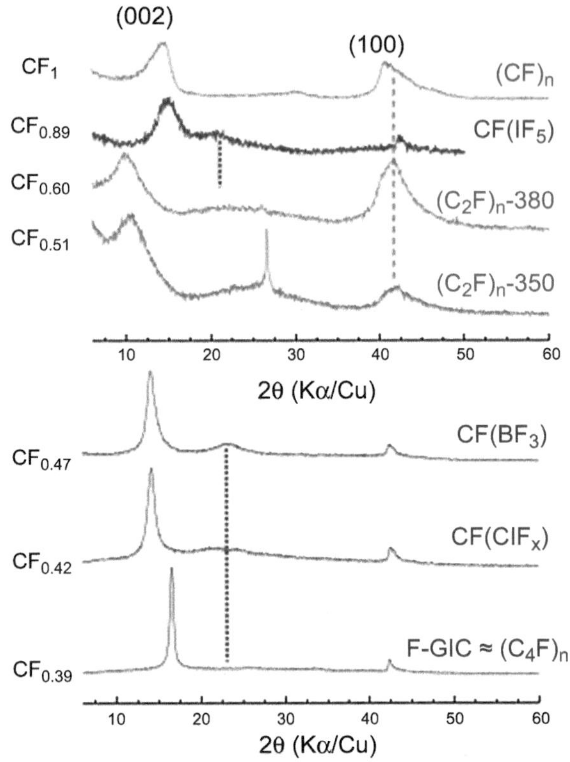

Figure 9.3 X-ray diffraction diagrams of various graphite fluorides.

interlayer spacing of about 0.88 and 0.62 nm, typical of $(C_2F)_n$ and $(CF)_n$ structural type. The second one at around 21 degrees ($d = 0.423$ nm) is related to the 002 reflection of a residual stage 1 of the MF_n-graphite intercalation compound (GIC) ($MF_n = IF_5$, IF_6^-, and IF_7 for $CF(IF_5)$, BF_4^- for $CF(BF_3)$, F^- for $(C_{2.5}F)_n$) which is formed at the beginning of the fluorination reaction process at room temperature [28]. The peak at around 41–42 degrees could be indexed simply as a 100 reflection with corresponding reticular distances between 0.214 and 0.221 nm. Such distances are associated with the C–C in-plane length (Table 9.3). The same $d_{C\text{-}C}$ values of 0.148 nm are found for $(C_2F)_n$ and $(CF)_n$ in accordance with a sp^3 carbon hybridization (corrugated sheets). On the contrary, the value of 0.142 nm for room temperature graphite fluorides ($CF(IF_5)$, $CF(BF_3)$, $CF(ClF_x)$, and $(C_{2.5}F)_n$) is related to sp^2 hybridization meaning that the planarity of the carbon sheet is maintained.

To change the C–F bonding and increase the fluorine content, the room temperature graphite fluorides were posttreated in pure F_2 gas. With increasing postfluorination temperature (T_{PF}), the hexagonal unit cell parameters change from 0.247 to 0.254 nm for a-axis and from 0.597 to 0.657 nm for c-axis. The evolution of the corresponding interlayer spacing as a function of the postfluorination temperature

Table 9.3 X-ray Diffraction Data of Various Graphite Fluorides

		Interlayer Distance d_{C-C} (nm)	$d_{C-C} = d_{100} \times 2/3$ (nm)	Carbon Hybridization
$(CF)_n$	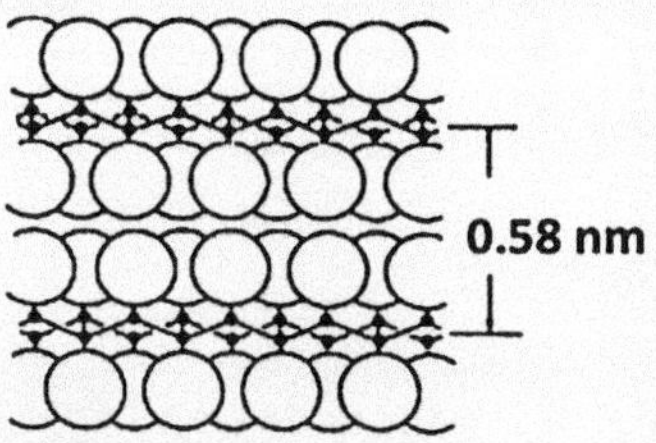	0.616	0.148	sp^3
$(C_2F)_n$	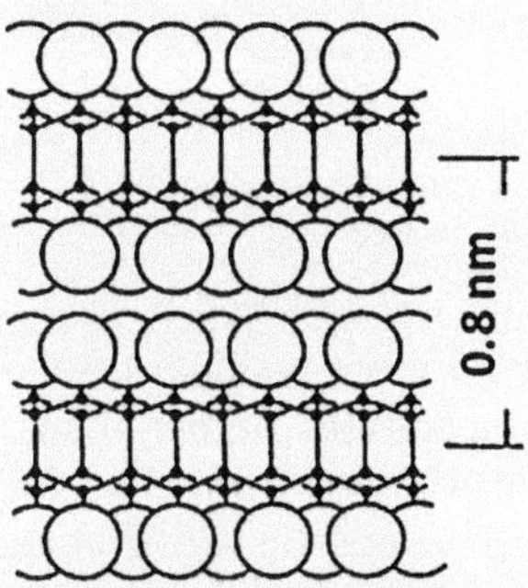	0.88	0.148	sp^3
$CF(IF_5)$	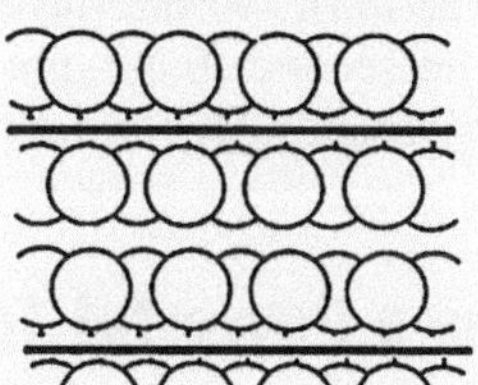	0.631	0.142	sp^2
$CF(BF_3)$		0.629	0.142	sp^2
$CF(ClF_x)$		0.626	0.142	sp^2
$(C_{2.5}F)_n$		0.514	0.142	sp^2

was recorded. XRD patterns are not shown here but Scheme 9.1 summarizes the processes. Up to 400°C, the C-C length is close to 0.142 nm, a value close to that of the pristine graphite. Such a result suggests that the planarity of the graphene layers is preserved for a postfluorination temperature as high as 400°C. Moreover, the broad peak at 21 degrees is still present, clearly indicating the persistence of the catalyst species most likely trapped between the graphene layers. For posttreatment temperatures between 400 and 500°C, a (001) line dissymmetry appears in the XRD diagrams along with two neighbor peaks centered at around 41.0 and 42.3 degrees corresponding to the 100 reflection. Such a doublet indicates that a new structure is formed with a larger d_{100} of 0.221 nm, characteristic of covalent C—F with, consequently, carbon sp^3 hybridization. Moreover, the total disappearance of the

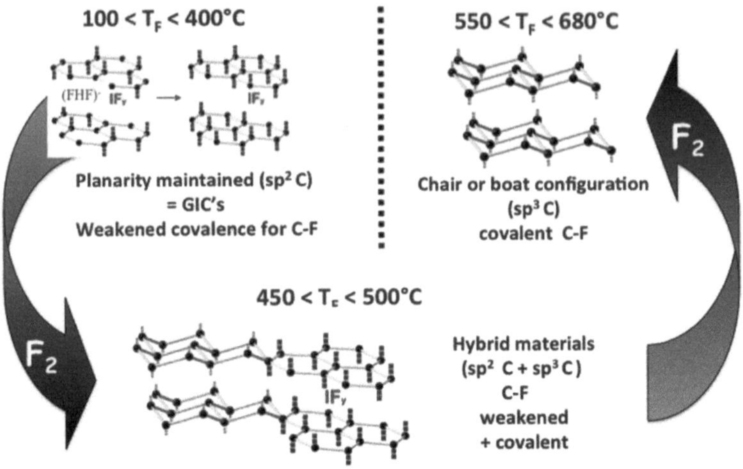

Scheme 9.1 Nature of the resulting sample using the catalytic fluorination followed by a postfluorination in F_2 gas.

reflection at around 21 degrees indicates that no more residual catalyst is trapped into the graphene interlayer spacing. For temperatures higher than 500°C, only one single structure remains, which is quite different from the $(CF)_n$ structure: even if the C—C bond length is similar for $CF(IF_5)$-600 (ie, $CF(IF_5)$ postfluorinated at 600°C) and $(CF)_n$ graphite fluoride ($d_{110} = 0.128$ nm), the interlayer spacing is higher for the first one (0.657 nm) compared to the second (0.616 nm). This confirms that the planarity is broken and an armchair structure is formed.

9.3 Nuclear Magnetic Resonance as a Powerful Tool for the Investigation of the C F Bonding

9.3.1 *Chemical Shifts*

The ^{19}F chemical shifts δ_{19F} are well correlated with a main covalent character of the C—F bonding: higher covalence is accompanied with a lower ^{19}F chemical shift (Fig. 9.4). To average or to decrease the strong ^{19}F—^{19}F homonuclear dipolar coupling and possible chemical shift anisotropy, which result in a substantial broadening of the lines, ^{19}F NMR experiments were performed using magic angle spinning (MAS) with spinning speed of 14 kHz (Fig. 9.5). The δ_{19F} value of −190 ppm was measured for covalent $(CF)_n$ [52] and $(C_2F)_n$ [53] types whereas a value of −160 ppm was obtained for room temperature graphite fluoride with weakened covalence [14–16]. For the case of graphite fluorides, C—F groups adjacent to carbons in nonsaturated fluorine-graphite intercalation compounds (F-GICs) exhibit a downfield chemical shift and saturated $(CF)_n$ exhibit upfield chemical shift. Only one bonding exists in covalent or weakened covalent fluorinated graphite but is distorted by the nature of the neighboring carbon (sp^2 or sp^3). This fact was confirmed by neutron diffraction [29]. It is to

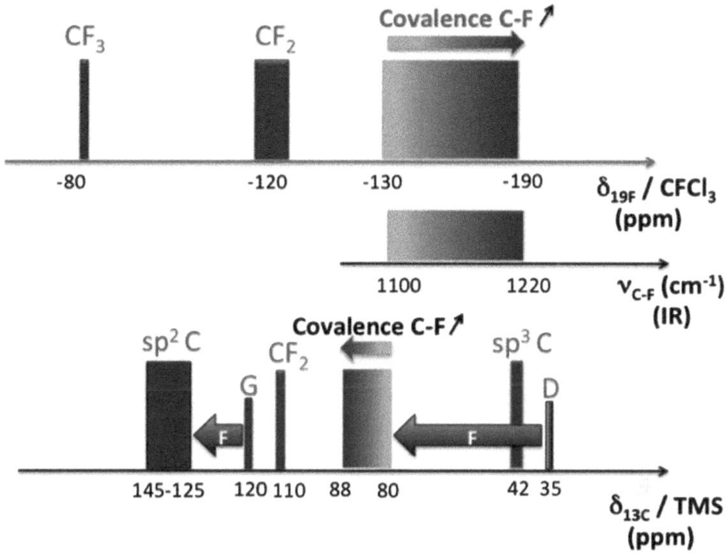

Figure 9.4 ^{19}F and ^{13}C chemical shift of C−F bonds, CF_2 and CF_3 groups. sp^3 *C* means nonfluorinated carbon atoms in $(C_2F)_n$ structural type. [13]C values of diamond (*D*) and graphite (*G*) are also reported. FTIR wave numbers of the C−F vibration are compared to ^{19}F chemical shifts.

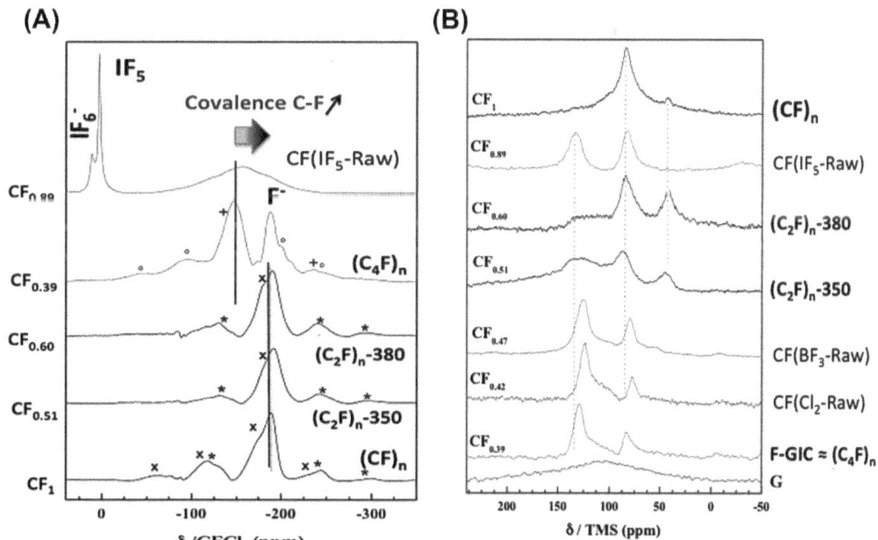

Figure 9.5 [19]F (A) and [13]C (B) nuclear magnetic resonance spectra of various graphite fluorides. The spinning rates are 10 and 14 kHz, respectively. ∗, +, and ∘ denote spinning sidebands.

note that narrow lines over 0–50 ppm range are assigned to IF_n intercalated species that are IF_5 and IF_6^-. For the case of $(C_{2.5}F)_n$, the line at −147 ppm is assigned to the fluorocarbon matrix, the lower δ_{19F} of the fluorocarbon matrix in $(C_{2.5}F)_n$ indicates larger hyperconjugation between C–F bonds and unfluorinated carbon atoms that is more pronounced weakening of the covalence [30]. In accordance with the works of Panich [31], who extensively studied the fluorine-GIC compounds, the second line is ascribed to intercalated fluorine ions. In addition to ^{19}F NMR, ^{1}H NMR experiments, not shown here, confirm a very low amount of ^{1}H nuclei; as a matter of fact, the corresponding resonance lines of FHF^- and $HClF^-$ at +13.5 and +6.3 ppm / TMS (tetramethylsilane), respectively, are very weak. Thus $(C_{2.5}F)_n$ compound represents mainly a fluorocarbon host matrix close to C_4F with additionally intercalated fluorine anions F^-. Similar to $(C_{2.5}F)_n$, $CF(IF_5)$, $CF(BF_3)$, and $CF(ClF_x)$ exhibit a weakened covalence and similarities with F-GICs because few residual catalysts are still intercalated. The weakening of the C–F covalence is also evidenced by ^{13}C chemical shifts; the value around 82 ppm is recorded for room temperature graphite fluorides whereas $(CF)_n$ and $(C_2F)_n$ exhibit a value close to 88 ppm. The higher the ^{13}C chemical shift, the stronger the covalence (Fig. 9.4).

NMR investigations on ^{13}C, ^{19}F, and ^{1}H of the structure and chemical bonding in raw and fluorinated detonation nanodiamonds (NDs) underline the formation of different fluorocarbon groups on the ND surface, which substitutes for hydrocarbon and hydroxyl groups [32–34]. The ^{13}C line with chemical shift of 35 ± 1 ppm for the raw NDs was attributed to the sp^3 carbons belonging to the ND core. ^{1}H to ^{13}C cross polarization allows the determination of surface-hydrogenated groups (CH, CH_2, and COH) and underlines the quasi-absence of a sp^2 carbon fullerene-like shell on the diamond surface [38].

9.3.1.1 *Hyperconjugation*

FTIR and ^{19}F MAS NMR spectroscopies have been used as they are excellent probes for the bonding determination of raw and postfluorinated $CF(IF_5)$ [14–16]. Between 400 and 2000 cm^{-1}, the FTIR spectra show vibration bands at 1135 and 1215 cm^{-1}. For $CF(IF_5)$, the weak band at 1135 cm^{-1} is attributed to weakened covalent C–F bond vibration [35], its small intensity is related to the conductor character of the sample. On spectra of $CF(IF_5)$ and $CF(IF_5)$-T_{PF} ($100 < T_{PF} < 500$°C, T_{PF} is the postfluorination temperature), two bands are observed, the intense one at 1215 cm^{-1} is assigned to covalent C–F bond [36]. Weakened covalent contribution appears only as a shoulder of the covalent one. The effect of the increase of the postfluorination temperature consists in the disappearance of the weakened covalent character which occurs simultaneously with the improvement of the covalence as evidenced for $CF(IF_5)$-T_{PF} with $T_{PF} > 500$°C. For $(CF)_n$ graphite fluoride, only the signal of covalent C–F is observed at 1215 cm^{-1}. In agreement with FTIR results, the line at −189 ppm on the room temperature ^{19}F MAS NMR spectra (Fig. 9.5A) corresponds to fluorine atoms involved in covalent C–F bonds. The second one at −119 ppm highlights the presence of CF_2 groups. The iodine fluoride content decreases with postfluorination temperature; at 400°C the content is low while at 530°C, iodine fluorides have been entirely removed from the fluorocarbon matrix. Room temperature graphite fluoride possesses weakened covalent C—F bonds as evidenced by the IR band at $1135\ cm^{-1}$ and the NMR line at -160 ppm. When the postfluorination temperature increases, the C—F bonding evolves toward a covalent character whatever the catalyst (IF_5 but also for the cases of BF_3 and ClF_x). This change is underlined by the shift of the C—F resonance line from -160 to -190 ppm. IR data always confirms this evolution because of the shift of the band at $1215\ cm^{-1}$ for covalent character. These data, correlated with XRD give the following process. The materials obtained after a postfluorination of room temperature graphite fluoride can be divided into three types as a function of the fluorination posttreatment temperature T_{PF}: (1) When T_{PF} is lower than 450°C, the samples contain intercalated iodine fluoride species, the C—F bonds are mainly weakened covalence because of the coexistence of nonfluorinated sp^2 C atoms and fluorinated sp^3 C atoms (hyperconjugation). (2) When T_{PF} is between 450 and 550°C, the catalyst species are eliminated, and the C—F bonds develop two natures: weakened covalence and covalent. (3) When T_{PF} is higher than 500°C, the fluorographite matrix contains no more residual catalyst. The C—F bonds become covalent but some sp^2 carbon hybridization still exists (10% determined by ^{13}C NMR). The physicochemical properties vary drastically from T_{PF} close to 400—450°C. The planarity of the carbon sheets is maintained (sp^2 carbon hybridization) for T_{PF} lower than 450°C. When T_{PF} increases, the C—F bonds develop a hybrid structure (both weakened and pure covalent C—F bonds coexist) as the carbon skeleton which consists of chair-type structure together with some planarity. The iodine fluoride species removal in the range of T_{PF} 400—450°C coincides with the sudden change of the C—F bond character. Then, the iodine fluoride species play an important role in the mechanisms because of their low mobility in the fluorocarbon interlayer space, the iodine fluoride species could hinder the diffusion of the F_2 molecules and shield the dangling bonds. In other terms, they could strengthen the planar configuration of the carbon sheets, ie, the hyperconjugation. Similar processes occur with room temperature graphite fluorides prepared with the other catalysts.

9.3.1.2 Curvature Effect

During fluorination, the C—F bond formation requires the change of the carbon hybridization from sp^2 to sp^3. The curvature prevents the formation of a pure sp^3 hybridization for the carbon atom, since it requires an important local strain. The residual sp^2-hybridized orbitals imply a weakening of the overlapping of the hybridized lobes of carbon and the fluorine atomic orbitals (Scheme 9.2). In other terms, this results in a weakening of the C—F bonding covalence. The change of the ^{19}F chemical shifts as a function of the diameter ϕ can be qualitatively explained by pyramidalization of the carbonaceous precursor. Because θ_p is related to $\sigma-\pi$ rehybridization and evolves as 1/d, the changes are more important on δ_{19F} and then on the covalence for the lower diameter and reach nearly the saturation for nanofibers with large diameter. Scheme 9.2 summarizes the above discussion. This covalence weakening affects the C—F bond strength and leads to lack of robustness of the C—F interaction within fluorofullerene clusters. A quasi-linear relationship of δ_{19F} with the IR wave number is

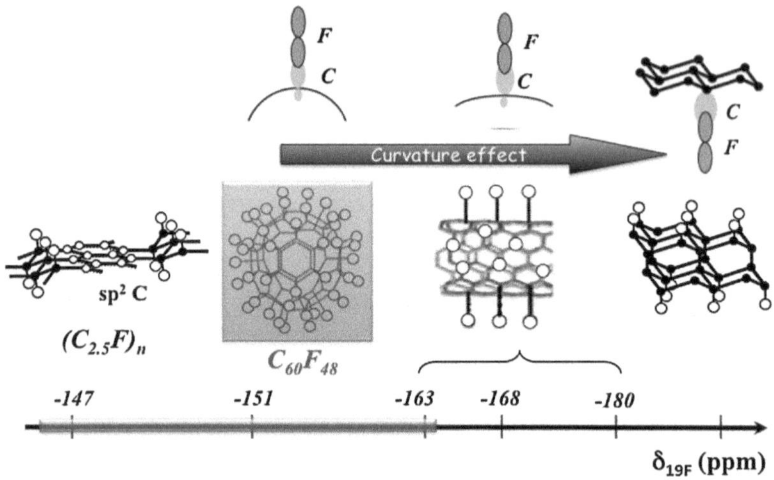

Scheme 9.2 Effect of curvature on the C–F bonding in fluorinated carbons.

observed for graphite fluorides, F-GIC, and fluorinated carbon nanotubes (Fig. 9.6) in accordance with either the hyperconjugation or the curvature effect. The lines of fluorine atoms involved in the C–F bond on the MAS ^{19}F spectra at 14 kHz [37] exhibit a chemical shift of −163, −168, and −180 ppm for fluorinated HiPCO (with $CF_{0.37}$ composition obtained by weight uptake), EA ($CF_{0.32}$), and Helix SWCNTs ($CF_{0.61}$), respectively. The chemical shift values for DWCNTs and MWCNTs fluorinated at 300°C ($CF_{0.42}$ and $CF_{0.43}$) are equal to −178 and −184 ppm. Before the discussion of those values, it is necessary to make the fluorination processes precise. For example,

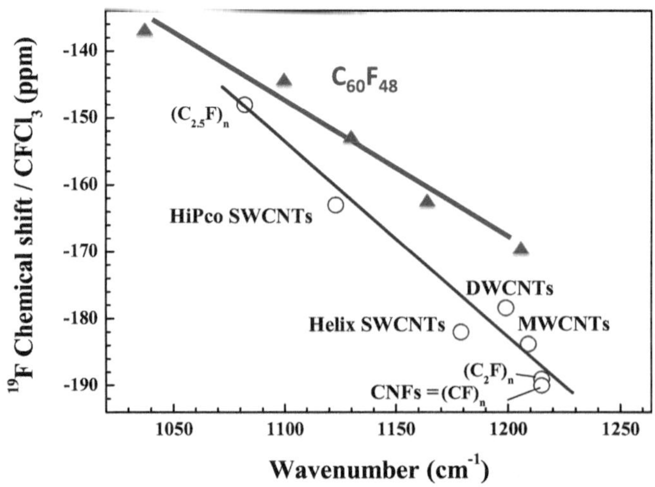

Figure 9.6 ^{19}F chemical shifts versus the wavenumber of the IR C–F vibration band.

as fluorinated DWCNTs are used as model compounds with larger diameter than SWCNTs, we can consider the chemical shift, IR wave numbers of the C−F bonds only for the case when fluorination concerns mainly the outer tubes. To check this point, Raman and ^{13}C NMR data were investigated. The radial breathing modes of the outer tubes disappeared after fluorination contrary to those of the inner tubes. The spectra are well fitted with two Lorentzian lines for fluorinated SWCNTs, the lines being assigned to sp^2 nonfluorinated carbon atoms in weak interaction with fluorine (131 ppm) and to covalent C−F bonds (84 ppm). Two additional lines are necessary for DWCNTs: a weaker one for carbon atoms in CF_2 groups (110 ppm) and a stronger line of nonfluorinated carbon atoms on the inner tube surface, which exhibit chemical shift close to that of graphite (120 ppm). In addition to Raman data, it indicates that fluorine atoms are located in the outer tube surface mainly in accordance with the works of Touhara [38,39]. The data set for C_{60} cannot be included in this relationship. It could indicate that the curvature is not the only parameter which affects the C−F bonding in fluorinated C_{60} and the steric hindrance between fluorine atoms should also be taken into account.

9.3.2 C−F Bond Length via Inverse ^{19}F−^{13}C CP-MAS

The C−F bond length of fluorinated carbons appears as a good indication of the C−F bonding, ie, covalence. A C−F bond length reduction is expected when the covalence increases. Together with other experimental methods, the C−F bond length can be estimated by NMR because this length is included in the expression of the dipolar interaction which can be determined using dipolar recoupling techniques such as rotational echo double resonance (REDOR) [40] or transferred echo double resonance [41,42]. Although the dipolar coupling information is lost when Hartmann-Hahn cross polarization (CP) associated to MAS is used, it can be reintroduced into the spectrum thanks to the inverse cross polarization (ICP) sequence (Fig. 9.7A) [43]. In comparison with the REDOR sequence usually used for the inter-nuclear distance measurements, the ICP sequence required easier experimental adjustment, the main difficulty is the Hartmann-Hahn matching. A second advantage is the absence of rotor-synchronized π pulses in ICP sequence unlike the REDOR sequence. Indeed, for the REDOR sequence and at high spinning speed the finite pulse length could result in an artifact. At high speed the π pulse then represent a significant fraction of the rotor period.

Using ICP, for short contact times, the amplitude of the CP signal is found to be oscillatory with a frequency ϕ related to the C−F bond length [44]. This behavior is observed only for carbon covalently bonded to fluorine and not for sp^3 carbon atoms (see Fig. 9.7B for fluorinated nanofibers). Using the ICP sequence, the integrated peak intensity of the carbon spectra was calculated as a function of contact time, thus revealing the CP dynamics (Fig. 9.7B); the extracted frequency was $\phi = 3976.0 \pm 18.6$ Hz. From $d_{C-F} = \sqrt[3]{10.0463/\varphi}$, $d_{C-F} = 0.136 \pm 0.001$ nm. Fourier transform of the ^{13}C magnetization evolution for C−F groups gives the Pake-like structure for all fluorination degrees of nanofibers (Fig. 9.7C). The C−F bond length (d_{C-F}) can be deduced from the Pake structure using the wings that are related to dipolar fluctuation. S1 is the separation between these wings. The bond

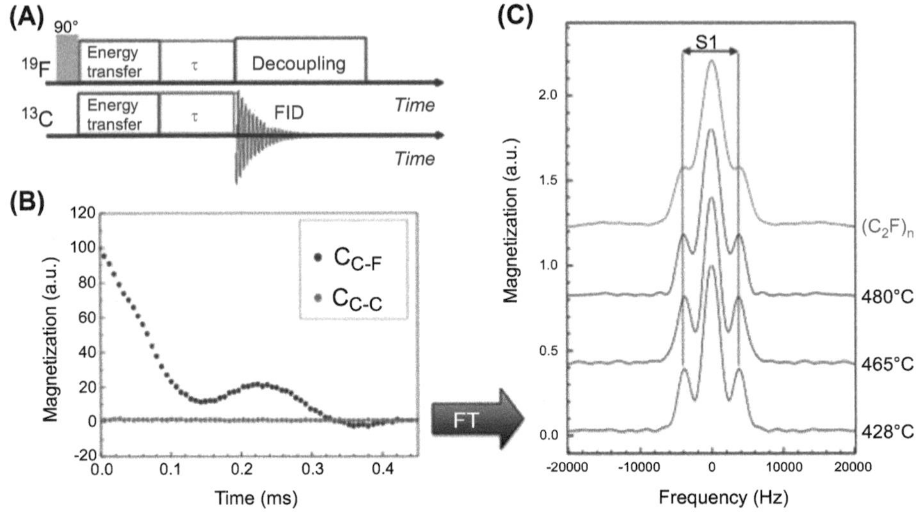

Figure 9.7 (A) Schematic diagram of the inverse cross polarization sequence. The evolution time is τ. (B) Time evolution of the ^{13}C magnetization for carbon atoms covalently bonded to fluorine (denoted C—F) and exclusively bonded to carbon (sp^3 C, marked C—C) of nanofibers fluorinated at 472°C with a spinning rate of 14.5 kHz at the $n = 1$ Hartmann-Hahn condition. (C) Fourier transform of the resulting oscillation as a function of the fluorination temperature. The curve of $(C_2F)_n$ graphite fluoride is added for comparison. The S1 value is equal to 7700 Hz for all the fluorinated nanofibers whatever the fluorination temperature. A C—F bond distance equal to 0.138 ± 0.002 nm is found in all studied cases. CNF-428, CNF-465, and CNF-480 mean carbon nanofibers fluorinated at 428, 465, and 480°C with F_2 gas.

length can be estimated from the equation $d_{C-F} = \sqrt[3]{20.0926/S1}$ (nm). It must be noted that these values as estimated by NMR could be overestimated because of possible molecular motions, which decrease the value of the second moment, inversely proportional to the C—F distance. The accuracy of the bond length value is an important data; the fitting procedure for the magnetization evolution gives us an error of 30 Hz for ϕ. It corresponds to a bond length error of 0.001 nm. From the Pake-like pattern, the error has been estimated to be 0.002 nm.

This method was first applied to $(CF)_n$ and $(C_2F)_n$ types high temperature graphite fluorides [45,13]. In both cases, a C—F bond length of 0.138 ± 0.002 nm was found in accordance with the value obtained from neutron diffraction data (0.136 nm) [30]. The same distance was found for CNFs whatever the reaction temperature in accordance with their covalent character [46]. The C—F bond lengths as seen by ICP are plotted as a function of the ^{19}F chemical shift (Fig. 9.8).

With a spinning speed of 12 kHz, a distance of 0.144 ± 0.002 and 0.1445 ± 0.002 nm is measured for the two samples with weakened covalence $(C_{2.5}F)_n$ and $(C_{2.5}F)_n$ postfluorinated at 350°C. IF_5 postfluorinated at 350°C exhibits a hybrid structure in which two kinds of C—F bond are present: weakened covalent and covalent; the majority being with weakened covalence. Two lines (−156.5

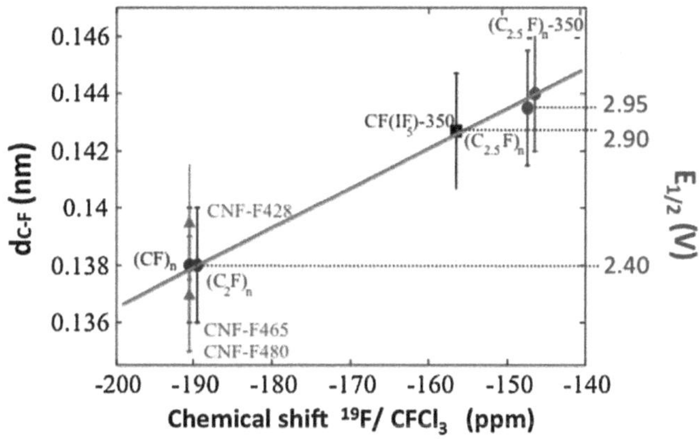

Figure 9.8 Dependence with δ_{19F} of the C—F bond length as seen by nuclear magnetic resonance and of the discharge potential for high temperature graphite fluorides $(CF)_n$ and $(C_2F)_n$, fluorinated nanofibers, $(C_{2.5}F)_n$ and room temperature graphite fluorides $CF(IF_5)$ and $(C_{2.5}F)_n$, postfluorinated at 350°C, $CF(IF_5)$-350, and $(C_{2.5}F)_n$-350, respectively.

and −178.5 ppm vs $CFCl_3$) are then observed for the ^{19}F NMR spectra. The higher the covalence, the shorter the bond length is. Two NMR distances are then expected. In reality, with a spinning speed of 12 kHz, the measure of the dipolar coupling leads to only one distance (0.143 nm) (Figs. 9.7 and 9.8), which is probably a weighted average distance between the weakened covalent ($d_{C-F} = 0.144 \pm 0.002$ nm) and the covalent NMR 1 $d_{C-F} = 0.138 \pm 0.002$ nm. This is in accordance with the ^{13}C spectra showing only one line with a chemical shift similar to the semi-covalent one (83 ppm/TMS).

The extracted C—F bond lengths (d_{C-F}) was equal to 0.147 ± 0.002 nm for fluorofullerene $C_{60}F_{\approx 48}$ (the error on the frequency value on the Pake-like structure leaded to an error of d_{C-F} close to 0.002 nm). Of course, the C—F bond lengths are average values of the various C—F bonds, which were confirmed by ^{19}F and ^{13}C NMR. Two reasons could explain why the C—F bond in fluorinated fullerenes is significantly longer than for covalent graphite fluorides: (1) Fluorinated C_{60} are independent molecules on which the fluorine atoms stand outside of fullerene cage. This could lead to steric hindrance between fluorine atoms. By analogy with graphite fluorides and F-GIC, the hyperconjugation results in a weakening of the C—F covalence as discussed before. (2) The curvature of the carbon lattice must be also taken into account.

Some discharge potentials in primary lithium battery ($E_{1/2}$) are plotted as a function of the ^{19}F chemical shift and compared to r_{CF} (Fig. 9.8). For samples with weakened covalence, ie, $(C_{2.5}F)_n$ and IF_5 both posttreated at 350°C, $E_{1/2}$ is significantly higher (2.95 and 2.90 V, respectively) than for covalent $(CF)_n$ graphite fluoride (2.4 V). Longer C—F bonds (0.144 ± 0.002 nm for $(C_{2.5}F)_n$ and IF_5 both posttreated at 350°C) needs less energy to be electrochemically broken than shorter ones in $(CF)_n$ (0.138 ± 0.002 nm). In other words, the longer the C—F bonds, the higher the discharge potentials.

9.3.3 Nuclear Magnetic Resonance Versus Near-Edge X-Ray Absorption Fine Structure

Solid state NMR and NEXAFS photoelectron spectroscopy are complementary techniques to investigate fluorinated graphites of any structural type. The covalence of the C—F bonds, as well as the fluorine content, can be extracted from the NMR spectra of ^{19}F and ^{13}C nuclei, regarding the chemical shifts and the integrated areas of various lines. In its turn, NEXAFS spectroscopy is an effective technique for obtaining data on electronic structure, symmetry, and energy distribution of unoccupied electron states in carbon-containing substances as well as on the coordination, the chemical state, and bonding of carbon atoms [47].

In this connection it is interesting to compare the data of NEXAFS and NMR techniques with a large panel for various fluorinated graphites having different fluorine contents and C—F bonding nature [48]. The C 1s and F 1s X-ray absorption spectra for a large panel of fluorinated graphites with different structures (Tables 9.1 and 9.2) as well as of reference systems were measured using total electron yield techniques with monochromatic synchrotron radiation and the facilities of the Russian—German beamline (RGBL) at the BESSY II (Berlin) [49] and the beamline D1011 of the MAX-IV laboratory (Lund) [50] under the same experimental conditions.

The C 1s absorption spectra of graphite fluorides $(C_yF)_n$ ($y = 1, 2, 2.5$) are compared in Fig. 9.9 with those of highly oriented pyrolytic graphite (HOPG), graph-CNDs, white graphite fluoride, and NDs. The most characteristic absorption peaks (resonances) *A* and *B-C* in the HOPG (graph-CNDs) spectrum are associated with the dipole-allowed transitions of the C 1s electrons to the empty π and σ states in the conduction band of the HOPG (graph-CNDs). These states (peaks *A* and *B-C*) are formed by the C $\pi 2p_z$ and sp^2-hybridized C $\sigma 2s + 2p_{x,y}$ orbitals oriented perpendicular and parallel to the plane of the carbon layer (graphene), respectively. Thus, they reflect the planar sp^2 hybridization of valence 2s and $2p_{x,y}$ states of carbon atoms and, hence, are the fingerprints of the trigonal coordination of carbon atoms in graphite-like structures.

A comparative analysis of the C 1s NEXAFS spectra of various fluorinated graphites with different structures (Tables 9.1 and 9.3) and C—F bonding with those of a large panel of fluorinated graphites with different structures (Tables 9.1 and 9.3) and C—F bonding of reference carbon compounds (HOPG, graph-CNDs, NDs) allows one to gain important information. The C 1s NEXAFS spectra of graphite fluorides $(CF)_n$ and $(C_2F)_n$ are completely different from that of the HOPG (graph-CNDs) but resemble the spectrum of NDs. This observation clearly indicates that the chemical bonding between carbon atoms in graphite fluorides and in diamond is due to covalent σ bonds, while π bonds are absent. The chemical bonding between carbon and fluorine atoms in the $(CF)_n$ and $(C_2F)_n$ is due to covalent C-C and C—F σ bonds, but π bonds are absent, as in the case of diamond. Upon fluorination, fluorine atoms form σ bonds with the C atoms with the participation of the C $2p_z$ states. Hence, upon fluorination, F atoms are attached to the basal plane of graphene sheets rather than substitute C in the plane. In this case, carbon atoms form new σ bonds with fluorine atoms through the participation of the C $2p_z$ states and planar graphene layers become corrugated.

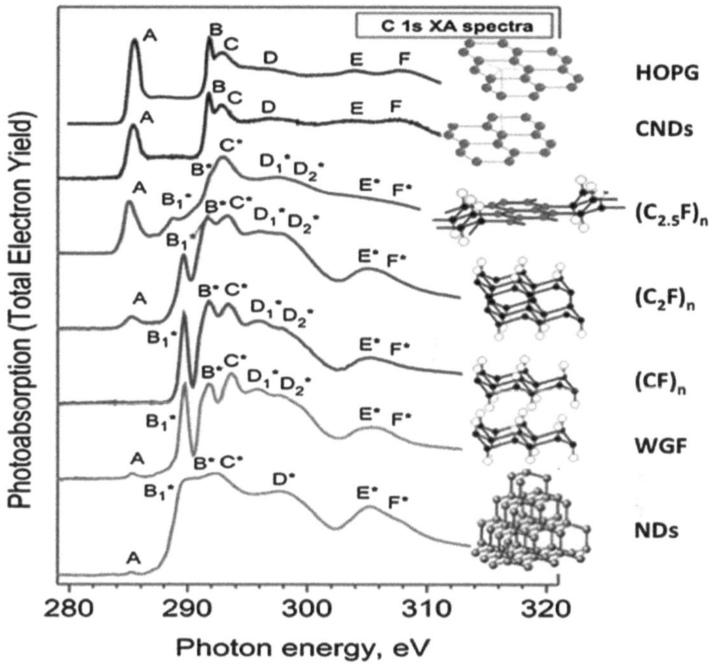

Figure 9.9 Comparison of C 1s near-edge X-ray absorption fine structure spectra of graphite fluorides $(CF)_n$, $(C_2F)_n$, and $(C_{2.5}F)_n$ with as well as those of reference systems highly oriented pyrolytic graphite (HOPG), graphitized carbon nanodiscs (CNDs), white graphite fluoride (WGF), and nanodiamonds (NDs). The C 1s X-ray absorption spectra of the graphite fluorides as well as of reference systems were measured using monochromatic synchrotron radiation and the facilities of the Russian—German beamline at the BESSY II (Berlin) [51] and the beamline D1011 of the MAX-IV laboratory (Lund)[52] under the same experimental conditions.

This inference is confirmed by a direct comparison of the C 1s and F 1s NEXAFS spectra of $(CF)_n$ (Fig. 9.10). The main absorption structures $B_1*—F*$ in these spectra are caused by the transitions of C 1s and F 1s electrons to the same set of unoccupied σ electron states formed by the C 2p and F 2p valence electron states. This results from a clear correlation of absorption structures in the C 1s and F 1s NEXAFS spectra of $(CF)_n$ aligned in energy using the energy separation ΔE (F 1s—C 1s) between the F 1s and C 1s core levels.

These findings suggest a hybridized (covalent) character of empty electron states in $(CF)_n$ and $(C_2F)_n$. In the case of $(C_{2.5}F)_n$, sp^2- and sp^3-bonded C atoms coexist, that is, there coexist regions with trigonal and tetrahedral coordination for the carbon atoms. The graphene sheets are still mainly planar and the C—F bonds exhibit weakened covalence because of the hyperconjugation. Finally, the postfluorination of room temperature graphite fluorides results in a transformation from sp^2- to sp^3-bonded carbon and in a change in the C—F bonding from weakened to pure covalence, as verified by NEXAFS study, in accordance with the NMR data.

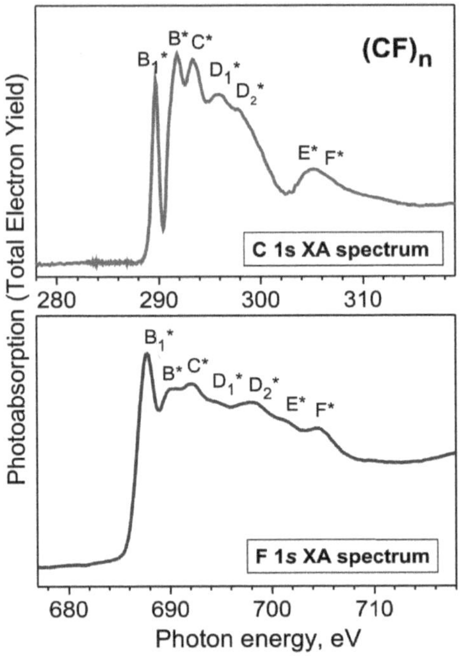

Figure 9.10 Comparison between the C 1s and F 1s near-edge X-ray absorption fine structure spectra of graphite fluoride $(CF)_n$. The spectra of $(CF)_n$ are aligned in energy using the energy separation ΔE (F 1s—C 1s) equal to ~399.0 eV [32].

9.3.4 Nuclear Magnetic Resonance Versus Electrochemistry (CF_x in Li Battery as a Local Probe of the C—F Bonding)

The discharge mechanism in Li/CF_x battery involves the breaking of C—F bonds. An intermediate graphite intercalation compound $C(Li^+ \cdot S\text{-}F^-)_x$, where S represents one or more solvent molecules coordinated, is formed and subsequently decomposes into the final discharge products, carbon and lithium fluoride. The solvated Li^+ ions are intercalated into graphite fluoride layers and the discharge voltage of a Li/CF_x cell is related to the solvation energy of Li^+ ions. Assuming that the intercalation energy is close whatever the CF_x materials, only the C—F bonding significantly changes (since solvation energy of Li^+ ion is equal). The comparison of the discharge potentials is then possible. Diffusion/migration resistance of Li^+ ion in the discharged products, ie, in graphene layers of CNFs and MWCNTs, among SWCNT bundles and among C_{60} and C_{70} molecules, may also act. The intercalation effect could be decreased, especially for very low applied current density. It is important to note that the discussed properties are not thermodynamic ones but obtained by kinetic measurements. Nevertheless, the C—F bonding may be discussed using discharge potentials.

The various samples used as electrode materials in primary lithium batteries are discharged using galvanostatic mode (Fig. 9.11); a similar current density was used,

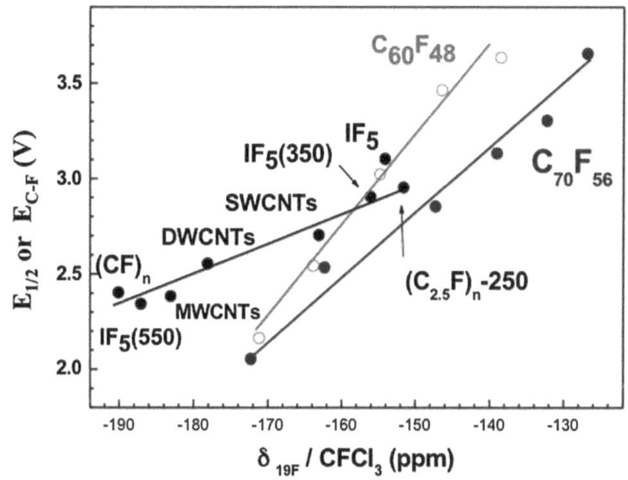

Figure 9.11 Correlation between the discharge potentials ($E_{1/2}$ or E_{C-F}) and ^{19}F NMR chemical shifts for a series of fluorinated C_{60}, C_{70}, nanocarbons, and graphites. IF_5-T_{PF} denoted graphite fluoride synthesized using a two-step process, fluorination at room temperature using a gaseous catalytic atmosphere $IF_5/HF/F_2$, and then posttreated at T_{PF} under pure F_2 gas ($T_{PF} = 350$ and 550°C). $(C_{2.5}F)_n$-250 is $(C_{2.5}F)_n$ posttreated at 250°C under pure F_2 gas. Composite electrodes were composed of fluorinated carbons (about 80% by weight, w/w), graphite powder (10%, w/w) to ensure electronic conductivity, and polyvinylidene difluoride (PVDF, 10%, w/w) as a binder. A PVDF microporous film wet with electrolyte, which was composed of a lithium salt ($LiClO_4$) dissolved in PC with 1 mol L^{-1} concentration, was sandwiched between the composite electrode and a lithium metal foil.

equal to 10 mAg^{-1}. Such low value is chosen to favor the electrochemical processes and to limit over-potential.

Whatever the number of walls, the nanotubes exhibit a well-defined plateau [53], related to reaction Eqn (9.1).

$$CF_x + xLi \rightarrow C + xLiF \quad (9.1)$$

The extracted potential at half of the total discharge capacity is easily measured on the plateau. On the contrary, for highly fluorinated fullerenes, no plateaus are observed and the derivative curves are necessary to underline the discharge potentials, which are called as C—F decomposition potentials (E_{C-F}). Five and six potentials are then obtained for $C_{60}F_{48}$ and $C_{70}F_{56}$, respectively. This result is in accordance with the previous studies in lithium batteries using liquid or solid electrolyte [54—57]. The electrochemical curves during the first discharge exhibited a reduction wave located at potentials higher from 1 to 1.5 V in comparison with nonfluorinated solid fullerene [58,59]. These preliminary studies revealed that the capacity is directly related to the initial level of fluorination. As for graphite fluorides, the reduction mechanism of fluorinated fullerenes results from the progressive defluorination, associated by the formation of LiF, which is thermodynamically favored. Two points must be noted: (1) the initial potential drop for fluorinated SWCNTs may be related to overvoltage due to their insulating behavior. When the electroreduction starts a few C–F bonds were converted in conductive carbon and LiF; the conduction of the electrode then slightly increased and the overvoltage decreased. Finally the potential reaches the stabilized value that is taken into account for the comparison; (2) ^{19}F NMR spectrum of fluorinated MWCNTs with $CF_{0.43}$ composition underlines large amount of CF_2 groups [37], which could hinder the electrochemical processes. This high relative content of CF_2 may explain the low capacity for this sample (320 $mAhg^{-1}$ instead of 571 $mAhg^{-1}$). A nearly linear dependence is observed for the discharge potential $E_{1/2}$ as a function of the ^{19}F NMR chemical shift, which is directly related to the covalence of the C–F bond and is then used to quantify the C–F bonding. The higher the curvature, ie, the lower the diameter, the higher the potential is. The relevance of our method is underlined when the values, $E_{1/2}$ and δ_{19F}, of graphite fluorides are added in Fig. 9.11. Because of the hyperconjugation, the covalence is weakened in the room temperature graphite fluoride (or posttreated in F_2 gas at low temperature). The potential of 3.1 V is then the highest of the present study, except for fullerenes which are a particular case. In the other side of the series, as expected, pure covalent high temperature graphite fluorides $(C_2F)_n$ and $(CF)_n$ and $CF(IF_5)$-T_{PF} posttreated at high temperature $CF(IF_5)$-550 exhibit the lowest potentials of 2.35 V. Similar potentials result from nanofibers indicating that the low curvature of this kind of nanotubes does not act anymore on the covalence. This is in accordance with the nondependence of the potential according to the fluorine content in fluorinated nanofibers; nearly the same potential has been measured whatever the fluorinated wall, outer for low F/C ratio or inner walls when the fluorination progress toward the core [60]. The quasi-linearity is verified in the overall series of fluorinated nanotubes and graphites. ^{19}F chemical shifts appear then useful to predict the discharge potentials in our electrochemical conditions. As fluorinated fullerenes contain several kinds of C–F bonds, depending on the position in pentagon or hexagon but also on their neighboring environment, this case must be separately discussed. Nevertheless, the comparison between $C_{60}F_{48}$ and $C_{70}F_{56}$ can be done. Once again, a linear dependence of the potentials with the chemical shift is observed in both cases. The data of $C_{70}F_{56}$ are slightly shifted toward the low potentials in comparison with $C_{60}F_{48}$. This could be related to the larger diameter of C_{70}. The elongated shape as compared to spherical C_{60} results in different covalence. The derivative curve of $C_{70}F_{56}$ exhibits a broad line centered at 2.85 V contrary to $C_{60}F_{48}$ (Fig. 9.11). It is also to note that the potential range is more extended for highly fluorinated fullerenes than for nanotubes and graphite with values included in between 3.6 and 2.0 V.

The discharge potentials in Li batteries are then correlated with the diameter of the outer tube, ie, the curvature of the carbon lattice. This has been done with precautions, ie, similar fluorine content, exclusive fluorination of the outer tubes, same electrochemical operating conditions. The higher the curvature, the higher the potential is because the covalence is weakened. This allows the potential of fluorinated nanotubes to be tuned in primary lithium batteries.

9.4 Tuning the C—F Covalence to Enhance the Applicative Properties

Three applications for which the C—F bonding plays a key role will be discussed: (1) use as electrode materials in primary lithium battery; (2) materials for microelectronics; (3) as lubricants.

Whatever the catalytic atmosphere ($MF_n/HF/F_2$, $MF_n = IF_5$, BF_3 or ClF_x) [15,61,62], the highest voltage plateau of the series was observed for the raw compound without postfluorination and can be correlated with the lowest C—F bond covalence. Fig. 9.12 displays the data for IF_5-containing catalysts. In the 250—550°C postfluorination temperature range, the covalent character of the C—F bonding increases with treatment temperature. This is indicated by the evolution of the ^{13}C NMR chemical shift of the carbon atoms which are in strong interaction with fluorine; the higher the chemical shift, the higher the C—F bond covalence is. The increase in C—F bond covalence results in a decrease in discharge potential because covalent C—F bonds require more energy to be broken during the discharge. In comparison, purely covalent high-temperature graphite fluoride with CF_1 composition exhibits a lower discharge voltage, equal to 1.96 V [1] or to 2.1 V in our electrochemical conditions.

Irrespective of the choice of catalyst, capacities were not always proportional to the F/C content. For the case of IF_5-T_{PF} series, although the fluorination level is nearly constant, the capacities of posttreated IF_5-T_{PF} increased with the treatment temperature (Fig. 9.13). These improvements are then rather correlated with the removal of the iodine fluoride species which occurs progressively during the thermal treatment. Capacities are in fact inversely proportional to the catalyst content indicating the strong correlation between these two parameters. Removal of the intercalated species is necessary to enhance electrochemical performance. So, in terms of commercial uses, the most interesting samples are those that have undergone postfluorination at

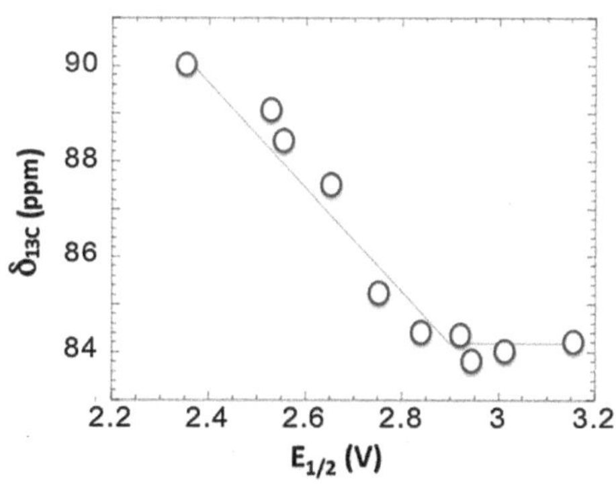

Figure 9.12 ^{13}C NMR chemical shift as a function of the discharge potential for raw and postfluorinated IF_5-T_{PF}.

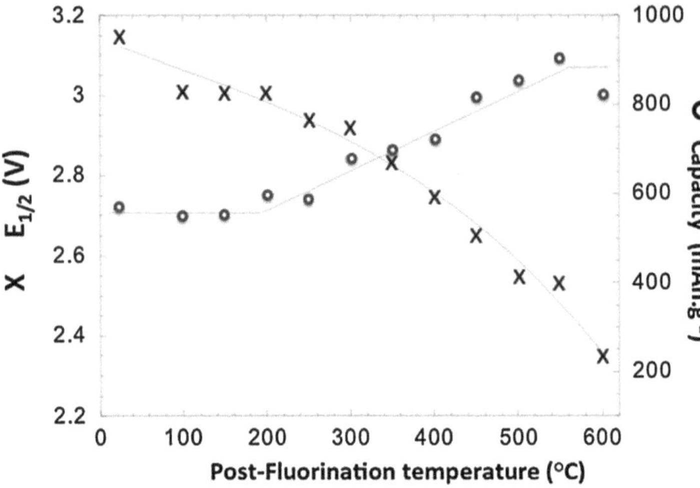

Figure 9.13 Discharge potentials and capacities as a function of the postfluorination temperature for IF_5-T_{PF} series.

temperatures higher than 400°C. Concerning energy and power density delivered, a good agreement is obtained for CF(IF_5)-400 due to simultaneous large discharge capacity and average potential, close to 700 $mAhg^{-1}$ and 2.8 V, respectively. For this sample, the best specific energy density and specific power density are 2149 and 902 Wkg^{-1}, respectively [63].

The covalence weakening affects the C–F bond strength and leads to lack of robustness of the C–F interaction within fluorocarbon lattice. Such an effect can be experimentally observed by the thermally activated decomposition of fluorinated fullerenes. Whereas a perfluorinated graphite plane remains unaffected, fluorofullerenes readily lose fluorine above 350°C in the absence of a fluorine atmosphere. Recently, fragmentation of C–F bonds was induced by electron irradiation of fluorinated graphene. Standard electron beam patterning processes can be used to engineer conductive and semiconductive structures with sizes ranging from a few micrometers down to a few tens of nanometers [64]. The fluorination of graphene was conducted prior to nanofabrication, by exposing natural graphite to F_2 atmosphere at 450°C, resulting in an F/C ratio of 0.28 as measured by mass uptake. Subsequently, few-layer fluorinated graphene flakes are obtained by mechanical cleavage of fluorinated graphite onto SiO_2 (300 nm)/p-doped Si substrate which acts as a back gate. The flakes are then identified by their low optical contrast, typically 2–4%. Electrical contacts were fabricated by standard electron beam lithography, evaporation of Cr/Au (5 nm/70 nm), and subsequent lift-off process (see the insets of Fig. 9.14). A topographical atomic force microscopy measurement of neighboring irradiated and nonirradiated fluorinated graphene regions evidenced that the electron irradiation does not cause any structural damage to the graphene material (see Fig. 9.14B). However, a significant change in height between the irradiated and nonirradiated regions was found. The height of the exposed region is 0.7 nm lower than the height of the unexposed

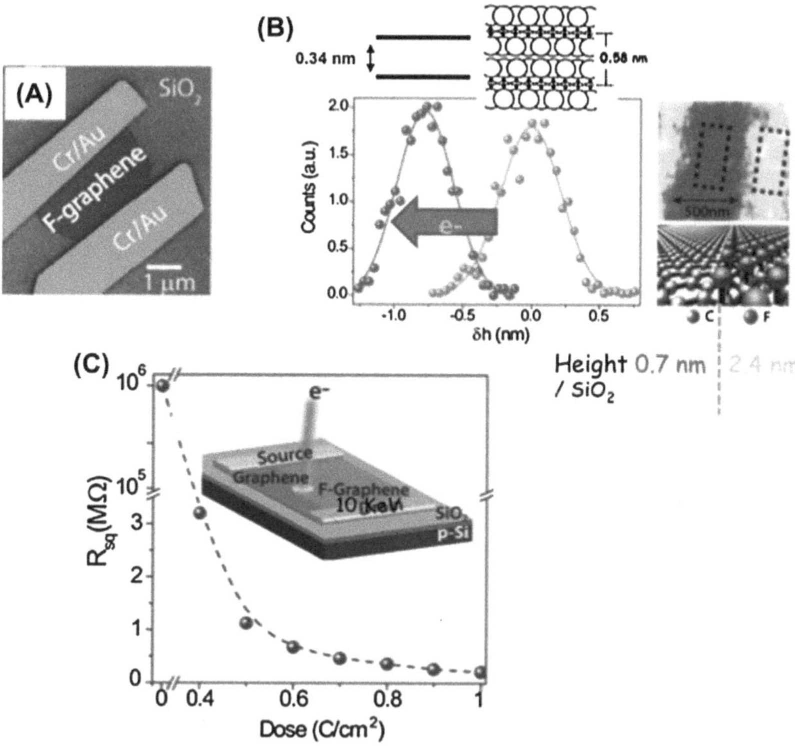

Figure 9.14 (A) Fluorinated graphene flakes are obtained by mechanical cleavage of fluorinated graphite onto SiO_2 (300 nm)/p-doped Si substrate which acts as a back gate. (B) AFM histogram data for the unexposed [orange (gray in print versions)] and exposed [red (dark gray in print versions)] fluorinated graphene extracted from the topographical AFM image shown. Yellow (light gray in print versions) corresponds to 2.4 nm measured from the SiO_2 substrate. The bottom panel illustrates the crystal structure of pristine graphene (gray) and fluorinated graphene [green (darker gray in print versions)]. (C) The measured sample resistance per square plotted against the electron irradiation dose (the *dashed black lines* is a guideline for the eyes). The inset shows an illustration of the device configuration under irradiation with a beam of electrons.

region, with the latter equal to 2.4 nm measured from the SiO_2. The experimentally observed reduction in height is in accordance with the electron irradiation assisted defluorination process because the interatomic layer spacing between atomically flat sp^2 graphene layers is smaller than the spacing between the three-dimensional atomic layers of sp^3 hybridized fluorinated graphene. The relative decrease in resistance per square upon electron irradiation of microstructures was at least seven orders of magnitude (from 1 TΩ to 100 kΩ), three orders of magnitude more conductive than what was previously reported in reduced graphene oxide patterning experiments [65].

Weakened covalence for the C–F bonds may favor both low friction coefficients and the formation of the tribofilm (see also Chapter 14 *Tribology of Fluorinated Carbons* of the present book).

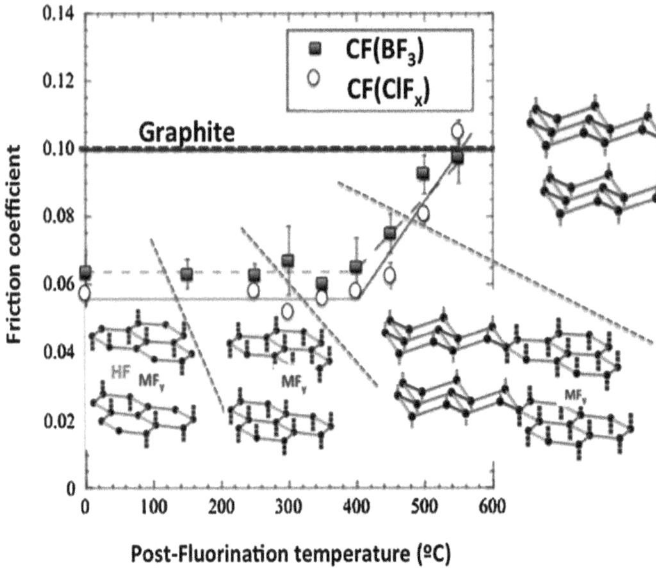

Figure 9.15 Intrinsic friction coefficients (after three cycles) of $CF(BF_3)$ and $CF(Cl_2)$ posttreated in F_2 gas.

The tribologic properties of the compounds are evaluated using an alternative ball-on-plane tribometer, the ball motion being a backward and forward motion on the static plane. The experimental details are given in Ref. 66. The friction coefficient μ is measured under air atmosphere with relative humidity of 50–55%. The values of friction coefficient measured after three cycles is considered to correspond to the intrinsic tribologic properties of the deposited film.

The friction coefficients obtained for the fluorinated graphite $CF(BF_3)$ and $CF(Cl_2)$ posttreated in F_2 gas at the beginning of the tribological test (3 cycles) are presented in Fig. 9.15 as a function of fluorination posttreatment temperature. All the compounds present good intrinsic friction properties, better than pristine graphite. The lowest friction coefficients are obtained for low and intermediate fluorination temperatures as μ remains in the same range ($\mu = 0.055$ and $\mu = 0.06$ for $CF(ClF_x)$ and $CF(BF_3)$ series, respectively) for initial compounds and materials posttreated up to 450°C. An increase of μ is observed for $T_{PF} > 500°C$ (0.09 and 0.1 for $CF(Cl_2)$ and $CF(BF_3)$ series, respectively).

The intrinsic tribologic properties of the tested graphite fluorides clearly demonstrate that the friction reduction process is not directly related to the fluorine content or the nature of the intercalated species (F^- or BF_4^-, IF_5 or IF_6^- in other works) because the friction coefficients appear the highest for the greatest F/C ratios and interlayer distances (Table 9.3). Low friction coefficients are obtained when a lot of graphene planes are maintained; such a case corresponds to partially fluorinated compounds (room temperature $< T_{PF} <$ 450°C), where graphitic domains can coexist with perfluorinated ones resulting in the weakening of the C–F bonding (hyperconjugation). A rapid increase of the friction coefficient is correlated to the increase of C−F covalent bond concentration. This induces a decrease of the graphitic domains extension evidenced by the decrease of La (deduced from the Raman spectra) or the disappearance of the Raman graphene bands. Long experiment results completed by Raman analyses of the tribofilm lead to the conclusion that the friction process induces a breaking of covalent C−F bonds and a partial rebuilding of graphitic structure in the perfluorinated compounds leading to an improvement of the tribo properties of the tribofilm. In the case of partially fluorinated materials, the presence of intercalated fluorides and the weakening of C−F bond covalence allow the released fluorine atoms to be trapped in the structure and/or to create new C−F bonds, which explains the good stability of the tribological performances of the tribofilm.

References

[1] N. Watanabe, T. Nakajima, H. Touhara, Graphite Fluorides, Elsevier, Amsterdam, 1988.
[2] H. Touhara, F. Okino, Carbon 38 (2000) 241−267.
[3] R.L. Fusaro, H.E. Sliney, ASLE Trans. 13 (1970) 56−65.
[4] A.S. Nazarov, V.G. Makotchenko, Inorg. Mater. 38 (2002) 278.
[5] R. Yazami, A. Hamwi, K. Guérin, M. Dubois, Y. Ozawa, J. Giraudet, F. Masin, Electrochem. Commun. 9 (2007) 1850−1855.
[6] P. Thomas, D. Himmel, J.L. Mansot, M. Dubois, K. Guérin, W. Zhang, A. Hamwi, Tribol. Lett. 34 (2009) 49−59.
[7] K. Delbé, P. Thomas, D. Himmel, J.L. Mansot, M. Dubois, K. Guérin, C. Delabarre, A. Hamwi, Tribol. Lett. 37 (2010) 31−41.
[8] R.L. Fusaro, H.E. Sliney, NASA Tech. Note D-5097, 1969.
[9] W. Zhang, M. Dubois, K. Guérin, P. Bonnet, E. Petit, N. Delpuech, D. Albertini, F. Masin, A. Hamwi, Carbon 47 (2009) 2763−2775.
[10] J. Giraudet, M. Dubois, K. Guérin, J.P. Pinheiro, A. Hamwi, W.E.E. Stone, P. Pirotte, F. Masin, J. Solid State Chem. 178 (2005) 1262−1268.
[11] J. Giraudet, M. Dubois, A. Hamwi, W.E.E. Stone, P. Pirotte, F. Masin, J. Phys. Chem. B 109 (2005) 175−181.
[12] J. Giraudet, M. Dubois, K. Guérin, A. Hamwi, F. Masin, J. Phys. Chem. Solids 67 (2006) 1100−1105.
[13] M. Dubois, J. Giraudet, K. Guérin, A. Hamwi, Z. Fawal, P. Pirotte, F. Masin, J. Phys. Chem. B 110 (2006) 11800−11808.
[14] M. Dubois, K. Guérin, J.P. Pinheiro, F. Masin, Z. Fawal, A. Hamwi, Carbon 42 (2004) 1931−1940.
[15] K. Guérin, J.P. Pinheiro, M. Dubois, Z. Fawal, F. Masin, R. Yazami, A. Hamwi, Chem. Mater. 16 (2004) 1786−1792.
[16] J. Giraudet, M. Dubois, K. Guérin, C. Delabarre, A. Hamwi, F. Masin, J. Phys. Chem. B 111 (2007) 14143−14151.
[17] S. Niyogi, M.A. Hamon, H. Hu, B. Zhao, P. Bhowmik, R. Sen, M.E. Itkis, R.C. Haddon, Acc. Chem. Res. 35 (2002) 1105−1113.
[18] S. Banerjee, T. Hemraj-Benny, S.S. Wong, Adv. Mater. 17 (2005) 17−29.
[19] J. Kürti, V. Zolyomi, M. Kertesz, G. Sun, New J. Phys. 5 (2003) 125-1−125-21.
[20] F. Chamssedine, K. Guérin, M. Dubois, E. Disa, E. Petit, Z. El Fawal, A. Hamwi, J. Fluorine Chem. 132 (2011) 1072−1078.

[21] M.-J. Jung, E. Jeong, S. Kim, S. Lee, J. Yoo, Y. Lee, J. Fluorine Chem. 132 (2011) 1127–1133.
[22] J. Im, M. Jung, Y. Lee, J. Colloid Interface Sci. 339 (2009) 31–35.
[23] P. Fulvio, S. Brown, J. Adcock, R. Mayes, B. Guo, X.-G. Sun, Chem. Mater. 23 (2011) 4420–4427.
[24] F. Leroux, M. Dubois, J. Mater. Chem. 16 (2006) 4510–4520.
[25] J. Parmentier, S. Schlienger, M. Dubois, E. Disa, F. Masin, T. Centeno, Carbon 50 (2012) 5135–5147.
[26] C.M. Ghimbeu, K. Guérin, M. Dubois, S. Hajjar-Garreau, C. Vix-Guterl, Carbon 84 (2015) 567–583.
[27] A. Hamwi, J. Phys. Chem. Solids 57 (1996) 677.
[28] T. Nakajima, N. Watanabe, Graphite, Fluorides, and Carbon- Fluorine Compounds, CRC Press, Boca Raton, FL, 1991, p. 84.
[29] Y. Sato, K. Itoh, R. Hagiwara, T. Fukunaga, Y. Ito, Carbon 42 (2004) 2897–2903.
[30] Y. Sato, K. Itoh, R. Hagiwara, T. Fukunaga, Y. Ito, Carbon 42 (2004) 3243–3249.
[31] A.M. Panich, Synth. Met. 100 (1999) 169–185.
[32] A.M. Panich, H.M. Vieth, A.I. Shames, N. Froumin, E. Osawa, A. Yao, J. Phys. Chem. C 114 (2010) 774–782.
[33] A.M. Panich, Crit. Rev. Solid State Mater. Sci. 37 (2012) 276–303.
[34] M. Dubois, K. Guérin, E. Petit, N. Batisse, A. Hamwi, N. Komatsu, J. Giraudet, P. Pirotte, F. Masin, J. Phys. Chem. C 113 (2009) 10371–10378.
[35] A. Hamwi, M. Daoud, J.C. Cousseins, Synth. Met. 26 (1988) 89–98.
[36] Y. Kita, N. Watanabe, Y. Fuji, J. Am. Chem. Soc. 101 (1979) 3832–3841.
[37] W. Zhang, M. Dubois, K. Guérin, P. Bonnet, H. Kharbache, F. Masin, A.P. Kharitonov, A. Hamwi, Phys. Chem. Chem. Phys. 12 (2010) 1388–1398.
[38] H. Muramatsu, Y.A. Kim, T. Hayashi, M. Endo, A. Yonemoto, H. Arikai, F. Okino, H. Touhara, Chem. Commun. (2005) 2002–2004.
[39] S. Kawasaki, Y. Kanamori, Y. Iwai, F. Okino, H. Touhara, H. Muramatsu, T. Hayashi, Y.A. Kim, M. Endo, J. Phys. Chem. Solids 69 (2008) 1203–1205.
[40] T. Gullion, J. Schaefer, J. Magn. Reson. 81 (1989) 196–200.
[41] A.W. Hing, S. Vega, J. Schaefer, J. Magn. Reson. 96 (1992) 205–209.
[42] A.W. Hing, S. Vega, J. Schaefer, J. Magn. Reson. A 103 (1993) 151–162.
[43] J. Giraudet, M. Dubois, K. Guerin, C. Delabarre, P. Pirotte, A. Hamwi, F. Masin, Solid State NMR 31 (2007) 131–140.
[44] P. Bertani, J. Raya, P. Reinheimer, R. Gougeon, L. Delmotte, J. Hirschinger, Solid State Magn. Res. 13 (1999) 219.
[45] J. Giraudet, M. Dubois, A. Hamwi, W.E.E. Stone, P. Pirotte, F. Masin, J. Phys. Chem. B 109 (2005) 175–181.
[46] F. Chamssedine, M. Dubois, K. Guérin, J. Giraudet, F. Masin, D.A. Ivanov, L. Vidal, R. Yazami, A. Hamwi, Chem. Mater. 19 (2007) 161–172.
[47] J. Stöhr, NEXAFS Spectroscopy, Springer-Verlag, Berlin, 1992.
[48] Y. Ahmad, M. Dubois, K. Guérin, A. Hamwi, Z. Fawal, A.P. Kharitonov, A.V. Generalov, A.Yu. Klyushin, K.A. Simonov, N.A. Vinogradov, I.A. Zhdanov, A.B. Preobrajenski, A.S. Vinogradov, J. Phys. Chem. C 117 (2013) 13564–13572.
[49] S.L. Molodtsov, S.I. Fedoseenko, D.V. Vyalikh, I.E. Iossifov, R. Follath, S.A. Gorovikov, M.M. Brzhezinskaya, Y.S. Dedkov, R. Puetner, J.S. Schmidt, V.K. Adamchuk, W. Gudat, G. Kaindl, Appl. Phys. A: Mater. Sci. Process. 94 (2009) 501–505.
[50] R. Nyholm, S. Svensson, J. Nordgren, A. Flodström, Nucl. Instr. Aand Mreth. A 246 (1986) 267–271.

[51] S.L. Molodtsov, S.I. Fedoseenko, D.V. Vyalikh, I.E. Iossifov, R. Follath, S.A. Gorovikov, M.M. Brzhezinskaya, Y.S. Dedkov, R. Puetner, J.S. Schmidt, Appl. Phys. A: Mater. Sci. Process. 94 (2009) 501−505.
[52] R. Nyholm, S. Svensson, J. Nordgren, A. Flodström, Nucl. Instrum. Methods Phys. Res. Sect. A 246 (1986) 267−271.
[53] M. Dubois, K. Guérin, Y. Ahmad, Electrochim. Acta 59 (2012) 485−491.
[54] A. Hamwi, C. Latouche, V. Marchand, J. Dupuis, R. Benoit, J. Phys. Chem. Solids 57 (1996) 991.
[55] N. Liu, H. Touhara, F. Okino, S. Kawasaki, Y. Nakacho, J. Electrochem. Soc. 143 (1996) 2267.
[56] Y. Matsuo, T. Nakajima, Electrochim. Acta 41 (1996) 15.
[57] F. Okino, S. Yajima, S. Suganuma, R. Mitsumoto, K. Seki, H. Touhara, Synth. Met. 70 (1995) 1447.
[58] R.G. Compton, R.A. Spackman, R.G. Wellington, L.H. Malcolm, J. Turner, J. Electroanal. Chem. 327 (1992) 337.
[59] Y. Chabre, D. Djurado, M. Armand, W.R. Romanov, N. Coustel, J.P. Mc Cauley, J.E. Fischer, A.B. Smith III, J. Am. Chem. Soc. 114 (1992) 764.
[60] R. Yazami, A. Hamwi, K. Guérin, M. Dubois, Y. Ozawa, J. Giraudet, F. Masin, Electrochem. Commun. 9 (2007) 1850.
[61] C. Delabarre, J. Giraudet, K. Guérin, M. Dubois, A. Hamwi, J. Phys. Chem. Solids 67 (2006) 1157−1161.
[62] C. Delabarre, K. Guérin, M. Dubois, J. Giraudet, Z. Fawal, A. Hamwi, J. Fluorine Chem. 126 (2005) 1078−1087.
[63] J. Giraudet, C. Delabarre, K. Guérin, M. Dubois, F. Masin, A. Hamwi, J. Power Sources 158 (2006) 1365−1372.
[64] F. Withers, T.H. Bointon, M. Dubois, S. Russo, M.F. Craciun, Nano lett. 11 (2011) 3912−3916.
[65] W. Zhongqing, W. Debin, K. Suenne, K. Soo-Young, H. Yike, M.K. Yakes, A.R. Laracuente, D. Zhenting, S.R. Marder, B. Claire, W.P. King, W.A. de Heer, P.E. Sheehan, E. Riedo, Science 328 (2010) 1373−1376.
[66] K. Delbé, P. Thomas, D. Himmel, J.-L. Mansot, M. Dubois, K. Guérin, C. Delabarre, A. Hamwi, Tribology Lett. 37 (2010) 31−41.

Preparation and Application of Fluorine–Carbon and Fluorine–Oxygen–Carbon Materials

10

Y. Matsuo
University of Hyogo, Himeji, Japan

Chapter Outline

10.1 Introduction

Fluorine and oxygen are the most and the second most electronegative elements and their sizes are small as shown in Table 10.1. Therefore, when they are inserted between the layers of graphite, various types of interactions are realized. In the case of fluorine, the resulting C–F bonding varies from ionic to covalent, while in the case of oxygen, various types of functional groups such as hydroxyl and epoxy ones can form. When fluorine is reacted with graphite at elevated temperatures, covalent type graphite

New Fluorinated Carbons: Fundamentals and Applications. http://dx.doi.org/10.1016/B978-0-12-803479-8.00010-3

Table 10.1 Physical Properties of Fluorine and Oxygen Atoms

	F	O
Van der Waals radius/nm	0.135	0.140
Ionic radius/nm	0.119	0.126
Electronegativity	4.0	3.5

fluorides in which sp^2 carbon-based planar sheets are converted to sp^3 carbon-based zigzag ones are formed. Two types of materials, poly(carbon monofluoride) $(CF)_n$ and poly(dicarbon monofluoride) $(C_2F)_n$ have been already well known as summarized elsewhere [1–3]. At lower temperatures, fluorine-intercalated graphite, C_xF. is formed in which sp^2 carbon-based planar sheets are maintained. The nature of C–F bonding in C_xF samples greatly changes, depending on their fluorine contents. As the fluorine content increases, it changes from ionic to "semi-ionic" or "semi-covalent." The latter interestingly provides the C1s and F1s X-ray photoelectron spectroscopy (XPS) peaks between those due to ionic and covalent C–F bondings [2,4] and gives shifted ^{19}F NMR peak [5], though its bond length is the same as that of the covalent one [6]. This slightly lower C–F bond order in C_xF than those in graphite fluoride is explained by the hyperconjugation involving the C-C bonds on the carbon sheets and C–F bond [6].

On the other hand, the oxidation of graphite provides a covalent type intercalation compound, graphite oxide (recently more frequently called as graphene oxide, hereafter abbreviated as GO). GO is synthesized by the reaction of graphite with strong oxidizing reagents such as $KMnO_4$ and $KClO_3$ in concentrated acids. It has been reported that hydroxyl and epoxy groups are mainly attached to carbon layers [7]; however, the structure of GO has not been fully understood yet, though many structure models have been proposed [8]. Since recently, this material has been attracting much attention for the precursor of graphene because it is easily exfoliated in polar solvents to form single sheets. Since covalent type graphite intercalation compounds are known only for fluorine/graphite and oxygen/graphite systems, interesting materials could be obtained in the system where both fluorine and oxygen exist with graphite.

In this chapter, electrochemical preparation of C_xF in an aqueous solution of hydrofluoric acid or in an all-solid cell using fluoride ion conductor, and the cathodic properties of resulting materials are first summarized. Then, preparation of carbon thin film containing a small amount of fluorine and oxygen for transparent and conducting electrode is also described.

10.2 Electrochemical Preparation of C_xF

10.2.1 Synthesis and Structure of C_xF

C_xF is usually prepared by the reaction of graphite with F_2 gas in the presence of a small amount of a catalyst, such as solid metal fluorides, gaseous halogen, or metal fluorides. The x value can reach 1.2 when high oxidation state metal fluorides, such as $KAgF_4$, are used under high fluorine gas pressure in anhydrous HF [9]. Electrochemical oxidation of graphite in fluoride ion–containing electrolyte solutions also provides C_xF. For example, stage 2 type $C_{12}F$–containing co-intercalated HF molecules have been obtained in an aqueous HF solution; however, decomposition of electrolyte solution prevents further oxidation of $C_{12}F$ and lower stage compounds are not formed [10]. C_xF is also formed during the electrochemical production of fluorine gas in KF·2HF melt. Impregnation of solid fluorides into the carbon electrode such as LiF can promote the formation of stage 2 type C_xF [11], otherwise graphite fluorides with low surface energy tend to form at the surface area. This leads to the loss of contact with electrolyte, resulting in the abrupt increase in the overpotential, which is known as an "anode effect" [1].

The properties and structure of C_xF greatly vary depending on the fluorine content. Fig. 10.1 shows the c-axis structure models of C_xF with various fluorine contents proposed by Nakajima et al. [4,12]. Fluoride ions with the diameter of 0.26–0.27 nm are inserted between carbon layers in stage 2 or higher samples as shown in Fig. 10.1(A); however, ionic type stage 1 C_xF has not been obtained and instead, two types of stage 1 samples with repeat distances along c-axis of 1.15 and 0.55 nm are formed. The former is the "bi-intercalation" type in which the ionically and covalently bonded fluorine layers (denoted as layers P and Q, respectively) are alternatively stacked and the latter is the covalent type stage 1 C_xF as shown in Fig. 10.1(B) and (C), respectively.

On the other hand, for the in-plane structure of C_xF, two types of superlattices have been observed. One is the hexagonal 2 × 2 [4,12] type obtained by X-ray and electron diffraction measurements for stage 2 or higher samples and the other is the

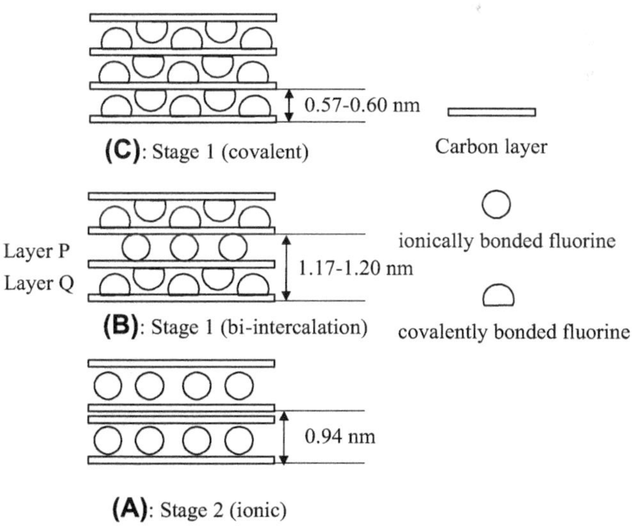

Figure 10.1 The c-axis structure models for C_xF with interlayer spacings of (A) 0.93, (B) 1.17, and (C) 0.57 nm.
Reproduced from T. Nakajima, M. Molinier, M. Motoyama, Carbon 29 (1991) 429–437, with permission from Elsevier.

orthorhombic $2 \times \sqrt{3}$ [13] type observed by atomic force microscopy (AFM) for a stage 1 sample. The proposed orientations of fluorine atoms on the graphite layers are as shown in Fig. 10.2.

10.2.2 Electrochemical Fluorination of Graphite in an Aqueous Solution of Hydrofluoric Acid [14,15]

Fig. 10.3 shows the cyclic voltammograms of graphite sheet in 47% HF aqueous solution at a scan rate of 0.5 mV/s. Three broad oxidation peaks at around 2.0, 2.48, and 2.6 V vs Pb/PbF_2 were observed when the potential was scanned to 3.0 V and only one reduction peak at 1.3 V was observed. The peak at 2.0 V is due to the formation of stage 2 type C_xF and the intercalated fluoride ions are reversibly de-intercalated at 1.8 V which was observed only when the upper limit potential was 2.2 V. The absence of corresponding reduction peaks for those at 2.4 and 2.6 V during cathodic scan means that covalent C—F bonding formed at these potentials and it is reduced at 1.3V.

Fig. 10.4 shows the variation of potential during electrochemical oxidation of graphite at various current densities. Plateaus at 2.0 and 2.5 V (denoted as plateaus I and II in Fig. 10.4, respectively) were observed at low current densities or at the beginning of the oxidation. The potential values of the plateau observed when the current density was high or after large amounts of charge passed through the cell increased with increase in the current density. At these plateaus, oxygen gas was vigorously generated, indicating that the charge passing through the cell was consumed for the oxidation of not only graphite but also water. This means that it is necessary to increase the anode potential to a high value at which O_2 evolution takes place by increasing the current density to oxidize graphite to higher levels. Fig. 10.5 shows the typical X-ray diffraction patterns of graphite oxidized at the plateaus of 2.0, 2.5, 3.2, and 3.5 V. The diffraction patterns of the sample obtained at higher potentials were identical to that obtained at 3.5 V. The repeat distances along c-axis of the obtained samples were 0.93, 0.89, 1.15, and 0.55 nm, respectively. The first and third values were ascribed to the phases shown in Fig. 10.1(A) and (B). On the other hand, the C_xF sample with an interlayer spacing of 0.89 nm has not been observed before. Based on the C/F ratio of four and covalent nature of C—F bonding, this phase is assigned to stage 2 covalent type C_xF as shown in Fig. 10.6(A). The c-axis repeat distance of the last one was the same as that shown in Fig. 10.1(C); however, the fluorine content was lower and it contained a considerable amount of oxygen. Considering that GO is prepared by the electrochemical oxidation of graphite in concentrated acids, it is reasonable to conclude that oxygen atoms are inserted into graphite together with fluorine. The position of oxygen and fluorine in the resulting materials is not clear; however, the short plateau at 2.5 V and the absence of diffraction peaks other than that at $2\theta = 16°$ suggest their homogenous distributions as shown in Fig. 10.6(B).

Fig. 10.7 shows the summary of the electrochemical fluorination of graphite in 47% HF aqueous solution. The threshold potentials for the C_xF samples with various interlayer spacings are 2.0, 2.5, 3.2, and 3.5 V.

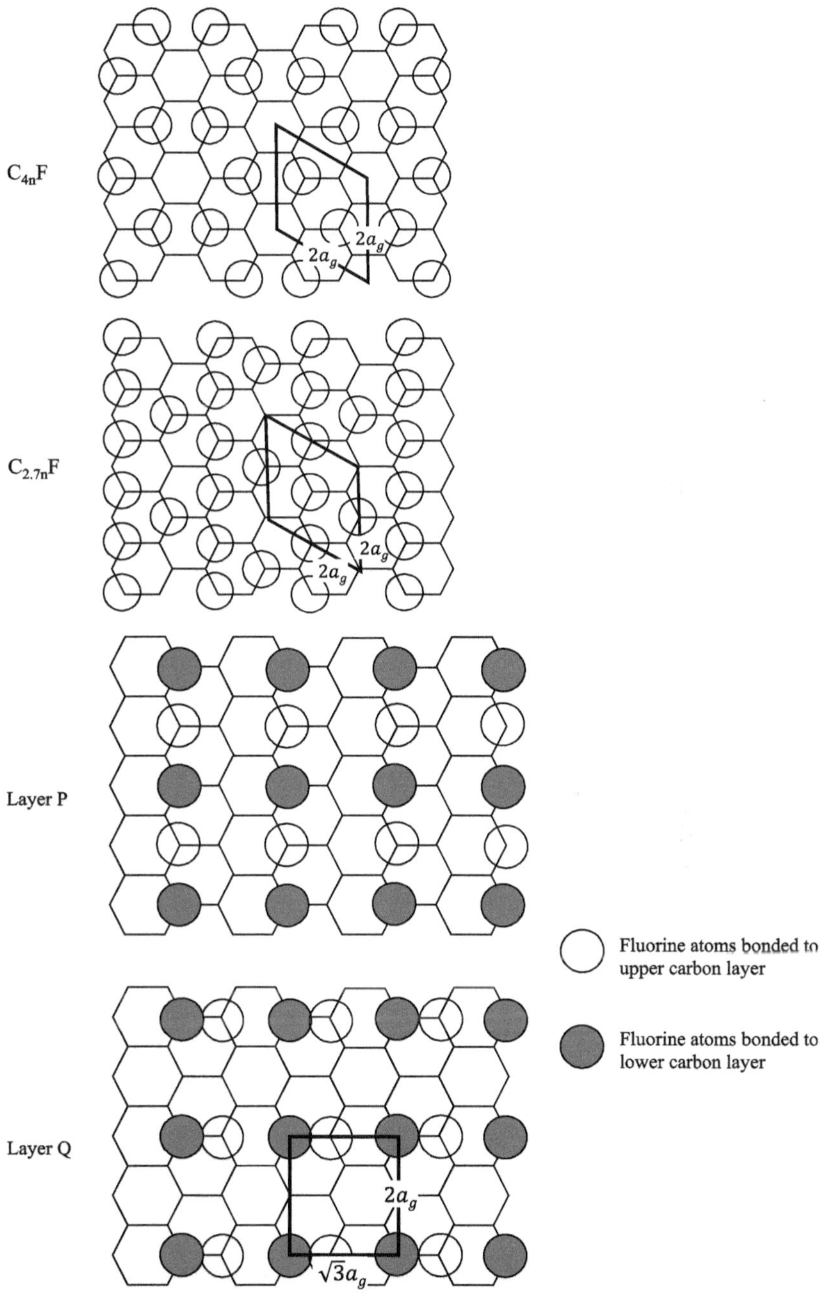

Figure 10.2 The in-plane structure models for various C_xF samples with superlattices of 2×2 and $2 \times \sqrt{3}$. a_g, lattice parameter of graphite; n, stage number.
Reproduced from T. Nakajima, M. Molinier, M. Motoyama, Carbon 29 (1991) 429–437 and N. Ikemiya, S. Hara, K. Ogino. T. Nakajima, Surf. Sci. Lett. 274 (1992) 524–528 with permission from Elsevier.

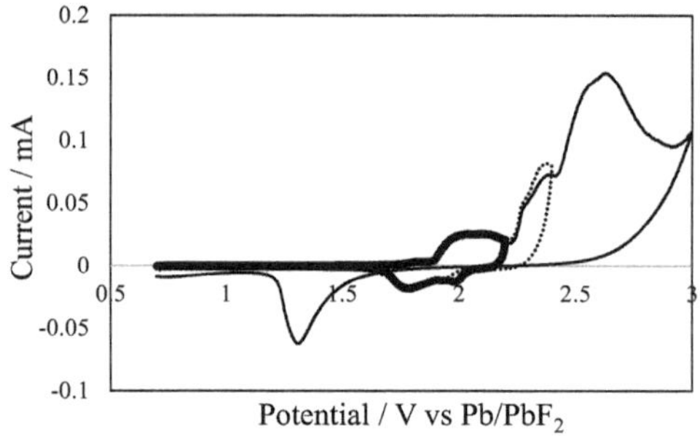

Figure 10.3 Cyclic voltammograms of graphite sheet electrode at 0.5 mV/s with various upper limit potentials.
Reproduced from Y. Matsuo, S. Hirata, M. Dubois, J. Fluorine Chem. 185 (2016) 36–41 with permission from Elsevier.

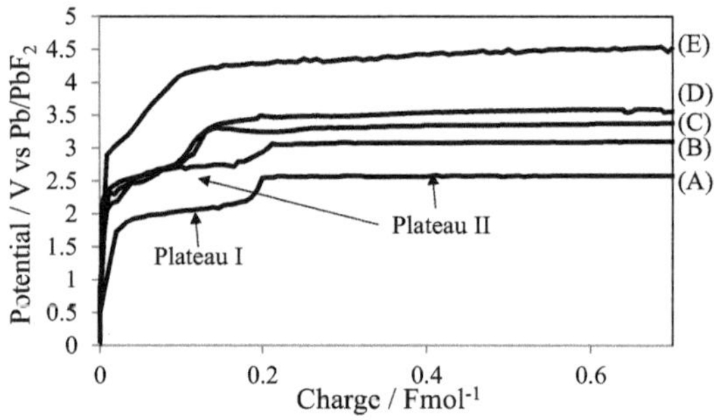

Figure 10.4 Variation of potentials during constant current oxidation of graphite at the current densities of 5, 50, 100, 200, and 800 mA/cm^2.

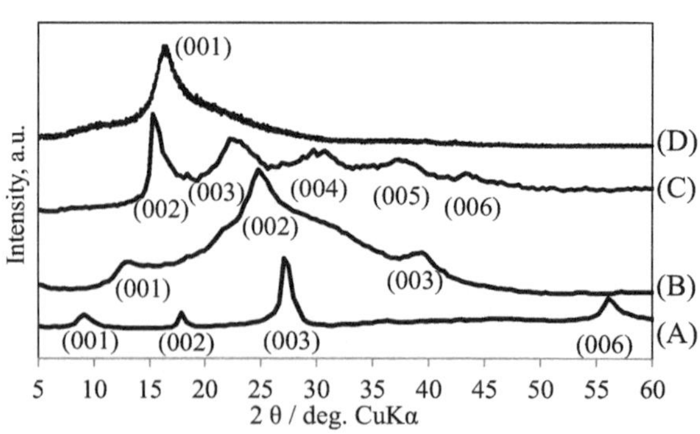

Figure 10.5 X-ray diffraction patterns of C_xF samples obtained at the plateaus of (A) 2.0, (B) 2.5, (C) 3.0, and (D) 3.6 V.

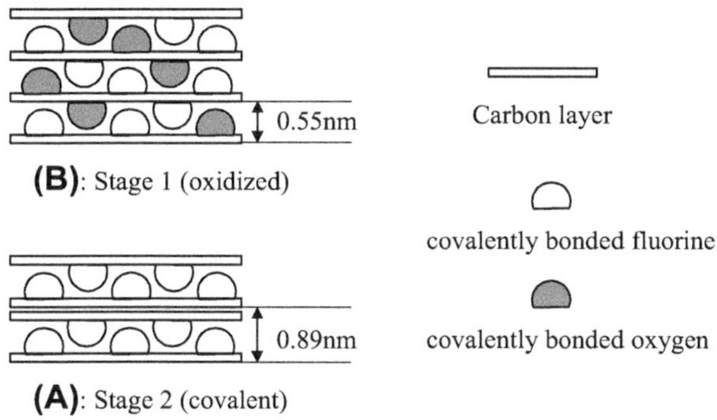

Figure 10.6 C-axis structure models of electrochemically prepared C_xF samples with interlayer spacings of (A) 0.89 and (B) 0.55 nm.

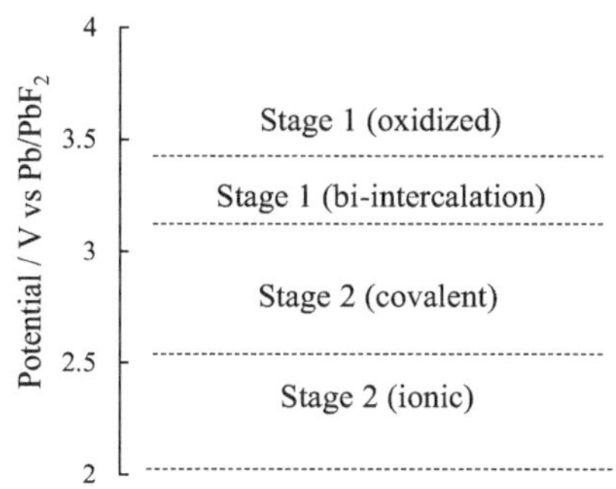

Figure 10.7 Threshold potentials for C_xF samples obtained by electrochemical oxidation of graphite in 47% HF aqueous solution.

10.2.3 Electrochemical Fluorination of Graphite in an All-Solid Cell [16]

To avoid the decomposition of water or HF in the reaction system, all-solid type electrochemical cell using a fast fluoride ion conductor, $PbSnF_4$ [17] as an electrolyte has been also tested. Fig. 10.8 shows the schematic illustration of the electrochemical cell. The edge of HOPG (highly oriented pyrolytic graphite) and Ni mesh counter electrodes were in contact with the solid electrolyte and the expected electrochemical reaction is shown in the figure. The fluoride ions formed at cathode move to the anode through the solid electrolyte and react with oxidized graphite to form C_xF. This type of cell was also used for the production of F_2 gas or fluorination of metals by using LaF_3 or $PbSnF_4$ as solid electrolytes [18,19]. The reaction was performed in the F_2 gas atmosphere, therefore chemical fluorination also occurred during the electrochemical fluorination of graphite.

The reaction condition and composition of the products prepared in this system are summarized in Table 10.2. The formation of insulating fluorides at the surface of electrode made it difficult to control the electrochemical reaction; however, the content of fluorine in C_xF tended to increase with increase in the reaction time and charge passing

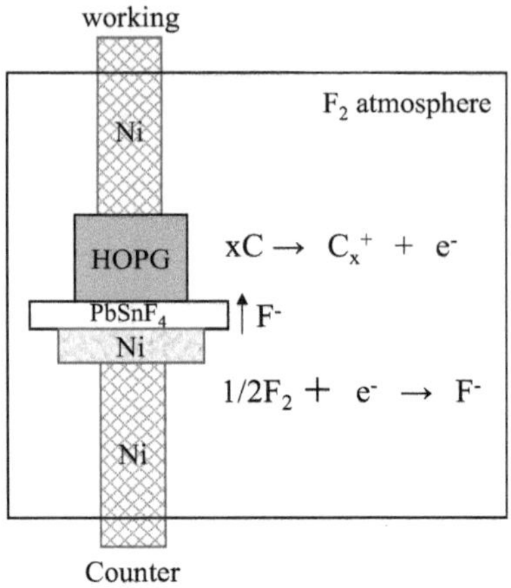

Figure 10.8 Schematic illustration of the electrochemical cell using solid fluoride ion conductor $PbSnF_4$. *HOPG*, highly oriented pyrolytic graphite.
Reproduced from Y. Matsuo, T. Nakajima, M. Motoyama, TANSO 160 (1993) 272–278 (in Japanese) with permission from the Carbon Society of Japan.

Table 10.2 Reaction Conditions and Composition of C_xF

Entry	Reaction Time/Day	Charge/10^3Fmol^{-1}	Composition[a]
1	12	0	$C_{4.6}F$
2	11	520	$C_{3.2}F$
3	1	0.22	$C_{16.5}F$
4	0.3	0.21	$C_{6.1}F$
5	5	0.09	$C_{9.1}F$
6	7	0.38	$C_{5.1}F$
7	6	0.65	$C_{4.6}F$
8	7	0.10	$C_{3.7}F$
9	5	1.1	$C_{3.6}F$
10	7	27	$C_{3.3}F$
11	4	2.1	$C_{3.0}F$
12	6	65	$C_{2.4}F$

[a]Determined from the weight change during reaction.

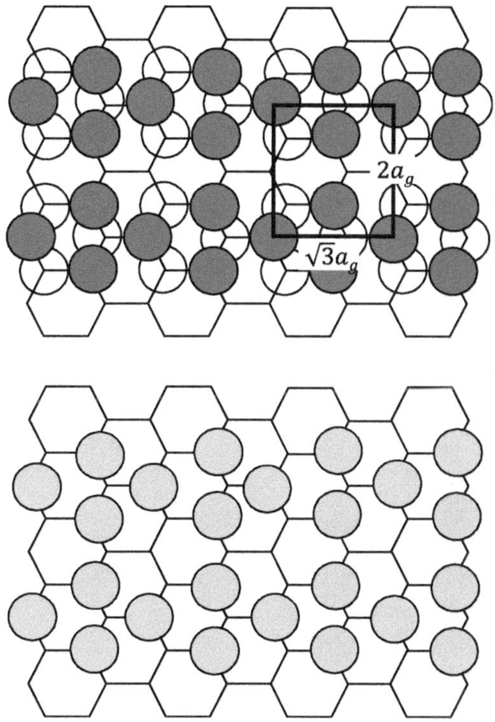

Figure 10.9 The in-plane structure model for $C_{1.8}F$ with superlattice of $2 \times \sqrt{3}$. Reproduced from Y. Matsuo, T. Nakajima, M. Motoyama, TANSO 160 (1993) 272–278 (in Japanese) with permission from the Carbon Society of Japan.

though the cell. Except for the samples prepared with short reaction time (entries 2–4), the fluorine contents were much higher than that obtained without electrochemical oxidation (entry 1; $C_{4.6}F$). Stage 1 bi-intercalation type $C_{2.4}F$ with the highest fluorine content was obtained. X-ray diffraction measurement indicated that this sample contained a phase with orthorhombic $2 \times \sqrt{3}$ in-plane superlattice. Considering the high fluorine content of this sample, a new arrangement of fluorine on carbon layer shown in Fig. 10.9 was proposed. In this model, the compositions of layers P and Q are C_8F_3 and C_8F_6, respectively and totally the composition is $C_{16}F_9$ ($C_{1.8}F$).

10.2.4 Cathodic Performance of C_xF

The C_xF samples with large fluorine contents are promising for the cathode materials of lithium primary battery with high energy densities. Typical discharge curve of C_xF prepared as described in Section 10.2.1 is shown in Fig. 10.10 [9], together with those of $(CF)_n$ prepared from natural graphite flake [20] and ball milled $(CF)_n$ [21]. The discharge voltage of $(CF)_n$ becomes higher when it is prepared from graphite with smaller L_c values (crystalline size along c-axis) [20] or it is ball milled [21]; however, it is still lower than 2.6 V. On the other hand, in the case of C_xF, reflecting its nature of

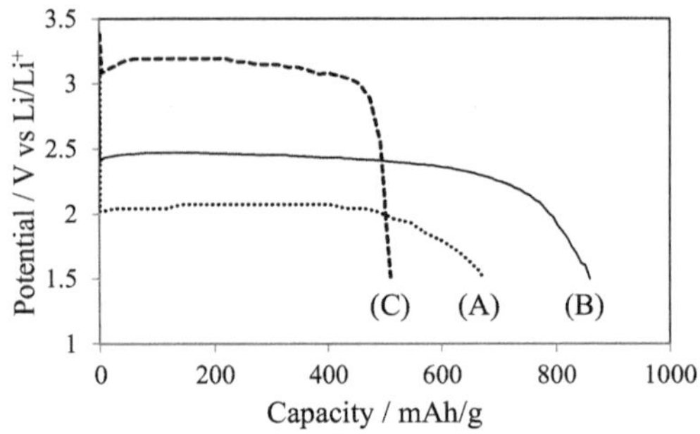

Figure 10.10 Discharge curves of (A) $(CF)_n$ prepared from natural graphite flake, (B) ball milled $(CF)_n$, and (C) $C_{1.4}F$ prepared by the reaction with F_2 gas using $KAgF_4$. Reproduced from N. Watanabe, T. Nakajima, R. Hagiwara, J. Power Sources 20 (1987) 87–92 and M.A. Reddy, B. Breitung, M. Fichtner, ACS Appl. Mater. Interfaces 5 (2013) 11207–11211 with permission from Elsevier and American Chemical Society.

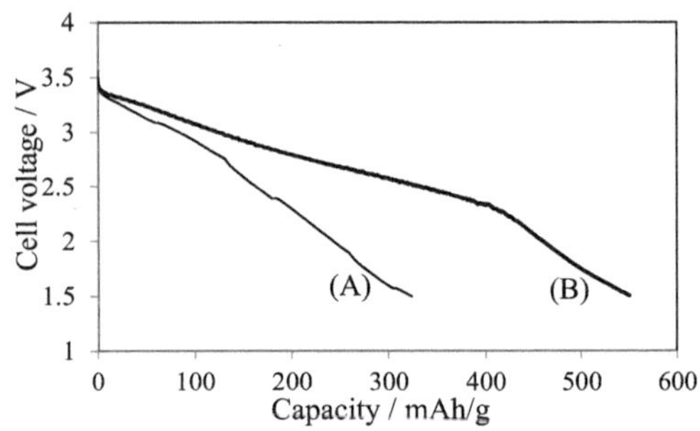

Figure 10.11 Discharge curves of (A) covalent type stage 2 C_xF and (B) oxidized stage 1 type C_xF. Reproduced from Y. Matsuo, S. Hirata, M. Dubois, J. Fluorine Chem. 185 (2016) 36–41 with permission from Elsevier.

C–F bonding, fluorine was more easily discharged and the discharge voltage exceeded 3 V. Accordingly, the energy density was similar to that of $(CF)_n$, though the capacity is slightly lower. The discharge curves of covalent stage 2 and oxidized stage 1 C_xF samples are shown in Fig. 10.11 [15]. The discharge voltage gradually decreased from 3.4 V and the capacity of the former was low; however, that of the latter reached a high value of 550 $mAhg^{-1}$. This discharge profile is very similar to that observed for fluorinated GO with a composition of $C_8O_{2.17}H_{0.63}F_{5.52}$ [22] and this means that not only fluorine but also oxygen in the oxidized C_xF sample is discharged.

10.3 Preparation of Transparent and Conducting Electrode From Graphene Oxide Containing Perfluoroalkyl Groups

10.3.1 Graphene Oxide for Transparent and Conducting Electrode

Recently, two dimensional and very thin graphene has been widely studied for transparent and conducting electrodes. Chemical vapor deposition method is usually employed for the preparation of electrodes with large surface area; however, high temperature and vacuum process are needed, which is not favorable for mass production. Therefore, the use of GO as the precursor is also promising because it is very easy to prepare thin film samples from the dispersion of GO. Many studies on the preparation of transparent and conducting electrode from the reduction of thin film of GO have been reported [23–33]. In these studies, GO thin films are prepared by spin coating, layer by layer, vacuum filtration methods, etc. and then the resulting samples are thermally, photochemically, or chemically reduced. To achieve high conductivity, high temperature is usually needed because it is difficult to remove all the oxygen atoms and defects are introduced during the reduction.

10.3.2 The Use of Fluorine in Fluoroalkyl Groups for the Modification of Inorganic Materials

Preparation of fluorine-containing inorganic materials by using fluoroalkyl groups as fluorine sources is attracting considerable attention. One method is the sol–gel one using trifluoroacetic acid, which has been developed by Fujihara [34]. Metal acetates are mixed with trifluoroacetic acid and then the solvent is removed. The resulting gel is heated to form metal fluorides. The other one is the reductive fluorination of inorganic oxides using polytetrafluoroethylene or polyvinylidene fluoride, which are proposed by Slater and Gover [35]. Inorganic oxides are mixed with these polymers and then heated, which results in the reduction of inorganic oxides by generated fluorine species. In this method, partially reduced oxyfluorides are obtained. Some active fluorine species would be generated during the thermal decomposition of fluoroalkyl groups such as $-(CF_2)-$ or CF_3-.

10.3.3 Introduction of Perfluoroalkyl Groups Into Graphene Oxide and Preparation of Carbon Thin Film Samples

Several methods have been proposed for the introduction of perfluroalkyl groups into GO, including acid/base, cation exchange, and silylation reactions [36]. In case of silylation, perfluoroalky groups are most strongly bonded with GO layers and they are not removed up to 300°C at which perfluoroalky are decomposed [37], and active fluorine species are expected to form during the thermal reduction of silylated GO thin films. The preparation procedure of carbon thin films containing a small amount of fluorine and oxygen is schematically shown in Fig. 10.12 [38]. Intercalation of *n*-butylamine molecules into GO was first performed and then the resulting sample was dispersed in toluene. The addition of 1H,1H,2H,2H-perfluorooctyltrichlorosilane resulted in the formation of silylated GO. Fig. 10.13 shows the X-ray diffraction patterns of GO before and after silylation. The diffraction peak at $2\theta = 12$ degree observed for GO shifted to lower angle of $2\theta = 5.9$ degree and silylated GO containing perfluoroalkyl groups with an interlayer spacing of 1.5 nm was obtained. The silylating reagents are bonded with the hydroxyl groups of GO via one or two O–Si bonds and C–O–C groups are not reacted, as shown in Fig. 10.13. To prepare thin film samples, it is necessary to exfoliate the silylated GO in appropriate solvents.

graphite

↓ Oxidation

GO

↓ n-butylamine
Toluene
Silylating reagents

Silylated GO

↓ Intercalation of n-hexadecylamine
Cyclohexane/chloroform

Dispersion of n-hexadecylamine-
intercalated silylated GO

↓ Cast on glass substrate

n-hexadecylamine-intercalated
silylated GO thin film

↓ Ethanol (Removal of n-hexadecylamine)

Silylated GO thin film

↓ Δ

Carbon thin film

Figure 10.12 Schematic illustration of the preparation procedure of carbon thin films containing a small amount of fluorine and oxygen from GO silylated with 1H,1H,2H,2H-perfluorooctyltrichlorosilane. GO, graphene oxide.

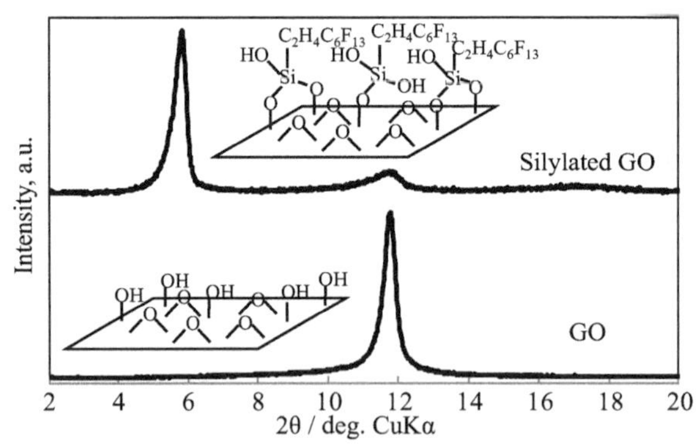

Figure 10.13 X-ray diffraction patterns of GO (A) before and (B) after silylated with 1H,1H,2H,2H-perfluorooctyltrichlorosilane. GO, graphene oxide.
Reproduced from Y. Matsuo, K. Iwasa, Y. Sugie, H. Usami, M. Kawaguchi, J. Fluorine Chem. 132 (2011) 669–672 with permission from Elsevier.

Exfoliation of the resulting silylated GO was realized by introducing *n*-hexadecylamine into it. The obtained sample well dispersed in chloroform/cyclohexane solution. Fig. 10.14 shows the AFM image of the sample obtained by casting the nanosheet solution of silylated GO on mica. The *n*-hexadecylamine molecules added for the

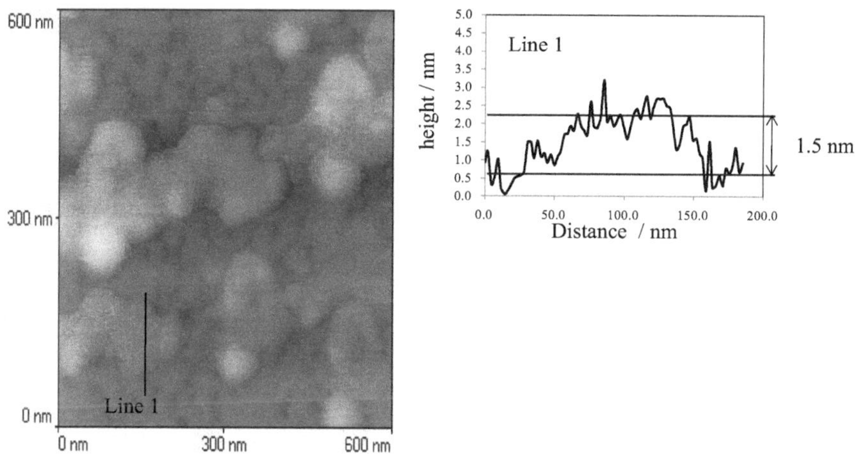

Figure 10.14 Atomic force microscopy image and height profiles of silylated graphene oxide nanosheets deposited on mica substrate.
Reproduced from Y. Matsuo, K. Iwasa, Y. Sugie, H. Usami, M. Kawaguchi, J. Fluorine Chem. 132 (2011) 669—672 with permission from Elsevier.

exfoliation were easily removed by washing with ethanol. The height of some of the particles was 1.5 nm, indicating that silylated GO was exfoliated into single sheets. Thin film samples with various thicknesses were obtained by casting the dispersion with various concentrations onto glass substrates. By heating the silylated GO thin films above 300°C under vacuum, carbon thin films were obtained.

10.3.4 Properties of Transparent and Conducting Electrodes From Silylated Graphene Oxide Containing Perfluoroalkyl Groups

Fig. 10.15 shows the sheet resistance of the carbon film prepared from silylated GO containing perfluoroalkyl groups as a function of the transmittance at 500 nm, together with that of the films prepared from GO silylated with octyltrichlorosilane in the same manner. In both cases, the resistance decreased with the decrease in the transparency. This behavior is commonly observed for carbons obtained from GO and this is because the resistance at the grain boundary became lower as the number of stacked carbon layers. The resistance of the former was always lower at the same transparency. Fig. 10.16 shows the XPS data of these samples in the O1s region. The peak intensity normalized by that of C1s peak area was considerably lower for the sample prepared from silylated GO containing perfluoroalkyl groups. This means that in the presence of perfluoroalkyl groups, GO was more effectively reduced. In the F1s region, a small peak at 686 eV was observed and Tauc's plot [39] made for the UV—vis spectra of these samples showed that more defects were introduced for this sample. These are ascribed to the active fluorine species generated by the decomposition of perfluoroalkyl groups. Higher degree of reduction is responsible for the lower sheet resistance.

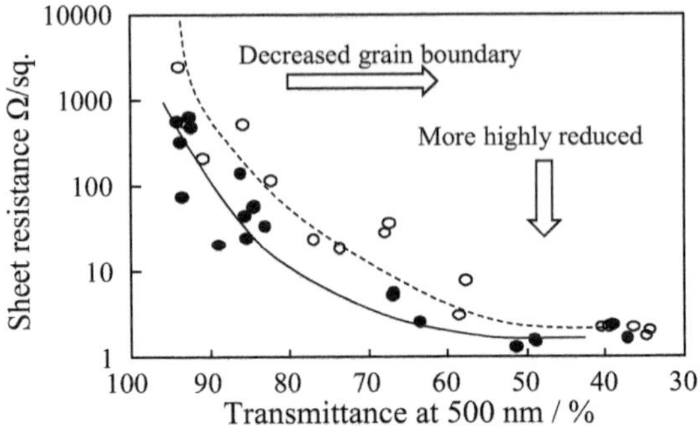

Figure 10.15 Variation of the sheet resistance of carbon films prepared from graphene oxide silylated with ○: 1H,1H,2H,2H-perfluorooctyltrichlorosilane and ●: octyltrichlorosilane as a function of their transmittance.
Reproduced from Y. Matsuo, K. Iwasa, Y. Sugie, H. Usami, M. Kawaguchi, J. Fluorine Chem. 132 (2011) 669−672 with permission from Elsevier.

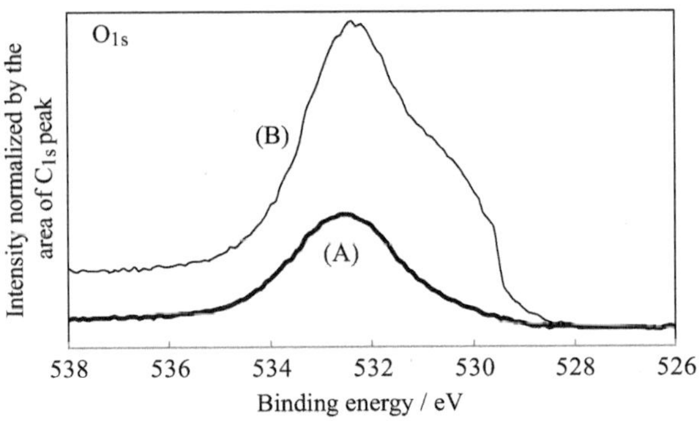

Figure 10.16 X-ray photoelectron spectroscopy spectra of carbon films prepared from graphene oxide silylated with (A) 1H,1H,2H,2H-perfluorooctyltrichlorosilane and (B) octyltrichlorosilane in O_{1s} region.
Reproduced from Y. Matsuo, K. Iwasa, Y. Sugie, H. Usami, M. Kawaguchi, J. Fluorine Chem. 132 (2011) 669−672 with permission from Elsevier.

References

[1] N. Watanabe, T. Nakajima, H. Touhara, Graphite Fluorides, Elsevier, Amsterdam, 1988.
[2] T. Nakajima, N. Watanabe, Graphite Fluorides and Carbon-fluorine Compounds, CRC Press, Boca Raton, 1991.
[3] T. Nakajima, Fluorinated carbon materials for energy conversion, in: T. Nakajima, B. Žemva, A. Tressaud (Eds.), Advanced Inorganic Fluorides-synthesis, Characterization and Applications-, Elsevier, Amsterdam, 2000, pp. 493−520.
[4] T. Nakajima, Synthesis, structure and physicochemical properties of fluorine-graphite intercalation compounds, in: T. Nakajima (Ed.), Fluorine-Carbon and Fluoride-Carbon Materials-Chemistry, Physics and Applications, Marcel Dekker, New York, 1995, pp. 1−32.

[5] J. Giraudet, M. Dubois, K. Gelin, C. Delabarre, A. Hamwi, F. Masin, J. Phys. Chem. B 111 (2007) 14143–14151.
[6] Y. Sato, K. Itoh, R. Hagiwara, T. Fukunaga, Y. Ito, Carbon 42 (2004) 3243–3249.
[7] W. Scholz, H.P. Boehm, Z. Anorg, Allg. Chem. 369 (1969) 327–340.
[8] T. Szabó, O. Berkesi, P. Forgó, K. Josepovits, Y. Sanakis, D. Petridis, I. Dékány, Chem. Mater. 18 (2006) 2740–2749.
[9] T. Nakajima, V. Gupta, Y. Ohzawa, H. Groult, Z. Mazej, B. Žemva, J. Power Sources 137 (2004) 80–87.
[10] H. Takenaka, M. Kawaguchi, M. Lerner, N. Bartlett, J. Chem. Soc. Chem. Commun. (1987) 1431–1432.
[11] T. Tojo, J. Hiraiwa, M. Doi, Y.-B. Chong, J. Fluorine Chem. 57 (1992) 93–99.
[12] T. Nakajima, M. Molinier, M. Motoyama, Carbon 29 (1991) 429–437.
[13] N. Ikemiya, S. Hara, K. Ogino, T. Nakajima, Surf. Sci. Lett. 274 (1992) 524–528.
[14] Y. Matsuo, M. Segawa, Y. Sugie, J. Fluorine Chem. 87 (1998) 145–150.
[15] Y. Matsuo, S. Hirata, M. Dubois, J. Fluorine Chem. 185 (2016) 36–41.
[16] Y. Matsuo, T. Nakajima, M. Motoyama, TANSO 160, 1993, pp. 272–278 (in Japanese).
[17] J. Reau, C. Lucat, J. Portier, P. Hagenmuller, L. Cot, S. Vilminot, Mat. Res. Bull. 13 (1978) 877–882.
[18] V.N. Bezmelnitsyna, A.V. Bezmelnitsyna, A.A. Kolmakov, J. Fluorine Chem. 77 (1996) 9–12.
[19] Y. Matsuo, T. Kakiya, Y. Sugie, J. Fluorine Chem. 101 (2000) 257–261.
[20] N. Watanabe, T. Nakajima, R. Hagiwara, J. Power Sources 20 (1987) 87–92.
[21] M.A. Reddy, B. Breitung, M. Fichtner, ACS Appl. Mater. Interfaces 5 (2013) 11207–11211.
[22] T. Nakajima, Y. Matsuo, Y. Morino, Denki Kagaku 61 (1993) 594–599 (in Japanese).
[23] S. Watcharotone, D.A. Dikin, S. Stankovich, R. Oiner, I. Jung, G.H.B. Dommett, G. Evmenenko, S.-E. Wu, S.-F. Chen, C.-P. Liu, S.T. Nguyen, R.S. Ruoff, Nano Lett. 7 (2007) 1888–1892.
[24] C. Gomez-Navarro, R. Weitz, R.T. Bittner, M. Scolari, A. Mews, M. Burghard, K. Kern, Nano Lett. 7 (2007) 3499–3503.
[25] S. Gilje, S. Han, M. Wang, R.B. Wang, K.L. Kaner, Nano Lett. 7 (2007) 3349–3353.
[26] X. Wang, L. Zhi, C. Müllen, Nano Lett. 8 (2008) 323–327.
[27] H. Becerril, J. Mao, Z. Liu, R.M. Stoltenberg, Z. Bao, Y. Chen, ASC Nano 2 (2008) 463–470.
[28] G. Eda, G. Fanchini, M. Chhowalla, Nat. Nanotech. 3 (2008) 270–274.
[29] G. Eda, C. Mattevi, H. Yamaguchi, H.-K. Kim, M. Chhowalla, J. Phys. Chem. C 131 (2009) 15768–15771.
[30] D. Li, M.B. Muller, S. Gilje, R.B. Kaner, G.G. Wallace, Nat. Nanotech. 3 (2008) 101–106.
[31] L.J. Cote, F. Kim, J. Huang, J. Am. Chem. Soc. 131 (2009) 1043–1049.
[32] C. Mattevi, G. Eda, S. Agnoli, S. Miller, K.A. Mkhyan, O. Celik, D. Mastrogiovanni, G. Granozzi, E. Garfunkel, M. Chhowalla, Adv. Funct. Mater. 19 (2009) 2577–2583.
[33] Z. Luo, Y. Lu, L.A. Somers, A.T.C. Johnson, J. Am. Chem. Soc. 131 (2009) 898–899.
[34] S. Fujihara, Sol-gel route to inorganic fluoride nanomaterials with optical properties, in: A. Tressaud (Ed.), Functinalized Inorganic Fluorides, Wiley & Sons, Chichester, 2010, pp. 307–326.
[35] P.R. Slater, R.K.B. Gover, J. Mater. Chem. 12 (2002) 291–294.
[36] Y. Matsuo, Hybrid intercalation compounds containing perfluoroalkyl groups, in: A. Tressaud (Ed.), Functionalized Inorganic Fluorides, Wiley & Sons, Chichester, 2010, pp. 469–488.

[37] U. Wanener, C. Rüssel, J. Non-Cryst. Solids 152 (1993) 167−171.
[38] Y. Matsuo, K. Iwasa, Y. Sugie, H. Usami, M. Kawaguchi, J. Fluorine Chem. 132 (2011) 669−672.
[39] J. Tauc, R. Grigorovici, A. Vancu, Phys. Status Solidi 15 (1966) 627−637.

Intercalation Chemistry and Application of B/C/N Materials to Secondary Batteries

11

M. Kawaguchi, K. Yamada, H. Ishikawa
Osaka Electro-Communication University, Neyagawa, Osaka, Japan

Chapter Outline

New Fluorinated Carbons: Fundamentals and Applications. http://dx.doi.org/10.1016/B978-0-12-803479-8.00011-5

11.1 Introduction

Boron/carbon/nitrogen (B/C/N) materials based on the graphite network are expected to expand the application of carbon-based materials to wider fields [1,2]. B/C/N materials are also classified to heteroatom-substituted carbon alloys among several kinds of carbon alloys [3,4]. Potential applications of these materials having the layered structures are the energy storage systems, such as secondary batteries [5,6] and capacitors [7], and energy conversion systems, such as photocatalysts [8] and fuel cells [9].

Intercalation behaviors have been already used for the secondary batteries, such as anodes of lithium (Li)-ion batteries. The chemical intercalation of Li into graphite was discovered in the 1950s [10]. Then, the electrochemically rechargeable intercalation/deintercalation of Li into/out of host materials was demonstrated [11], which has been practically used as Li-ion batteries (LIBs).

Recently, secondary batteries in the next generation, such as sodium (Na) ion [12] and magnesium (Mg) batteries [13], have been proposed. There are several merits to use electrochemical systems of Na and Mg. Nakanishi has suggested that these systems include not only the low cost and environmental benefits but also the safety on the cathode which has been already investigated [14]. Concerning the anodes, however, these metals cannot be easily intercalated into graphite. Present authors have recently reported that Na and Mg can be intercalated into B/C/N materials by using the electrochemical method [15] and the two-bulb method [16], respectively. These results suggest that reductive intercalation occurs for B/C/N materials easier than for graphite, and might be applied to the anodes of secondary batteries in the next generation, such as Na-ion batteries (SIBs) [17]. On the other hand, oxidative intercalation of anion, such as PF_6^-, into graphite has been investigated [18]. Both reductive and oxidative intercalation into graphite can be applied to the dual graphite batteries [18,19]. From the results described above, B/C/N materials are also possible candidates of anodes of dual batteries.

In this chapter, we summarize the reductive intercalation of metals into B/C/N materials and the application of the intercalation to the anodes of Li- and SIBs in a various kinds of electrolyte solutions including fluoride-related electrolytes. Results of modification of B/C/N materials by using fluorine-based oxidative reagents, such as hydrofluoric acid, and a combination of B/C/N materials and graphite as electrodes of dual carbon alloy batteries have been included in this chapter. Intercalation of metals into boron/carbon (B/C) materials based on the graphite network has also been discussed.

11.2 Preparation of Boron/Carbon/Nitrogen and Boron/Carbon Materials

B/C/N materials were first reported by Kosolapova et al. [20]. They prepared the materials by using a solid–gas reaction of a mixture of carbon and boron powders under a nitrogen or ammonia atmosphere. Since then, the chemical vapor deposition (CVD) method, solid–gas reaction, and precursor pyrolysis method have been used for the preparation of B/C/N, B/C, and carbon/nitrogen (C/N) materials. Particularly, the

CVD method is one of the more effective methods to prepare B/C/N, B/C, and C/N materials because the materials are in a metastable state and the CVD method can create the metastable state during the preparation of the materials. Since graphite and hexagonal boron nitride (h-BN) are thermodynamically stable, B/C/N materials tend to transform to graphite and h-BN when they are heated at temperatures higher than 2270 K.

Materials with compositions of BC_2N [21,22], BC_3N [23], BCN [23], BC_6N [6], BC_3 [24], BC_8 [17], C_3N_4-type [25], C_xN [26], and others have been prepared by using the CVD method. Fig. 11.1 shows a distribution map of compositions of B/C/N, B/C, and C/N materials prepared so far [27]. There are areas of compositions near stoichiometric ones of BCN, BC_2N, BC_3N, BC_4N, BC_6N, etc. These materials with nonstoichiometric compositions have been usually prepared. For example, materials with compositions of BC_xN_y, where $x = 2.0-2.4$ with $y = 0.8-0.9$ [15] and $x = 6.1$ with $y = 0.91$ [6] have been prepared by the CVD method. These compositions of materials were analyzed by the usual combustion method for carbon and nitrogen, and by dissolution of the material in an acid solution followed by the emission spectroscopy using induction coupled plasma for boron. The material with the composition $BC_{2.0\sim2.4}N_{0.8\sim0.9}$ is called BC_2N in this chapter. Fig. 11.2 shows a possible atomic arrangement in the monolayer of BC_2N [28]. Boron, carbon, and nitrogen are in the same layer in which boron and nitrogen connect to each other to make strong B–N bonds. Based on the polarization rule proposed by Shimoyama [29], B/C/N materials are expected to have semiconducting properties. In fact, prepared B/C/N materials showed semiconducting properties [23,30–32]. As is indicated in Fig. 11.1, boron

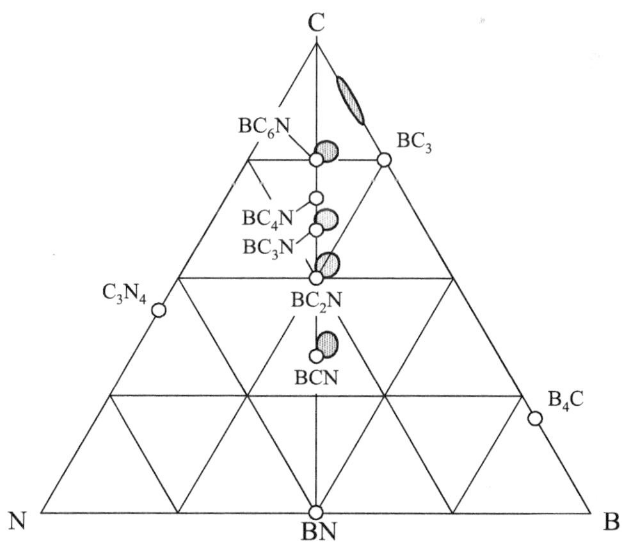

Figure 11.1 Distribution map of compositions of boron/carbon/nitrogen, boron/carbon and carbon/nitrogen materials.
Reproduced with permission from M. Kawaguchi, TANSO No267 (2015) 84–93. Copyright 2015, The Carbon Society of Japan.

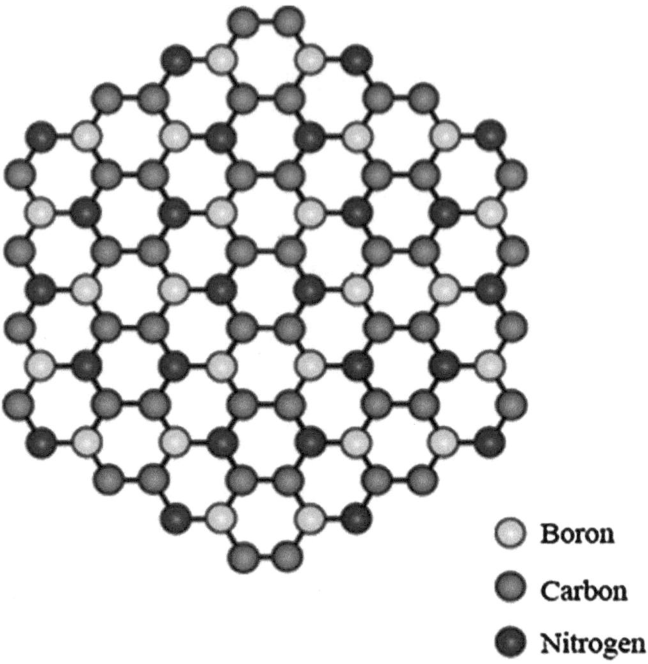

Figure 11.2 A possible atomic arrangement in the monolayer of BC_2N [28].

contents are usually larger than stoichiometric compositions, probably because boron atoms themselves can be alloyed in the graphene layer and act as the crystallization reagent [17,33,34], while nitrogen atoms themselves cannot be easily alloyed in the graphene layer. By using the CVD method, materials having different compositions can be made by modifying reaction conditions, such as the molar ratio of starting materials and the reaction temperature.

Fig. 11.3 indicates X-ray powder diffraction patterns of the materials with the compositions $BC_{4.7}N_{0.75}$, prepared at 1770 K; $BC_{7.9}$, prepared at 1170 K; $BC_{6.9}$, prepared at 1770 K; $C_{54}N$, prepared at 1770 K; and carbon, prepared at 1770 K. $BC_{4.7}N_{0.75}$ was prepared by a CVD reaction using boron trichloride (BCl_3) and acetonitrile (CH_3CN) as starting materials with a molar ratio BCl_3:CH_3CN = 1:1 at a temperature of 1770 K. When the reaction condition with a molar ratio BCl_3:CH_3CN = 2:1 at the same temperature 1770 K was used, the material with a composition $BC_{2.2}N_{0.76}$ was obtained [17]. The materials with compositions of $BC_{7.9}$ and $BC_{6.9}$ were prepared from the CVD reaction of BCl_3 and ethylene [17]. $BC_{7.9}$ could be a single phase with a graphite-like layered structure, while $BC_{6.9}$ was a mixture of BC_x having the graphite-like layered structure and B_4C as a by-product. The materials with compositions of $BC_{4.7}N_{0.75}$ and $BC_{7.9}$ are called BC_5N and BC_8, respectively, in this chapter. Sharp 002 diffraction lines of BC_8 and BC_5N in comparison with those of $C_{54}N$ and carbon suggest that boron atoms in the graphene layer have a role of crystallization of materials [17,33,34].

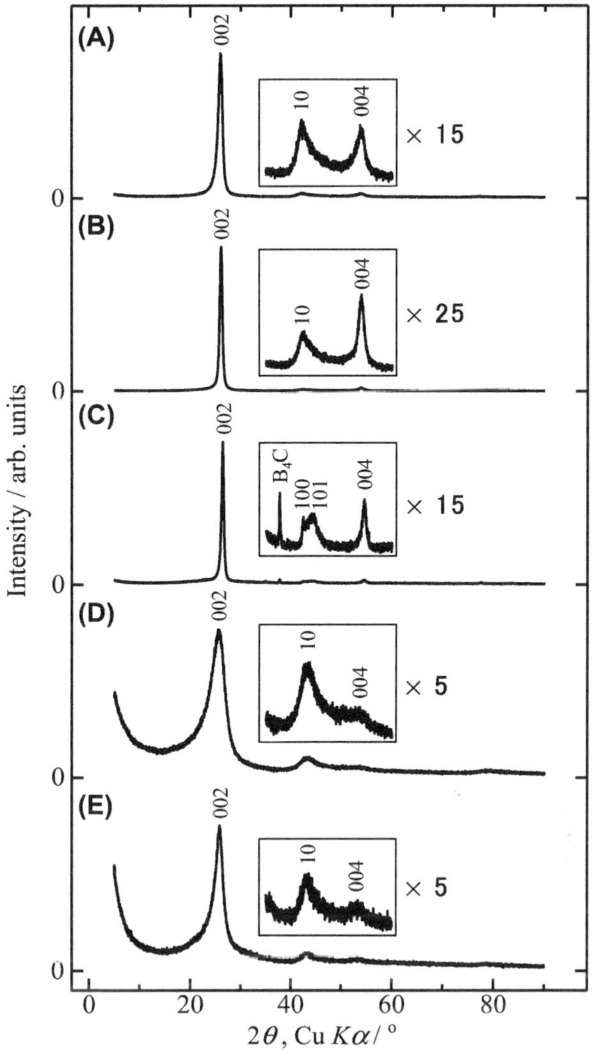

Figure 11.3 X-ray diffraction patterns of (A) boron/carbon/nitrogen powder prepared at 1770 K, (B) boron/carbon (B/C) powder prepared at 1170 K, (C) B/C powder prepared at 1770 K, (D) carbon/nitrogen powder prepared at 1770 K, and (E) carbon powder prepared at 1770 K used in this study as original host materials.

11.3 Intercalation of Li Into Boron/Carbon/Nitrogen and Boron/Carbon Materials and Its Application to Anode of Li-Ion Batteries

LIBs have been increasingly produced, having been used for mobile phones, PCs, and electric vehicles. Li-intercalated graphite (Li-GIC) is used as the anode of LIBs, while

Li itself cannot be used as the anode because of the formation of whiskers during discharge/charge cycles. Li can be smoothly intercalated into graphite to form stage structures, making it possible for LIBs to work safely. The composition of fully intercalated Li-GIC is LiC_6 [35], which corresponds to the reversible capacity of 372 mAh/g. It would be useful if a host material could be intercalated with Li to make an intercalation compound whose reversible capacity is more than 372 mAh/g. In this section, the intercalation of Li into B/C/N and B/C materials by using the electrochemical method is summarized.

11.3.1 Electrochemical Intercalation of Li Into Boron/Carbon/Nitrogen Materials

Fig. 11.4 shows the electrochemical discharge/charge curves of BC_5N, which were prepared at 1770 K (Fig. 11.3A), by the galvanostatic method. The electrolyte solution was 1M-$LiPF_6$ dissolved in ethylene carbonate (EC) and diethyl carbonate (DEC) with a volume ratio of 1:1, which is indicated to be 1 M-$LiPF_6$/(EC + DEC) in this chapter. The terms discharge and charge mean the electrochemical intercalation of Li into B/C/N materials and deintercalation of Li out of the materials, respectively. The intercalation of Li into BC_5N began at a potential higher than that of graphite, probably because of the role of boron described in Section 11.6.2. The solid electrolyte interface (so-called SEI [36]) formed at around 0.7 V only on the first discharge, which was similar to the case of graphite. Reversible discharge/charge capacity of BC_5N during the second through fifth cycles was c.330 mAh/g, which was also similar to the value in the case of graphite by using the same experimental condition.

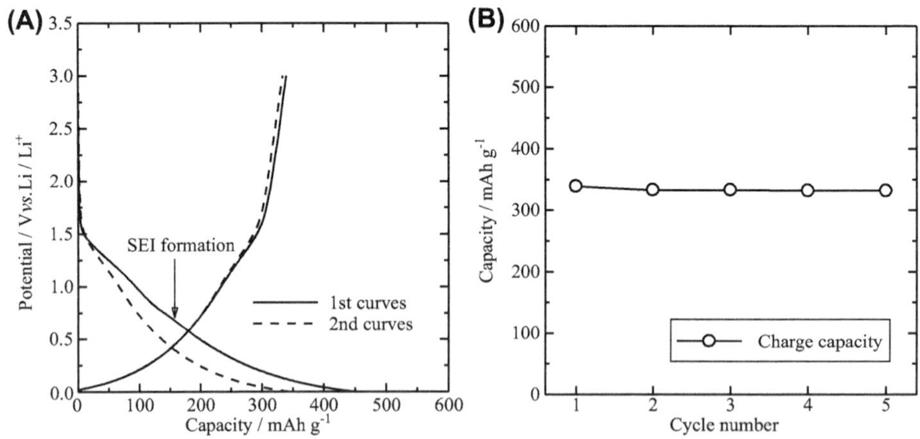

Figure 11.4 Discharge/charge curves (A), and charge (deintercalation) capacities (B) of BC_5N by the galvanostatic method using a current density of 100 $\mu A/cm^2$ in 1 M-$LiPF_6$/(EC + DEC) electrolyte solution. *DEC*, diethyl carbonate; *EC*, ethylene carbonate; *SEI*, solid electrolyte interface.

11.3.2 Electrochemical Intercalation of Li Into Boron/Carbon/Nitrogen Material Treated With Hydrofluoric Acid

To check the influence of boron content in B/C/N material on the electrochemical properties, we treated $BC_{4.7}N_{0.75}$ (so-called BC_5N) with hydrofluoric acid (50%) at 300 K for 24 h. After the treatment, the composition $BC_{4.7}N_{0.75}$ changed to a composition of $BC_{5.0}N_{0.77}$, which suggests that a part of boron atoms were eliminated by the treatment with hydrofluoric acid as follows:

$$BC_xN_y + 3\alpha HF \rightarrow B_{1-\alpha}C_xN_y + \alpha BF_3 + (3\alpha/2)H_2$$

In fact, gas evolution was observed during the treatment. The material with the composition $BC_{5.0}N_{0.77}$, which was obtained by the hydrofluoric treatment, is called HF-treated BC_5N in this chapter. X-ray diffraction (XRD) pattern of HF-treated BC_5N was almost the same as the untreated BC_5N (Fig. 11.3A), which means that the crystal structure did not change after the treatment with hydrofluoric acid.

Fig. 11.5A and B shows discharge/charge curves and charge (de-intercalation) capacities with cycle numbers, respectively, of HF-treated BC_5N by the galvanostatic method. Charge capacities (363–369 mAh/g) were slightly larger than that of untreated BC_5N (332–339 mAh/g). This might be explained by the following: holes in the graphene layer could be made by the elimination of boron through the treatment with hydrofluoric acid, which could make the insertion of Li through the holes on the basal plane of HF-treated BC_5N layer. This treatment of using hydrofluoric acid can be one of the useful methods to modify the layered structures of B/C/N and B/C materials.

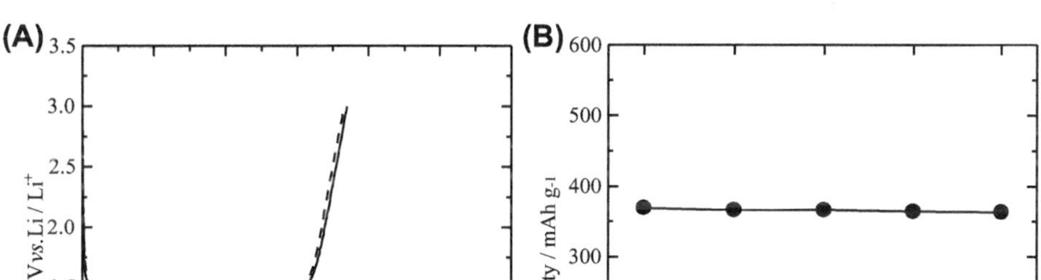

Figure 11.5 Discharge/charge curves (A), and charge (deintercalation) capacities (B) of HF-treated BC_5N by the galvanostatic method using a current density of 100 $\mu A/cm^2$ in 1 M-$LiPF_6$/(EC + DEC) electrolyte solution. *DEC*, diethyl carbonate; *EC*, ethylene carbonate.

11.3.3 Electrochemical Intercalation of Li Into Boron/Carbon Materials

Electrochemical intercalation of Li into B/C materials has been studied [37–39]. Way and Dahn reported the reversible capacity of 437 mAh/g for $B_{0.17}C_{0.83}$ ($BC_{4.9}$), which was prepared by CVD reaction of BCl_3 with benzene at 1170 K [37]. Shirasaki et al. reported that the reversible capacities depended on the boron contents of B/C materials prepared by CVD reaction of BCl_3 with acetylene at 1410 K (Fig. 11.6) [38]. The result indicated the B/C material containing 13 at% boron ($BC_{6.7}$) showed the highest reversible capacity.

Fig. 11.7 shows the electrochemical discharge/charge curves of BC_8, which was prepared at 1170 K (Fig. 11.3B), by the galvanostatic method in 1M-$LiPF_6$/(EC + DEC) electrolyte solution [39]. The intercalation of Li into BC_8 began at a potential higher than that of graphite, which was similar to the case of Li intercalation into BC_5N and has been described in Section 11.3.1. In this case, however, the reversible capacity was 450 mAh/g which is 1.2 times larger than that of graphite (372 mAh/g).

The results that intercalation of Li into BC_5N and BC_8 occurs at high potentials could be due to the role of boron in the materials. There are two roles of boron in the materials [27]. One is that boron plays a role in the crystallization of the materials with a graphite-like structure, which has been described in Section 11.2. The other is that boron modifies the electronic structure of materials. The electron deficiency of boron atom in B/C/N material lowers the energy level at the bottom of the conduction band [15] in comparison with that of graphite. In other words, B/C/N materials have the electron affinity larger than that of graphite. This might be the reason why Li,

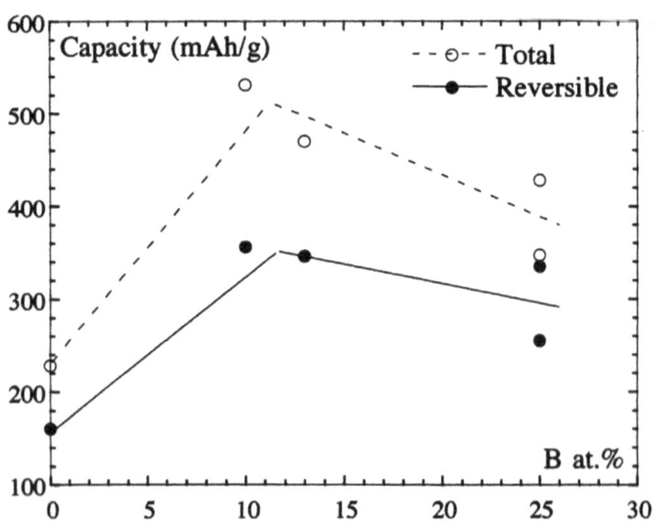

Figure 11.6 The capacities of Li–carbon cells as a function of the boron content. Reproduced with permission from T. Shirasaki, A. Derré, K. Guérin, S. Flandrois, Carbon 37 (1999) 1961–1964. Copyright 1999 Elsevier.

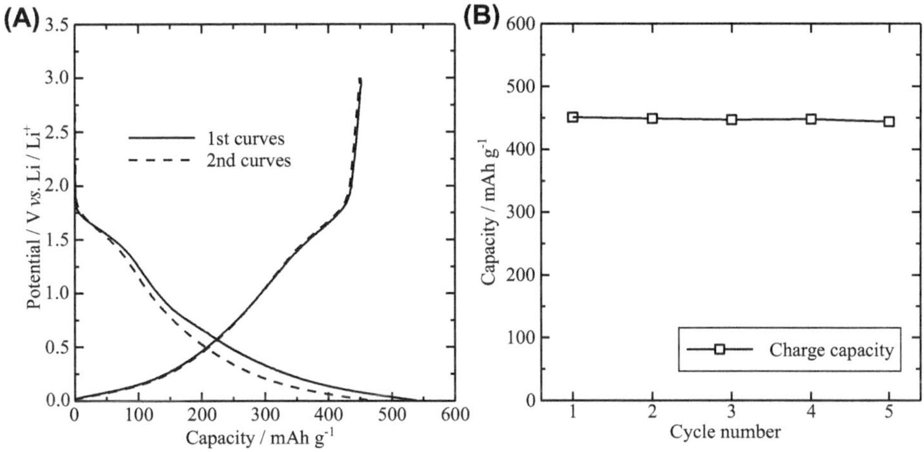

Figure 11.7 Discharge/charge curves (A), and charge (deintercalation) capacities (B) of BC_8 by the galvanostatic method using a current density of 100 μA/cm^2 in 1 M-$LiPF_6$/(EC + DEC) electrolyte solution. *DEC*, diethyl carbonate; *EC*, ethylene carbonate.

having large ionization potential, can be intercalated into B/C/N and B/C materials easier than the cases of graphite; in other words, Li can be intercalated into B/C/N and B/C materials at higher potentials than the case of graphite.

11.4 Intercalation of Na and Mg Into Boron/Carbon/Nitrogen Materials

Since Li, Na, and Mg are less electropositive metals among the group 1 and 2 metals, they are not easily intercalated into graphite through a chemical reaction. However, Li can be intercalated into graphite by using the electrochemical method. Na and Mg cannot be easily intercalated into graphite through the electrochemical method as well as the chemical reaction. It would be interesting to investigate the intercalation of these metals into host materials. In this section, the intercalation of Na and Mg into B/C/N materials is summarized.

11.4.1 Intercalation of Na Into BC_2N by Chemical Reaction

Since Na cannot easily be intercalated into graphite, only a mixture of sixth and seventh stage compounds [40] or an eighth stage compound [41] was made. By using BC_2N as a host material, Na can be intercalated into BC_2N through the chemical reaction by the two-bulb method to form the first stage compound whose *d*-spacing was 430 pm [42]. On the other hand, no intercalation of Na into graphite was observed under the same reaction condition. Therefore, Na can be intercalated into BC_2N more effectively than graphite. This would be due to a strong interaction between the guest Na and the host BC_2N.

11.4.2 Intercalation of Mg Into BC_2N by Chemical Reaction

Although Mg has never been intercalated into graphite, it has been intercalated into BC_2N through the chemical reaction to form a second stage compound [16]. The intercalation reaction was carried out by the vapor-phase reaction of Mg with BC_2N at 920 K. Based on the XRD and electron spectroscopy for chemical analysis (ESCA) study, the intercalation compound Mg-BC_2N obtained above was estimated to be the second stage compound whose distance of Mg-intercalated layer could be 365 pm, assuming that the diameter of the Mg^{2+} ion is similar to that of the Li^+ ion. This is the first example in which Mg was intercalated into a material based on the graphite network. On the other hand, almost no intercalation of Mg into graphite was observed under the same reaction condition. Therefore, Mg can be intercalated into BC_2N more effectively than graphite.

11.5 Intercalation Mechanism of Metals Into Boron/Carbon/Nitrogen Materials

Na and Mg can be intercalated into BC_2N by the chemical reaction as described in Section 11.4, while these metals cannot be easily intercalated into graphite. It would be important to consider an intercalation mechanism by which the intercalation of Na and Mg into BC_2N could be explained. In this section, we consider the intercalation mechanism based on the relationship between the electronic structures of B/C/N materials and the ionization potentials of metals.

11.5.1 Electronic Structure of Boron/Carbon/Nitrogen Materials

X-ray absorption (XAS) and X-ray emission spectroscopies are useful tools to investigate electronic states of materials. Unoccupied orbitals (conduction bands) of materials can be observed on XAS spectrum, particularly on X-ray absorption near-edge structure (XANES). Fig. 11.8 shows XANES in the *CK* region of BC_2N, BC_6N, noncrystalline carbon, and graphite. Unoccupied orbital is constituted by π^* and σ^* peaks of the layered materials having an sp^2 hybridized structure [15]. The lower energy part of π^* peaks, which includes the bottom of the conduction band of the material, is enlarged in Fig. 11.8B. The bottom of the conduction band of BC_2N was at a lower energy than graphite by about 2 eV. The difference between the vacuum level and the bottom of the conduction band is the electron affinity of the material. Therefore, an electron affinity of BC_2N is 2 eV larger than that of graphite. Since the electron affinity of graphite is 4.6 eV [43], the electron affinity of BC_2N should be 6.6 eV.

11.5.2 Relation Between Electronic Structures of Host Materials and Ionization Potentials of Metals

When an intercalation reaction, for example, of an alkali metal with a host material occurs, the alkali metal supplies an electron to the host material (Fig. 11.9) [15]. In this case, it is important to know the relationship between the ionization potential

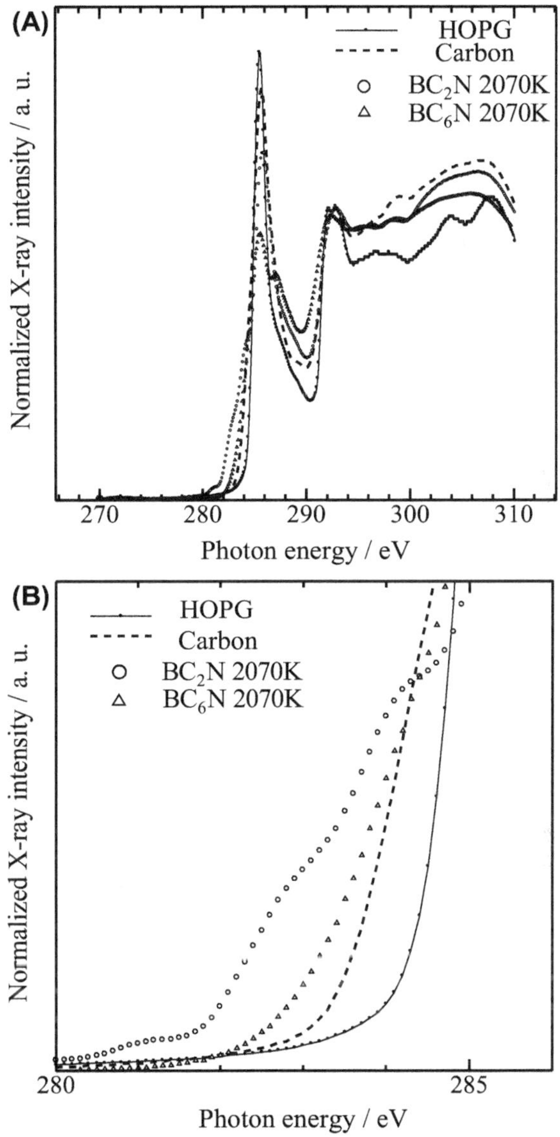

Figure 11.8 X-ray absorption near-edge structure in the C*K* region of BC_2N prepared at 2070 K, compared with those of highly oriented pyrolytic graphite (HOPG), noncrystalline carbon, and BC_6N prepared at 2070 K. Incident angle: 45°. (A) Whole spectra, (B) spectra enlarged of the lower energy part in (A).

Reproduced with permission from M. Kawaguchi, K. Ohnishi, K. Yamada, Y. Muramatsu, J. Electrochem. Soc. 157 (2010) P13–P17. Copyright 2010, The Electrochemical Society.

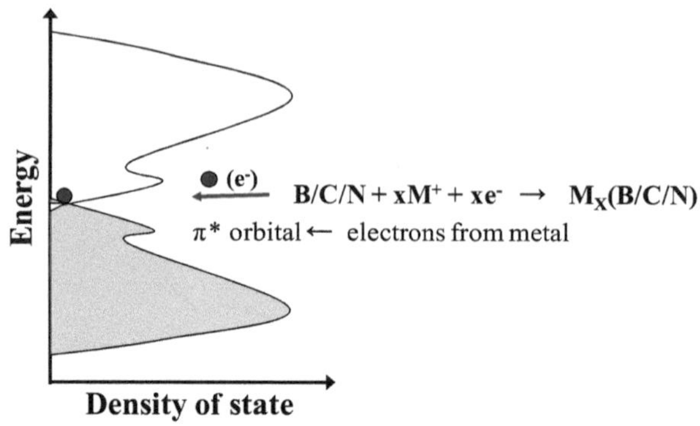

Figure 11.9 A schematic relationship between intercalation of alkali metal (M) and an electronic structure of host material (boron/carbon/nitrogen).
Reproduced with permission from M. Kawaguchi, K. Ohnishi, K. Yamada, Y. Muramatsu, J. Electrochem. Soc. 157 (2010) P13–P17. Copyright 2010, The Electrochemical Society.

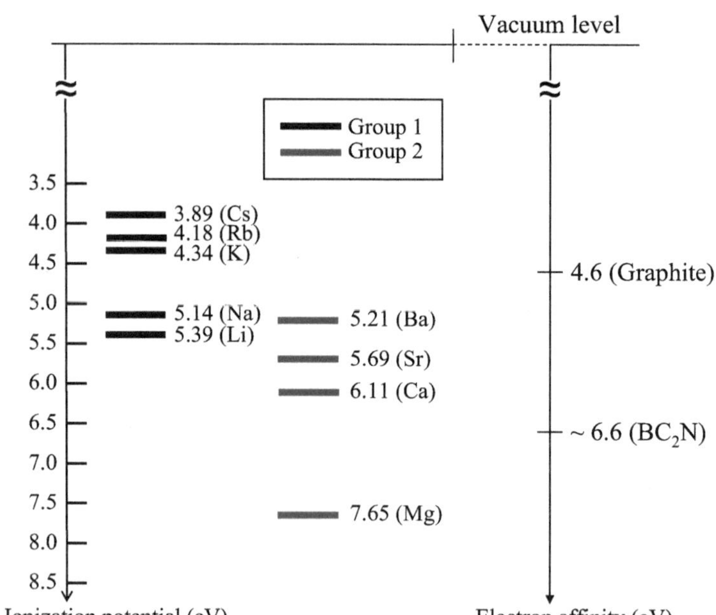

Figure 11.10 Relationship between ionization potentials of group 1 and 2 metals and the electron affinities of the host materials.
Reproduced with permission from M. Kawaguchi, A. Kurasaki, Chem. Commun. 48 (2012) 6865–6988. Copyright 2012, Royal Society of Chemistry.

of the metals and the electron affinities of the host materials (Fig. 11.10) [16]. The heavier alkali metals, potassium (K), rubidium (Rb), and cesium (Cs) can be easily intercalated into graphite. In these cases, the ionization potentials of K (4.34 eV), Rb (4.18 eV), and Cs (3.89 eV) lie below the electron affinity of graphite (4.6 eV) [43].

On the other hand, Li is not easily intercalated into graphite through a vapor-phase reaction, but can be intercalated by electrochemical methods, and Na is rarely intercalated into graphite [40,41], probably because the ionization potentials of Li (5.39 eV) and Na (5.14 eV) lie above the electron affinity of graphite. The alkaline earth metals, calcium (Ca), strontium (Sr), and barium (Ba), are also less electropositive with ionization potentials of 6.11, 5.69, and 5.21 eV, respectively. These are intercalated into graphite in a manner similar to that of Li. However, Mg has not been intercalated into graphite, probably because of its high ionization potential (7.65 eV).

As described in Section 11.5.1, the bottom of the conduction band of BC_2N was observed at a lower energy than that of graphite according to the results of XANES. The difference between the bottoms of the BC_2N and graphite conduction bands was about 2 eV. The electron affinity of BC_2N may therefore be stronger than that of graphite (4.6 eV). Probably because of the strong electron affinity of BC_2N compared to the less electropositive Na and Mg, these metals could be intercalated into BC_2N to form lower-stage intercalation compounds. Particularly in the case of Mg, the intercalated divalent ion Mg^{2+} could make a strong lattice energy in the intercalation compound with a charge transfer [27].

11.6 Intercalation of Na Into Boron/Carbon/Nitrogen and Boron/Carbon Materials and Its Application to Anode of Na-Ion Batteries

Na and Mg ion batteries are considered to be the best candidates for secondary batteries in the next generation [44], because Na and Mg are rich on earth and adequate as anodes having low redox potentials. In addition, Mg is a divalent metal so that batteries using Mg would have higher volumetric energy densities. However, unfortunately, the electrochemical intercalation of Mg into a host material for anodes has not been reported yet.

In the case of SIBs, anode materials showing higher potentials than Na metal is needed because of the high reactivity of Na metal deposited on the anode. Fig. 11.11 shows the comparison between new SIBs and present LIBs (LIB) proposed by Nakanishi et al. [14]. Fortunately, cathode materials which show high voltages have been already developed. As described in Section 11.4.1, Na can be intercalated into B/C/N materials. It would be useful to study the intercalation of Na into B/C/N materials by using electrochemical method. In this section, the electrochemical intercalation of Na into B/C/N and B/C materials is summarized and its application to the anode of SIBs is discussed.

11.6.1 Electrochemical Intercalation of Na Into Boron/Carbon/Nitrogen and Boron/Carbon Materials

On the basis of the successful intercalation of Na intercalation into BC_2N, an attempt to intercalate Na into B/C/N materials through the electrochemical method was made.

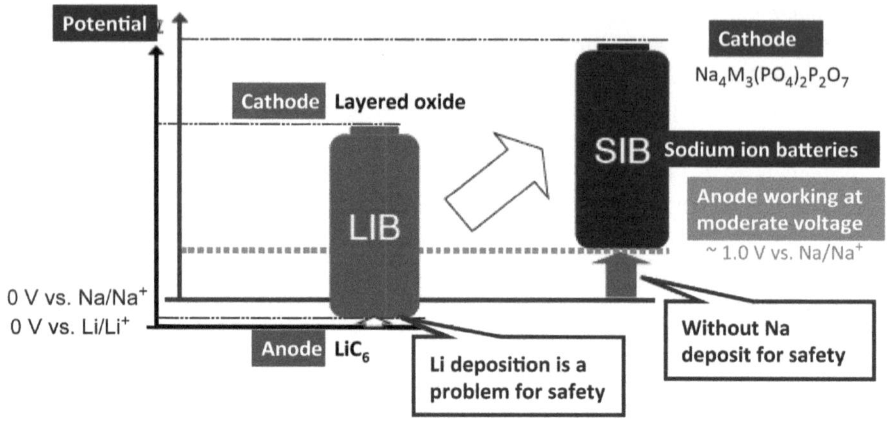

Figure 11.11 Comparison between new Na-ion batteries (SIBs) and present Li-ion batteries (LIBs). The article was originally written in Japanese.
Reproduced with permission from S. Nakanishi, M. Nose, H. Nakayama, Electrochemistry 83 (2015) 182−187 (in Japanese). Copyright 2015, The Electrochemical Society of Japan.

Fig. 11.12 shows discharge/charge curves of BC_5N electrode by the galvanostatic method in 1 M-$NaPF_6$/(EC + DEC) electrolyte solution. On the first discharge, at least two slights were observed. One at around 0.5 V versus Na/Na^+ might be due to the formation of SEI. On the second discharge and thereafter, the slight at 0.5 V could not be observed, but a slight at 0.8 V and another slight at around 0 V versus Na/Na^+ were observed. The reversible capacities of c.190 mAh/g were higher than that of graphite and similar to that of carbon.

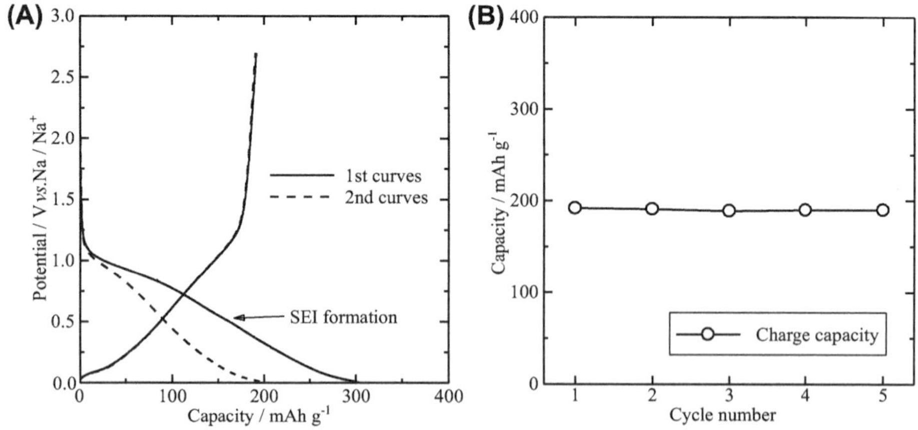

Figure 11.12 Discharge/charge curves (A), and charge (deintercalation) capacities (B) of BC_5N by the galvanostatic method using a current density of 100 μA/cm^2 in 1 M-$NaPF_6$/(EC + DEC) electrolyte solution. *DEC*, diethyl carbonate; *EC*, ethylene carbonate; *SEI*, solid electrolyte interface.

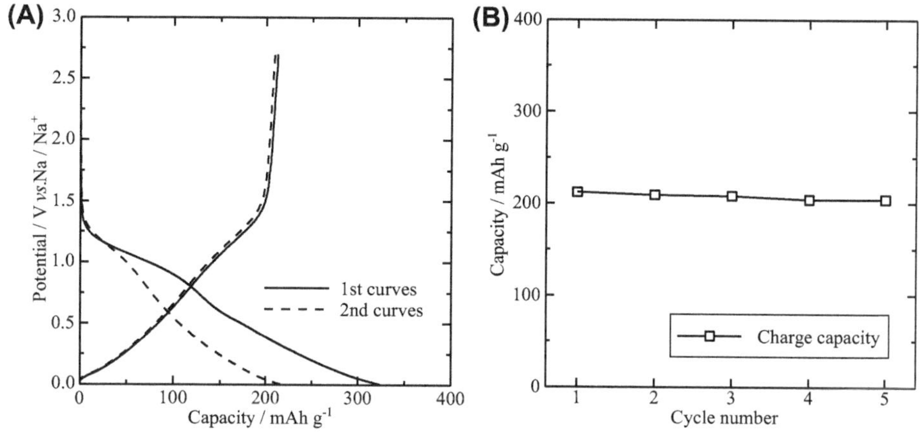

Figure 11.13 Discharge/charge curves (A), and charge (deintercalation) capacities (B) of BC_8 by the galvanostatic method using a current density of 100 μA/cm^2 in 1 M-$NaPF_6$/(EC + DEC) electrolyte solution. *EC*, ethylene carbonate; *DEC*, diethyl carbonate.

Fig. 11.13 shows first discharge/charge curves of BC_8 prepared at 1170 K. The interesting point is that the B/C material prepared at a lower temperature showed a higher potential at the beginning of the discharge (intercalation of Na^+ ion) than that of carbon [17]. Generally B/C material containing a larger amount of boron showed a higher discharge potential. This result was consistent with that observed in the case of B/C/N material (example: Fig. 11.12) and might be explained by the energy scheme, which is described in Section 11.6.2.

11.6.2 Mechanism of Electrochemical Intercalation of Na Into Boron/Carbon/Nitrogen and Boron/Carbon Materials

On the electrochemical intercalation of Na into BC_5N, second stage compound was made whose *d*-spacing was 403 pm (Fig. 11.14A) when the potential was kept at 0.7 V versus Na/Na^+ by the constant current and the constant voltage (CCCV) method. Fig. 11.14B shows the XRD pattern of Na-intercalated BC_2N when the potential was kept at 0.003 V versus Na/Na^+ by the CCCV method. The sharp diffraction lines indicated that the intercalation material was a first stage compound whose *d*-spacing was 440 pm. From these results, the slights at 0.8 and 0 V versus Na/Na^+ observed on the discharge (Fig. 11.12) could be due to the formation of second and first stage compounds, respectively. On the other hand, it is known that mainly insertion reaction occurs in the case of Na into nanopores of carbon [45]. Therefore, the electrochemical reaction of Na into BC_5N proceeds in a different way from carbon. The formation of stage structure, in the case of B/C/N and B/C materials used as anodes, is one of the advantages for the safety of batteries.

As described in Section 11.6.1, the intercalation of Na^+ ion into the B/C/N and B/C materials having higher boron content began at higher potentials, suggesting that

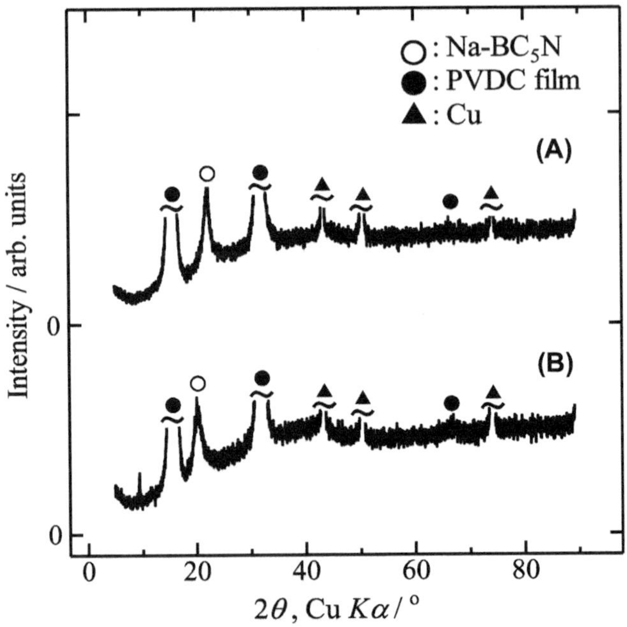

Figure 11.14 X-ray diffraction patterns of Na-intercalated BC_5N (Na-BC_5N) prepared by the CCCV method. Potentials were fixed at (A) 0.7 V and (B) 0.003 V versus Na/Na^+ in 1 M-$NaPF_6$/(EC + DEC) electrolyte solution. *CCCV*, constant current and the constant voltage; *DEC*, diethyl carbonate; *EC*, ethylene carbonate.

boron content in the B/C/N and B/C materials tended to determine the potentials on the discharge/charge cycles. The fact that intercalation of Na^+ ion into reduced B/C/N materials (B/C/N^-) occurred at higher potentials could be due to a coordination of the B/C/N^- layers to Na^+ ion at lower energy on its conduction band, which is shown in Fig. 11.15 as an energy scheme [17]. In fact, the XANES in the C*K* region (Fig. 11.8) indicated that the bottoms of the conduction bands of B/C/N materials were at energies lower than those of noncrystalline carbon and graphite. Molecular orbital calculation suggested that the conduction band of B/C/N material is constructed by the hybridization of carbon 2s and 2p orbitals with boron 2s and 2p orbitals [46]. Based on these experimental and calculated results, we found that the electron deficiency of boron atom in B/C/N material decreases the energy level in the bottom of the conduction band in comparison with that of graphite. In other words, B/C/N materials have an electron affinity larger than that of graphite. This might be the reason why Na, having a large ionization potential, can be intercalated into B/C/N materials easier than into graphite. This is another important role of boron in B/C/N materials. It is expected that Na^+ ion can also be electrochemically intercalated into the B/C/N^- layers during the discharge at potentials over the redox potential of Na (deposition of Na metal), resulting in the materials showing high reversible capacities in the anode matrix of SIBs.

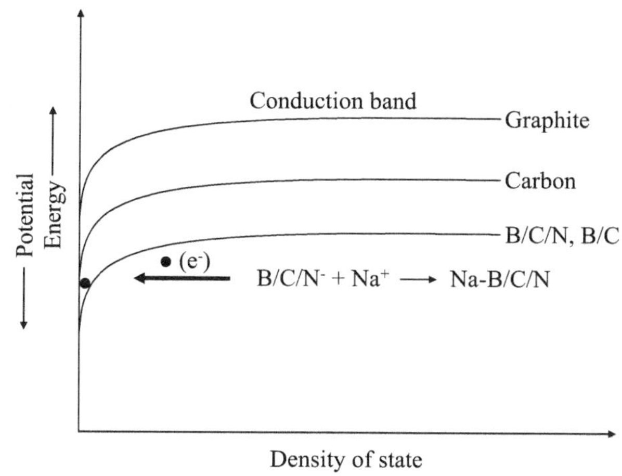

Figure 11.15 Energy scheme of the bottom of conduction band and the electrochemical potential for B/C/N and B/C materials, noncrystalline carbon, and graphite. *B/C*, boron/carbon; *B/C/N*, boron/carbon/nitrogen.
Reproduced with permission from K. Yamada, H. Ishikawa, C. Kamiwaki, M. Kawaguchi, Electrochemistry 83 (2015) 452–458. Copyright 2015, The Electrochemical Society of Japan.

11.7 Application of Boron/Carbon/Nitrogen Materials to Dual Carbon Alloy Batteries

Since graphite can be intercalated with both reductive metals, such as Li and Na, and oxidative species, such as hexafluorophosphate anion PF_6^- [18], dual graphite batteries have been investigated [18,19]. Fig. 11.16 shows the schematic diagram of the dual carbon battery proposed by Steel and Dahn [18]. Ishihara indicated that the capacity of dual carbon batteries depended on the weight ratio of the anode and cathode, which is shown in Fig. 11.17. The results indicated that maximum capacity of 150 mAh/g was obtained with the anode/cathode ratio of 20 in 1 M n-butyl pyridinium PF_6 dissolved in a mixed solvent of propylene carbonate (PC) and methyl ethyl carbonate (MEC) (1 M-n-$BPPF_6$/(PC + MEC)). In this kind of battery, the combination of graphite and other host material can be used if the other host material has larger capacities than that of graphite. Dual batteries using carbon alloys for anode and cathode are called dual carbon alloy batteries in this chapter.

Since Li and Na can be effectively intercalated into B/C/N materials, B/C/N materials can be used as anodes of dual carbon alloy batteries. It will be interesting particularly to discover a dual carbon alloy battery in the electrochemical Na system, which would include several merits and be one of the candidates of batteries in the next generation. The present authors have checked the intercalation/deintercalation of PF_6^- into/out of graphite by using the electrochemical method in 1 M-$NaPF_6$/(EC + DEC) (Fig. 11.18). The PF_6^- ion can be reversibly intercalated into graphite, although the reversible capacity (20 mAh/g) is not so large and almost the same as that measured in 1 M-$LiPF_6$/(EC + DEC). Fig. 11.19 shows the first trial of discharge/charge curves

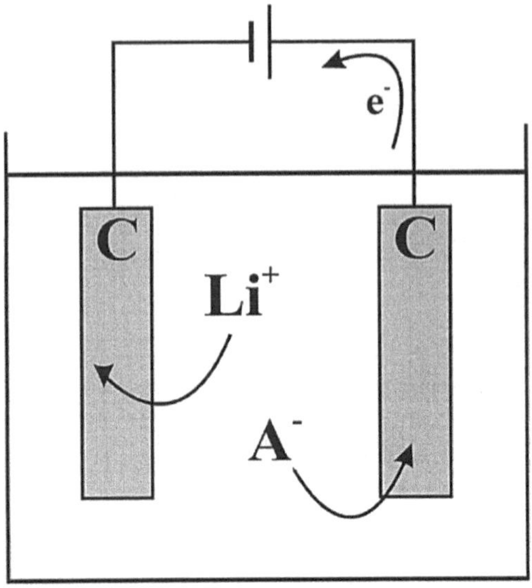

Figure 11.16 Schematic diagram of the dual carbon battery.
Reproduced with permission from J.A. Seel, J.R. Dahn, J. Electrochem. Soc. 147 (2000) 892–898. Copyright 2000, The Electrochemical Society.

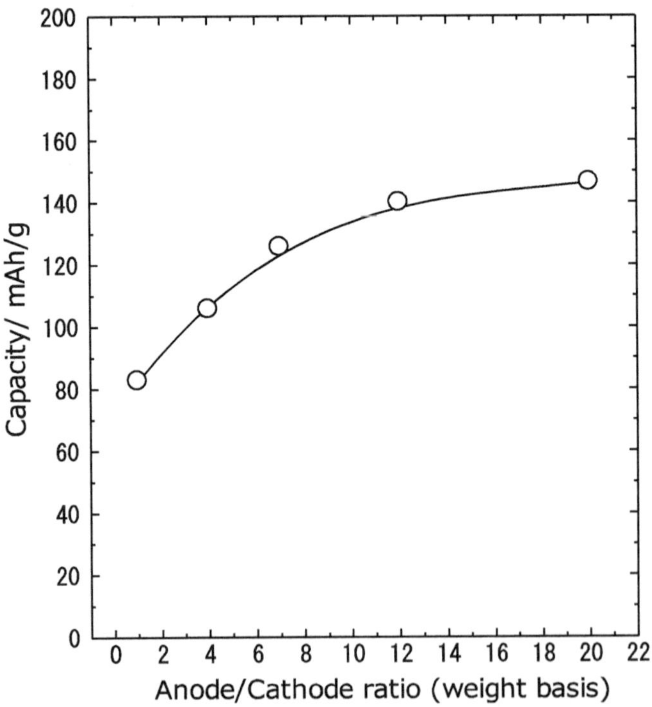

Figure 11.17 Capacity of the PF_6^- intercalation as a function of the amount of active carbon anode. Electrolyte: 1M n-BPPF$_6$ in PC–MEC (2:1), cathode: graphite, 10 mg. *MEC*, methyl ethyl carbonate; *PC*, propylene carbonate.
Reproduced with permission from T. Ishihara, Y. Yokoyama, F. Kozono, H. Hayashi, J. Power Sources 196 (2011) 6956–6959. Copyright 2011 Elsevier.

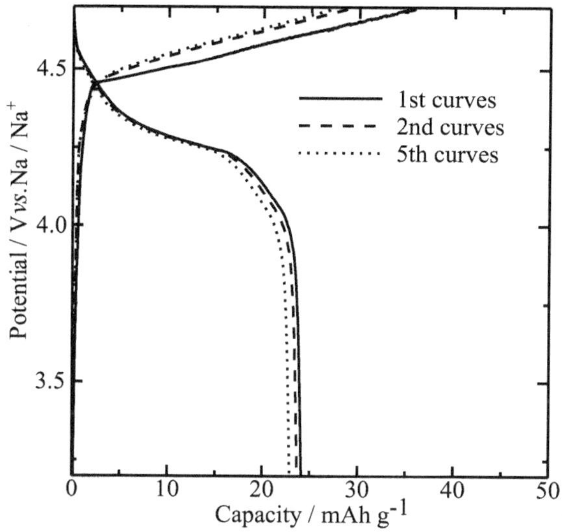

Figure 11.18 Discharge/charge curves of graphite by the galvanostatic method using a current density of 100 μA/cm^2 in 1 M-NaPF$_6$/(EC + DEC) electrolyte solution. *DEC*, diethyl carbonate; *EC*, ethylene carbonate.

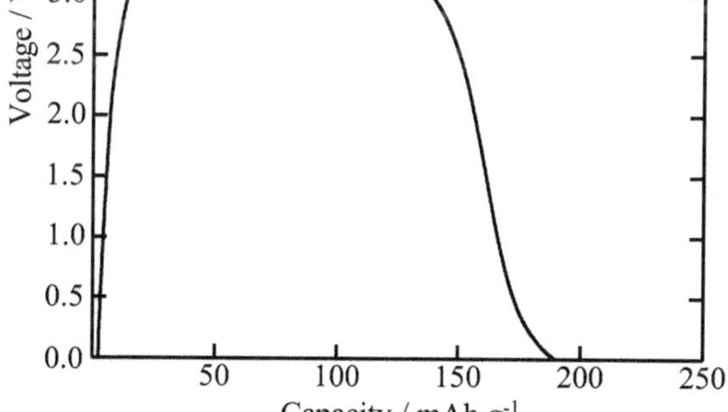

Figure 11.19 Discharge/charge curves of dual carbon alloy battery with a combination of BC_5N anode and graphite cathode by the galvanostatic method using a current density of 100 μA/cm^2 in 1 M-NaPF$_6$/(EC + DEC) electrolyte solution. *DEC*, diethyl carbonate; *EC*, ethylene carbonate.

of dual carbon alloy batteries with a combination of a BC_5N anode and graphite cathode in 1 M-$NaPF_6$/(EC + DEC) [47]. The maximum capacity of the BC_5N anode was 190 mAh/g when the anode/cathode ratio was 9. The observed high voltage (3.7 V) on average is one of the advantages on dual carbon alloy batteries. The optimum condition for the carbon alloy batteries will be further investigated.

11.8 Summary

Intercalation chemistry and application of B/C/N and B/C materials to secondary batteries are summarized in this chapter. Na and Mg can be intercalated into B/C/N and B/C materials by using the electrochemical method and chemical reaction, while these metals cannot be easily intercalated into graphite. The intercalation of these metals has been explained by the relation between the ionization potentials of metals and the electron affinities of host materials. Intercalation of Li and Na into B/C/N and B/C materials can be applied to anodes of Li- and SIBs. Application of the materials to dual carbon alloy batteries has also been proposed for the future. In conclusion, B/C/N and B/C materials are promising materials in the fields of energy systems, particularly in the energy storage systems.

Acknowledgment

This work was partly supported by MEXT, KAKENHI (15H03852), Japan.

References

[1] M. Kawaguchi, N. Bartlett, Syntheses, structures, and intercalation chemistry of B/C/N materials based on the graphite network, in: T. Nakajima (Ed.), Fluorine-carbon Fluoride-carbon Materials, Marcel Dekker, New York, 1995, pp. 187–238.
[2] M. Kawaguchi, Adv. Mater. 9 (1997) 615–625.
[3] E. Yasuda, TANSO No170 (1995) 317–320 (in Japanese).
[4] E. Yasuda, M. Inagaki, K. Kaneko, M. Endo, A. Oya, Y. Tanabe (Eds.), Carbon Alloys, Elsevier, Oxford, 2003.
[5] M. Morita, T. Hanada, H. Tsutsumi, Y. Matsuda, M. Kawaguchi, J. Electrochem. Soc. 139 (1992) 1227–1230.
[6] M. Kawaguchi, Y. Imai, N. Kadowaki, J. Phys. Chem. Solids 67 (2006) 1084–1090.
[7] H. Konno, T. Ito, M. Ushiro, K. Fushimi, K. Azumi, J. Power Sources 195 (2010) 1739–1746.
[8] M. Kawaguchi, Y. Ishida, CARBON2013, Rio de Janeiro, Brazil, 2013.7.14–19, Abstract No.113.
[9] J. Ozaki, N. Kimura, T. Anahara, A. Oya, Carbon 45 (2007) 1847–1853.
[10] A. Hérold, Bull. Soc. Chim. Fr. 187 (1955) 999–1012.
[11] R. Yazami, P.H. Touzain, J. Power Sources 9 (1983) 365–371.

[12] S. Komaba, T. Itabashi, M. Watanabe, H. Groult, N. Kumagai, J. Electrochem. Soc. 154 (2007) A322–A330.
[13] Y. Orikasa, T. Masese, Y. Koyama, T. Mori, M. Hattori, K. Yamamoto, T. Okado, Z.D. Huang, T. Minato, C. Tassel, J. Kim, Y. Kobayashi, T. Abe, H. Kageyama, Y. Uchimoto, Sci. Rep. 4 (2014) 5622 (1-6).
[14] S. Nakanishi, M. Nose, H. Nakayama, Electrochemistry 83 (2015) 182–187 (in Japanese).
[15] M. Kawaguchi, K. Ohnishi, K. Yamada, Y. Muramatsu, J. Electrochem. Soc. 157 (2010) P13–P17.
[16] M. Kawaguchi, A. Kurasaki, Chem. Commun. 48 (2012) 6865–6988.
[17] K. Yamada, H. Ishikawa, C. Kamiwaki, M. Kawaguchi, Electrochemistry 83 (2015) 452–458.
[18] J.A. Seel, J.R. Dahn, J. Electrochem. Soc. 147 (2000) 892–898.
[19] T. Ishihara, Y. Yokoyama, F. Kozono, H. Hayashi, J. Power Sources 196 (2011) 6956–6959.
[20] T.Y. Kosolapova, G.N. Makarenko, T.I. Serebryakova, E.V. Prilutskii, O.T. Khorpyakov, O.I. Chernysheva, Poroshk. Metall. 1 (1971) 27–33.
[21] J. Kouvetakis, T. Sasaki, C. Shen, R. Hagiwara, M. Lerner, K.M. Krishnan, N. Bartlett, Synth. Met. 34 (1989) 1–7.
[22] T. Sasaki, M. Akaishi, S. Yamaoka, Y. Fujiki, T. Oikawa, Chem. Mater. 5 (1993) 695–699.
[23] M. Kawaguchi, T. Kawashima, T. Nakajima, Chem. Mater. 8 (1996) 1197–1201.
[24] J. Kouvetakis, R.B. Kaner, M.L. Sattler, N. Bartlett, J. Chem. Soc. Chem. Commun. (1986) 1758–1759.
[25] M. Kawaguchi, S. Yagi, H. Enomoto, Carbon 42 (2004) 345–350.
[26] T. Nakajima, M. Koh, Carbon 35 (1997) 203–208.
[27] M. Kawaguchi, TANSO No267 (2015) 84–93.
[28] A.Y. Liu, R.M. Wentzcovitch, M.L. Cohen, Phys. Rev. B 39 (1989) 1760–1765.
[29] I. Shimoyama, Y. Baba, T. Sekiguchi, Carbon 71 (2014) 1–10.
[30] R.B. Kaner, J. Kouvetakis, C.E. Warble, M.L. Sattler, N. Bartlett, Mater. Res. Bull. 22 (1987) 399–404.
[31] M.O. Watanabe, S. Itoh, K. Mizushima, T. Sasaki, J. Appl. Phys. 22 (1995) 2880–2882.
[32] M.R. Uddin, S. Majety, J. Li, Y. Lin, H.X. Jiang, J. Appl. Phys. 115 (2014) 093509 (1–7).
[33] B.M. Way, J.R. Dahn, T. Tiedje, K. Myrtle, M. Kasrai, Phys. Rev. B 46 (1992) 1697–1702.
[34] T. Shirasaki, A. Derre, M. Menetrier, A. Tressaud, S. Flandrois, Carbon 38 (2000) 1461–1467.
[35] G.S. Parry, D.E. Nixon, K.M. Lester, B.C. Levene, J. Phys. C 2 (1969) 2156–2158.
[36] J.O. Besenhard, M. Winter, J. Yang, W. Biberacher, J. Power Sources 54 (1995) 228–231.
[37] B.M. Way, J.R. Dahn, J. Electrochem. Soc. 141 (1994) 907–912.
[38] T. Shirasaki, A. Derré, K. Guérin, S. Flandrois, Carbon 37 (1999) 1961–1964.
[39] M. Kawaguchi, K. Yamada, H. Ishikawa, C. Kamiwaki, CARBON2015, Dresden, Germany, 2015.7.12–17, Abstracts: Lectures 130.
[40] A. Métrot, D. Guerard, D. Billaud, A. Hérold, Synth. Met. 1 (1980) 363–369.
[41] R.C. Asher, J. Inorg. Nucl. Chem. 10 (1959) 238–249.
[42] K. Yamada, A. Kurasaki, M. Kawaguchi, TANSO No249 (2011) 161–167 (in Japanese).
[43] N. Bartlett, B.W. McQuillan, Graphite chemistry, in: M.S. Whittingham, A.J. Jacobson (Eds.), Intercalation Chemistry, Academic Press, New York, 1982, pp. 19–53.
[44] X. Kang, Chem. Rev. 114 (2014) 11503–11618.
[45] P. Thomas, D. Billaud, Electrochim. Acta 47 (2002) 3303–3307.
[46] Y. Muramatsu, Adv. Quantum Chem. 42 (2003) 353–361.
[47] K. Yamada, M. Kawaguchi, Unpublished Data.

Structures of Highly Fluorinated Compounds of Layered Carbon

12

R. Hagiwara [1], *Y. Sato* [1,2]
[1]Kyoto University, Kyoto, Japan; [2]National Institute of Advanced Industrial Science and Technology, Tsukuba, Japan

Chapter Outline

12.1 Introduction

Fluorination of graphite to synthesize highly fluorinated binary compounds with layered structures, commonly known as graphite fluorides, is performed mostly by two methods. Direct fluorination of graphite by elemental fluorine (ambient pressure) at elevated temperatures [1–3] is one of them to give poly(carbon monofluoride) $(CF)_n$, and/or poly(dicarbon monofluoride) $(C_2F)_n$.

$(CF)_n$ is a colorless solid obtained by fluorination of graphite. In $(CF)_n$, all the double bonds of carbon atoms in an original graphite sheet are cleaved to be covalently bound to fluorine atoms giving a two-dimensional sp^3 carbon array forming cyclohexanes linked in chair-form (Fig. 12.1) [4–6]. Every carbon atom is bound to a fluorine

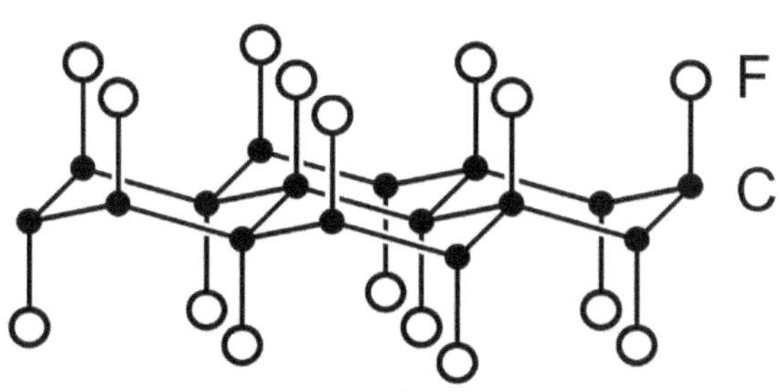

Figure 12.1 Structure of $(CF)_n$ [4–6].

New Fluorinated Carbons: Fundamentals and Applications. http://dx.doi.org/10.1016/B978-0-12-803479-8.00012-7

atom alternatively above and below the array of carbon atoms. Fluorine atoms form a closely packed layer which endues an extremely low surface energy to this polymer. The sheets are stacked to form a layered structure. The interaction between the two $(CF)_n$ sheets occurs between closely packed sheets of fluorine atoms which are much weaker than that between the sheets of carbon atoms in graphite and enables this material to be applied as an excellent solid lubricant. The separation of the sheets is approximately 0.60 nm which decreases down to ~0.585 nm as the minimum with the improvement of crystallinity. The crystallinity of $(CF)_n$ formed depends on that of the starting carbon material as well as the reaction temperature and pressure of F_2. Fluorination of graphite with high crystallinity is completed at relatively high temperature, around 600°C. The reaction is more or less accompanied by the formation of fluorocarbon gas. Graphite with lower crystallinity such as petroleum coke graphitized at low temperature is fluorinated at lower temperatures to give $(CF)_n$ with lower crystallinity. Elevation of fluorination temperature improves the crystallinity by the elimination of the part of lower crystallinity as fluorocarbon gas, giving a crystalline $(CF)_n$ as a residue although the yield of the material is decreased.

$(C_2F)_n$ is a gray solid obtained by fluorination of graphite under atmospheric pressure of elemental fluorine at 350–400°C for a few weeks. Unlike $(CF)_n$, $(C_2F)_n$ is obtained only from highly crystalline graphite. In $(C_2F)_n$, all the double bonds of carbon atoms in an original graphite sheet are cleaved to be connected to adjacent carbon sheets via C–C bonds giving a double-decked array of cyclohexanes linked in chair-form. Fluorine atoms are bound to every other carbon atoms in each deck to form a two-dimensional close packing of fluorine atoms above and below. The layer separation is about 0.90 nm. For the connectivity of two adjacent carbon sheets, there are two possibilities, AB- and AA′-type (Fig. 12.2). Focusing the carbon framework, the former is a piece of cubic and the latter is hexagonal diamond sliced in atomic level perpendicular to [111] and [001] direction, respectively of the lattice. Touhara et al. concluded that the AB-type is more reasonable than the AA′-type, following to the stacking sequences of carbon sheets in typical acceptor-graphite intercalation compounds (GICs) with stage-2 structures where adjacent graphite layers not intercalated hold the stacking of original graphite [6]. However, it should be noted that $(C_2F)_n$ is a particular case with the three-dimensional network of sp^3-hybridized carbon atoms half of which are forming C–C covalent bond between the adjacent sheets while such bonding is absent in the gallery not intercalated in general stage-2 GICs. The carbon sheet structure, AB- or AA′-type, should be experimentally checked by the spectroscopic and diffraction analyses. In both cases, it is easily anticipated that the occlusion of the adjacent carbon layers is essential to form C–C bonds between the two layers by the solid-state reaction. Graphite with low crystallinity in which the carbon layer stacking is turbostratic never gives $(C_2F)_n$, only $(CF)_n$ being formed instead. Even with a highly crystalline graphite, $(CF)_n$ is more or less included as a by-product giving F/C atomic ratio higher than 0.5.

Another type of layered carbon fluorides is obtained by the room-temperature reactions of graphite and elemental fluorine in the presence of fluoroacid catalysts, such as HF, AsF_5, IF_5, and OsF_6, which have been classified as fluorine–GICs (C_xF), distinguished from $(CF)_n$ and $(C_2F)_n$ prepared by direct fluorination of graphite at elevated

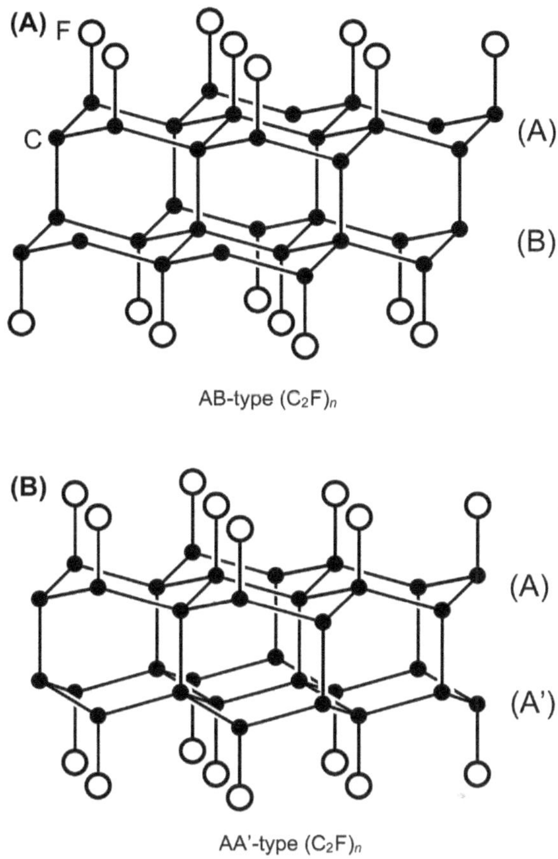

Figure 12.2 Structure models of $(C_2F)_n$ [6]. (A) AB-type $(C_2F)_n$; (B) AA′-type $(C_2F)_n$.

temperatures [7]. Contamination of a trace of metal fluorides as fluorometallate anions in C_xF is not avoidable, but the amount of such contaminants in highly fluorinated compounds is much smaller than that of the fluorine atoms directly bound to the carbon atoms [8,9]. Discussion on the structures of C_xF ($x < 6$) in the previous studies has been based on the diffraction and spectroscopic data such as X-ray diffraction (XRD), X-ray photoelectron spectroscopy (XPS), IR, and NMR (nuclear magnetic resonance). The a-lattice parameters of the C_xF ($x = 1.3-6$) based on the graphite hexagonal lattice are 0.246–0.248 nm [10,11], slightly larger than that of original graphite (0.2456 nm), but smaller than those of $(CF)_n$ and $(C_2F)_n$ having sp^3-hybridized buckled carbon sheets (0.262 and 0.257 nm, respectively). The carbon atoms unbound to fluorine atoms in C_xF give the ^{13}C NMR peak at 135 ppm [8] and the C1s XPS peak at the binding energy, E_B, of 284–285 eV [11,12], suggesting that they maintain the graphitic sp^2 hybridization. On the other hand, the carbon atoms bound to fluorine atoms in C_xF give the C1s XPS peak at E_B of 287–288 eV [11], being lower than those observed for $(CF)_n$ and $(C_2F)_n$ (around 290 eV [2]). The F1s XPS peak of C_xF is also

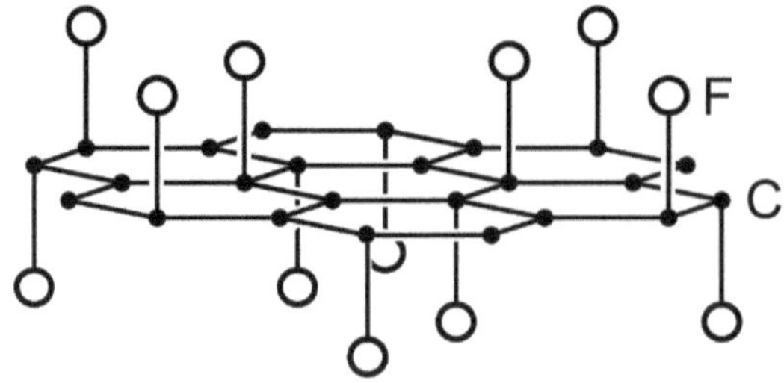

Figure 12.3 Proposed structural model of stage-1 $C_{2.0}F$ [8].

found at E_B of around 687 eV, lower than that of $(CF)_n$ and $(C_2F)_n$ (around 690 eV). The IR absorption band ascribed to the C—F stretching mode in C_xF is observed at around 1100 cm^{-1} [8], while those for $(CF)_n$ and $(C_2F)_n$ are observed at higher wave numbers around 1220 cm^{-1} [2]. Based on these results a structural model with planar carbon sheets has been proposed initially for C_xF (Fig. 12.3). In this model, both carbon atoms bound and unbound to fluorine atoms maintain the sp^2 hybridization, and the C—F bonds possess "semi-ionic" character with the bond order of lower than that of the typical covalent C—F bond [7,8,11]. However, estimation of the C—F bond order and the electronic states of the carbon atoms bound to fluorine atoms in C_xF based on XPS data seems to be difficult, since E_B's of the C1s and F1s peaks are strongly affected by the chemical environments such as the nearest neighbor atoms and/or groups around, as has been reported for other various kinds of fluorocarbon polymers [13]. The chemical shift of the ^{13}C NMR peak ascribed to the carbon atoms bound to fluorine atoms in C_xF, 88 ppm [8], is almost identical to that observed for $(CF)_n$, 89 ppm [14], which does not seem to support the proposed semi-ionic nature of C—F bonds in C_xF. The chemical shift of the ^{19}F NMR peak ascribed to the fluorine atoms in C_xF is also explainable without assuming the semi-ionic nature [15,16]. Transmission electron microscopic images and the electronic properties of stage-1 C_xF suggests that the carbon sheets in this compound are partially buckled due to the formation of covalent C—F bonds [17]. The planarity of the carbon sheets proposed for C_xF has never been experimentally proved.

In this chapter, structures of these highly fluorinated compounds of layered carbon are discussed mostly based on the results of neutron diffraction experiments [18,19]. Elucidating local structures and C—F bond nature of these fluorinated compounds should also be beneficial to understand fluorination mechanism of graphene monolayer recently attracting much interest, since chemical and physical properties of graphene are significantly affected by formation of C—F bonds [20,21].

12.2 Experimental

The details of preparative conditions of $(CF)_n$, $(C_2F)_n$, and C_xF used for structural analysis in the present study are given in Refs. [18,19]. These samples were prepared so that they keep as high crystallinity as possible although they are not intrinsically crystalline materials.

Neutron diffraction measurements were carried out in the high intensity total scattering spectrometer (HIT-II) installed at the pulsed neutron source in the High Energy Accelerator Research Organization (KEK, Tsukuba, Japan). Each powder sample was put in a vanadium tube cell with an inner diameter of 8.0 mm and a wall thickness of 0.025 mm. The details of the data processing are given in Refs. [18,19].

12.3 Results and Discussion

12.3.1 Structures of $(CF)_n$ and $(C_2F)_n$

The interlayer distances, d_{001}, and the a-lattice parameters derived for $(CF)_n$ and $(C_2F)_n$ from the XRD patterns (Fig. 12.4) are listed in Table 12.1, together with their F/C atomic ratios determined by elemental analysis. These structural parameters clearly indicate that the main components are $(CF)_n$ and $(C_2F)_n$, respectively, although their observed F/C atomic ratios are slightly larger than the stoichiometric values (1.00 and 0.50, respectively). It would be mainly caused by the peripheral groups such as $>CF_2$ and $-CF_3$ formed on the surfaces. Formation of small $(CF)_n$-like domains undetectable by XRD is not avoidable for the preparation of $(C_2F)_n$ as described above.

Structure factors, $S(Q)$, of $(CF)_n$ and $(C_2F)_n$ derived from neutron diffraction data are shown in Fig. 12.5, compared with those of graphite and poly(tetrafluoroethylene) (PTFE). Note that Q denotes $4\pi(\sin\theta)/\lambda$ where θ and λ represent the wave length of neutron beam and the diffraction angle, respectively. The small angle scattering is observed for both the graphite fluorides at Q of lower than that of the 001 Bragg

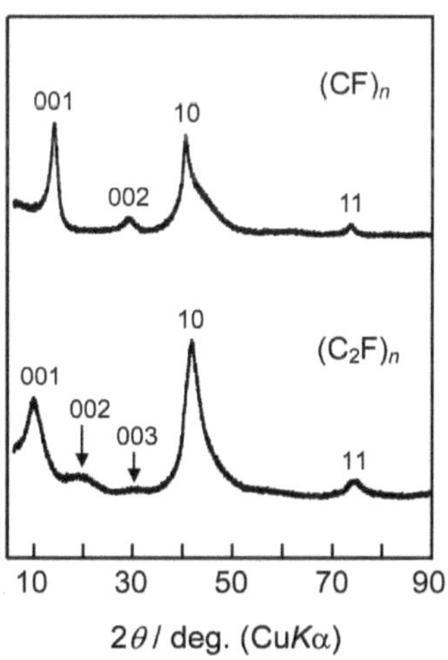

Figure 12.4 X-ray diffraction patterns of $(CF)_n$ and $(C_2F)_n$.

Table 12.1 Structural Characteristics of the Samples $(CF)_n$ and $(C_2F)_n$ in the Present Work

Sample	F/C Atomic Ratio	d_{001}/(nm)	$a(10)^a$/(nm)	$a(11)^a$/(nm)
$(CF)_n$	1.190	0.623	0.258	0.258
$(C_2F)_n$	0.696	0.872	0.252	0.256

[a] *a*(*hk*) denotes the *a*-lattice parameter estimated from the *hk* diffraction peak in the X-ray diffraction pattern.

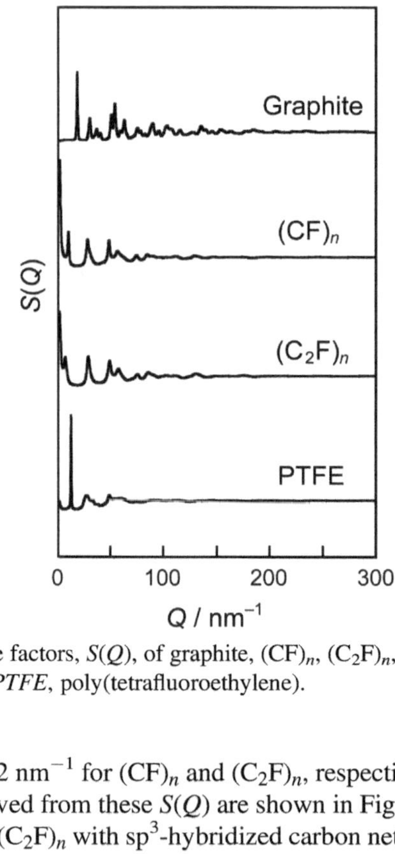

Figure 12.5 The structure factors, $S(Q)$, of graphite, $(CF)_n$, $(C_2F)_n$, and PTFE derived from the neutron diffraction data. *PTFE*, poly(tetrafluoroethylene).

diffraction (10.0 and 7.2 nm^{-1} for $(CF)_n$ and $(C_2F)_n$, respectively). Radial distribution functions, $RDF(r)$, derived from these $S(Q)$ are shown in Fig. 12.6. Due to the layered structures of $(CF)_n$ and $(C_2F)_n$ with sp^3-hybridized carbon networks and covalent C—F bonds, the $RDF(r)$ profiles of the $(CF)_n$ and $(C_2F)_n$ are similar to each other, while they significantly differ from that of graphite. A poorly resolved peak profile compared to those of $(CF)_n$ and $(C_2F)_n$ is found at the atomic correlation distance, r, above 0.3 nm in $RDF(r)$ of PTFE that also possesses covalent C—F bonds on sp^3-hybridized carbon chains. The peaks corresponding to the pair correlations for C—F and C—C in $RDF(r)$ of $(CF)_n$, $(C_2F)_n$, and PTFE are well approximated by Gaussian distribution functions as shown in Fig. 12.7. The C—F and C—C bond lengths, r_{C-F} and r_{C-C}, respectively, determined from the positions of these Gaussian peaks are listed in Table 12.2 and

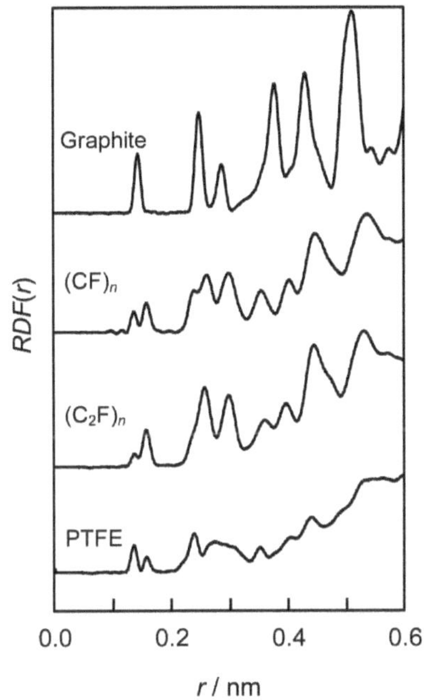

Figure 12.6 Radial distribution functions, *RDF*(*r*), of graphite, $(CF)_n$, $(C_2F)_n$, and PTFE. *PTFE*, poly(tetrafluoroethylene).

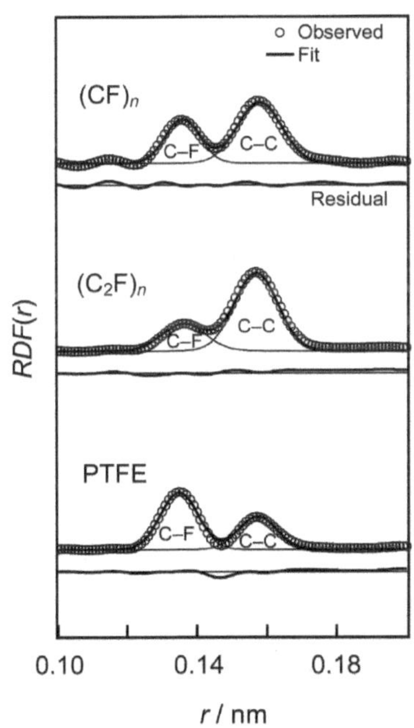

Figure 12.7 Radial distribution functions, *RDF*(*r*), of $(CF)_n$, $(C_2F)_n$, and PTFE from 0.1 to 0.2 nm (expanded). *PTFE*, poly(tetrafluoroethylene).

Table 12.2 C–F and C–C Bond Lengths in $(CF)_n$, $(C_2F)_n$, and PTFE

	r_{C-F}/(nm)		r_{C-C}/(nm)	
Sample	Present Work	References	Present Work	References
$(CF)_n$	0.136	0.141 [5][a]	0.158	0.154 [5][a]
$(C_2F)_n$	0.136	0.141 [6][a]	0.157	0.153 [6][a]
PTFE	0.135	0.136 [22]	0.157	0.158 [22]

PTFE, poly(tetrafluoroethylene).
[a]Estimated as the sum of the typical covalent radii of carbon and fluorine atoms.

compared to the values reported previously [5,6,22]. It is clearly shown that both $(CF)_n$ and $(C_2F)_n$ possess r_{C-F} and r_{C-C} of close to those in PTFE.

The structural parameters such as the *a*-lattice parameters, atomic coordinates and thermal factors in the proposed structural models of $(CF)_n$, AB-, and AA′-type $(C_2F)_n$ in Fig. 12.2 were refined so as to give the best possible agreement between the atomic pair distribution functions, *G*(*r*), calculated and that experimentally obtained from *S*(*Q*). During the refinement, the space symmetry of the monolayer in each model ($P\overline{3}$ m1 for $(CF)_n$ and AB-type $(C_2F)_n$, $P\overline{6}$ m2 for AA′-type $(C_2F)_n$ see Fig. 12.8) was maintained, and d_{001}, r_{C-F}, and r_{C-C} were fixed as they have been determined and listed in Table 12.2. Although $(CF)_n$ and $(C_2F)_n$ would actually take random stacking of the monolayers to cause the absence of the *hkl* ($hl \neq 0$ and/or $kl \neq 0$) diffraction peaks in the XRD patterns, only four types of the stacking sequences, αα, αα′, αβ,

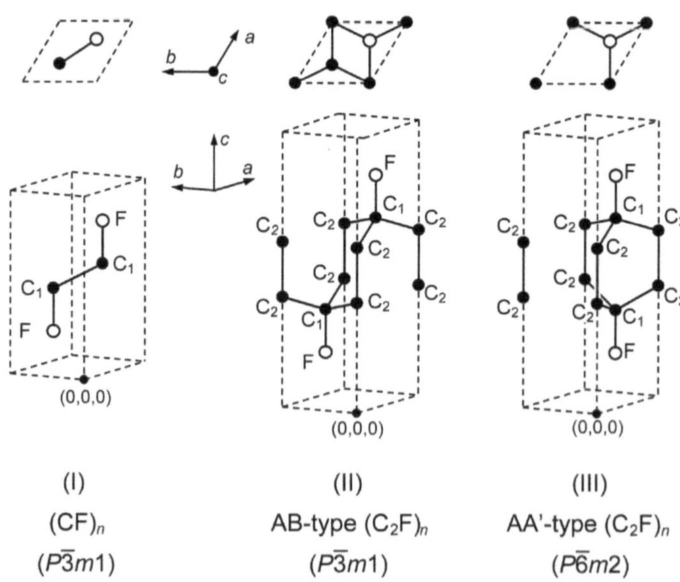

Figure 12.8 The unit cells of the monolayers of (I) $(CF)_n$, (II) AB-type $(C_2F)_n$, and (III) AA′-type $(C_2F)_n$.

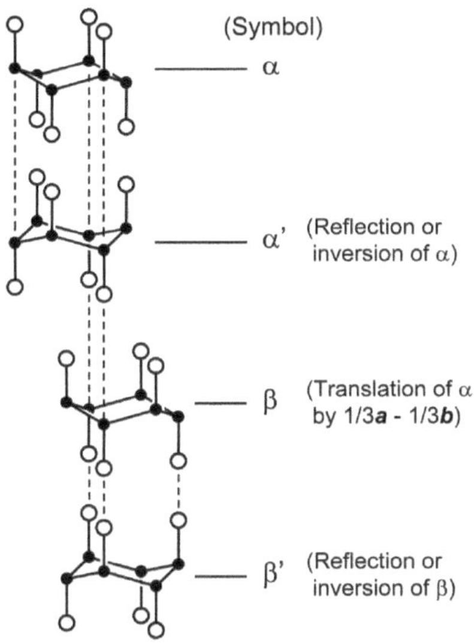

Figure 12.9 The spatial relations between the stacked layers assumed in this chapter. In this figure, the stacking of the C_6 rings in $(CF)_n$ is shown as an example.

and $\alpha\beta'$ (Fig. 12.9), were assumed here to simplify the calculation of $G(r)$. The layers α' (or β') in the layer stacking of $(CF)_n$ and AB-type $(C_2F)_n$ assumed here denotes the mirror images of the layers α (or β) on the ab-plane at $z = 0$, while that for the AA′-type $(C_2F)_n$ model means the inverse of the layer α (or β) on the origin of the unit cell (0,0,0). Consequently, the degree of agreement between the $G(r)$ curves calculated and observed for each compound at r of shorter than the d_{001} value does not strongly depend on the type of the stacking sequences, suggesting that the contribution of the interlayer correlations in the profile of $G(r)$ at such small r would be much less than that of the intralayer correlations. Fig. 12.10 shows the best-fit curve on $G(r)$ of $(CF)_n$ (I; derived from the $(CF)_n$ model with $\alpha\alpha'$ stacking sequences) and those on $G(r)$ of $(C_2F)_n$ (II and III; derived from AB- and AA′-type $(C_2F)_n$ models, respectively, with $\alpha\alpha$ stacking sequences). The disagreement between the $G(r)$ of $(CF)_n$ and the calculated curve I at r of around 0.23 nm is explained by the F—F correlation in the peripheral $>CF_2$ groups which is not taken into account for the calculation. $G(r)$ calculated for AB- and AA′-type $(C_2F)_n$ models clearly differ from each other at r above the distance between the second-nearest C—C pairs (0.26 nm), since the C_6 ring formed between a pair of carbon sheets in a monolayer in the former model is a chair form while that in the latter is a boat form (Fig. 12.8). The curve II is better fitted to the $G(r)$ of $(C_2F)_n$ than the curve III at and above 0.3 nm, indicating that the AB-type model is more reasonable than AA′-type model for the structure of $(C_2F)_n$.

The structural parameters obtained for $(CF)_n$ and $(C_2F)_n$ by $G(r)$-based refinement are listed in Table 12.3. The parameters z_i change as the c-lattice parameter ($=d_{001}$) varies with the experimental conditions such as the crystallinities of the original carbon host and fluorination temperature, but the distance between the layers of fluorine atoms

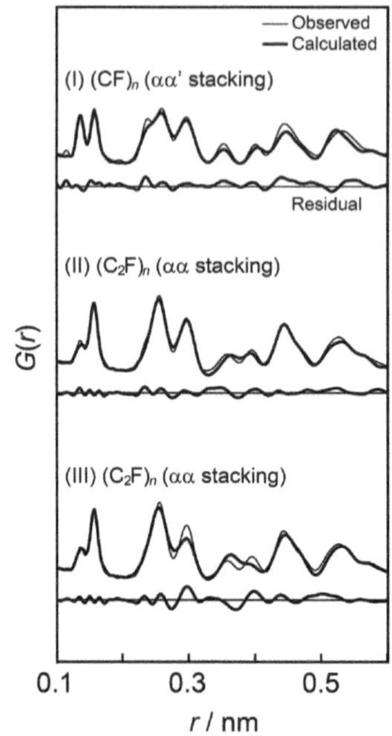

Figure 12.10 The best-fit results on $G(r)$ observed for $(CF)_n$ and $(C_2F)_n$.

Table 12.3 The Refined Structural Parameters of $(CF)_n$ and $(C_2F)_n$

Sample	Model	Lattice Constants/(nm)	Atomic Coordinates	Bond Angles/ (degrees)
$(CF)_n$	αα stacking ($P\bar{3}\,m1$)	a: 0.261 c: 0.623[a]	C: $(1/3,2/3,z_1)$, $(2/3,1/3,1-z_1)$ F: $(1/3,2/3,z_2)$, $(2/3,1/3,1-z_2)$ $z_1 = 0.461$, $z_2 = 0.243$	F–C–C: 108 C–C–C: 111
$(C_2F)_n$	AB-type αα stacking ($P\bar{3}\,m1$)	a: 0.257 c: 0.872[a]	C_1: $(1/3,2/3,z_1)$, $(2/3,1/3,1-z_1)$ C_2: $(0,0,z_2)$, $(0,0,1-z_2)$ F: $(1/3,2/3,z_3)$, $(2/3,1/3,1-z_3)$ $z_1 = 0.351$, $z_2 = 0.410$, $z_3 = 0.195$	F–C_1–C_2: 109 C_2–C_1–C_2': 110 C_1–C_2–C_2': 110 C_1–C_2–C_2': 109

[a]Fixed as d_{001} determined from X-ray diffraction patterns.

above and below the carbon layer (0.320 and 0.532 nm for $(CF)_n$ and $(C_2F)_n$, respectively) would be independent of such factors. The *a*-lattice parameter of $(CF)_n$ (0.261 nm) corresponding to the distance between the nearest pair of fluorine atoms is larger than that of $(C_2F)_n$ (0.257 nm). The bond angles ∠F−C−C and ∠C−C−C in $(C_2F)_n$ are a little closer to that at the regular tetrahedral site (109°28′) than those in $(CF)_n$. It is expected that carbon atoms composing the two-dimensional network of C−C bonds in $(CF)_n$ are slightly shifted from the regular tetrahedral site to reduce the repulsive force working between the nearest pairs of fluorine atoms, but such distortion of the lattice is less likely to occur in $(C_2F)_n$ with the three-dimensional network of C−C bonds.

12.3.2 Structure of C_xF

Fig. 12.11 shows the XRD patterns of three C_xF samples with different C/F atomic ratios, *x*, of 2.47, 2.84, and 3.61. Three peaks found in each pattern are ascribed to the 001, 10, and 11 diffraction based on the hexagonal lattice. The interlayer distances, d_{001}, are 0.574, 0.559, and 0.520 nm for C_xF with $x = 2.47$, 2.84, and 3.61, respectively, and their *a*-lattice parameters are equally 0.247 nm based on the XRD data. Assuming that the carbon sheets in these samples maintain the planarity, r_{C-C} is estimated to be 0.143 nm from the *a*-lattice parameter of 0.247 nm. The d_{001}-plot against $1/x$ (F/C atomic ratio) for these three samples shown in Fig. 12.12 well agrees with those of the compounds obtained in the previous works [10,11,23]. F1s and C1s XPS spectra of the C_xF samples shown in Fig. 12.13 have the characteristic peaks ascribed to fluorine atoms bound to carbon atoms at around 687 eV, carbon atoms

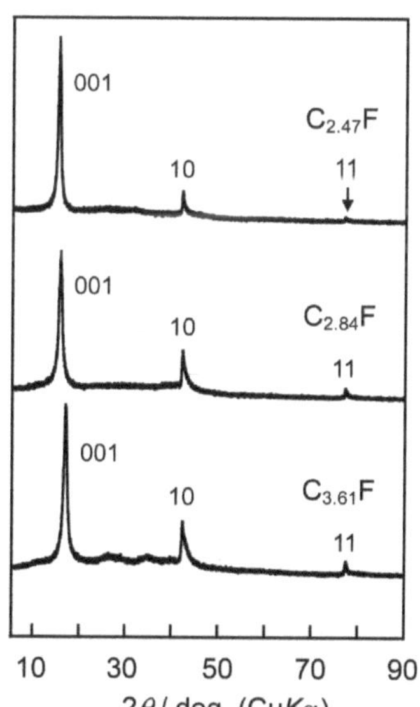

Figure 12.11 X-ray diffraction patterns of the stage-1 C_xFs.

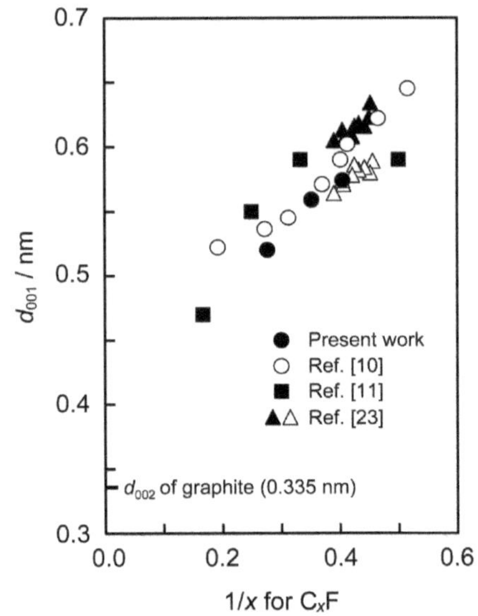

Figure 12.12 The plot of d_{001} of stage-1 C_xF on $1/x$.

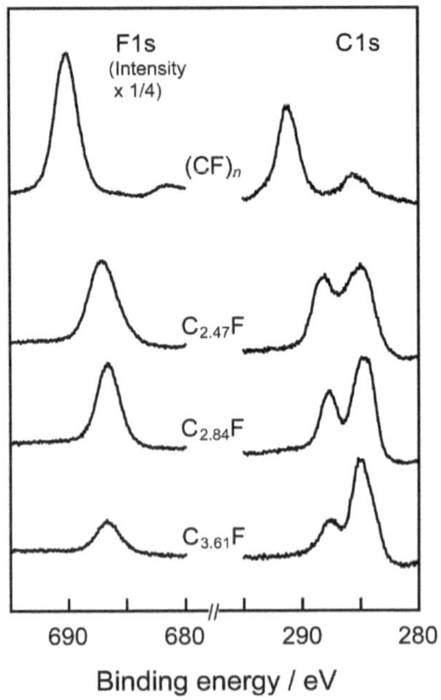

Figure 12.13 F1s and C1s X-ray photoelectron spectroscopy spectra of the stage-1 C_xF samples.

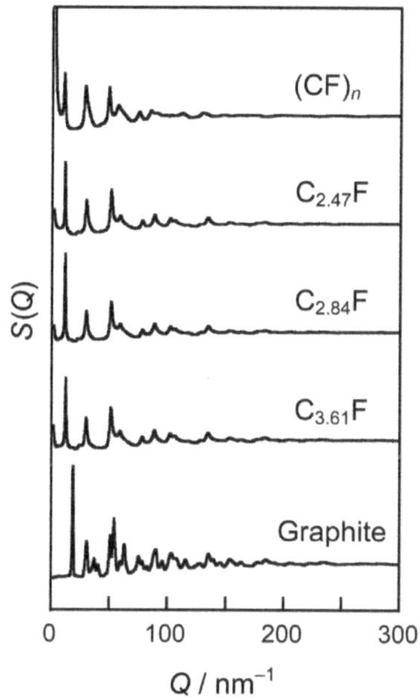

Figure 12.14 The structure factors, $S(Q)$, of graphite and the stage-1 C_xF samples derived from the neutron diffraction data.

bound and unbound to fluorine atoms at around 288 and 285 eV, respectively. Relative intensities of these peaks vary with the C/F atomic ratios, x of C_xF.

Structure factors, $S(Q)$, of the C_xF samples derived from neutron diffraction data are shown in Fig. 12.14, compared with that of graphite. The absence of the $00l$ ($l > 1$) and hkl ($hl \neq 0$ and/or $kl \neq 0$) Bragg diffraction peaks for hexagonal lattice in $S(Q)$ is explained by the random stacking of the fluorinated carbon sheets spreading the *ab* direction along the *c*-direction in these compounds. Radial distribution functions, $RDF(r)$, derived from $S(Q)$ of these C_xF samples, are shown in Fig. 12.15 compared with those of $(CF)_n$ and graphite. The peaks corresponding to C—F and C—C bonds in $RDF(r)$ of C_xF overlap with each other to give a single peak at around 0.14—0.15 nm, while those in $RDF(r)$ of $(CF)_n$ are well separated. These $RDF(r)$ without any peaks at r of 0.170—0.222 nm clearly show that r_{C-F} in C_xF is smaller than the value previously expected (0.17 nm [7]) while the interatomic distance of the second-nearest C—F pair in the compounds, $r_{\underline{C}-C-\underline{F}}$, is larger than 0.222 nm. These r_{C-F} and $r_{\underline{C}-C-\underline{F}}$ are not explained by the structural model with planar carbon sheets (Fig. 12.16A), in which the bond lengths are related by Eq. (12.1);

$$(r_{\underline{C}-C-\underline{F}})^2 = (r_{C-F})^2 + (r_{C-C})^2 \tag{12.1}$$

where r_{C-C} is 0.143 nm as derived above. On the other hand, a structural model with covalent C—F bonds in which the carbon sheets are partially buckled around the sp^3-hybridized carbon atoms (Fig. 12.16B) explains simultaneously r_{C-F} of smaller than

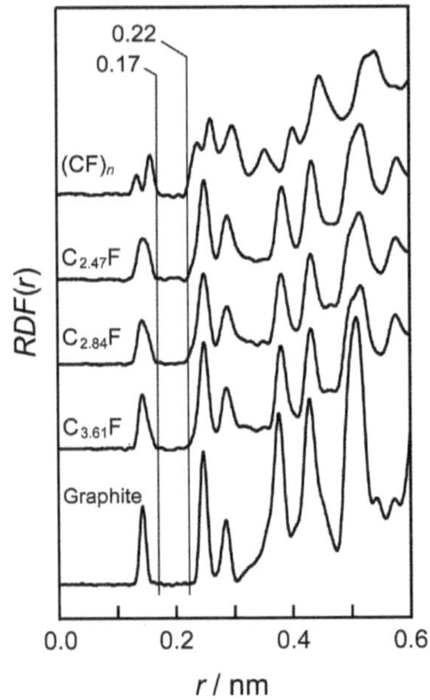

Figure 12.15 Radial distribution functions, $RDF(r)$, of graphite, the stage-1 C_xF samples and $(CF)_n$.

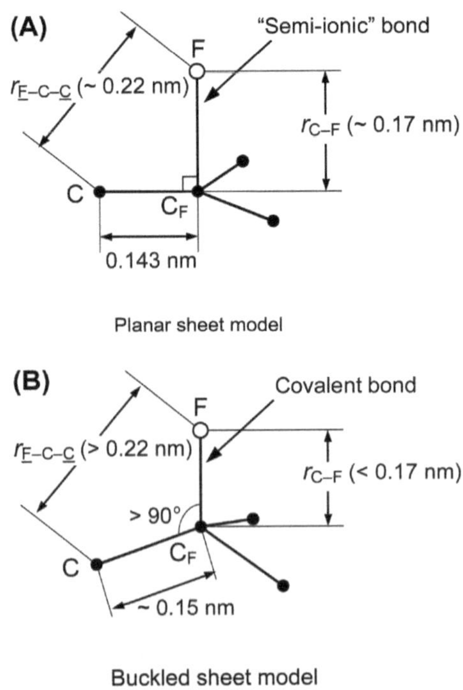

Figure 12.16 Models of local structure around a C–F bond in C_xF with (A) planar carbon sheets and (B) partially buckled carbon sheets.

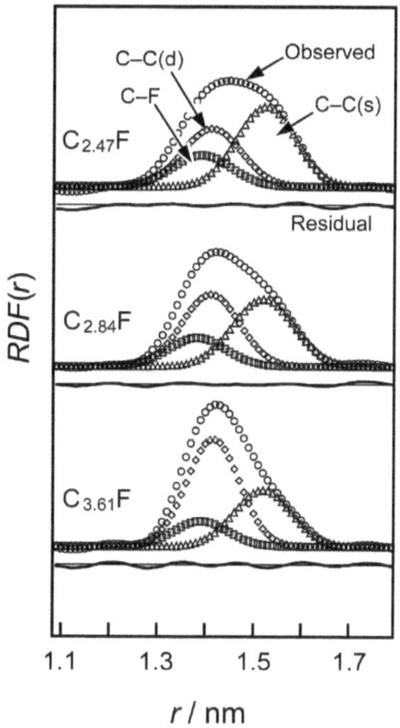

Figure 12.17 Radial distribution functions, *RDF*(r), of the stage-1 C_xFs.

0.17 nm and $r_{\underline{C}-C-\underline{F}}$ of larger than 0.222 nm without any contradictions. In the following discussion, carbon atoms bound and unbound to fluorine in C_xF are denoted as C_F and C_C, respectively. Assuming that the difference between the bond lengths of C_F-C_F and C_F-C_C in C_xF is negligibly small, these bonds will be denoted by C–C(s) so as to be distinguished from the graphitic C_C-C_C bonds, C–C(d).

The *RDF*(r) profiles of the C_xF samples at around 0.13–0.16 nm were approximated by Gaussian distribution functions. The C–F peak area in each *RDF*(r) was fixed so as to match coordination number of fluorine atoms around a carbon atom in average, CN_{C-F}, which corresponds to F/C atomic ratio of the compound, $1/x$. At the same time, coordination number of carbon atoms around a fluorine atom in average, CN_{F-C}, was necessarily fixed to one. As shown in Fig. 12.17, each *RDF*(r) profile is well approximated with three Gaussian peaks at 0.140, 0.142, and 0.153 nm which correspond to the bond lengths of r_{C-F}, $r_{C-C(d)}$, and $r_{C-C(s)}$ in C_xF, respectively. As listed in Table 12.4, r_{C-F} and $r_{C-C(s)}$ determined here slightly differ from those in $(CF)_n$ and $(C_2F)_n$ (0.136 and 0.157–0.159 nm, respectively). The numbers of C–C(d) and C–C(s) bonds around a carbon atom in average, $CN_{C-C(d)}$ and $CN_{C-C(s)}$, respectively, derived from the peak areas of these Gaussian peaks are listed in Table 12.5. The sum of $CN_{C-C(d)}$ and $CN_{C-C(s)}$ in each sample is about three, as is expected for the two-dimensional network of carbon sheets in which each carbon atom is bound to other three carbon atoms around.

Table 12.4 C—F and C—C Bond Lengths in the Layered Carbon Fluorides

Sample	r_{C-F}/(nm)	r_{C-C}/(nm)
C_xF ($x = 2.47, 2.84, 3.61$)	0.140	C—C(d): 0.142
		C—C(s): 0.153
$(CF)_n$	0.136	0.158
$(C_2F)_n$	0.136	0.157

Table 12.5 Nearest Neighbor Coordination Numbers of the C_xF Samples

Sample	CN_{C-F}	$CN_{C-C(d)}$	$CN_{C-C(s)}$	$CN_{C-C(d)} + CN_{C-C(s)}$	$CN_{C-C(s)}/2CN_{C-F}$
$C_{2.47}F$	0.41[a]	1.26	1.75	3.01	2.1
$C_{2.84}F$	0.35[a]	1.50	1.52	3.02	2.1
$C_{3.61}F$	0.28[a]	1.96	1.13	3.09	2.0

[a]Fixed so as to agree with the F/C atomic ratio ($=1/x$ of C_xF).

The numbers of C—C(s) and C—F bonds, N[C—C(s)] and N[C—F], were compared to discuss the arrangement of fluorine atoms on a single graphene layer in C_xF. The ratio N[C—C(s)]/N[C—F] is given as $(CN_{C-C(s)} \times 1/2)/CN_{C-F}$ in which the coefficient 1/2 is to avoid double counting of $N[X-Y]$ where atoms X and Y are not distinguished from each other. Four types of tetrahedral sites surrounded by $(3\text{-}n)C_C + nC_F + F$ ($n = 0, 1, 2, 3$ for type 0, 1, 2, 3, respectively) are possible to occupy for a C_F atom in C_xF (Fig. 12.18). Let one assume a structural model of stage-1 C_xF consisting of N_1 fluorine atoms and xN_1 carbon atoms, in which all the C_F atoms take the type-n_1 tetrahedral sites ($(3\text{-}n_1)C_C + n_1C_F + F$). Numbers of C_F and C_C atoms, $N[C_F]$ and $N[C_C]$, in this model are N_1 and $(x\text{-}1)N_1$, respectively, the former corresponding to N[C—F]. Each C_F atom at type-n_1 tetrahedral sites is bound to other n_1 of C_F atom(s) and to $(3\text{-}n_1)$ of C_C atom(s), therefore numbers of C_F—C_F and C_F—C_C bonds, N[C_F—C_F] and N[C_F—C_C], respectively, are derived as follows;

$$N[C_F{-}C_F] = N[C_F] \times n_1 \times 1/2 = n_1N_1/2 \quad (12.2)$$

$$N[C_F{-}C_C] = N[C_F] \times (3 - n_1) = (3 - n_1)N_1 \quad (12.3)$$

regarding both C_F—C_F and C_F—C_C bonds as C—C(s) bonds of longer than the graphitic C—C(d) bonds, N[C—C(s)] is given as the sum of N[C_F—C_F] and N[C_F—C_C].

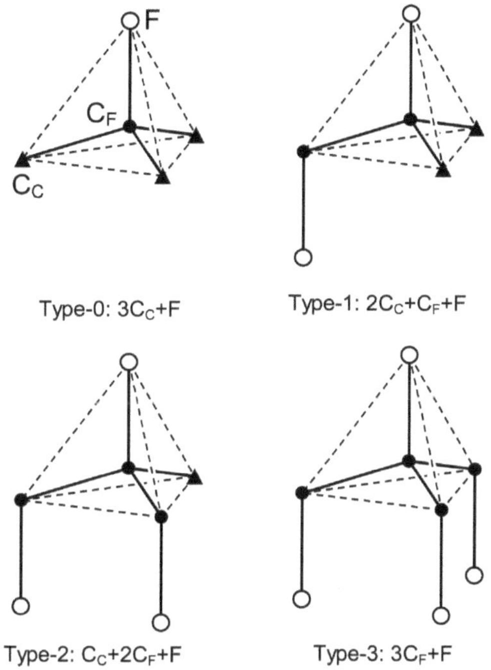

Figure 12.18 Possible tetrahedral sites for the C_F atoms in C_xF.

$$N[\mathrm{C{-}C(s)}] = n_1N_1/2 + (3 - n_1)N_1 = (6 - n_1)N_1/2 \quad (12.4)$$

Therefore N[C−C(s)]/N[C−F] is given as follows

$$N[\mathrm{C{-}C(s)}]/N[\mathrm{C{-}F}] = ((6 - n_1)N_1/2)/N_1 = (6 - n_1)/2. \quad (12.5)$$

Among the four possible sites for C_F atoms in C_xF, the type 3 ($3C_F + F$) of almost equivalent with the position of a carbon atom in $(CF)_n$ does not explain the differences in E_B's of the C1s XPS peaks between these compounds. Leaving the type-3 sites out of account as the major C_F positions in C_xF, N[C−C(s)]/N[C−F] of 2.0−2.1 estimated for the three C_xF samples (Table 12.5) are reasonably explained by the structures with most of the C_F atoms taking the type-2 sites ($C_C + 2C_F + F$).

No in-plane superlattice, ie, the ordered arrangement of fluorine atoms, is formed in C_xF used in the present study, as has been suggested by the XRD patterns and $S(Q)$ exhibiting only the *hk* diffraction peaks ascribed to the graphite-like hexagonal lattice besides the 001 peak. These diffraction data represent only the averaged information on the disordered structure of C_xF with partially buckled carbon sheets on which graphite-like C−C(d) bonds ($r_{C-C(d)} = 0.142$ nm) and the elongated C−C(s) bonds ($r_{C-C(s)} = 0.153$ nm) coexist. Therefore the graphite-like a-lattice parameter of the compound estimated from the *hk* diffraction peaks is larger than that of original graphite by only about 0.01 nm. The electric conductivities of these partially buckled

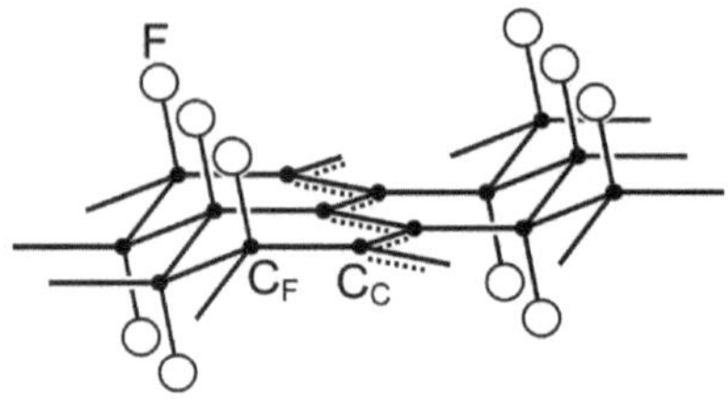

Figure 12.19 The conjugated π bonds remaining on the partially buckled carbon sheet in C_xF.

carbon sheets in C_xF are lower than that of planar graphene sheets, and strongly depends on the C/F atomic ratio, as has been previously reported [17]. However, even the conductivity of C_xF with x around two (5×10^{-8}–1×10^{-7} S/cm [10]) is much larger than those of electric insulators such as $(CF)_n$ and $(C_2F)_n$ ($<10^{-15}$ S/cm [2]), suggesting that the conductive network of the conjugated π bonds (corresponding to C–C(d) bonds) is partially maintained on the carbon sheets (Fig. 12.19). Such an arrangement of the C–C(d) bonds on the carbon sheets in stage-1 $C_{2.0}F$ is impossible if the major C_F positions are $3C_C + F$ and $2C_C + C_F + F$ tetrahedral sites, as shown in Fig. 12.20A and B, respectively. Each C_C atom in the former model is a radical with an unpaired electron and each pair of the neighboring C_C atoms in the latter model are bound to each other forming an isolated double bond. On the other hand, the network of the conjugated π bonds is partially maintained on the carbon sheets by allocating the C_F atoms the $C_C + 2C_F + F$ tetrahedral sites (Fig. 12.20C).

This new model of C_xF having both the conjugated π bonds on the carbon sheets and covalent C–F bonds does not conflict with the ^{13}C NMR spectra of the compound in which the peak appearing at 88 ppm ascribed to the sp^3-hybridized C_F atoms is clearly separated from the peak at 135 ppm ascribed to the sp^2-hybridized C_C atoms [8]. However, other spectroscopic results such as IR and XPS spectra previously obtained for C_xF suggest that the C–F bond nature in this compound is different from those in $(CF)_n$ and $(C_2F)_n$. The IR absorption bands ascribed to the C–F stretching mode appears at a lower wave number (around 1100/cm [8]) than those observed for $(CF)_n$ and $(C_2F)_n$ (around 1220/cm [2]). The C1s and F1s XPS peaks corresponding to C_F and F atoms in this compound shift toward lower E_B's (288 and 687 eV, respectively; Fig. 12.13) than those observed for $(CF)_n$ and $(C_2F)_n$ (290 and 690 eV for C1s and F1s peaks, respectively [2]). On the other hand, E_B of the C1s XPS peak ascribed to the C_C atoms in C_xF (around 285 eV; Fig. 12.13) is a little higher than that of the original graphite (284.3 eV), but much lower than that of the sp^3-hybridized C_C atoms in $(C_2F)_n$ (288.5 eV [2]). Based on the structural model of C_xF proposed here in which each C_F atom is connected to the network of the conjugated π bonds, these spectroscopic results can be reasonably explained by the hyperconjugation between C–C bonds on the carbon sheets and covalent C–F bonds. Such hyperconjugation involving the C–F σ bond has been reported for benzyl fluorides such as $C_6H_5CH_2F$ and $C_6H_5CFMe_2$ [24–28]. The electrons involved in the C–F σ bonds in C_xF are slightly delocalized by this hyperconjugation, resulting in the small decrease of the bond order. Therefore the C–F bond length in this compound

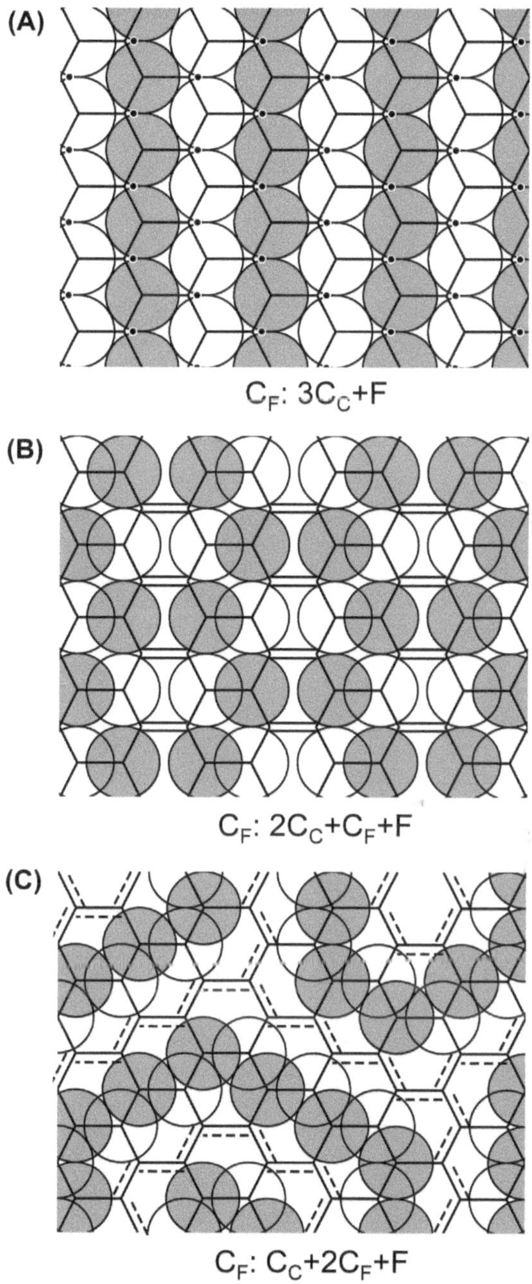

Figure 12.20 The arrangement of fluorine atoms on a carbon sheet in stage-1 $C_{2.0}F$ in which each CF atom takes the tetrahedral site of (A) $3C_C + F$, (B) $2C_C + C_F + F$, and (C) $C_C + 2C_F + F$. *Gray* and *white circles* represent the fluorine atoms above and below the carbon sheets, respectively. *Dots* in the model (A) represent the dangling bonds.

determined here (0.140 nm) is slightly longer than those in $(CF)_n$ and $(C_2F)_n$ (0.136 nm). The DFT calculations carried out for the C–F bond length perpendicular to the benzene ring in p-$NH_2C_6H_4CF_3$ was 0.1400 nm [29], which was exactly the same as was experimentally determined for C–F bond length of C_xF in the present study.

References

[1] V.O. Ruff, O. Bretschneider, F. Ebert, Z. Anorg. Allg. Chem. 217 (1934) 1–18.
[2] N. Watanabe, T. Nakajima, H. Touhara, Graphite Fluorides, Elsevier, Amsterdam, 1988.
[3] N. Watanabe, T. Nakajima, Graphite Fluorides and Carbon–Fluorine Compounds, CRC Press, Boca Raton, 1991.
[4] R.J. Lagow, R.B. Badachhape, J.L. Wood, J.L. Margrave, J. Chem. Soc. Dalton. Trans. (1974) 1268–1273.
[5] M. Takashima, N. Watanabe, J. Chem. Soc. Jpn. (1975) 432–436.
[6] H. Touhara, K. Kadono, Y. Fujii, N. Watanabe, Z. Anorg. Allg. Chem. 544 (1987) 7–20.
[7] T. Nakajima, Synthesis, structure, and physicochemical properties of fluorine–graphite intercalation compounds, in: T. Nakajima (Ed.), Fluorine–Carbon and Fluoride–Carbon Materials, Chemistry, Physics, and Applications, Marcel Dekker, New York, 1995.
[8] T. Mallouk, B.L. Hawkins, M.P. Conrad, K. Zilm, G.E. Maciel, N. Bartlett, Philos. Trans. R. Soc. Lond. A314 (1985) 179–187.
[9] H. Takenaka, M. Kawaguchi, M. Lerner, N. Bartlett, J. Chem. Soc. Chem. Commun. (1987) 1431–1432.
[10] T. Mallouk, N. Bartlett, J. Chem. Soc. Chem. Commun. (1983) 103–105.
[11] R. Hagiwara, M. Lerner, N. Bartlett, J. Chem. Soc. Chem. Commun. (1989) 573–574.
[12] I. Palchan, M. Crespin, H. Estrade-Szwarckopf, B. Rousseau, Chem. Phys. Lett. 157 (1989) 321–327.
[13] G. Nansé, E. Papirer, P. Fioux, F. Moguet, A. Tressaud, Carbon 35 (1997) 175–194 (and the references therein).
[14] T.R. Krawietz, J.F. Haw, J. Chem. Soc. Chem. Commun. (1998) 2151–2152.
[15] A.M. Panich, T. Nakajima, S.D. Gorena, Chem. Phys. Lett. 271 (1997) 381–384.
[16] J. Giraudet, M. Dubois, K. Guérin, C. Delabarre, A. Hamwi, F. Masin, J. Phys. Chem. B 111 (2007) 14143–14151.
[17] M.S. Dresselhaus, M. Endo, J.P. Issi, Physical properties of fluorine– and fluoride–graphite intercalation compounds, in: T. Nakajima (Ed.), Fluorine–Carbon and Fluoride–Carbon Materials, Chemistry, Physics, and Applications, Marcel Dekker, New York, 1995.
[18] Y. Sato, K. Itoh, R. Hagiwara, T. Fukunaga, Y. Ito, Carbon 42 (2004) 2897–2903.
[19] Y. Sato, K. Itoh, R. Hagiwara, T. Fukunaga, Y. Ito, Carbon 42 (2004) 3243–3249.
[20] J.E. Johns, M.C. Hersam, Acc. Chem. Res. 46 (2013) 77–86.
[21] S. Yuan, M. Rösner, A. Schulz, T.O. Wehling, M.I. Katsnelson, Phys. Rev. Lett. 114 (2015) 047403.
[22] B.R. Schwartz, G.R. Mitchell, Polymer 35 (1994) 3139–3148.
[23] Y. Sato, T. Kume, R. Hagiwara, Carbon 41 (2003) 351–357.
[24] E.T. McBee, I. Serfaty, T. Hodgins, J. Am. Chem. Soc. 93 (1971) 5711–5715.
[25] H. Schmidt, A. Schweig, Tetrahedron Lett. 14 (1973) 981–984.

[26] W. Adcock, A.N. Abeywickrema, Aust. J. Chem. 33 (1980) 181–187.
[27] G. Distefano, A. Modelli, M. Guerra, D. Jones, S. Rossini, J. Mol. Struct. 174 (1988) 177–182.
[28] R. Benassi, F. Taddei, J. Mol. Struct. 418 (1997) 59–71.
[29] O. Exner, S. Böhm, New J. Chem. 32 (2008) 1449–1453.

Lithium–Graphite Fluoride Battery—History and Fundamentals

13

T. Nakajima
Aichi Institute of Technology, Toyota, Japan

Chapter Outline

13.1 Development of Li/(CF)$_n$ Battery

Fluorine gas (F_2) is produced by electrolysis of molten KF•2HF using a carbon anode at c. 90°C. Large overpotential is usually observed for carbon anode. Fluorinated carbon film with a low surface energy is formed on carbon anode during the electrolysis, often leading to the occurrence of anode effect. Anode effect in which high cell voltage and no current flow are observed is caused by the fluorinated carbon film with low surface energy which extremely reduces the wettability of

New Fluorinated Carbons: Fundamentals and Applications. http://dx.doi.org/10.1016/B978-0-12-803479-8.00013-9

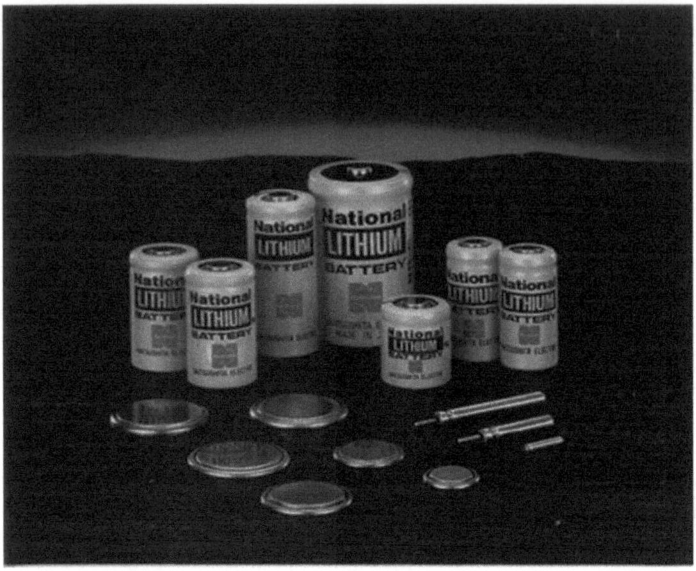

Figure 13.1 Lithium–graphite fluoride batteries [6].

electrolyte with carbon anode [1]. In our laboratory at Kyoto University, the research on anode effect led to the study on the synthesis, structure, and physical and chemical properties of fluorinated carbon materials, that is, graphite fluorides [1–5]. Many compounds such as fluorides, chlorides, oxides, sulfides, etc. were examined as cathodes of primary lithium battery in the 1960s. However, these trials were not successful. It was finally found that graphite fluoride is the best cathode material [1–5]. Based on the research on graphite fluorides made in our laboratory at Kyoto University, primary lithium–graphite fluoride battery (primary $Li/(CF)_n$ battery) was commercialized by Matsushita Battery Industrial Co., Ltd. in 1973 as shown in Fig. 13.1 [6]. The $Li/(CF)_n$ battery shows excellent properties such as high and flat discharge potential, large discharge capacity and energy density, long storage life, and so on. In our laboratory, synthetic and electrochemical studies were extensively made on the cell reaction of $Li/(CF)_n$ battery, discharge characteristics of graphite fluorides prepared from various carbon materials, structural factors of graphite fluoride governing the discharge characteristics, and new graphite fluoride with high performance, prepared from a unique carbon with submicronic thickness (submicronic layered carbon) via graphite oxide (recently called graphene oxide) [1–5]. The $Li/(CF)_n$ battery is the first commercial lithium battery, giving a strong impact to research and development of high energy density batteries, and leading to the commercialization of primary Li/MnO_2 battery in 1975 and rechargeable lithium batteries (lithium ion batteries) in 1991. In this chapter, these results on $Li/(CF)_n$ battery are summarized.

Recent topic is that $Li/(CF)_n$ battery was used for a minor planet explorer, "Hayabusa." After 7 years of journey in the space (2003–10), Hayabusa came back to the earth and a capsule containing sand or rock powder obtained on the minor planet, "Itokawa" was collected by electric wave radiation using Li/$(CF)_n$ battery. Thus Li/$(CF)_n$ battery is quite stable and long storage life.

13.2 Synthesis and Properties of Graphite Fluorides

13.2.1 Graphite Intercalation Compounds

Graphite fluoride is a kind of graphite intercalation compounds (GICs). Table 13.1 shows classification of GICs, which are classified in two categories, that is, those with covalent and ionic bonds between host graphite and guest species. GICs with covalent bonds are graphite fluorides and graphite oxide (recently called graphene oxide). Only fluorine and oxygen having high electronegativities give covalent type GICs in which fluorine or oxygen is directly bonded to carbon atom of graphene layers. Atoms or molecules with low ionic potentials such as alkali metals and those with high electron affinities such as halogen atoms, fluorides, chlorides, or acids are inserted into graphite, yielding a large number of ionic type GICs. Ionic type GICs consist of donor and acceptor types. Alkali and alkaline earth metals with low ionic potentials are intercalated into graphite by giving an electron from alkali and alkaline earth metals to host graphite (LiC_6, KC_8, SrC_6, CaC_6, etc.). This type GICs are called donor type because electron is donated from guest to host graphite. On the other hand, electron is accepted by intercalated guest species from graphite in the acceptor type GICs (C_8Br, C_8AsF_6, C_8SbF_6, C_7FeCl_3, etc.).

13.2.2 Synthesis of Graphite Fluorides

There are two crystal forms of $(CF)_n$ and $(C_2F)_n$ in graphite fluorides as shown in Fig. 13.2 [5]. They consist of covalent carbon-fluorine bonds. $(CF)_n$ is a white compound; however, $(C_2F)_n$ is black probably because a trace of sp^2 carbon atoms is contained due to the relatively low formation temperature. $(CF)_n$ and $(C_2F)_n$ are both synthesized by the fluorination of carbon materials with elemental fluorine (F_2) at 300–600°C. $(CF)_n$ type graphite fluoride is usually prepared from various

Table 13.1 Graphite Intercalation Compounds (GICs)

Covalent Type GICs		**$(CF)_n$, $(C_2F)_n$ (Graphite Fluorides)**
		$C_8O_2(OH)_2$ (Graphite Oxide (Graphene Oxide))
Ionic type GICs	Donor type	LiC_6, KC_8, RbC_8, CsC_8, CaC_6, SrC_6, BaC_6, SmC_6, EuC_6, YbC_6, etc.
	Acceptor type	$C_{x>2}F$, C_9IF_5, C_8Br, C_8ICl, C_6HNO_3, $C_{12}TiF_5$, C_8AsF_6, C_8SbF_6, C_7FeCl_3, C_6CuCl_2, C_9AlCl_3, $C_{18}AlBr_3$, etc.

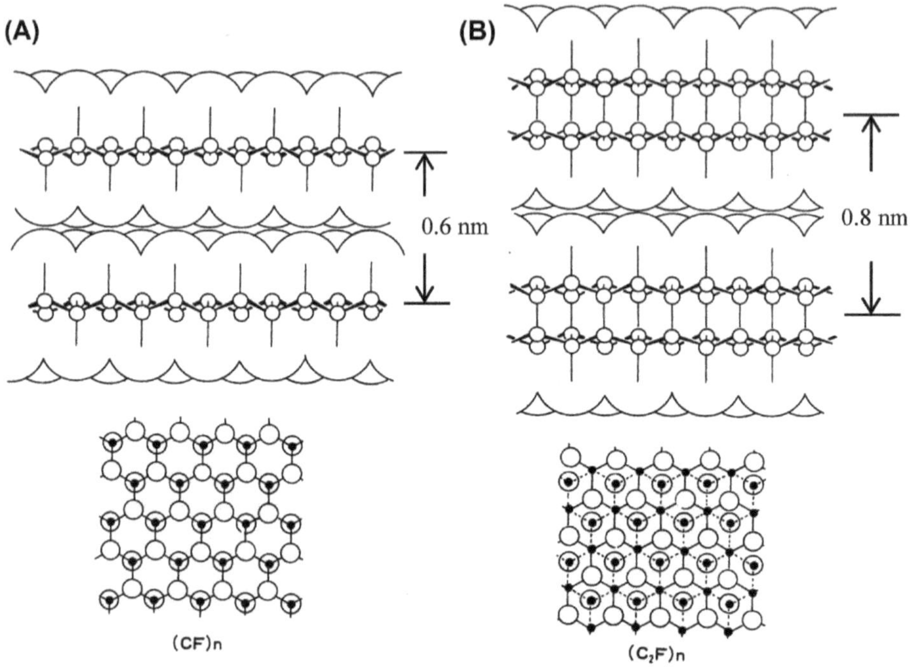

Figure 13.2 Structures of $(CF)_n$ (A) and $(C_2F)_n$ (B) [5].

carbon materials. On the other hand, $(C_2F)_n$ is formed under limited conditions. Natural graphite with high crystallinity gives $(CF)_n$ at 570–600°C and $(C_2F)_n$ at 350–400°C. Between 400 and 570°C, mixtures of $(CF)_n$ and $(C_2F)_n$ are obtained [2,5].

Natural graphite + F_2 →
- $(CF)_n$ (570–600°C)
- $(CF)_n + (C_2F)_n$ (400–570°C)
- $(C_2F)_n$ (350–400°C)

On the other hand, a low crystalline carbon such as petroleum coke yields only $(CF)_n$ at 300–590°C [2,5].

$$\text{Petroleum coke} + F_2 \rightarrow (CF)_n \ (300\text{–}590°C)$$

Synthetic graphite such as petroleum coke graphitized at 2700°C gives $(CF)_n$ at 570–600°C and mixtures of $(CF)_n$ and $(C_2F)_n$ at 337–570°C [2,5].

Petroleum coke graphitized at 2700°C + F_2 →
- $(CF)_n$ (570–600°C)
- $(CF)_n + (C_2F)_n$ (<50%) (480–570°C)
- $(CF)_n + (C_2F)_n$ (50–80%) (337–480°C)

Thus $(C_2F)_n$ is formed by the fluorination of natural and synthetic graphites with high crystallinity at relatively low temperatures. $(C_2F)_n$ content is higher in graphite fluoride mixture prepared from high crystalline natural graphite. Formation of fluorocarbon gases such as CF_4, C_2F_6, and so on is increased by carbon–carbon bond breaking at high temperatures above 600°C.

13.2.3 Surface Properties of Graphite Fluoride

Graphite fluorides are highly hydrophobic materials due to the covalent carbon-fluorine bonds. Surface free energies of solids are obtained by contact angle measurements using organic solvents and water. Table 13.2 shows the surface free energies of graphite fluoride in comparison with those of graphite sheet and polytetrafluoroethylene (PTFE) tablet [1–5]. Surface free energies were calculated from the contact angles of water and formamide. Surface free energies of $(CF)_n$ and PTFE are much smaller than that of graphite sheet. Surface free energy of flaky $(CF)_n$ is similar to that of PTFE. $(CF)_n$ tablet prepared from powdery sample shows the much smaller value of 2 mJ m^{-2}, which may be due to the effect of CF_2 and CF_3 groups existing at the edge of fluorinated graphene layers.

Heats of immersion of graphite fluorides in organic solvents are shown in Table 13.3 [1,2]. Heats of immersion for graphite fluorides and PTFE are much smaller than those for graphite, showing their hydrophobic property. Heats of immersion of $(CF)_n$ are 32–42 mJ m^{-2} smaller than those of PTFE, but those of $(C_2F)_n$ are similar to the values for PTFE.

Graphite fluorides take the layered structures in which CF or C_2F layers are stacked along *c*-axis. Graphite fluorides are therefore good solid lubricants [2,3]. Their low surface free energies provide low friction coefficients.

Table 13.2 Contact Angles of Water and Formamide, and Surface Free Energies of Graphite Fluoride, Graphite Sheet, and PTFE [1–3,5]

Sample	Contact Angle (degree) at 30°C		Surface Free Energy (mJ m^{-2})
	Water	Formamide	
Graphite sheet	95	81	44
$(CF)_n$ flake	117	97	18
$(CF)_n$ tablet	143	128	2
PTFE tablet	109	95	20

PTFE, polytetrafluoroethylene.

Table 13.3 Heats of Immersion ($-\Delta H_i$) of Graphite Fluorides, PTFE, and Graphite ($mJ\ m^{-2}$) [1,2]

	$(CF)_n$	$(C_2F)_n$	PTFE	Graphite
1-Buthanol	36	54	56	114
Methanol	42	–	–	102
Cyclohexane	32	–	–	–
CCl_4	35	54	–	115
1-Nitropropane	–	–	54	–
1-Cyclobutane	–	63	56	106
n-Heptane	–	–	58	–

PTFE, polytetrafluoroethylene.

13.3 Cell Reaction of Lithium–Graphite Fluoride Battery

13.3.1 Electromotive Forces of Li/CF$_x$ Cell Calculated on the Basis of Simple Cell Reaction

Thermodynamic functions of graphite fluoride were obtained by Wood et al. [7] as a function of x in CF_x to calculate the electromotive force (EMF) of Li/$(CF)_n$ cell. Measured samples were graphite fluorides, $CF_{1.125}$ ($(CF)_n$), $CF_{0.597}$ ($(C_2F)_n$), and fluorine-intercalated graphite $CF_{0.237}$ (C_4F) prepared at room temperature in the presence of HF. Heat capacities of these samples were measured using cryostat calorimeter at 5–310 K. Heats of combustion were also obtained using bomb calorimeter at 298K. Heat capacities and standard entropies of graphite and CF_x samples are given in Tables 13.4 and 13.5 [2,7]. Standard enthalpy changes of formation of CF_x samples obtained from heats of combustion are given in Table 13.6. In these data, one mol of CF_x is defined as CF_x containing one mol of carbon [2,7].

Calculation of EMF of Li/$(CF)_n$ cell was made using these data, based on the simple cell reaction: $CF_x + x\text{Li} \rightarrow C + x\text{LiF}$, in which LiF and C are directly formed without any intermediate product [2,7].

(a) EMF of Li/$CF_{0.237}$ cell

From JADRANSKI NAFTOVOD (JANAF) data,
$\Delta H_f°(\text{LiF, s}) = -616.93\ \text{kJ mol}^{-1}$,
$S_{298}(\text{LiF, s}) = 35.66\ \text{JK}^{-1}\ \text{mol}^{-1}$,
$S_{298}(\text{Li, s}) = 29.10\ \text{JK}^{-1}\ \text{mol}^{-1}$,
$S_{298}(\text{C, s}) = 5.69\ \text{JK}^{-1}\ \text{mol}^{-1}$.
Using $\Delta H_f°(CF_{0.237}, \text{s}) = -25.44\ \text{kJ mol}^{-1}$, the enthalpy change for $CF_{0.237}(s) + 0.237\text{Li}(s) \rightarrow C(s) + 0.237\text{LiF}(s)$ is $-120.77\ \text{kJ mol}^{-1}$ ($= \Delta H_{298}$).

Table 13.4 Heat Capacities (C_p) of Graphite and CF_x Samples [2,7]

Temperature (K)	Graphite (JK^{-1} mol^{-1})	$CF_{0.237}$ (JK^{-1} mol^{-1})	$CF_{0.597}$ (JK^{-1} mol^{-1})	$CF_{1.125}$ (JK^{-1} mol^{-1})
15	0.0427	0.2600	0.1469	0.3175
25	0.1255	0.5199	0.3612	0.7363
50	0.5063	1.448	1.347	2.449
75	1.046	2.615	3.064	5.185
100	1.658	3.852	5.086	8.468
150	3.229	6.300	9.011	14.51
200	4.937	8.735	12.55	19.45
250	6.816	11.06	15.65	23.70
300	8.590	13.32	18.54	27.57

Table 13.5 Standard Entropies (S_{298}) of Graphite and CF_x Samples at 298 K [2,7]

Sample	S_{298} (JK^{-1} mol^{-1})
C (graphite)	5.74
$CF_{0.237}$	11.22
$CF_{0.597}$	14.71
$CF_{1.125}$	23.64

Table 13.6 Standard Enthalpy Changes of Formation ($\Delta H_f°$) of CF_x Samples at 298 K [2,7]

Sample	$\Delta H_f°$ (kJ mol^{-1})
$CF_{0.237}$	−25.43
$CF_{0.597}$	−101.74
$CF_{1.125}$	−195.73

The entropy change for this cell reaction is $-3.98\ JK^{-1}\ mol^{-1}$ ($= \Delta S_{298}$) using $S_{298}(CF_{0.237}, s) = 11.22\ JK^{-1}\ mol^{-1}$, $S_{298}(Li, s) = 29.10\ JK^{-1}\ mol^{-1}$, $S_{298}(LiF, s) = 35.66\ JK^{-1}\ mol^{-1}$ and $S_{298}(C, s) = 5.69\ JK^{-1}\ mol^{-1}$.

Therefore $\Delta G_{298} = -119.5$ kJ mol^{-1} ($\Delta G = \Delta H - T\Delta S$).

From $\Delta G_{298} = -0.237FE_{298}$, E_{298} is 5.23 V ($E_{298} = 5.23$ V, $(dE/dT)_{298} = -0.2$ mV K^{-1}).

Table 13.7 Calculated Electromotive Forces and Temperature Coefficients for Li/CF$_x$ ($0.597 \leq x \leq 1.125$) Cell at 298 K [2,7]

x in CF$_x$	ΔG_{298} (kJ mol^{-1})	E_{298} (V)	$(dE/dT)_{298}$ (mV K^{-1})
0.597	−265.0	4.60	−0.09
0.7	−309.9	4.59	−0.09
0.8	−353.5	4.58	−0.09
0.9	−397.1	4.57	−0.10
1.0	−440.7	4.57	−0.10
1.125	−495.2	4.56	−0.10

(b) EMF of Li/CF$_x$ ($0.597 \leq x \leq 1.125$) cell

Enthalpy change of formation and entropy for CF$_x$ are obtained by interpolation of the data for $CF_{0.597}$ and $CF_{1.125}$ as a function of x in CF$_x$.

Using the following data:

$CF_{0.597}$: $\Delta H_f^\circ = -101.74$ kJ mol^{-1}, $S_{298} = 14.71$ JK^{-1} mol^{-1},

$CF_{1.125}$: $\Delta H_f^\circ = -195.73$ kJ mol^{-1}, $S_{298} = 23.64$ JK^{-1} mol^{-1},

$\Delta H_f^\circ(CF_x, s)$ and $S_{298}(CF_x, s)$ are calculated:

$\Delta H_f^\circ(CF_x, s) = (4.53 - 178.01x)$ kJ mol^{-1},

$S_{298}(CF_x, s) = (4.62 + 16.91x)$ JK^{-1} mol^{-1}.

From these and JANAF data, the thermodynamic functions for $CF_x(s) + xLi(s) \rightarrow C(s) + xLiF(s)$ are obtained:

$\Delta H_{298} = -(438.92x + 4.53)$ kJ mol^{-1},

$\Delta S_{298} = (1.07 - 10.35x)$ JK^{-1} mol^{-1},

$\Delta G_{298} = -(435.83x + 4.85)$ kJ mol^{-1}.

ΔG_{298}, E_{298}, and $(dE/dT)_{298}$ calculated as a function of x in CF$_x$ are summarized in Table 13.7 [2,7]. The calculated EMF values are in the range of 4.56−4.60 V, which are higher than the observed open circuit voltages (OCVs), 3.1−3.5 V by more than 1 V as described in the next section. This suggests that discharge reaction of Li/(CF)$_n$ cell is different from the simple reaction: $CF_x + xLi \rightarrow C + xLiF$. For this discrepancy between the calculated and observed OCVs, Whittingham proposed that cell reaction was the formation of ternary graphite intercalation compound ($CF + xLi \rightarrow CLi_xF$) [8].

13.3.2 Thermodynamic Data of Real Cell

OCV ($E_{CF0.93}$ (V vs Li/Li$^+$)) of Li/$CF_{0.93}$ cell was measured in 1 mol L^{-1} $LiClO_4$-propylene carbonate (PC) between 18 and 30°C after 5% discharge. The result is shown in Fig. 13.3 [9], from which thermodynamic functions for cell reaction are calculated as given in Table 13.8 [9]. The observed $E_{CF0.93}$ was much lower than the calculated value, 4.57 V. The entropy decrease in the real cell is larger than the calculated value, and the Gibbs free energy and enthalpy changes are both smaller. These suggest that the cell reaction is not direct formation of LiF and carbon, but discharge product is a complex.

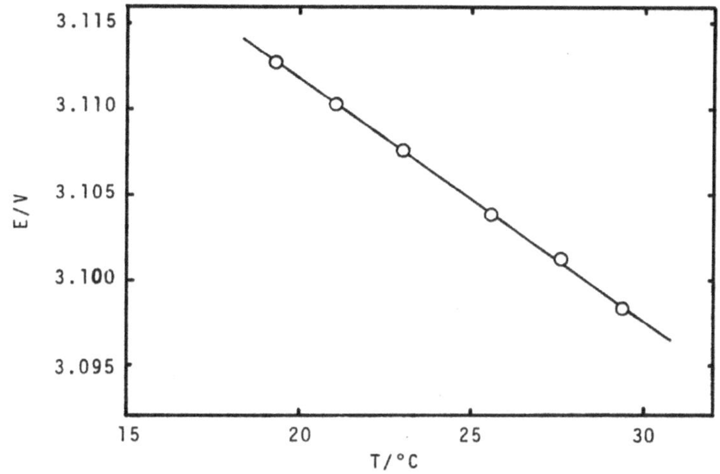

Figure 13.3 Change in open circuit voltage of Li/$CF_{0.93}$ cell as a function of temperature [9].

Table 13.8 Observed and Calculated Thermodynamic Data for Li/$CF_{0.93}$ Cell at 25°C [9]

	Observed	Calculated[a]
$E_{CF0.93}$ (V vs Li/Li^+)	3.10	4.57
$(\partial E_{CF0.93}/\partial T)_p$ (mV K^{-1})	−1.44	−0.10
ΔS (JK^{-1} mol^{-1})	−129	−9.0
$T\Delta S$ (kJ mol^{-1})	−38.4	−2.7
ΔG (kJ mol^{-1})	−278	−410
ΔH (kJ mol^{-1})	−316	−413

[a]Calculated Based on the Cell Reaction: $CF_{0.93}$(s) + 0.93Li(s) → C(s) + 0.93LiF(s).

OCV measurements of Li/$CF_{1.00}$ and Li/$CF_{0.61}$ cells were performed in 1 mol L^{-1} $LiClO_4$-PC, ɣ-butyrolactone (BL), and dimethyl sulfoxide (DMSO) at 25°C and in 1 mol L^{-1} $LiClO_4$–sulfolane (TMS) at 30°C (m.p. of TMS: 28.5°C) after 25% discharge [10]. Table 13.9 shows Gibbs free energies of solvation for Li^+ ion calculated by modified Born's equation, electrode potentials of Li/Li^+ reference, and OCVs in four solvents [10]. Potential of Li/Li^+ reference electrode decreases with increasing solvation energy for Li^+ ion. The electrode potentials of graphite fluoride cathode relative to Li/Li^+ in PC are shown in Fig. 13.4 [10]. If the cell reaction is direct formation of LiF and C (CF + Li → LiF + C), that is, desolvation and formation of LiF proceed fast, OCVs should be the same in four different solvents, in other words, electrode potentials of graphite fluoride should be lower by 0.01, 0.13, and 0.21 V in

Table 13.9 Gibbs Free Energy Changes of Solvation for Li^+, Electrode Potentials of Li/Li^+ Reference, and OCVs of $Li/CF_{1.00}$ and $Li/CF_{0.61}$ cells in four solvents [10]

Solvent	DMSO	BL	TMS	PC
ΔG_s (J(mol-Li^+)$^{-1}$)	-5.00×10^5	-4.80×10^5	-4.58×10^5	-4.56×10^5
Li/Li^+ (V)	−0.45	−0.25	−0.02	0.00
OCV ($Li/CF_{1.00}$)	3.50	3.30	3.28	3.28
OCV ($Li/CF_{0.61}$)	3.47	3.39	3.27	3.26

OCV, open circuit voltage; *DMSO*, dimethyl sulfoxide; *BL*, ɣ-butyrolactone; *TMS*, $LiClO_4$—sulfolane; *PC*, propylene carbonate.

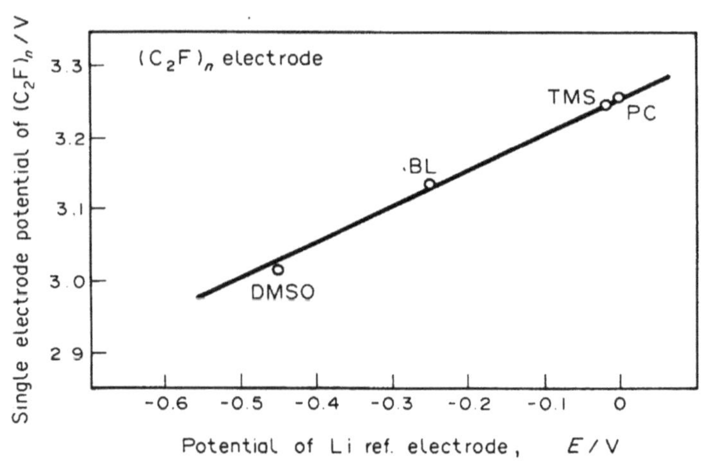

Figure 13.4 Electrode potentials relative to Li/Li^+ in propylene carbonate of graphite fluoride cathode in four solvents [10].

TMS, BL, and DMSO, respectively, than those in Fig. 13.4. However, measured OCV value increases with increasing solvation energy for Li^+ ion as given in Table 13.9. Analysis of discharge product revealed that graphite fluoride cathode consisted of carbon, LiF, and organic solvent [1,2]. These results suggest that solvent molecules participate in the cell reaction, that is, cathode product containing solvent molecules determines the electrode potential of graphite fluoride cathode.

13.3.3 Cell Reaction

Based on the thermodynamic data shown in the previous section, the following cell reaction was proposed [10,11]. Solvated Li^+ ion is inserted into fluorinated graphene layers to form cathode product $C_x \cdot F^- \cdot Li^+ \cdot zS$, which may have some stability due to

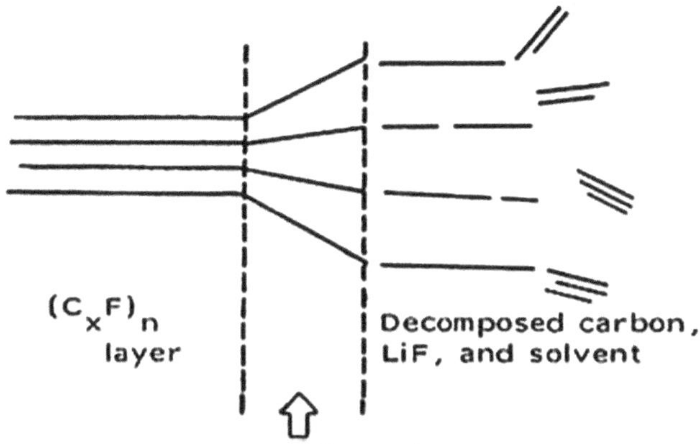

Figure 13.5 Schematic illustration of discharge reaction of graphite—fluoride cathode [11].

the strong carbon-fluorine bond and high solvation ability of Li^+ ion as given in Eqs. (13.1—13.3). The discharge product, $C_x{\cdot}F^-{\cdot}Li^+{\cdot}zS$, determines the cathode potential and gradually decomposes to carbon, LiF, and solvent by chemical reaction (Eq. (13.4)). The discharge reaction of graphite fluoride is illustrated in Fig. 13.5 [11].

$$\text{Anode reaction: } Li + yS \rightarrow Li^+{\cdot}yS + e^- \text{ (S: solvent molecule)} \tag{13.1}$$

$$\text{Cathode reaction: } C_xF + Li^+{\cdot}yS + e^- \rightarrow C_x{\cdot}F^-{\cdot}Li^+{\cdot}zS + (y\text{-}z)S \tag{13.2}$$

$$\text{Cell reaction: } C_xF + Li + yS \rightarrow C_x{\cdot}F^-{\cdot}Li^+{\cdot}zS + (y\text{-}z)S \tag{13.3}$$

$$\text{Decomposition of cathode product: } C_x{\cdot}F^-{\cdot}Li^+{\cdot}zS \rightarrow xC + LiF + zS \tag{13.4}$$

Large exothermic heat is observed when $Li/(CF)_n$ battery is discharged. This may be due to the decomposition of cathode product releasing heat, that is, crystallization of LiF. It is also supported by the analysis that final discharge product contains low crystalline carbon, LiF, and solvent [12].

13.4 Structural Factors of Graphite Fluoride Governing Discharge Characteristics

13.4.1 Discharge Capacity

Discharge capacity is governed by chemical composition, that is, the amount of fluorine atoms bonded to basal plane of graphene layers. There are two crystal forms in graphite fluorides, $(CF)_n$ and $(C_2F)_n$. $(CF)_n$ has larger capacity than $(C_2F)_n$ because of the higher fluorine content. Discharge capacities of $(CF)_n$ and $(C_2F)_n$ prepared

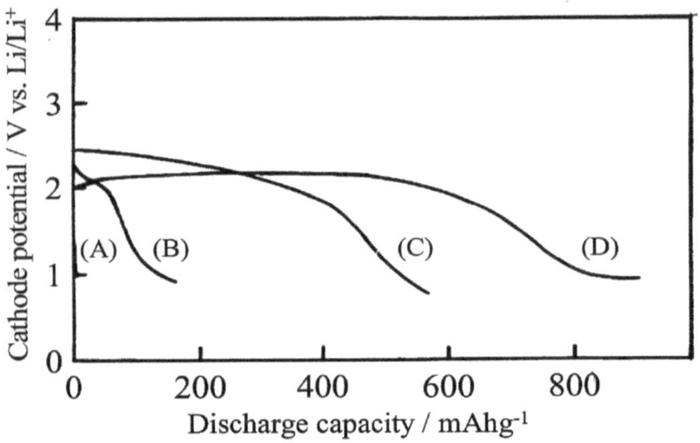

Figure 13.6 Discharge capacities of PTFE (A), fluorinated coronene $C_{24}F_{36}$ (B), $(CF)_n$ prepared from petroleum coke (C), and $(CF)_n$ prepared from natural graphite (D), obtained in 1 mol L^{-1} $LiClO_4$-PC (25°C, 0.5 mA cm^{-2}) [11]. *PTFE*, polytetrafluoroethylene.

from natural graphite were approximately 710 and 600 mA hg^{-1}, respectively in 1 mol L^{-1} $LiClO_4$-PC at 25°C (current density: 0.5 mA cm^{-2}, cut-off potential: 1.5 V vs Li/Li^+) [11,13].

Fluorine atoms bonded to basal plane are electrochemically discharged. But those bonded to edge plane as CF_2 and CF_3 groups are stable against electrochemical reduction [14]. Therefore $(CF)_n$ having large graphene layers gives a large capacity. Fig. 13.6 shows discharge capacities of $(CF)_n$ samples prepared from natural graphite and petroleum coke, fluorinated coronene, $C_{24}F_{36}$, and PTFE [11]. Natural graphite with high crystallinity has much larger crystallite size along *a*-axis, that is, much larger graphene layers than petroleum coke. Therefore the amount of fluorine atoms bonded to graphene layers is larger in $(CF)_n$ prepared from natural graphite than petroleum coke. Discharge capacity of $(CF)_n$ prepared from natural graphite is larger than that from petroleum coke as shown in Fig. 13.6. Fluorinated coronene contains a small amount of dischargeable fluorine atoms. PTFE with a linear chain structure is not electrochemically discharged. The order of capacities are as follows: $(CF)_n$ prepared from natural graphite (D) > $(CF)_n$ prepared from petroleum coke (C) > fluorinated coronene (B) > PTFE (A).

13.4.2 Discharge Potential

Main factors influencing the discharge potential of graphite fluoride cathode are surface area and interlayer spacing of fluorinated graphene layers. Fluorination of carbon materials enlarges surface area of graphite fluoride by fluorine insertion into graphene layers partly accompanying carbon-carbon bond breaking as given in Table 13.10 [5]. BET surface areas of graphite fluoride prepared from raw and graphitized petroleum cokes are 30–40 times larger than those of pristine materials. Surface areas of graphite

Table 13.10 Surface Areas of Carbon Materials and Graphite Fluoride [5]

Carbon Material	Fluorination Temperature (°C)	Surface Area (m^2g^{-1})	
		Carbon Material	Graphite Fluoride
Petroleum coke (raw)	380–500	6.4	206–240
Petroleum coke (heat-treated at 2800°C)	560	2.4	98
Natural graphite (290–883 μm)	475–600	0.6	95–115
Natural graphite (61–74 μm)	460–555	1.4	102–105

fluoride prepared from natural graphite flake and powder are 160–190 and 75 times larger than those of original samples, respectively. Increase of surface area brings about increase in electrode area of graphite fluoride cathode, which reduces current density and then makes the cathode potential high because discharge of the battery is made at a constant current.

Discharge of graphite fluoride cathode proceeds with insertion of solvated Li^+ ions. Large overpotential of graphite fluoride cathode is caused by the delay of Li^+ ion diffusion in the cathode product [15]. Discharge potential and overpotential are affected by interlayer spacing of fluorinated graphene layers, $d_{(001)}$ value and half width of (001) diffraction line, $\beta_{(001)}$ (the repeat distance along *c*-axis is indexed as (001) or (002) according to the stacking structure of fluorinated graphene layers [16]). As $d_{(001)}$ increases, interlayer space between fluorinated graphene layers increases, which makes easy the transfer of solvated Li^+ ions. The $\beta_{(001)}$ is inversely proportional to the thickness of graphite fluoride crystallite along *c*-axis. With increasing $\beta_{(001)}$, thickness of crystallite size of fluorinated graphene layers decreases, which makes easy the expansion of fluorinated graphene layers by the insertion of solvated Li^+ ions. As shown in Fig. 13.7, cathode overpotential decreases with increasing $d_{(001)}$ and $\beta_{(001)}$ [13]. Natural and synthetic graphites yield $(CF)_n$ with high crystallinity having relatively small $d_{(001)}$ and $\beta_{(001)}$ values. Overpotentials are high for these $(CF)_n$ samples, A, B, and C. Carbon with low crystallinity such as non-heat-treated petroleum coke provides $(CF)_n$ with low crystallinity having large $d_{(001)}$ and $\beta_{(001)}$. Furthermore the $d_{(001)}$ and $\beta_{(001)}$ values of $(CF)_n$ samples, E, F, G, and H increase with decreasing fluorination temperature. The overpotentials are low in $(CF)_n$ samples prepared from non-heat-treated petroleum coke as shown in Fig. 13.7. Discharge potential of $(CF)_n$ prepared from petroleum coke is therefore higher than that prepared from natural graphite (Fig. 13.6) [11].

Potential drop of graphite fluoride cathode is normally observed at the beginning of discharge since graphite fluoride is an electric insulator. To improve this phenomenon, slight decomposition of graphite fluoride was attempted in Cl_2 atmosphere [17]. Partial decomposition graphite fluoride improves the potential drop without decrease in the discharge capacity.

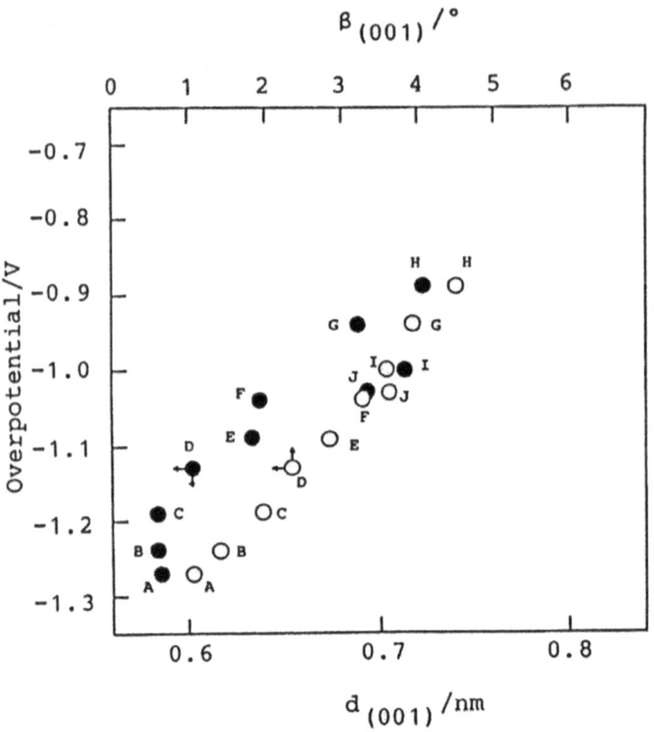

Figure 13.7 Overpotentials of $(CF)_n$ cathodes in 1 mol L^{-1} $LiClO_4$-PC at 25% discharge as functions of d(001) (●) and half width of (001) diffraction line, β(001) (○) (25°C, 0.5 mA cm^{-2}) [13]. Raw materials and fluorination temperatures of $(CF)_n$ samples: (A) natural graphite, 590°C, (B and C) petroleum coke heat-treated at 2800, 600, and 510°C, respectively, (D) exfoliated graphite, 590°C, (E, F, G, and H) non-heat-treated petroleum coke, 420, 360, 340, and 290°C, respectively, (I) commercial $(CF)_n$ prepared from petroleum coke.

13.5 Discharge Characteristics of Graphite Fluoride Prepared From a New Carbon With Submicronic Thickness (Submicronic Layered Carbon), Obtained by Thermal Decomposition of Graphite Oxide (Graphene Oxide)

13.5.1 Preparation of Submicronic Layered Carbon

Since discharge of graphite fluoride cathode proceeds by the insertion of solvated Li^+ ion into fluorinated graphene layers, cathode overpotential would decrease if thickness of fluorinated graphene layers is reduced. Exfoliated graphite is usually prepared by thermal decomposition of partially decomposed GIC with ionic bonds. Thickness of stacked graphene layers is reduced by expansion of graphite due to

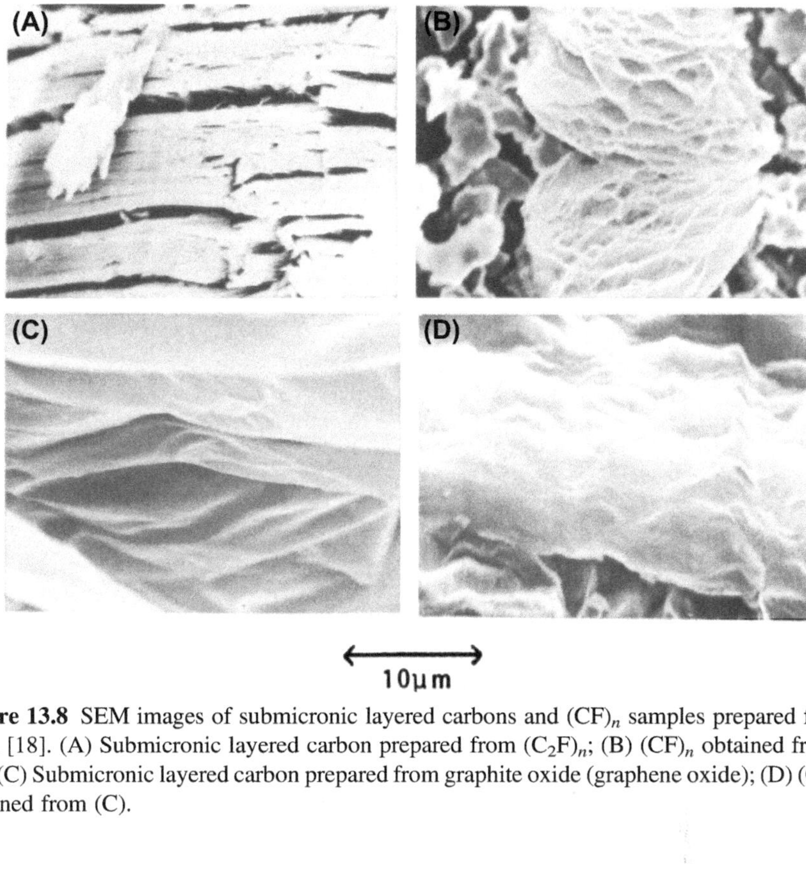

Figure 13.8 SEM images of submicronic layered carbons and $(CF)_n$ samples prepared from them [18]. (A) Submicronic layered carbon prepared from $(C_2F)_n$; (B) $(CF)_n$ obtained from (A); (C) Submicronic layered carbon prepared from graphite oxide (graphene oxide); (D) $(CF)_n$ obtained from (C).

release of residual intercalated species at a high temperature. Thermal decomposition of GICs with covalent bonds, graphite fluoride, and graphite oxide (graphene oxide) may be better method to prepare a new carbon with smaller *c*-axis thickness than exfoliated graphite. Fig. 13.8 shows new carbons with submicronic thickness (submicronic layered carbons) prepared from $(C_2F)_n$ (A) and graphite oxide (C), and $(CF)_n$ samples prepared from them (B and D) [18]. Fig. 13.9 also shows a submicronic layered carbon obtained by thermal decomposition of $(CF)_n$. The *c*-axis thickness is much less than 1 μm.

Submicronic layered carbon is easily prepared from thermal decomposition of graphite oxide (graphene oxide) as shown in Fig. 13.8C [18–20]. There are three methods for preparation of graphite oxide, that is, Brodie, Staudenmaier, and Hummers and Offeman methods [2]. Among them, Hummers and Offeman method using $KMnO_4$ [21] is suitable for the preparation of submicronic layered carbon. Natural graphite (10 g) and $NaNO_3$ (5 g) is put into H_2SO_4 (230 mL), and then $KMnO_4$ (5–40 g) is added slowly to the solution under stirring [18–20]. After that, water is gradually poured into the solution. When the temperature of the solution is elevated to about 90–100°C, graphite oxide is formed. Water is further added without isolation

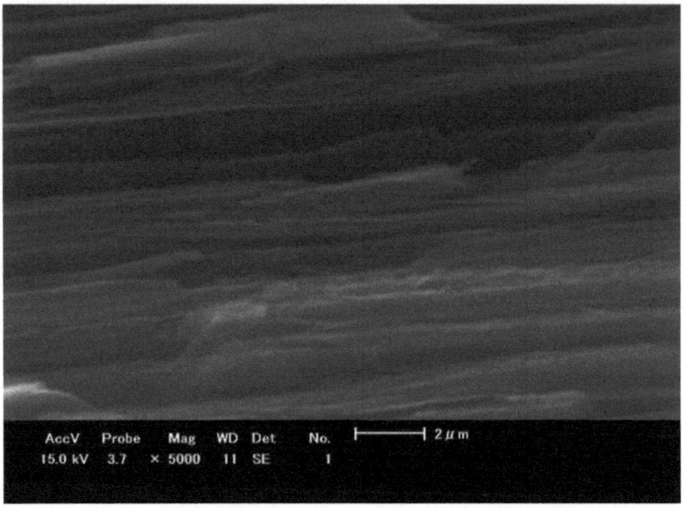

Figure 13.9 SEM image of submicronic layered carbon prepared from $(CF)_n$ (thermal decomposition: 700°C, 2 h).

of graphite oxide. When the temperature reaches to 140–180°C, graphite oxide is decomposed, giving submicronic layered carbon. The obtained submicronic layered carbon is washed in a funnel several 10 times until neutralization of filtrate, and dried under vacuum at 60°C for one day. The dried submicronic layered carbon is heat-treated at 400°C for 2 h to reduce oxygen. The oxygen content thus obtained was 12 wt% [20]. The crystallite sizes of submicronic layered carbons prepared from graphite oxide are shown in Fig. 13.10 [20]. The crystallite sizes along *c*-axis (L_c) are in the range of 10–20 nm much smaller than >100 nm of original natural graphite. Crystallite size along *a*-axis, L_a is always larger than L_c. Fluorination of submicronic

Figure 13.10 Crystallite sizes of submicronic layered carbons prepared from graphite oxide (graphene oxide) as a function of amount of $KMnO_4$ [20]. L_a: crystallite size along *a*-axis, L_c: crystallite size along *c*-axis.

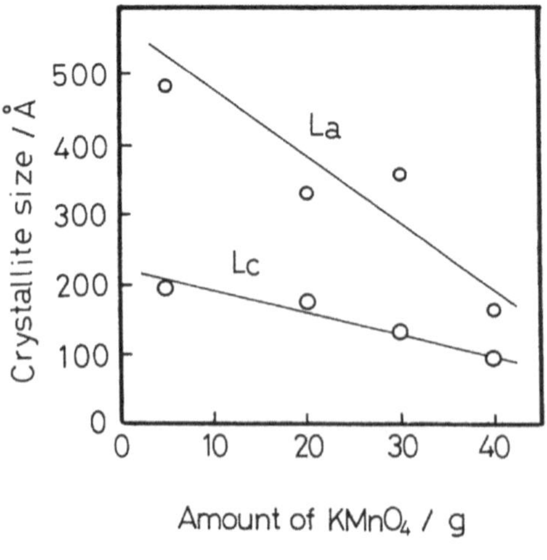

Table 13.11 Surface Areas of Carbon Materials and $(CF)_n$ Samples [20]

Sample	Surface Area ($m^2 g^{-1}$)	
	Carbon Material	$(CF)_n$
Natural graphite	0.74	96
Submicronic layered carbon via graphite oxide	13	155
	20	154

layered carbon at 400–600°C yields $(CF)_n$ [18–20]. Fluorination temperature to obtain $(CF)_n$ is highly lowered. Fluorinated products at 300–350°C contain $(C_2F)_n$ [18–20]. Surface areas of submicronic layered carbons and obtained $(CF)_n$ samples are both larger than those of natural graphite and $(CF)_n$ as given in Table 13.11 [20].

13.5.2 Discharge Characteristics of $(CF)_n$ Prepared From Submicronic Layered Carbon

Discharge curve of $(CF)_n$ obtained from submicronic layered carbon is shown in Fig. 13.11 in comparison with those of $(CF)_n$ and $(C_2F)_n$ prepared by direct fluorination of natural graphite [18]. The discharge potential is higher than those of $(CF)_n$ and $(C_2F)_n$, and discharge capacity is also much larger than that for $(CF)_n$. The discharge capacity was 789 mA hg^{-1} much larger than 674 mA hg^{-1} of $(CF)_n$ prepared by direct

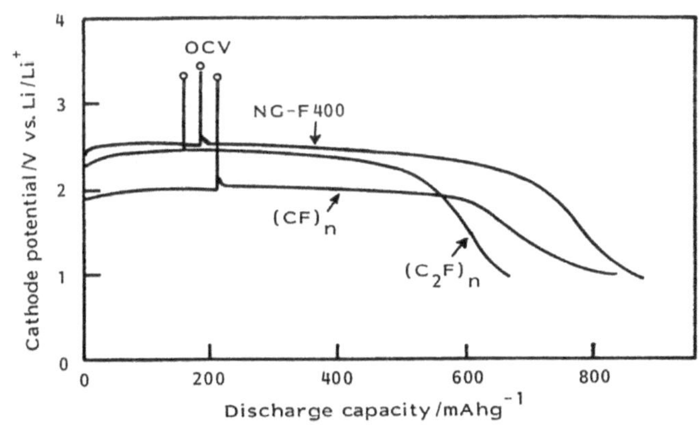

Figure 13.11 Discharge curves of $(CF)_n$ samples prepared from submicronic layered carbon and natural graphite in 1 mol L^{-1} $LiClO_4$-PC (25°C, 0.5 mA cm^{-2}) [19]. NG-F400: $(CF)_n$ prepared at 400°C from submicronic layered carbon via graphite oxide, $(CF)_n$: prepared at 610°C from natural graphite, $(C_2F)_n$: prepared at 350°C from natural graphite.

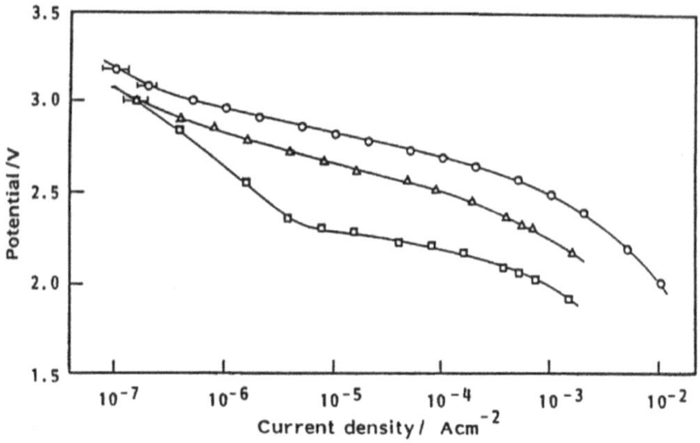

Figure 13.12 Galvanostatic polarization curves of graphite fluoride cathodes at 25% discharge, obtained in 1 mol L^{-1} $LiClO_4$-PC at 25°C [18]. −○−○−○−: $(CF)_n$ prepared at 400°C from submicronic layered carbon via graphite oxide,−Δ−Δ−Δ−: $(C_2F)_n$ prepared from natural graphite at 350°C,−□−□−□−: $(CF)_n$ prepared at 600°C from natural graphite.

fluorination of natural graphite (cut-off potential:1.5 V vs Li/Li^+) [19]. Fig. 13.12 shows discharge potentials in the wide range of current densities [18]. The discharge potentials of $(CF)_n$ prepared from submicronic layered carbon are higher than those for $(C_2F)_n$ and $(CF)_n$ in the whole range of current densities. Capacity reduction at high current densities is also small in the $(CF)_n$ prepared from submicronic layered carbon as shown in Fig. 13.13 [18]. These may be due to large surface area of $(CF)_n$ prepared from submicronic layered carbon and easy insertion and transfer of Li^+ ion in the cathode product.

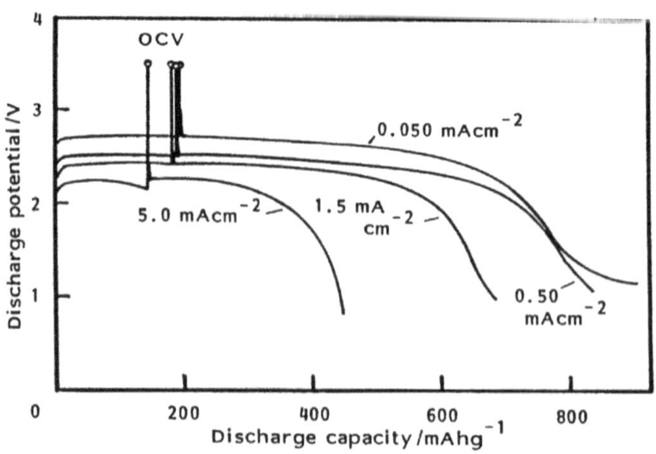

Figure 13.13 Discharge curves of $(CF)_n$ prepared at 400°C from submicronic layered carbon via graphite oxide at various current densities, obtained in 1 mol L^{-1} $LiClO_4$-PC at 25°C [18].

13.6 Conclusions

Li/(CF)$_n$ battery is the first lithium battery commercialized in 1973, having excellent properties such as large capacity and energy density, high stability, long storage life, and so on. These are derived from the highest electronegativity of fluorine and high stability of graphite fluoride cathode. Graphite fluoride is stable in air with high hydrophobicity because of the covalent carbon-fluorine bonds. Development of Li/(CF)$_n$ battery was the beginning of high energy density batteries. Practical use of Li/(CF)$_n$ battery led to the commercialization of Li/MnO_2 battery in 1975 and lithium ion battery in 1991. Nowadays, these batteries are applied to many kinds of electronic devices, hybrid cars, electric vehicles, and so on. In particular, development of high energy density batteries with high rate charge/discharge and high safety is urgently requested [22].

Rererences

[1] N. Watanabe, T. Nakajima, H. Touhara, Graphite Fluorides, Elsevier, Amsterdam, 1988 (and references therein).

[2] T. Nakajima, N. Watanabe, Graphite Fluorides and Carbon-fluorine Compounds, CRC Press, Boca Raton, Florida, 1991 (and references therein).

[3] T. Nakajima, N. Watanabe, ChemTech 20 (1990) 426–430 (and references therein).

[4] T. Nakajima, J. Fluorine Chem. 100 (1999) 57–61 (and references therein).

[5] T. Nakajima, J. Fluorine Chem. 149 (2013) 104–111 (and references therein).

[6] Cover of "National Technical Report".

[7] J.L. Wood, A.J. Valerga, R.B. Badachhape, J.L. Margrave, R & D Tech. Report, ECOM-0105-F, March, 1972. Contract DAAB 07-72-C-0105.

[8] M.S. Whitingham, J. Electrochem. Soc. 122 (1975) 3139–3141.

[9] K. Ueno, N. Watanabe, T. Nakajima, J. Fluorine Chem. 19 (1982) 323–332.

[10] N. Watanabe, R. Hagiwara, T. Nakajima, H. Touhara, K. Ueno, Electrochim. Acta 27 (1982) 1615–1619.

[11] N. Watanabe, T. Nakajima, R. Hagiwara, J. Power Sources 20 (1987) 87–92.

[12] H. Touhara, H. Fujimoto, N. Watanabe, A. Tressaud, Solid State Ionics 14 (1984) 163–170.

[13] N. Watanabe, R. Hagiwara, T. Nakajima, J. Electrochem. Soc. 131 (1984) 1980–1984.

[14] K. Naga, Y. Ohzawa, T. Nakajima, Electrochim. Acta 51 (2006) 4003–4010.

[15] R. Hagiwara, T. Nakajima, N. Watanabe, J. Electrochem. Soc. 135 (1988) 2128–2133.

[16] Y. Sato, K. Itoh, R. Hagiwara, T. Fukunaga, Y. Ito, Carbon 42 (2004) 2897–2903.

[17] R. Hagiwara, T. Nakajima, K. Nogawa, N. Watanabe, J. Appl. Electrochem 16 (1986) 223–228.

[18] T. Nakajima, R. Hagiwara, K. Moriya, N. Watanabe, J. Electrochem. Soc. 133 (1986) 1761–1766.

[19] T. Nakajima, R. Hagiwara, K. Moriya, N. Watanabe, J. Power Sources 20 (1987) 93–98.

[20] T. Nakajima, A. Mabuchi, R. Hagiwara, N. Watanabe, J. Electrochem. Soc. 135 (1988) 273–277.

[21] W.S. Hummers Jr., R.E. Offeman, J. Am. Chem. Soc. 80 (1958) 1339.

[22] T. Nakajima, H. Groult (Eds.), Advanced Fluoride-based Materials for Energy Conversion, Elsevier, Waltham, MA, USA, 2015.

Fluorinated Nanocarbons for Lubrication

14

Ph. Thomas[1], *Ph. Bilas*[1], *A. Molza*[1], *L. Legras*[2], *J.L. Mansot*[1], *K. Guérin*[3,4], *M. Dubois*[3,4]
[1]Université des Antilles, Pointe à Pitre, France; [2]EDF R&D MAI, Les renardières, Moret-sur-Loing, France; [3]Université Blaise Pascal, Aubiere, France; [4]CNRS, Aubière, France

Chapter Outline

New Fluorinated Carbons: Fundamentals and Applications. http://dx.doi.org/10.1016/B978-0-12-803479-8.00014-0

14.1 Introduction to Tribological Applications

14.1.1 Brief Introduction of Tribology

Tribology is the part of science concerned with friction, wear, and lubrication processes [1–4]. Friction corresponds to energy losses associated to relative displacement of moving pieces in various mechanical assemblies. Friction has negative impact on yields of engines and more generally on energy converters (see Fig. 14.1).

Wear corresponds to the loss of material generated during the relative movement of the pieces in mechanical assemblies. Its major negative impacts concern the lifetime of the mechanical assemblies subjected to wear.

To reduce the negative impacts of friction and wear, lubrication is used. It consists in the separation of the sliding surfaces by a thin lubricant film (solid, liquid, or gaseous) to avoid direct contacts between asperities of the moving antagonist surfaces.

The Stribeck curve [5], presented in Fig. 14.2, allows four different liquid lubrication regimes to be defined. They depend on the viscosity (η) of the lubricant, the load carried by the contacting and moving surfaces (F_N), and their relative sliding speed (V).

The most severe regime, which will be the regime of interests for the carbon nanoparticle–fluorinated derivatives, is the boundary regime for which the lubricant film is quasi absent and additives have to be added to the lubricating base oil to achieve friction and wear reduction.

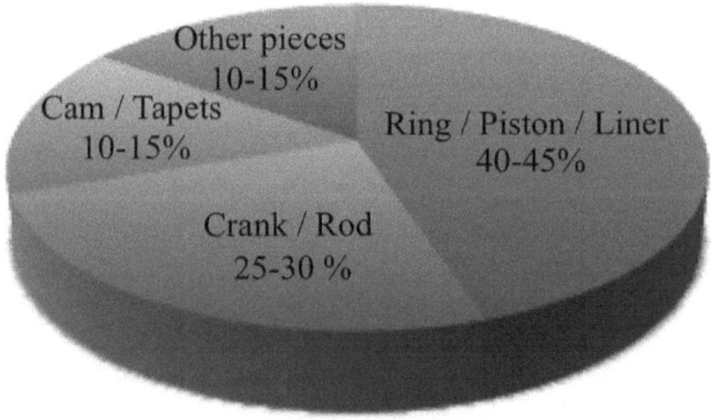

Figure 14.1 Distribution of the energy losses by friction in the various parts of a diesel engine.

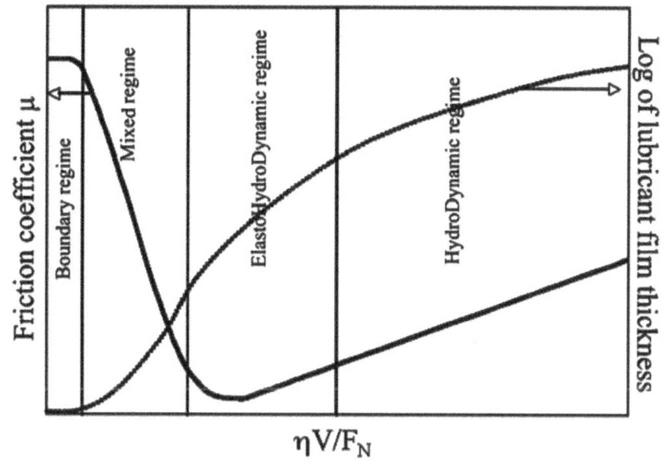

Figure 14.2 Stribeck curve representing the evolutions of the friction coefficient and of the lubricant film thickness as a function of the $\eta V/F_N$ parameter.

14.1.2 Introduction to the Experimental Device for Friction and Wear Studies in the Boundary Regime

To measure friction and wear, specific devices, called tribometers, are used. The schematic representation of a reciprocating sphere on plane tribometer, used for friction and wear investigations, is presented in Fig. 14.3. Depending on the tribometer type, the contact geometry can be sphere/plane, cylinder/plane, cylinder/cylinder, or plane/plane. In the case of the sphere/plane or cross cylinder geometries, the contact is self-adjusted and high pressures can be generated in the contact (1 GPa) for relatively small normal applied load (12 N) for balls or cylinders of 10 mm diameters.

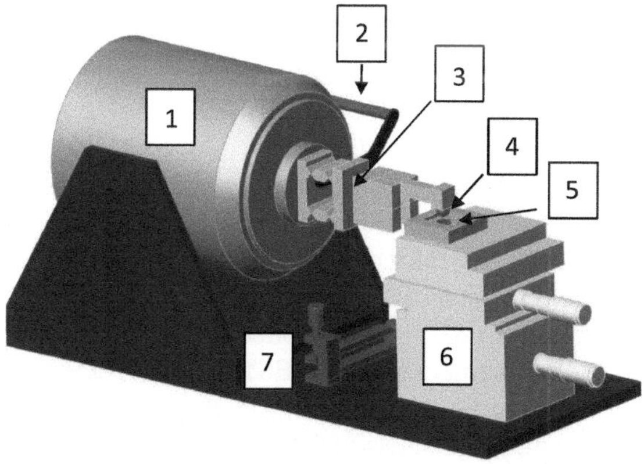

Figure 14.3 Schematic representation of an alternative sphere/plane tribometer: (1) motor, (2) displacement transducer, (3) triaxial force transducer, (4) sphere, (5) plane, (6) x, y, z adjustable sample holder, (7) elastic device for normal load application.

Depending on the tribometer type, the normal load can be applied by inertial (mass), pneumatic, or elastic devices (a spring). The normal and lateral (friction) forces are measured either by piezoelectric transducers or strain gauges devices. Control of the tribometer, data acquisition, and treatments are done using computers.

14.1.3 Short Overview of Action Mechanisms of Conventional Additives in Boundary Lubrication Regime

14.1.3.1 Friction Reduction Additives

The unctuous additives, first introduced by Hardy [6], are amphiphilic molecules such as organic acids, esters, and alcohol molecules. The friction reduction action of monolayers of a great variety of these molecules was first investigated by W.A. Zisman [7].

He pointed out the relationship between the boundary friction coefficient and the monolayer thickness adsorbed on the sliding surfaces. He especially demonstrated, in the case of linear fatty acids, that the friction coefficient decreases as a function of the monolayer thickness, ie, the molecular length tending to an asymptotic value of 0.1 for acids larger than dodecanoic acid. These results led to the developments of monomolecular layer lubrication models [8—11] establishing quantitative relationship between the friction coefficient and the molecular film thickness.

However, it is important to note that after few cycles of friction such monolayers are destroyed either by bond breakings or chemical reactions with the contacting surfaces leading to the built up of new interface materials, with physical properties and thicknesses different from the initial monolayer [12], and finally its progressive elimination from the sliding contact. If the molecular layer is formed from a dispersion of molecules into a hydrocarbon base oil embedding the sliding contact, the friction can remain stable at a relatively low value as far as continuous adsorption of amphiphilic molecules on the sliding surfaces occurs [11], which is the role of the lubricating oil.

14.1.3.2 Antiwear Additive Action Mechanisms

In the boundary lubrication regime, the action of conventional antiwear additives occurs via the built up of a tribologic film resulting from complex chemical reactions between the additive molecules and the sliding surfaces in contact [13—15]. The antiwear action mechanism of a conventionally used antiwear additive, zinc dialkyl dithiophosphate (ZDDP), has been extensively studied [9—16]. The elemental composition and the structure of the so-called tribofilm or antiwear film was established using complementary techniques such as analytical transmission electron microscopy (ATEM), Auger electron spectroscopy (AES), X-ray photoelectron spectroscopy (XPS), and X-ray absorption near edges structures (XANES) [16,17]. They reveal that the chemical reaction involved for the built up of the film in the physical contact conditions (pressure raising 1 GPa and shear stress of 10^5 s^{-1}) are readily different from thermal degradation. The collected results lead to an accurate description of the antiwear film which is mainly composed of amorphous Fe/Zn poly(thio)phosphate embedding some nanocrystallites of ZnO and ZnS.

Although ZDDP revealed itself as a highly efficient antiwear additive during the last decades, its action mechanism presents two limitations:

- As far as the antiwear film needs reactions between the additive and the surfaces, the film built up is not immediate (depending on the reactions kinetics). During this film establishment period, called induction period, the wear rate is as high as the additive free oil.
- The development and use, for various mechanical applications, of new materials presenting weak chemical reactivity with the additive do not allow the antiwear film to be produced.

To overcome these limitations, alternative ways based on the use of nanosized particles of tribologic film precursors are now proposed to build up the protective tribologic film without chemical reaction of the additive with the contacting surfaces [18–21].

14.1.3.3 The Solid Nanoadditives Approach

This recent approach is based on the direct feeding of the sliding interface with nanoparticles of tribo-active phases (MoS_2, graphite, hexagonal boron nitride, etc.) or precursors of friction and wear reduction phases, such as encapsulated nanoparticles, carbon nano-onions or nanotubes, metal dichalcogenides or fullerenes,..., dispersed in lubricant base oils or greases to produce the tribofilm without reaction with the substrate surfaces [18–26]. The use of nanosized particles is required to obtain a perfect feeding of the sliding interface. This approach allows composite particles to be synthesized, which can result in multiple action additives with associated properties such as friction reduction, antiwear, anticorrosion, etc.

14.1.4 Tribologic Applications of Fluorinated Derivatives of Graphitized Carbon Nanoparticles

Graphite is well known to present good intrinsic friction properties, attributed to its lamellar structure inducing easy shearing along directions parallel to the basal graphene sheets. It has been commonly used as solid lubricant or as complementary additive in specific oils and greases which already contain classical antiwear and friction-reduction additives.

Due to the excellent tribologic properties of graphite, graphitized carbon nanoparticles appeared as good candidates for the solid nanoadditives strategy application. As far as fluorinated derivatives of graphite, such as CF_x, developed and tribologically studied in the 1960s [27–29] revealed improved antiwear and friction reduction compared to graphite, it appears natural to envisage a new family of tribo-active nanoparticles consisting in fluorinated derivatives of graphitized carbon nanoparticles. The tribologic properties and tribologic application of these last particles are the subject of the present chapter. The modulation of the C–F bond strength should modify the tribologic performances of the materials. The covalence of the C–F bond is known to be weakened when the curvature of the initial carbon matrix increases [30]. So the investigation of the friction properties is carried out on fluorinated nanocarbons presenting different morphologies (0D spherical, 1D tubular, and 2D discotic structures).

14.2 Fluorination

General trends about fluorination of carbonaceous (nano)materials using molecular fluorine are given in Chapter 9. Curvature, graphitization, and high surface effects are discussed in the latter. This part will focus on the particular case of carbonaceous nanomaterials with different morphologies. To avoid effect of amorphous regions, such as preferential fluorination and decomposition under fluorine gas, the starting materials were post-treated under inert atmosphere at high temperature in the 1800–2700°C range according to the sample. Details are given in Table 14.1.

The 0D character of the graphitized carbon blacks (denoted GCBs) is emphasized by the spherical shape of the nanoparticles with average diameter of 44 nm (Fig. 14.4A). Moreover, GCBs mainly consist in aggregated spherical particles. Their specific surface area (SSA) was found to be equal to 46 m^2/g [39]. The size of the crystalline domains, ie, the coherence length along the c-axis (Lc) determined by X-ray diffraction, is 7 nm.

The morphology of the carbon nanofibers (CNFs) (Fig. 14.4B) is characterized by quite cylindrical shapes with diameters in the 100–200 nm range and length in the 5–10 mm range. Fig. 14.4B points out the 1D character of these compounds. SSA and Lc of these fibers are equal to 19 m^2/g and 39 nm, respectively [35].

The micrograph collected on 2D material shows flat particles presenting nanodisc and nanocone shapes with diameters ranging from 0.8 to 3.2 mm and 0.5 to 2.8 mm, respectively, and typical thicknesses lying in the 15–65 nm range (Fig. 14.4C). Cones

Table 14.1 Fluorination Conditions According to the Starting Material

Starting Material	Reactive Species	T_F Range (°C)	Notation	F/C Range (±0.02)	References
GCBs[c]	F_2	340–420	GCB-MT_F	0.08–1.04	[31]
	F•	320–480	GCB-AT_F	0.37–0.83	[31]
CNFs[a]	F_2	380–480	CNF-MT_F	0.06–1.04	[32–35]
	F•	420–500	CNF-AT_F	0.12–0.91	[33–35]
CNDs[b]	F_2	450–520	CND-MT_F	0.14–1.0	[36–38]
	F•	500–550	CND-AT_F	0.7–0.95	[37,38]

[a]High purity (>90%) carbon nanofibers, 2–20 μm in length, were supplied by courtesy of the MER Corporation, Tucson, Arizona. They were obtained by chemical vapor deposition and heat treated at 1800°C in an argon atmosphere to enhance their graphitization degree.
[b]The commercially available product for nanocones/discs, provided by N-Tec Corporation, consists of a mixture of 70 wt.% of nanodiscs (CND), 20 wt.% of nanocones, and 10 wt.% of amorphous carbons. Due to the content of nanodiscs, the mixture is called CND. The carbon nanocones and nanodiscs were produced by pyrolysis of heavy oil using the Kvaerner carbon black and hydrogen process [41].
[c]Graphitized carbon blacks (GCBs), developed by Superior Graphite, Chicago, Illinois, consist of aggregates of nanospheres of 40–50 nm resulting from graphitization process by heat treatment at 2000°C in inert gas [39].

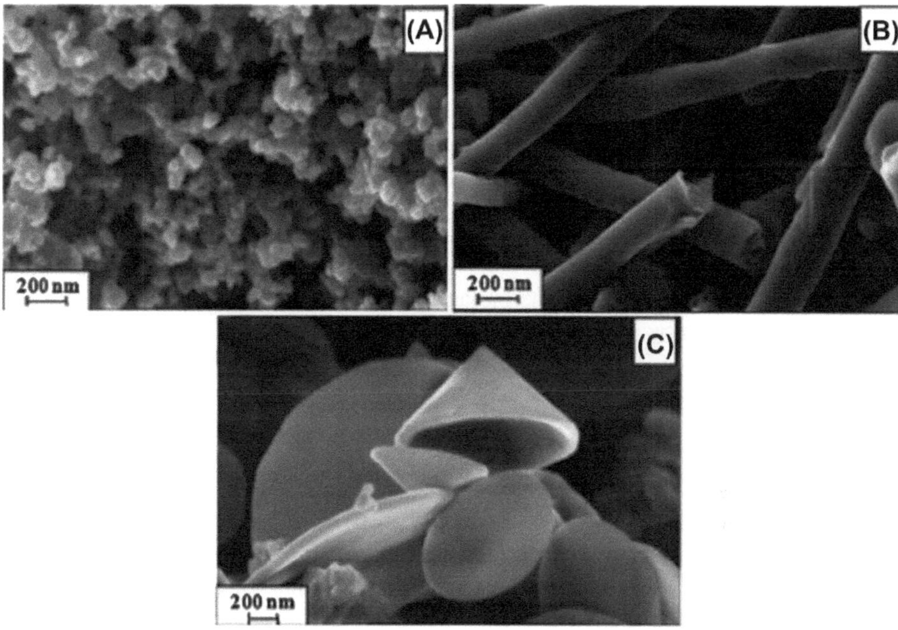

Figure 14.4 Scanning electron microscopy images of the fluorinated nanocarbons with F/C $\approx$ 1: (A) graphitized carbon blacks consist in spherical particles (0D), (B) carbon nanofibers consist in concentric cylinders of graphite sheets or single graphite sheet rolled in around itself (1D) and (C) a mixture of carbon nanodiscs and nanocones are formed by conical/discotic graphene stacking (2D).

exhibit different apex angles equal to 112.98, 83.68, 60.08, 38.98, and 19.28° [40,41]. Lc is around 2 nm which means that the as-product exhibited low degree of order. An annealing treatment under argon at 2700°C was then performed (the resulting sample is denoted carbon nanodisc (CND)). It improves the crystallinity since Lc reaches a much higher value of 43 nm. SSA of CNDs is equal to 29 m^2/g.

To tailor the dispersion of fluorine atoms into the carbonaceous lattice, two fluorination routes were carried out according to reactive species: direct with F_2 and controlled with atomic fluorine. "Controlled" means that the amount and location of fluorine atoms may be finely adjusted. Direct fluorination was carried out with pure fluorine gas flow in a nickel (or monel) reactor. The reaction temperatures T_F were ranged according to the starting material (see Table 14.1). The duration of each experiment was of 3 h, except for nanofibers (16 h). Table 14.1 summarizes both the fluorination conditions and the notation as a function of the starting nanomaterials. In the case of controlled fluorination, terbium tetrafluoride was used as fluorinating agent. It was obtained from TbF_3 (Aldrich, 99.9%) in pure F_2 gas at 500°C. For the fluorination by TbF_4, a nickel (or monel) closed reactor was used to preserve the defined fluorine amount (atomic and/or molecular) released by the thermal decomposition of TbF_4. A two-temperature oven was used: the part containing TbF_4 was heated at 450°C whatever the experiment whereas nanocarbons were heated at temperatures T_F (Table 14.1). A reaction time of 16 h was used. Prior to the heating, a primary vacuum (10^{-2} atm) was applied to the reactor. The reactions involved during the fluorination are the following ones:

$$TbF_4 \rightarrow TbF_3 + F\bullet \text{ or } 1/2F_2$$

$$C + xF\bullet \text{ or } x/2F_2 \rightarrow CF_x$$

The fluorination level "x" (ie, F/C molar ratio) of nanocarbons fluorinated by F_2 or TbF_4 was determined by weight uptake and NMR (Table 14.1).

14.3 Structural Characterization and C—F Bonding

According to X-ray diffraction and solid state NMR, the fluorination of both CNDs and CNFs using TbF_4 decomposition resulted in the formation of an unique highly fluorinated phase, ie, $(CF)_n$ type, whatever the fluorination temperature. This fluorinated phase was always mixed with some residual non-fluorinated nanocarbons and a high fluorination level can be obtained (x close to the unity) without hyperfluorination contrary to the F_2 process. The higher homogeneity of the materials obtained using this controlled process in comparison with the direct fluorination using fluorine gas can be explained by the low kinetic of decomposition of TbF_4 that allows a continuous addition of fluorine to carbon matrix and a more progressive fluorination. For the case of GCBs, $(C_2F)_n$ structural type was never formed whatever the fluorination way, $(CF)_n$ type being the only phase formed.

The fluorination of 0D, 1D, and 2D nanocarbons resulted in fluorinated compounds presenting similar nanostructuration whereas structural differences were observed for initial materials, ie, different coherence lengths. This was related to the conversion of the carbon hybridization from sp^2 to sp^3 and the structural changes from graphite to fluorinated phases either $(CF)_n$ or a $(C_2F)_n/(CF)_n$ mixture. For a given fluorine content, the structural parameters La and Lc were quite similar (Fig. 14.5). As an example, Lc values were 7, 13, and 43 nm for initial GCBs, CNFs, and CNDs, respectively. When F/C was 0.75, the Lc values were 5, 9, and 15 nm for GCB-M410, CNF-M450, and CND-M500, respectively. ^{13}C NMR confirmed the structural phase formed during fluorination. As a matter of fact, in the F/C/C/F stacking sequence of $(C_2F)_n$, half of the carbon atoms were fluorinated. The other atoms, also with sp^3 hybridization, were linked only to other carbon atoms, as in a diamond phase. The typical ^{13}C NMR line of those ^{13}C nuclei was observed at 42 ppm (vs tetramethylsilane) and appeared as a good indicator of the presence of $(C_2F)_n$ phase. Such line was observed only for CNDs and CNFs fluorinated with F_2. Moreover, ^{19}F NMR highlighted the nature of the interaction between carbon and fluorine atoms, ie, the C—F bonding. The comparison of samples with similar fluorination level obtained by direct and controlled fluorination using TbF_4 indicated also different fluorination mechanisms.

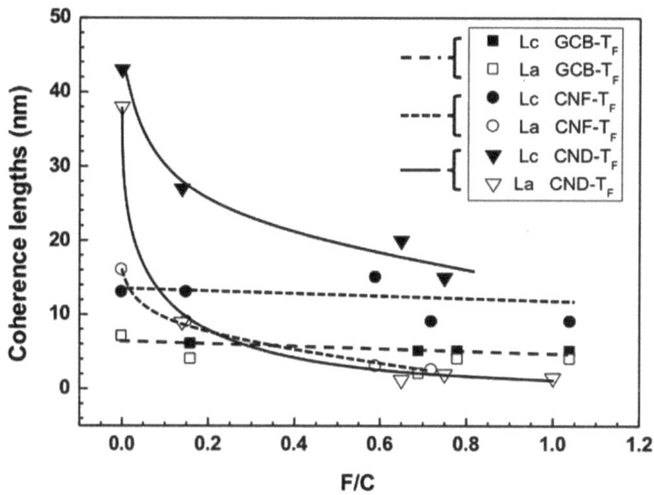

Figure 14.5 For the cases of GCB-MT$_F$, CNF-MT$_F$, and CND-MT$_F$ series, Lc and La coherence lengths calculated by XRD and by Raman spectroscopy, respectively. *Lines* are guides for eyes. *CND*, carbon nanodisc; *CNF*, carbon nanofiber; *GCB*, graphitized carbon black; *XRD*, X-ray diffraction.

First of all, whatever the sample, the main line at −190 ppm was assigned to covalent C—F bonds meaning that the C—F bonding is similar regardless the starting material and the fluorination route [42—45]. Thanks to high spinning rate (in magic angle spinning conditions), a second line at −175 ppm was observed, present only for fluorinated CNDs and CNFs as for $(C_2F)_n$ type graphite fluoride. This line was unambiguously assigned to fluorine nuclei present only in the $(C_2F)_n$ structure: ^{19}F in weak interaction with nonfluorinated sp^3 carbon atoms. The structure of $(C_2F)_n$ can be described using an F/C/C/F stacking sequence for fluorocarbon sheets whereas for $(CF)_n$ the sequence is F/C/F/F/C/F. All carbon atoms exhibit sp^3 hybridization. Half of the carbon atoms were fluorinated in $(C_2F)_n$ while the others being linked to carbon atoms. By analogy with the hyperconjugation effect in fluorine-graphite intercalation compounds (F-GICs) that involves C—C in nonfluorinated parts and C—F bonds [46], the covalence of the C—F bonds located close to these diamond-like atoms could be weakened in the same manner. The ^{19}F chemical shift was then of −175 ppm (vs $CFCl_3$) instead of −190 ppm for pure covalent C—F. In the case of F-GICs, −170 ppm was recorded. The line at −175 ppm, which was observed, thanks to a high spinning rate, can be considered as an indicator of the presence of $(C_2F)_n$ structural type. $(C_2F)_n$ type was maintained in a wide temperature range for the direct fluorination of CNFs and CNDs whereas $(CF)_n$ type is formed during the controlled fluorination using TbF_4 whatever the temperature. This was due both to progressive generation of atomic fluorine by thermal decomposition of TbF_4 and to its high reactivity and diffusion. It is to be noted that the presence of an isotropic line at −130 ppm indicated the presence of some CF_2 groups. CF_2 contents are generally well correlated with the amount of CF_3 groups. For each spectrum of nanocarbons, narrow lines were present in the −60/−90 ppm range corresponding to CF_3 groups and are superimposed with spinning sidebands of C−F and CF_2 lines. The intensity of the lines related to the amounts of CF_2 and CF_3 groups was equivalent for compounds presenting similar fluorine content whatever the precursor.

Disorder is induced during the fluorination. Such phenomenon may be evidenced by Raman diffusion that focuses on the extension of the graphitic monocrystalline domains perpendicularly to the c-axis. Because the C−F vibration is not active in Raman, this technique gives information about the non-fluorinated carbonaceous lattice, which could be, however, affected by the neighboring fluorinated parts. Most of the Raman spectra for the fluorinated nanocarbons exhibited three bands as classically observed for carbonaceous materials. The vibration band at 1580 cm^{-1} was related to the well-known G band characteristic of graphite structure. The two bands near 1345 and 1620 cm^{-1} were assigned to the D and D′ modes and appeared when disorder is present in the graphitic lattice, characterizing the impurities and/or defects. In all fluorinated series, the D and D′ bands relative intensities increased as a function of the fluorination temperatures showing a size reduction of the remaining graphitic domains (Fig. 14.5). The D band downshift also recorded in the three families was attributed both to size reduction of the graphitic domains and to the creation of covalent C−F bonds [47−51].

The size reduction of the remaining graphitic domains in the fluorinated compounds was expressed as the coherence length La corresponding to their spatial extension perpendicular to the c-axis. The La values deduced from the I_D/I_G intensity ratios using the formula $La = C(\lambda)/(I_D/I_G)$, where $C(\lambda) = C_0 + \lambda C_1$ is equal to 10.3 nm with an excitation wavelength of 514.5 nm, C_0 and C_1 are nearly constant in the 400−700 nm wavelength range ($C_0 = -12.6$ nm and $C_1 = 0.033$ nm) [52], are evidenced in Fig. 14.5. The decrease of the spatial extension of the graphitic domains from initial carbon phases to fluorinated derivatives highlights the progressive conversion of the graphite lattice to fluorinated phases. This behavior was also observed for GCB-AT_F, CNF-AT_F, and CND-AT_F series prepared with TbF_4 [31−38].

Changes of the morphologies were observed during the fluorination especially at the highest fluorination temperatures. Direct fluorination may induce some partial delamination of external fluorinated layers (underlined by the *arrows* in Fig. 14.6A), whereas such phenomena do not occur in the case of controlled fluorination (Fig. 14.6B).

The fluorination started on the defective or unsaturated carbons sitting at outlayers of the nanofiber tips although their amount was quite low since the sample was post-treated at 1800°C under argon atmosphere. The fluorination could then progress toward the core through an opening of some external walls to permit the volume expansion resulting from the accommodation of fluorine atoms and the carbon hybridization change from sp^2 to sp^3. Scanning electron microscopy (SEM) image of CNF-A420 in Fig. 14.7A supports this proposed model which is schematically shown in accordance with Ref. [53]. Nevertheless, using F_2 gas at 480°C (Fig. 14.7B), this process occurred with a partial exfoliation resulting in some cracks due to the expansion [32].

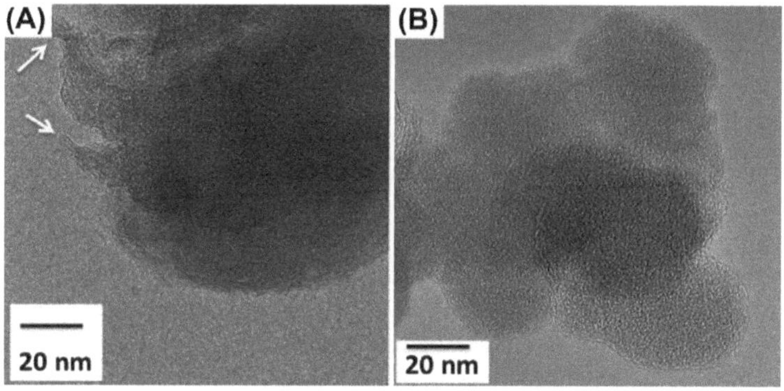

Figure 14.6 Transmission electron microscopy images of graphitized carbon blacks fluorinated with: (A) F_2 (F/C = 0.89) and (B) TbF_4 (F/C = 0.83).

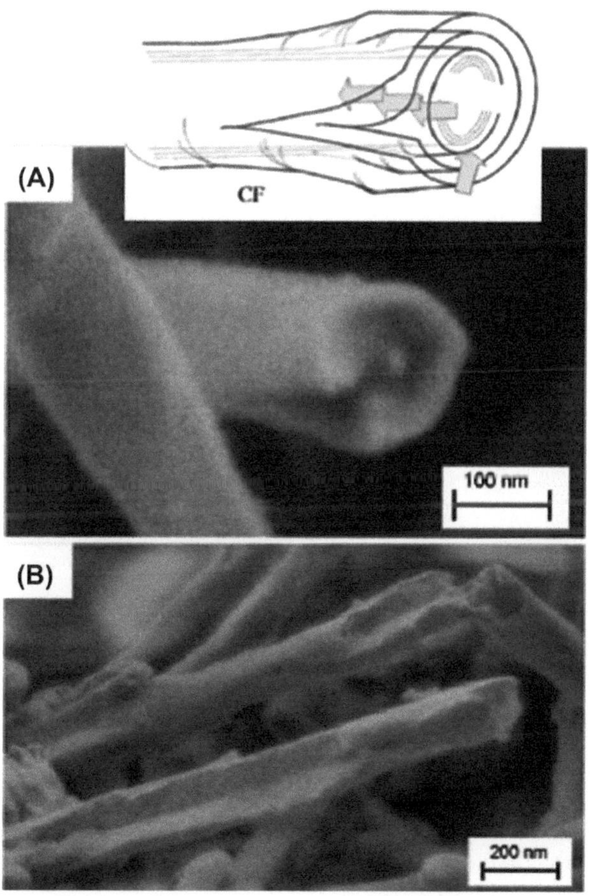

Figure 14.7 Scanning electron microscopy image of carbon nanofibers fluorinated with TbF_4 at 420°C (A) and F_2 at 480°C (B). Scheme of the fluorination mechanism is added.

14.4 Dispersion of Fluorinated Parts in the Carbon Matrix

When the fluorine content is not the maximum, ie, F/C is lower than the unity, some residual non-fluorinated parts still exist. Different locations of non-fluorinated and fluorinated regions were observed according to the fluorination route.

Fluorination requires both the change of the carbon hybridization from sp^2 to sp^3 and accommodation of fluorine atoms. Both processes result in drastic volume expansion. Atomic force microscopy (AFM) underlines that highly fluorinated nanodiscs were swelled without cracks because of the accommodation of fluorine atoms [38]. AFM (Fig. 14.8) revealed that swelling was more pronounced on the edges and the maximum swelling was observed in the overall perimeter when F_2 is the reactive species. The edge seems then to be the preferential pathway of fluorine atoms/molecules for the fluorination. Using atomic fluorine released by TbF_4, the swelling is more homogenous showing a first difference according to the reactive species. For carbon blacks and nanofibers, the structures are closed and the fluorine diffusion started in sheet defects. The small size of the GCB particles associated with their high defect density, induced by the geometric constraints of the graphene layers, let suppose a high reactivity of this material with fluorine. As a consequence, the fluorination temperature is quite low because of relatively easy diffusion of fluorine in the structure (see Table 14.1). Previous works [32] allowed us to demonstrate that for CNFs, the first steps of their direct fluorination resulted essentially in fluorination of the external (surface) graphene layers. When the fluorination temperature increased, fluorination progressed toward the inner layers of the nanostructure. Comparison in Table 14.1 evidences that higher fluorination temperatures are needed for nanodiscs as far as sheet edges are the starting points of the fluorination and that fluorine atoms must diffuse in the interlayer van der Waals gap.

Transmission electron microscopy (TEM) bright field image of CNFs fluorinated with TbF_4 at 480°C ($F/C = 0.70 \pm 0.02$), shown in Fig. 14.9C, underlines that the

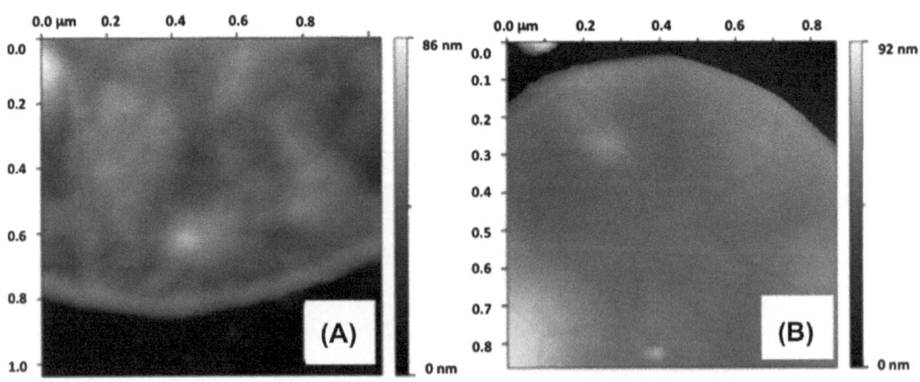

Figure 14.8 AFM images of CND-M500 (A), and CND-A500 (B).

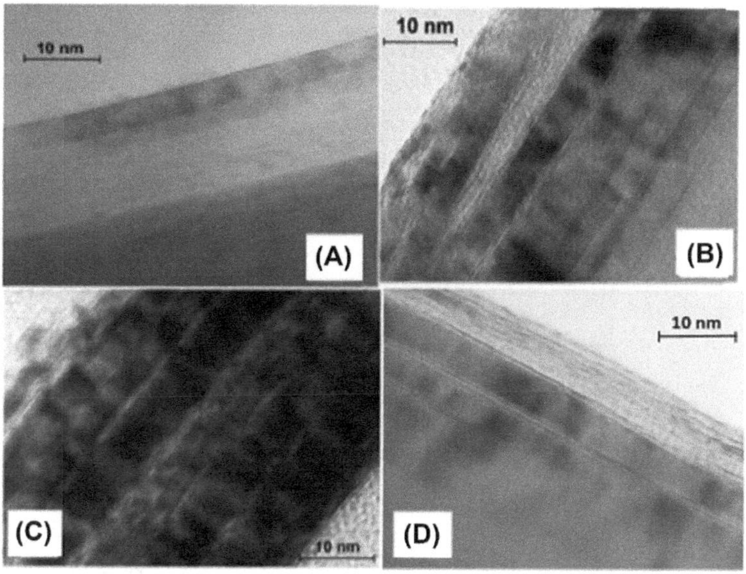

Figure 14.9 Transmission electron microscopy bright field images of the raw carbon nanofibers (A) and sample fluorinated at 420°C (B) and 480°C (C) using TbF_4 as fluorinating agent. Sample fluorinated with F_2 at 420°C (D).

layered structure was maintained even for high temperature and high fluorine level (0.65–0.70). However, in small domains, due to the accommodation of fluorine atoms within the graphene layers (Fig. 14.9B and C), the interlayer distance increased from the one of raw CNFs (0.34 nm, Fig. 14.9A) up to the classical value of $(CF)_n$-type graphite fluoride structure (d-spacing equal to 0.60 nm [54]). The less ordered layers, which were more spatially separated due to the accommodation of fluorine atoms, were uniformly dispersed using the fluorination at 480°C by TbF_4 (Fig. 14.9C). In the case of the direct reaction with F_2, such disordered layer dispersion has not been observed at 480°C, but some exfoliation occurred. Moreover, for the lower fluorination temperature at 420°C for the direct ($F/C = 0.39 \pm 0.02$) and controlled fluorination (Fig. 14.9B), the distorted layers were located near the nanofiber surface [32]. This TEM study revealed at the nanoscale the difference between the two fluorination routes. By the occurring of short fragments after the fluorination, TEM also underlined that the nanofiber length decreased at the higher temperatures, ie, 480 and 500°C, contrary to the low fluorination temperature (420°C). As for the direct fluorination with F_2 [32], breaking of the nanofibers occurred in addition to their swelling; the average diameter increases from 150 nm for the raw CNFs to 153, 184, and 225 nm for the fluorinated samples CNF-A420, CNF-A480, and CNF-A500, respectively (these values are estimated taking into account at least 40 measurements on TEM images).

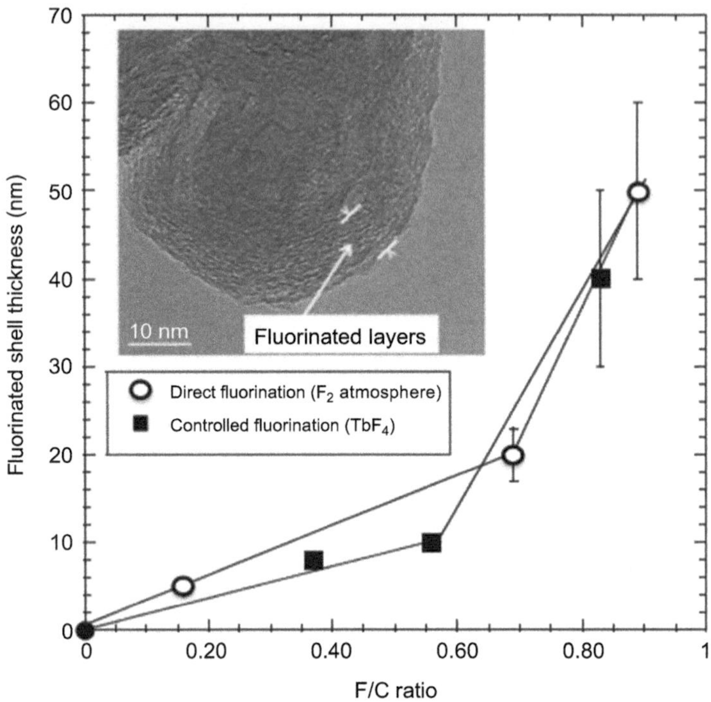

Figure 14.10 Evolution of the thickness of the fluorinated shell as a function of F/C ratio for fluorinated graphitized carbon blacks (GCBs) obtained from direct and controlled fluorination. HRTEM micrograph of GCB-A410 ($CF_{0.56}$) is given as a representative example [46].

The diameter of initial GCBs varies from 10 to 50 nm, the average diameter being 30 nm. The high resolution transmission electron microscopy (HRTEM) images recorded for fluorinated compounds revealed that the average diameter of the particles increases up to 50 nm for both GCB-M380 ($CF_{0.69}$) and GCB-M400 ($CF_{0.89}$) compounds.

For the case of GCBs, the difference between the sample obtained from controlled and direct fluorination is less defined. The thickness of the fluorinated shell is estimated from HRTEM (see Fig. 14.10 for a representative example). The external layers are first fluorinated for low F/C ratios, and then the fluorination process progresses toward inner ones. Fluorination progression is clearly pointed out in Fig. 14.10 which presents the evolution of the thickness of the fluorinated shell as a function of the fluorine content. A linear increase of the thickness of the shell is observed for $F/C < 0.6$ for both direct and controlled fluorination. For high fluorination rates, the values are similar because fluorination occurs up to the core of the particles [55].

After that the similarities (C−F bonding, increase of the disorder) and differences (fluorination mechanisms, dispersion of the fluorinated parts in the carbon lattice, morphologies) were highlighted, the tribological properties will be discussed and explained.

14.5 Macrotribologic Properties of Fluorinated Nanocarbons

14.5.1 Sample Preparation and Tribologic Tests Conditions

The friction properties of the compounds were determined using a ball-on-plane tribometer, consisting of an AISI 52100 steel ball describing an alternative motion on a static AISI 52100 steel plane. Prior to friction tests, the planes were polished to obtain an average peak-to-peak roughness of 350 nm to improve the adherence of the tested tribofilms. The balls were used as delivered and the average peak-to-peak roughness is 90 nm. The metallurgical characteristics as well as 2D (R_a) and 3D (S_a) roughness data, obtained with an optical Altisurf profilometer (resolution 10 nm), are summarized in Table 14.2.

A normal load of 10 N was applied to the ball which moved on the static plane surface at a speed of 6 mm/s. According to Hertz's theory [56], the contact diameter was 140 μm and the mean contact pressure was 0.65 GPa.

After ultrasonic cleaning of balls and planes in acetone and ethanol baths in order to eliminate pollutants and remaining abrasive particles, few milligrams of powdery materials were deposited on the plane surface. After ball-on-plane contact, a drop of pentane (boiling point 36°C) was added to improve the particle feeding of the sliding interface and accelerate the formation of a homogeneous film between the contacting surfaces.

The friction properties were investigated with the tribometer shown in Fig. 14.3 under air atmosphere (relative humidity: 50 ± 5%) at 298 ± 1 K. The friction experiments allowed us to investigate:

- The intrinsic friction properties of the deposited films, ie, before structural transformations induced by the friction process. The intrinsic friction coefficient was determined at three cycles after total evaporation of pentane (few cycles).
- The evolution of the friction coefficient as a function of the cycle number by means of long-term experiments in order to investigate the friction properties changes induced by the transformations of the initial materials during the friction process.

Table 14.2 Metallurgy and Roughness Parameters of Balls and Planes

	Balls	Planes
Metallurgy	AISI 52100 stainless steel	AISI 52100 stainless steel
Vickers hardness	850	850
Reduced Young's modulus E^*	115 GPa	115 GPa
Diameter	9.5 mm	10 mm
Roughness	$R_a = 90 \pm 10$ nm	$R_a = 350 \pm 10$ nm
	$S_a = 120$ nm	$S_a = 120$ nm

14.5.2 Intrinsic Friction Properties of Fluorinated Derivatives of Graphitized Carbon Nanofibers

The evolution of the intrinsic friction coefficient in air (recorded three cycles after total pentane evaporation) of fluorinated derivatives of graphitized CNFs obtained by fluorination under F_2 atmosphere (direct fluorination) and fluorination with TbF_4 (controlled fluorination) are presented in Fig. 14.11 as a function of the fluorine content.

As it can be seen, intrinsic friction coefficient decreases from 0.13 for the pristine CNFs down to 0.08 for a fluorination rate F/C of 0.15. Then the intrinsic friction coefficient does not evolve whatever was the F/C ratio. To investigate this surprising result, elemental and structural characterization by ATEM and nanometric scale friction studies were carried out on fluorinated CNFs with F/C = 0.15.

14.5.2.1 Sample Preparation of Fluorinated Carbon Nanofibers for Structural Transmission Electron Microscopy and Nanotribological Studies (Friction Force Microscopy) and Experimental Setup

Two milligrams of fluorinated CNFs (F/C = 0.15) were introduced into epoxy resin contained in a small volume polyethylene vessel with conic bottom for embedding. Fibers were concentrated at the bottom by a few minutes centrifugation (3000 g).

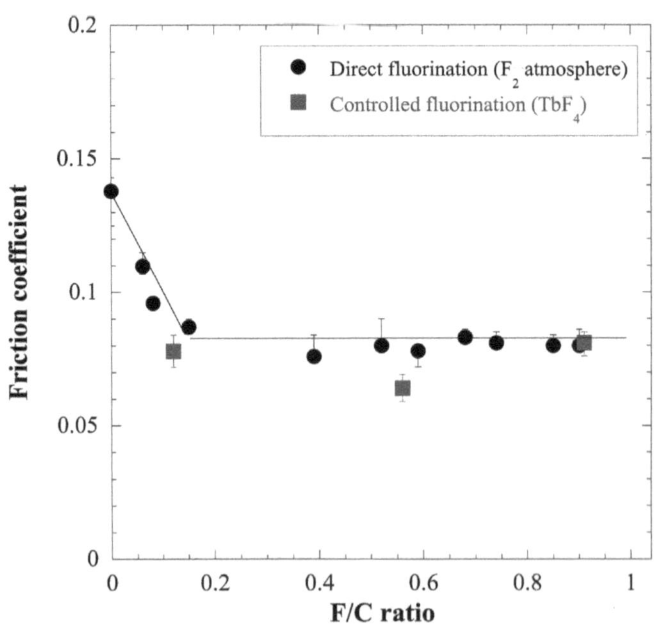

Figure 14.11 Evolution of intrinsic friction coefficient (recorded three cycles after pentane evaporation) of fluorinated carbon nanofibers as a function of F/C ratio.

Polymerization of the resin was performed at 37°C for 48 h in an oven. For TEM investigations, thin sections (60 nm thick) of the epoxy/fiber block was prepared using an ultramicrotome (Leica Ultracut). For friction force microscopy (FFM) studies, the remaining block was introduced in a dual beam scanning electron microscope (FEI Helios Nanolab from EDF—R&D, Department MMC and MAI) in order to cut the fibers protruding from the epoxy. The amorphous layer induced by focused ion etching was removed in a plasma cleaner using oxygen and argon atmosphere. The block is then directly used for SEM and FFM studies.

SEM studies were carried out on an FEI Quanta 250 ESEM equipped with a tungsten cathode running at an accelerating voltage of 5 kV. SEM investigations were carried out in low vacuum conditions with water vapor pressure of 130 Pa.

TEM investigations of the structure and composition of the nanofibers were carried out on the prepared thin sections using a TEM (FEI Tecnaï F 20 X twin) equipped with an FEG running at 200 kV. Elemental analyses and profiles were acquired using X-ray emission spectrometry with SiLi diode energy dispersive detection (EDS) EDAX. Elemental mapping was carried out by electron energy loss spectrometry (EELS) and energy filtered imaging using a post column Gatan quantum energy filter.

Morphology (AFM) and friction behavior at nanoscale (FFM) studies were carried out using a PICO scanning probe microscope from Molecular Imaging Corp. The cantilever was mounted on a piezoelectric tube, which allowed the tip to scan the sample surface. All experiments were performed in constant force mode. To calibrate the microscope and quantify the forces, the specific method developed in previous works [57,58] was used. A normal stiffness of the cantilever of 0.55 ± 0.05 N/m was measured using the method developed by Sader et al. [59]. The pull-off force was negligible and remained constant during all the experiments. Experiments were carried out under air atmosphere at 25°C with a relative humidity of 55%. The fast scan direction was perpendicular to the long axis of the cantilever and the sliding speed was 5.5 μm/s. No wear was revealed at the end of the tests.

14.5.2.2 Characterization of the Fluorinated Carbon Nanofibers Used for Nanotribologic Investigations

Fig. 14.12 compares images collected on dispersed fluorinated CNFs at different scale, using SEM (Fig. 14.12A), TEM (Fig. 14.12B), and AFM (Fig. 14.12C and D). Fig. 14.12C emphasizes the facetted surface of the fibers indicating that CNFs are constituted of joining flat graphitic nanodomains whose relative orientation allows them to accommodate the average curvature of the fiber's surface.

Fig. 14.13A presents a low magnification bright field transmission electron micrograph collected on a thin section prepared from fluorinated CNFs embedded in epon block. The embedding epoxy resin which covers the main part of the picture appears in light gray. Longitudinal sections and cross-sections of some fibers are easily identified as *dark gray areas* and we can note also the presence of few broken fibers (*white arrows*). Numerous holes (*white areas*) are also present in the epon matrix and are attributed to the fact that some fibers were not cut by the diamond knife during the sections preparation. In Fig. 14.13B, the secondary electron image collected on the

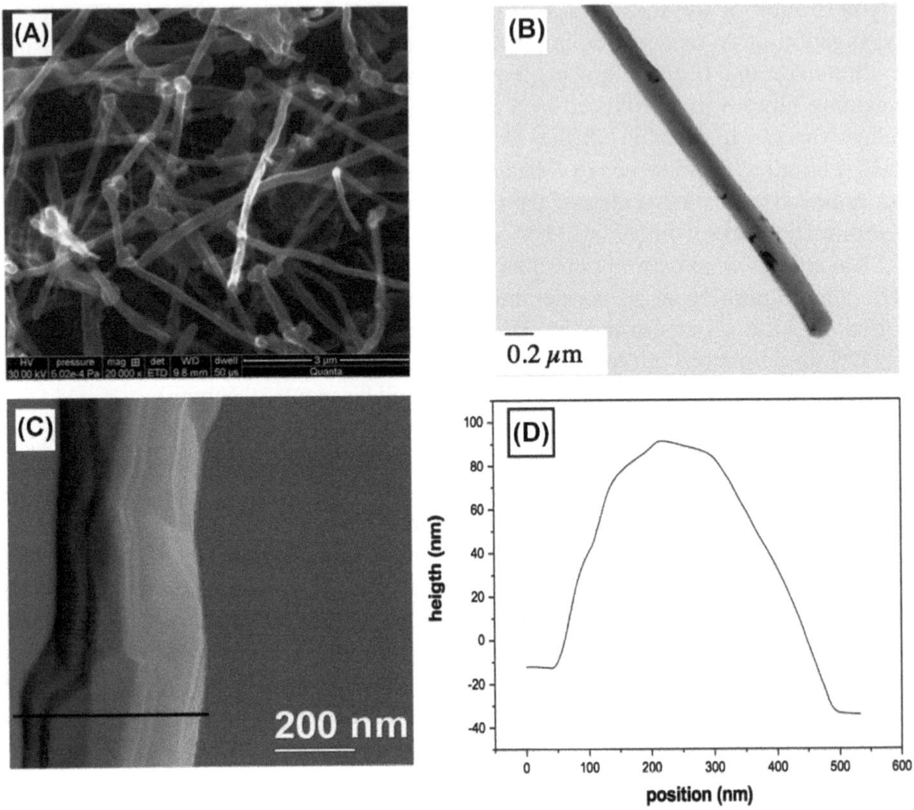

Figure 14.12 Morphology of the fluorinated carbon nanofibers (F/C = 0.15). (A) Secondary electron image showing fibers with diameters in the range 100–200 nm, (B) Transmission electron microscopy low magnification image collected on a nanofiber deposited on a thin carbon layer, the fiber presents a *quasi cylindrical shape*, (C) Topographic AFM image collected on a nanofiber revealing its facetted surface induced by the graphitization process, (D) Topographic profile recorded along the *black line* in Fig. 14.12C.

surface of the remaining block after the diamond cutting exhibits *dark areas* corresponding to regions where fibers are concentrated. Some scratches due to defaults of the diamond knife cutting edge appear as longitudinal straight lines. The magnified image of a fibers concentrated region presented in Fig. 14.13C reveals the topography of the epon block surface and shows the presence of numerous uncut fibers emerging over few micrometers from the block surface.

The distribution of fluorine atoms in the structure of CNFs in the particular case of F/C = 0.15 can be determined from TEM analysis of the ultramicrotomy section. Fig. 14.14A presents a typical TEM micrograph recorded on a fiber cross-section. Three areas can be distinguished: the core (1), the intermediary layer (2), and the external shell (3). The core (Fig. 14.14B) extending over 100 nm appears constituted of numerous randomly oriented nanocrystalline (graphitic) domains as far as

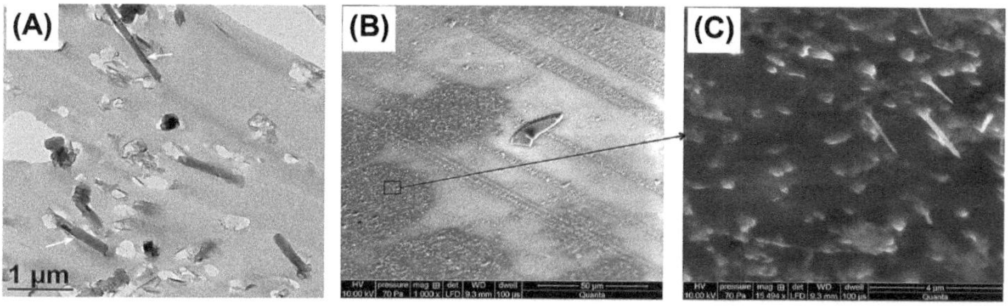

Figure 14.13 Morphologies of an ultramicrotomy section of fibers embedded in epon and of the surface of the corresponding remaining epon block: (A) Transmission electron microscopy micrograph of an ultramicrotomy section of nanofibers embedded in epon. Cross-sections and longitudinal sections of fibers are clearly visible. We can note the presence of some broken fibers (*white arrows*). Numerous holes are present in the resine part indicating that lots of fibers were not cut by the diamond knife, (B) Low magnification secondary electron image recorded on the surface of the remaining epon block after the preparation of ultramicrotomy sections showing, in *dark gray*, the region where nanofibers are concentrated in the resine, (C) Magnified image of the region underlined by the *black rectangle* in (B) which clearly reveals protruding fibers not suitable for AFM studies.

(002) graphite lattice fringes clearly appear in the micrograph. The intermediary layer (2) (Fig. 14.14E) which extends over 50−150 nm around the core reveals highly organized graphitic domains in which the (002) graphite lattice fringes are easily seen.

The contrast changes and the enhancement of the inter-reticular distances visible in the external shell of the fiber's cross-section (Fig. 14.14B and E) point out changes of both structure and chemical composition. The inter-reticular distances measured on the high-resolution electron micrograph raise 0.56 ± 0.02 nm close to the 0.58 nm of the $(CF)_n$ AA structure [54,59−65]. The EDX and EEL spectra (Fig. 14.14C and D) recorded at the periphery of the fiber's section confirm the presence of carbon and fluorine in the external shell. The fluorine concentration distribution (Fig. 14.14F), collected by EDXS analyses along the line profile corresponding to the superimposed black line in Fig. 14.14A, reveals that fluorine is only present in the surrounding shell whose thickness ranges between 30 and 40 nm.

The evolution of the epon block surface after surface [focused ion beam (FIB)] treatments is underlined in Fig. 14.15. Fig. 14.15A presents a SEM micrograph in SE mode recorded onto epon block surface after diamond knife cutting. Fig. 14.15B shows the surface state after ion milling in the dual beam. The vertical lines are due to curtaining effect resulting from ion milling. Fluorinated CNFs concentrated in some areas are clearly visible in the ion-milled region and less protruding fibers are detected. However, ion milling induces some superficial perturbations (amorphization, gallium ion implantation, etc.). This problem is overcome by argon/oxygen surface etching in the plasma cleaner. The SEM image collected in secondary electron mode with an accelerating voltage of 5 kV of the resulting clean surface is presented in Fig. 14.15C. CNFs are always clearly visible and the AFM topographic image and

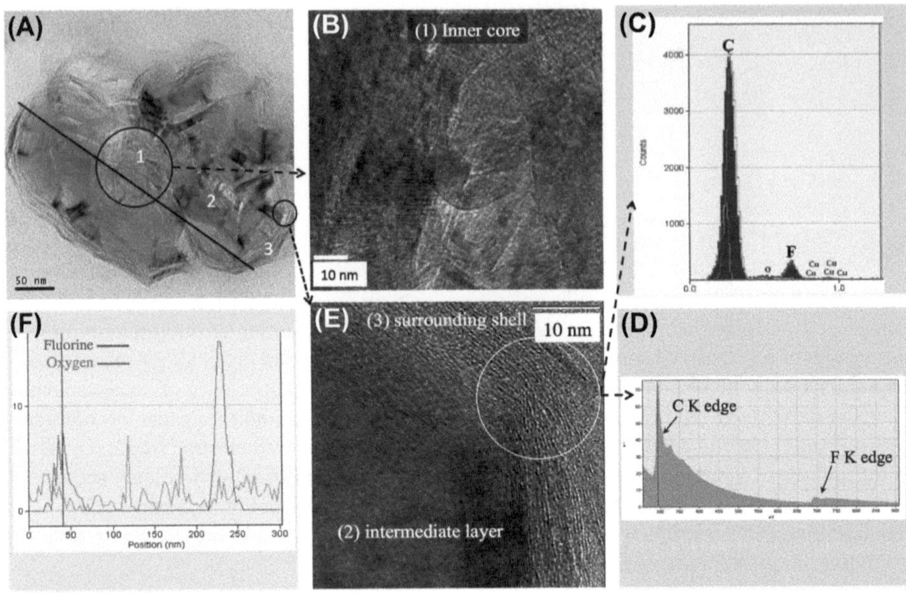

Figure 14.14 Transmission electron microscopy (TEM) results acquired on a fluorinated CNF (F/C = 0.15) cross section: (A) Low magnification TEM bright field image of the section in which three regions can be distinguished: (1) the inner core, (2) the intermediary layer, (3) the external shell, (B) HREM image corresponding to the inner core (1). It appears to be composed of randomly oriented graphitic nanodomains in which graphitic (002) lattices are clearly visible, (C) and (D) EDX and Electron Energy Loss spectra, recorded on the area underlined by the *white circle* in (E), revealing the presence of fluorine in the probed area, (E) HREM image recorded on the intermediary layer (2) and the external shell (3). The intermediary layer (2) presents two planar joining graphitic crystals in which the graphite (002) lattices are clearly visible. The surrounding shell shows a more ill-organized structure in which lattices with interreticular distances of 5.6 ± 0.2Å are clearly visible and attributed to the $(CF)_n$ carbon perfluorinated phase, (F) Fluorine and oxygen distribution profile acquired, using EDXS, along the superimposed *black line* in (A). The fluorine distribution clearly demonstrates that fluorine is only present in the surrounding shell of 30–40 nm thickness.

topographic profile shown in Fig. 14.15D confirms that some fibers slightly protrude from the surface.

This last characteristic of the sample will imply the development of a specific analysis of the lateral force images collected on the sample in order to obtain reliable and quantitative friction results.

14.5.2.3 Nanotribological Study of Friction Properties of Various Parts of the Partially Fluorinated Nanofibers

Fig. 14.16A presents a topographic image of a single fiber. The height profile acquired along the superimposed *white line* reveals a regular shape corresponding to a 200 nm horizontal flat surface surrounded by two surfaces tilted with an angle of

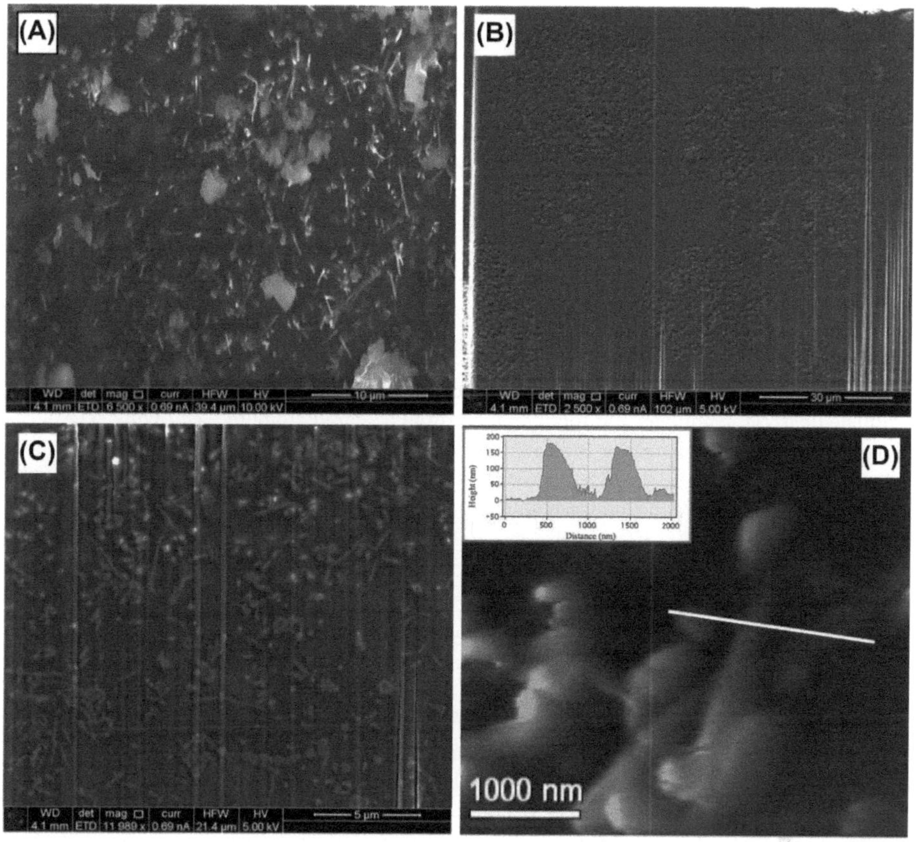

Figure 14.15 Evolution of the epon block surface after the consecutive surface treatments: (A) SE image collected on the surface of the remaining block surface after the preparation of ultramicrotomy sections, (B) SE micrograph of the surface collected after cutting of the emerging fibers by FIB. Less protruding nanofibers are detected, (C) SE micrograph of the surface recorded after the final oxygen/argon plasma cleaning applied to remove FIB-induced amorphized layer, (D) AFM topographic image recorded on the plasma cleaned area and height profile acquired along the *white line* in the image. It is clearly shown that the FIB nano-machining and plasma cleaning lead to sectioned fibers slightly protruding from the surface (around 150 nm).

$\theta = 0.30 \pm 0.02$ rad. This shape profile can be attributed both to the different etching rates of epoxy resin, fluorinated carbon phases and graphitic phases, and convolution of the real shape with the tip profile.

This fiber was scanned at several applied loads F_N^D from 12 to 60 nN and the shape remained unchanged during the whole experiments.

The fiber orientation and scan line are misaligned from $\xi = 0.64$ rad. In this case the horizontal calibrated driving lateral forces used to calculate the friction coefficient are $\left(F_0^D - F_{LL}^D\right)\sin\xi$ (see Fig. 14.16B) and $\left(F_{LR}^D - F_0^D\right)\sin\xi$ (not shown). The expression of the friction coefficient becomes:

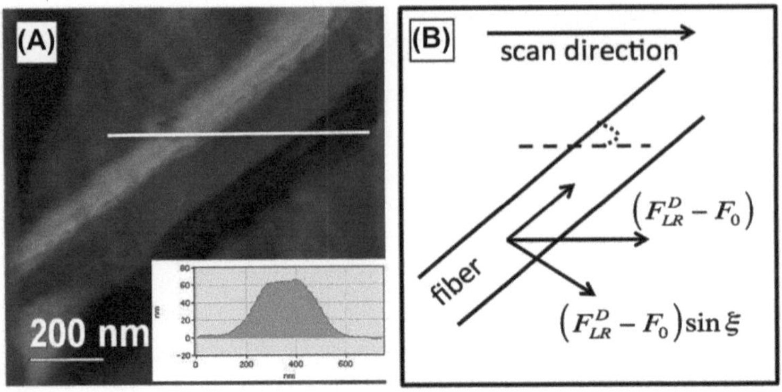

Figure 14.16 (A) AFM topographic image recorded on the longitudinal section of a single fluorinated carbon nanofibers and topographic profile acquired along the *white line*. The topography of the etched surface is attributed to the etching rates which are different for the various components of the surface (higher etching rates for epoxy resin and fluorinated part than for graphite), (B) Vector diagram showing the component of the driving force.

$$\mu = \frac{-F_{\mathrm{N}}^{\mathrm{D}}(1 + tg^2\theta) + \sqrt{\left[F_{\mathrm{N}}^{\mathrm{D}}(1 + tg^2\theta)\right]^2 + 4\beta^2 tg^2\theta \sin^2\xi}}{2\beta tg^2\theta \sin\xi} \tag{14.1}$$

To calculate the friction coefficient, special attention is paid to scan and use the same area on the fiber for every normal applied load. As previously mentioned, classical method, using conventional friction loop, cannot be applied to extract friction curve $F_{\mathrm{F}}(F_{\mathrm{N}})$ from lateral friction images. Then Eq. (14.1) is used to calculate the friction coefficient. The quantitative data are deduced from the experimental lateral values extracted along the line reported on the corresponding topographic image (Fig. 14.17A). The evolution of the calculated friction coefficient is displayed in Fig. 14.17C. Each point represents the average of 100 values calculated with the experimental values collected along constant inclination surface. The *dashed line* in Fig. 14.17B is shown as example.

The friction coefficient evolution as a function of sliding distance exhibits three plateaus, where friction coefficient remains constant, separated by strongly evolving regions. The plateaus correspond to sample areas where the tip is sliding at a constant speed on a surface with a constant inclination. The correlation to the topographic profile allows us to determine that the lowest calculated (see Eq. (14.1)) friction coefficient (0.20 ± 0.05) corresponds to the graphitized core of the fiber. The friction coefficient of the fluorinated surrounding shell, which corresponds to the second plateau, is 0.35 ± 0.05. The highest friction coefficient recorded outside the fiber corresponds to the tribological properties of the embedding resin. The strong evolutions of the friction between the plateaus of constant friction values are attributed to the "edge effects" (change of sliding slope) convolved with the tip radius.

To interpret the nanotribologic behavior of the various parts of the nanofiber, the preceeding results are compared to those recorded on reference compounds. For this

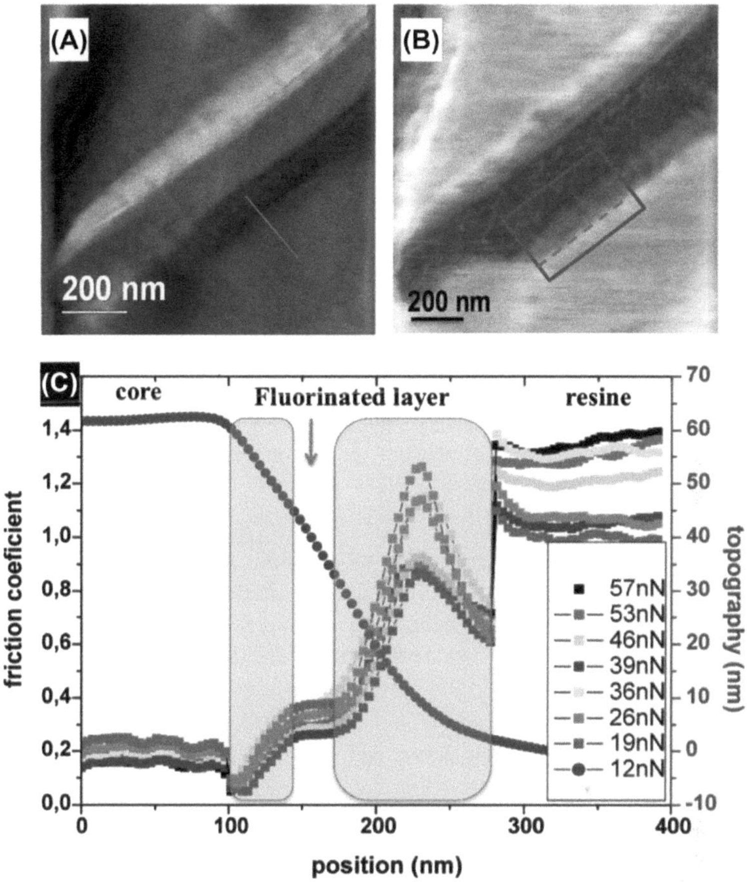

Figure 14.17 (A) AFM topographic image recorded on a slightly protruding fiber, (B) Lateral force image acquired on the fiber, (C) Topographic profile and evolution of the friction coefficient.

purpose, nanotribologic experiments are performed on monocrystalline highly oriented pyrolytic graphite (HOPG) and corresponding perfluorinated derivative.

14.5.2.4 Nanotribologic Behavior of Mono Crystalline Highly Oriented Pyrolytic Graphite and Corresponding Perfluorinated Derivative

For these investigations, perfluorinated HOPG was prepared by direct reaction, at 600°C during 3 days, of gaseous fluorine on an HOPG block of 10 × 50 × 1 mm dimensions [66,67]. Since oxygen/argon plasma cleaning was performed on the fibers, this treatment is applied as well to mono crystalline HOPG and its perfluorinated derivative.

The nanotribological tests were performed in air on pristine and plasma-treated samples of HOPG and its perfluorinated derivative by FFM. The normal loads

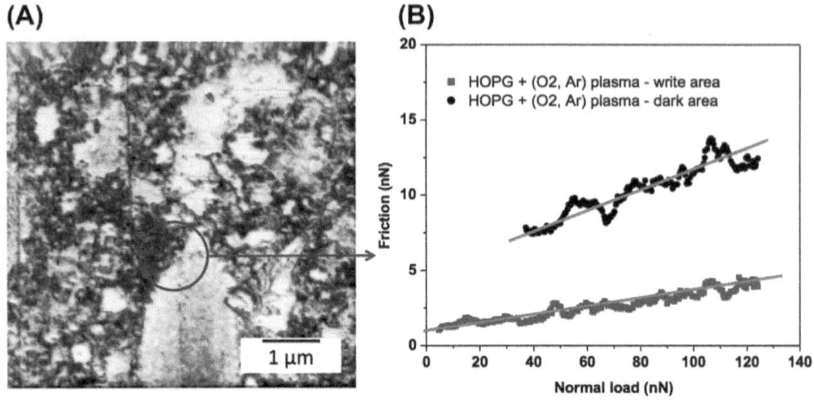

Figure 14.18 (A) Lateral image of plasma-treated HOPG obtained during the eighth scan, (B) The friction coefficients are respectively 0.07 ± 0.02 on the *dark areas* and 0.02 ± 0.02 on the *white areas*. *HOPG*, highly oriented pyrolytic graphite.

applied during the tests ranged from 0 to 120 nN. The results are presented in Fig. 14.18 for HOPG and Fig. 14.19 for perfluorinated HOPG. The evolution of the friction forces as a function of the total normal load (applied load + adhesion force) reveals a linear relationship between the friction and normal load for the

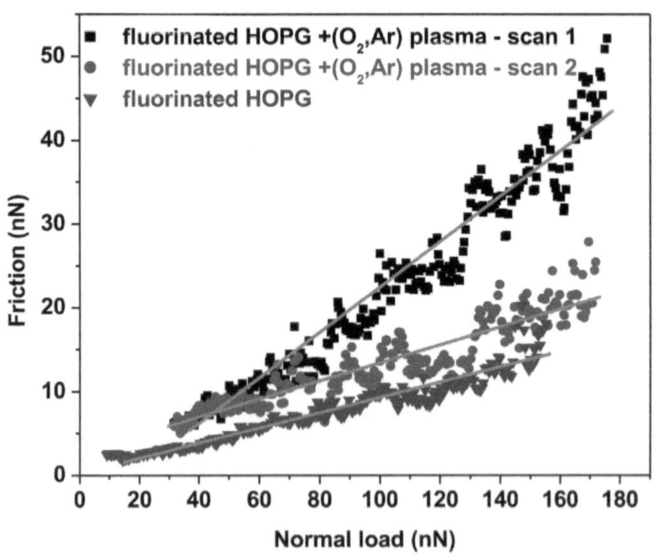

Figure 14.19 Evolution of the friction force as a function of the normal load for native perfluorinated HOPG and plasma-treated perfluorinated HOPG. During the first scan on the surface, the friction coefficient is 0.24 ± 0.03. It decreases down to 0.13 ± 0.02 at the second scan and remains constant during the following scans. *HOPG*, highly oriented pyrolytic graphite.

two compounds, which allows us to quantitatively deduce the friction coefficients defined as the slopes of the F_F versus F_N curves. The error is the standard deviation deduced from 10 experiments.

For pristine HOPG, the friction coefficient is close to 0.012 ± 0.002 whereas it is 10 times higher ($\mu = 0.10 \pm 0.02$) for its perfluorinated derivative. These results are in good concordance with those previously obtained by Kwon et al. [68].

For the plasma-treated HOPG sample, the friction properties evolve during the tip motion indicating that the plasma exposition modifies the surface. The friction properties, acquired on the same area at 120 nN, stabilize after seven scans. Fig. 14.18A shows the image obtained during the eighth scan. This one reveals *dark and bright areas* indicating that the sample presents locally different friction behaviors. The evolution of the friction force as a function of the normal load recorded on the *dark and bright areas* are presented in Fig. 14.18B. The deduced friction coefficients are 0.07 ± 0.02 and 0.02 ± 0.02 for the *dark and bright areas*, respectively. The plasma treatment of HOPG samples leads to friction coefficient higher than pristine HOPG which can explain the friction coefficient recorded on the graphitized fiber's core.

For the plasma-treated perfluorinated HOPG sample, the slopes of the friction curves (Fig. 14.19) show that during the first scan the friction coefficient is 0.24 ± 0.03. This one decreases down to 0.13 ± 0.02 during the second scan and then remains constant. These results indicate that the surface modifications induced by the plasma treatment only affect an ultrathin surface layer, this one being removed during the first scan to reveal the native surface.

It is important to note that several scans were applied on the nanofibers before the final image recording leading as for plasma-treated reference compounds to stabilized friction properties of the sample. The results collected on the fibers on the reference pristine and plasma-treated HOPG and perfluorinated HOPG compounds revealed in all cases that the friction coefficient of pristine graphitic structure is lower than the fluorinated one. This leads to the conclusion that friction reduction mechanism is not due to the bulk properties of fluorinated carbon.

14.5.2.5 Macroscopic Friction Reduction Mechanism of Fluorinated Carbon Nanofibers

Nanotribologic experiments carried out on various areas of partially fluorinated CNFs revealed that the fluorinated shell exhibited higher friction coefficient than graphite core as fluorinated HOPG presents also higher friction coefficient than pristine HOPG. These results imply that friction is not due to the bulk properties of fluorinated carbon. Coming back to the friction coefficient as a function of F/C ratio, it is interesting to note that above F/C = 0.15, the friction coefficient stabilizes up to F/C = 1. This indicates that above a critical F/C ratio, fluorination process or fluorinated layer thickness do not induce any changes. As far as it is not due to bulk effect, it is reasonable to attribute the friction reduction process to fiber's surface interactions. As far as fluorination progresses from the fiber's surface toward the inner part, the fluorination will first affect the surface energy up to the obtention of a

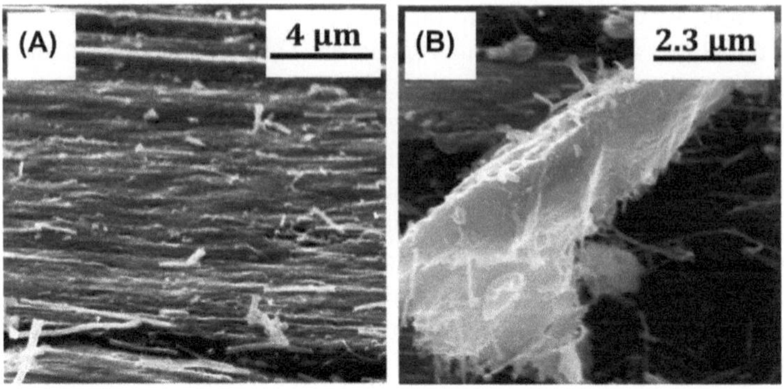

Figure 14.20 Secondary electron images collected (A) on the surface of the tribofilm and (B) on a fragment of the tribofilm extracted from the wear scar. The tribofilm appears constituted of assembly of individual nanofibers.

critical fluorinated thickness corresponding to F/C = 0.15 above which fiber's surface energy remains unchanged whatever was the F/C ratio. Friction reduction would then be due to interaction among individual fibers. To confirm the hypothesis, the tribofilm was extracted from the friction surfaces and its internal structure was studied by SEM.

Fig. 14.20 presents secondary electron micrographs recorded on the tribofilm and on one fragment extracted from the wear scar. The pictures clearly reveal that the tribofilm is mainly composed of nanofibers lying parallel to the sliding direction (Fig. 14.20A). This result strongly supports the proposed friction mechanism implying interaction between fibers.

As a conclusion, friction reduction process with CNFs results from interaction among fibers during sliding. The additional friction reduction resulting from fluorination can be attributed to the lowering of fiber surface energy by the formation of an external fluorinated layer. The interactions among fibers consequently decrease up to the stabilization of the surface energy associated with a critical thickness of the fluorinated layer (50 nm) corresponding to a fluorination rate of F/C = 0.15. As for higher fluorination rate where the surface energy is no more affected, the interactions among fibers remain constant and consequently lead to constant value of friction coefficient whatever is the F/C ratio.

14.5.3 Intrinsic Tribologic Properties of the Various Fluorinated Carbon Nanoparticles

The evolution of the intrinsic friction coefficient in air (recorded three cycles after total pentane evaporation) of the tested nanoparticles fluorinated under F_2 atmosphere (direct fluorination) are presented in Fig. 14.21 as a function of the fluorine content.

The intrinsic tribologic properties of pristine materials are similar in the case of CNFs and carbon blacks ($\mu = 0.13 \pm 0.02$ for CNFs and $\mu = 0.12 \pm 0.03$ for

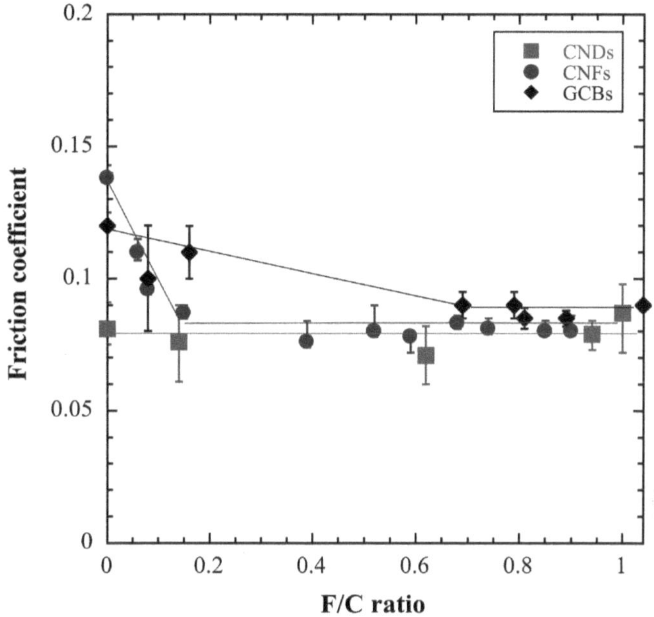

Figure 14.21 Evolution of intrinsic friction coefficient (recorded three cycles after pentane evaporation) of fluorinated carbon nanodiscs (CNDs), nanofibers (CNFs), and carbon blacks (GCBs) as a function of the fluorine content (direct fluorination).

GCBs) whereas the friction coefficient is lower ($\mu = 0.08 \pm 0.01$) in the case of CNDs. This is probably due to the morphology of the particles and their ability to feed the contact vicinity, the flat structure of CNDs being the most favorable.

Fig. 14.21 also underlines two different friction behaviors depending on the nature of the initial nanocarbons. Fluorination strongly improves the friction properties in the case of CNFs and GCBs: μ decreases for low or intermediate fluorine contents to raise an asymptotic value close to 0.08 for F/C superior to 0.15 and 0.6 for CNFs and GCBs, respectively. In the case of CNDs, the influence of fluorination is less significant as the friction coefficient remains in the 0.07–0.08 range.

The correlation of the tribologic results with TEM analyses allows us to explain the friction behavior of GCBs. As in the case of CNFs, TEM micrographs show that fluorination first affects the external layers of the particles and then progresses toward inner ones [55,69]. For F/C < 0.6, the decrease of the friction coefficient is attributed to the thickening of the surrounding fluorinated layer resulting in a decrease of the surface energy of the particles and in consequence in a lowering of interparticle interactions. For F/C > 0.6, the friction coefficient remains stable because the surrounding fluorinated layer is thick enough (≈ 30 nm) to stabilize the surface energy of the particles. These results strongly suggest that the friction reduction mechanisms of fluorinated GCBs are similar to the CNFs one and involve surface effects in the early stage of friction. In the case of fluorinated CNDs, the fluorination process occurs via the edge of the nanodiscs and in consequence does not only affect the external layers of the particles as fluorine atoms diffuse in the whole structure. The stability of the friction coefficient as a function of the fluorination rate suggests that the friction reduction mechanisms of CNDs involve bulk effects [70].

When the initial nanoparticles are fluorinated using TbF_4 fluorinating agent (controlled fluorination), the friction properties are improved compared to direct fluorination, as shown in Fig. 14.22. The lowest friction coefficient, close to 0.065, is obtained for fluorination contents superior to 0.4 for all the tested materials. This decrease of μ is probably due to different interparticle interactions induced by the fluorination process. In the case of controlled fluorination, the diffusion of fluorine radicals leads to a continuous addition of fluorine in the carbon matrix and a more progressive and deeper fluorination. The accumulation of fluorine in the outer shell is reduced preserving the external layers from damaging, observed for direct fluorination, up to high fluorination rates [31].

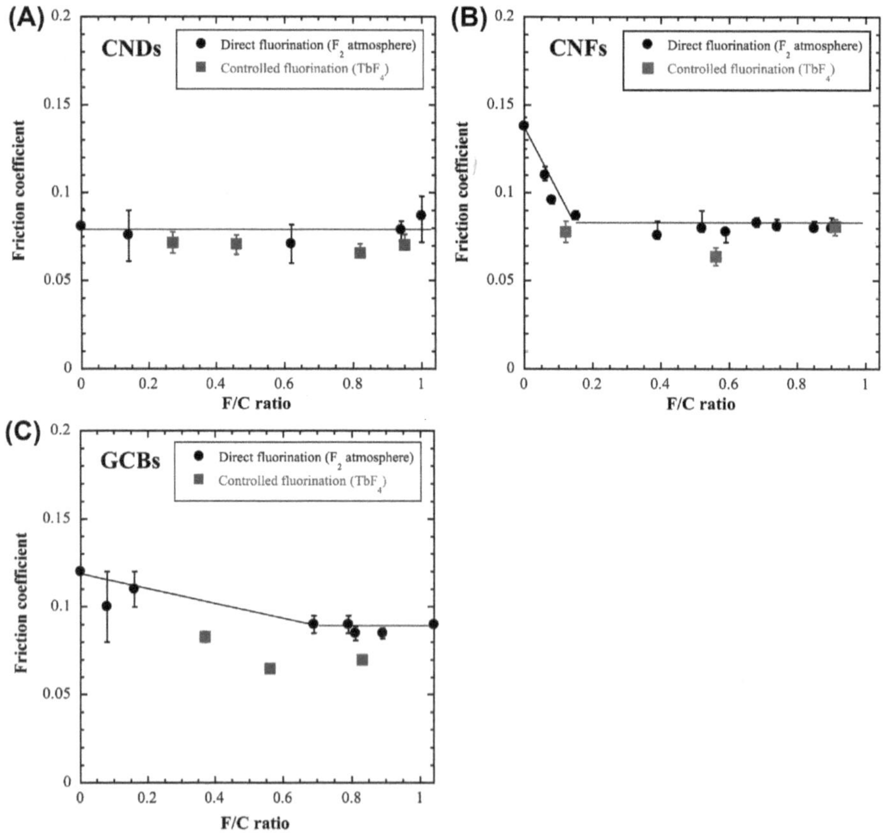

Figure 14.22 Evolution of intrinsic friction coefficient (recorded three cycles after pentane evaporation) of fluorinated carbon nanodiscs (CNDs) (A), nanofibers (CNFs) (B), and carbon blacks (GCBs) (C) as a function of the fluorine content in the case of direct and controlled fluorination.

14.5.4 Influence of the Friction Process on the Tribologic Behavior and the Structure of the Nanoparticles

Fig. 14.23 presents the evolution of the friction coefficient of the tested initial nanoparticles and fluorinated derivatives (direct fluorination) recorded after 500 cycles of friction, in comparison with the intrinsic ones, as a function of the fluorination rate. The influence of the friction process on the friction coefficient is underlined for pristine and weakly fluorinated CNDs (Fig. 14.23A) and CNFs (Fig. 14.23B) as μ decreases for F/C < 0.4. No evolution of the friction coefficient is observed for higher F/C ratios. In the case of fluorinated carbon blacks (Fig. 14.23C), the friction coefficient decreases during the tribologic test down to very low values ($0.06 < \mu < 0.07$) independent of the fluorination rate. This drastic improvement of the friction properties occurs during the first 200 cycles and then a stabilization of μ is observed up to the end of the

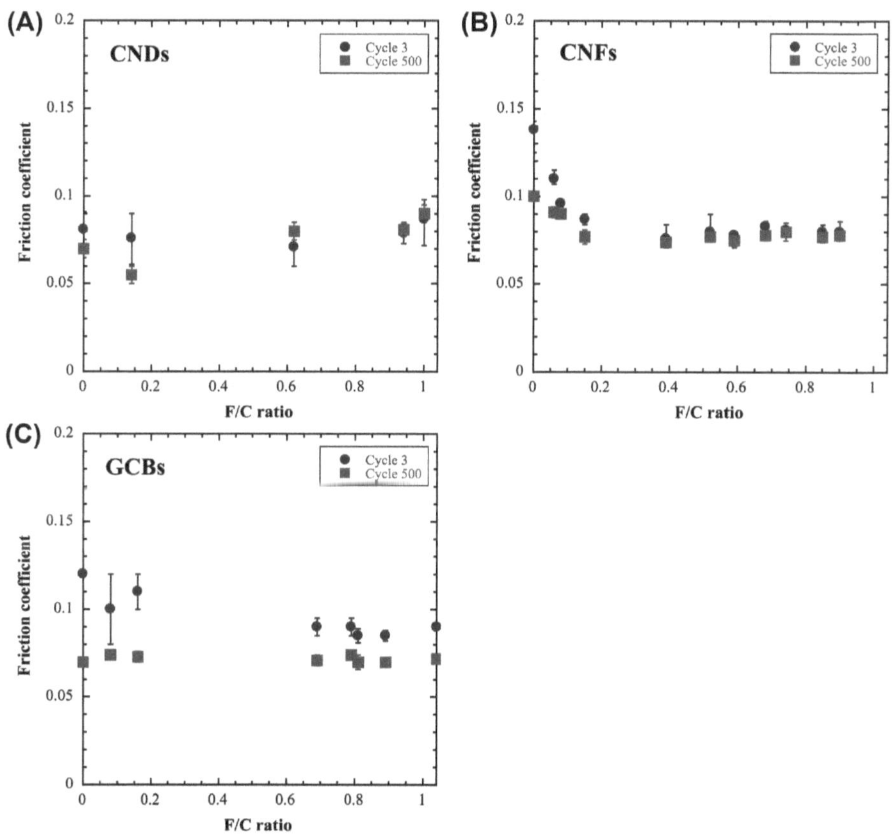

Figure 14.23 Evolution of friction coefficient recorded after 500 cycles of friction in comparison with intrinsic ones for fluorinated carbon nanodiscs (CNDs) (A), carbon nanofibers (CNFs) (B), and carbon blacks (GCBs) (C) as a function of the fluorine content in the case of direct fluorination.

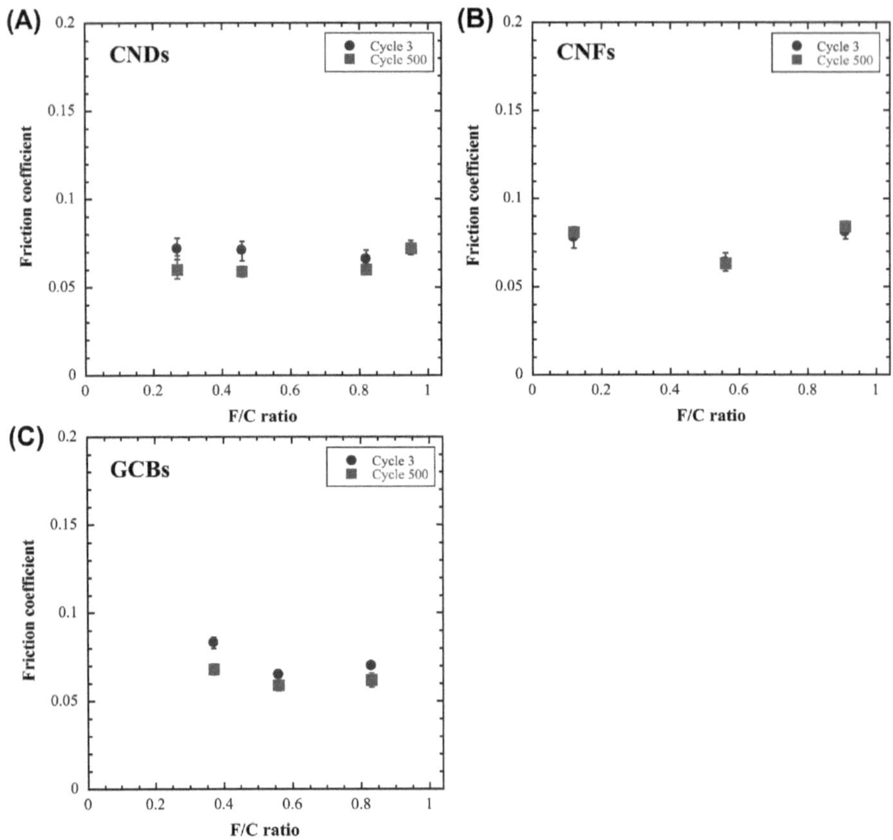

Figure 14.24 Evolution of friction coefficient recorded after 500 cycles of friction in comparison with intrinsic ones for fluorinated carbon nanodiscs (CNDs) (A), nanofibers (CNFs) (B), and carbon blacks (GCBs) (C) as a function of the fluorine content in the case of controlled fluorination.

test [55]. The stability of the friction coefficient suggests that the friction interfaces are similar for all the tested fluorinated GCBs whatever the initial fluorine contents were. It has to be noticed that the influence of the friction process is more significant for low fluorine contents for all fluorinated carbon materials.

In the case of controlled fluorination, the influence of the friction process is emphasized for fluorinated CNDs and GCBs (Fig. 14.24A and C). Very low friction coefficient values, close to 0.06 are obtained, suggesting structural evolution of the friction interfaces during the friction process. In the case of fluorinated CNFs, the friction coefficient remains stable during the friction test (Fig. 14.24B).

To confirm the hypothesis of structural evolution of the nanoparticles suggested by the friction properties observed for long-term experiments, Raman spectroscopy (HR 800 Horiba multi channel spectrometer fitted with a Peltier cooled CCD detector for spectra recording, see experimental conditions in Ref. [55]) and SEM investigations

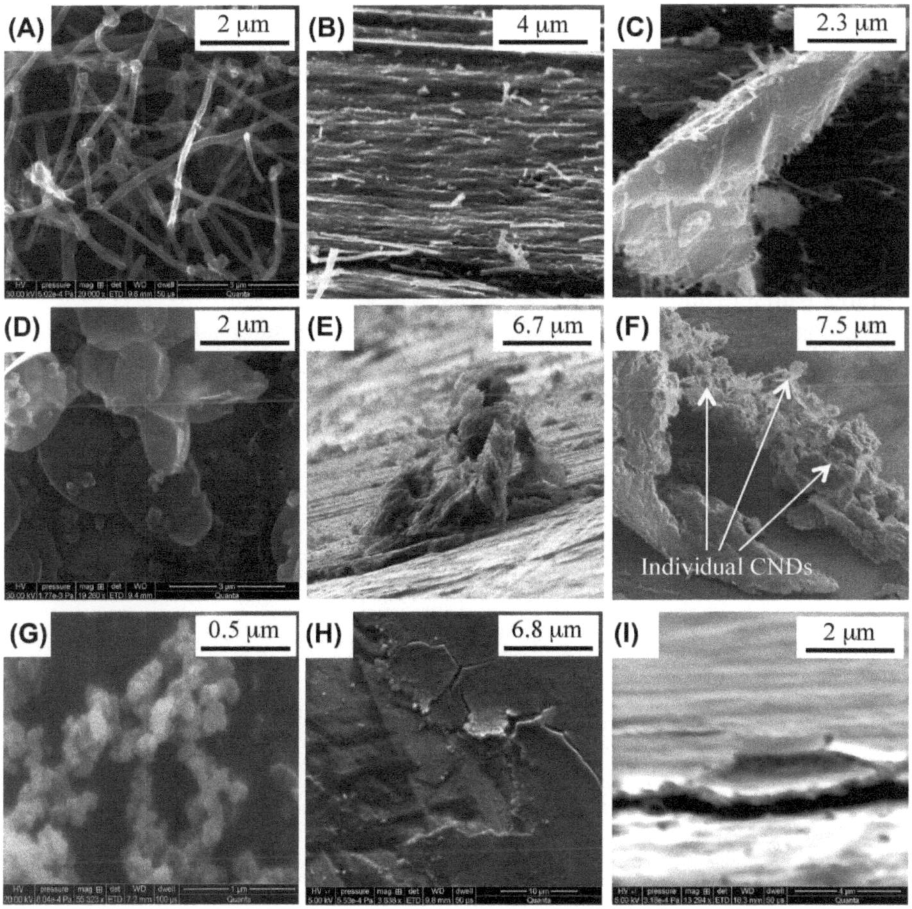

Figure 14.25 Scanning electron microscopy micrographs recorded on initial carbon nanofibers (CNFs) (A), carbon nanodiscs (CNDs) (D), carbon blacks (GCBs) (G); (B), (E), (H) aspect of the corresponding tribofilms; (C), (F), (I) particles of the tribofilms extracted from the wear scars.

(FEI Quanta 250 ESEM) were performed on initial fluorinated nanocarbons and associated tribofilms (direct fluorination).

SEM micrographs of initial nanocarbons, corresponding tribofilms, and particles of tribofilms extracted from the wear scars are presented in Fig. 14.25. In the case of CNFs, Fig. 14.25B and C emphasizes the fibrous nature of the tribofilm, constituted of individual nanofibers embedded in an ill-organized matrix. The disordered matrix is also observed in the case of CNDs (Fig. 14.25E and F) and CBGs (Fig. 14.25H) coating some individual particles (Fig. 14.25F and I). Note that the same structure of the tribofilms is observed for corresponding fluorinated particles.

Raman spectroscopy is a powerful technique to underline structural evolution of the particles induced by the friction process. In the case of fluorinated carbons, the investigation of the evolution of the D band position gives information about the presence of fluorine atoms in the carbonaceous structure. Indeed, the fluorination of graphitic carbons leads to a significant downshift of the D band as a function of increasing fluorine content as well as an increase of the D band intensity compared to the G one. This behavior is attributed to both size reduction of the graphitic domains induced by fluorination and formation of covalent C—F bonds [47—50]. Fig. 14.26 points out this downshift of the D band for initial CNFS, CNDs, and GCBs. For all the tested materials, Raman spectra recorded on tribofilms emphasizes a shift of the D band toward high wave numbers during the friction process. This behavior is attributed to a partial defluorination of the particles.

The correlation of tribologic results and SEM and Raman analyses suggests that fluorination improves the friction properties in the early stage of friction. Structural evolution of the particles is observed during the friction process. In particular for CNDs and GCBs, the tribofilms are composed of fluorinated carbon phases dispersed in a carbon matrix. The friction coefficient remains stable for long-term tribologic

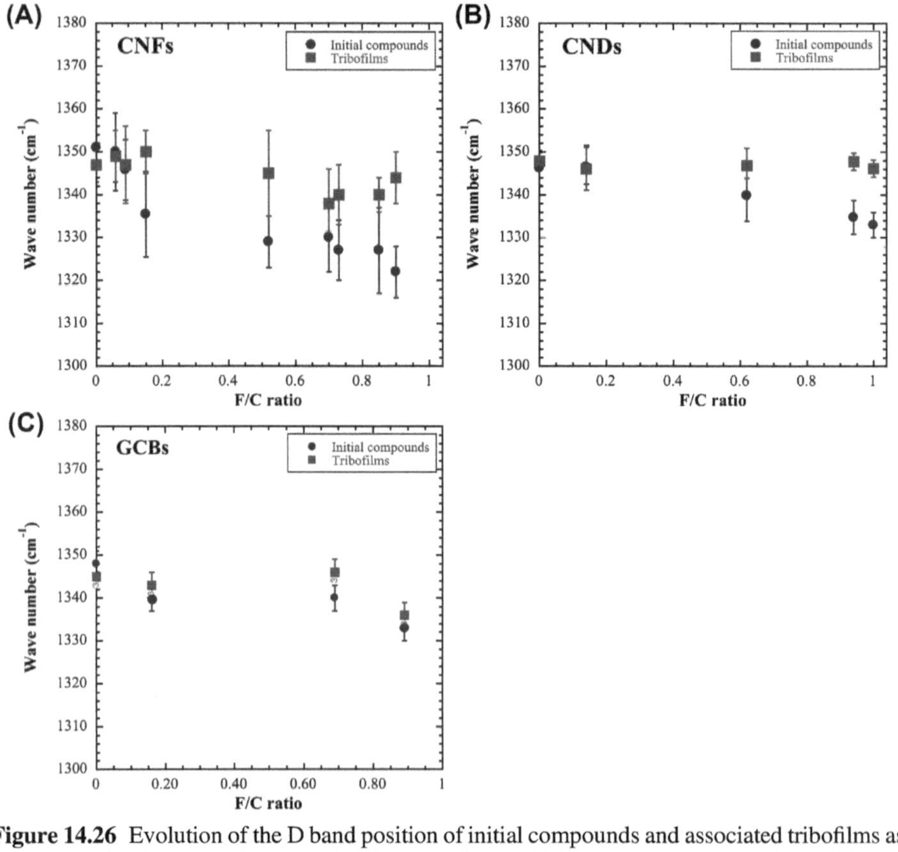

Figure 14.26 Evolution of the D band position of initial compounds and associated tribofilms as a function of F/C ratio in the case of carbon nanofibers (CNFs) (A), carbon nanodiscs (CNDs) (B), and carbon blacks (GCBs) (C).

experiments and does not depend on the fluorine content. The friction reduction properties are consequently mainly due to the carbon matrix. This behavior is less significant in the case of CNFs because an orientation of the fibers along the sliding direction is induced by the friction process [69]. As a consequence, the initial morphology of the fibers is no longer maintained.

14.6 Conclusion

The study of the tribologic properties of fluorinated nanocarbons shows that fluorination improves the lubricating performances of the particles, in particular for 0D (GCBs) and 1D (CNFs) materials. A drastic decrease of the friction coefficient is observed and attributed to a lowering of the surface energy of the particles due to the fluorination of external layers. When the fluorinated shell is thick enough to stabilize the surface energy, the friction coefficient does not evolve any more. The friction reduction mechanism is consequently attributed to surface effects in the case of fluorinated GCBs and CNFs. The behavior is different in the case of 2D compounds (CNDs). The improvement of the intrinsic friction properties due to fluorination is weak because fluorination occurs via the edges of CNDs and concerns the whole layers of the discs.

The tribologic performances of nanocarbons also depend on the fluorination process. The friction properties are improved when the nanoparticles are fluorinated using a fluorinating agent such as TbF_4 (controlled fluorination) by comparison with direct fluorination under F_2 atmosphere. This is probably due to the more homogeneous and progressive diffusion of fluorine atoms in the graphitic structure induced by controlled fluorination.

Long-term tribologic experiments lead to evolutions of the structure and the composition of the tribofilms, underlined by Raman and SEM investigations. The friction process leads to an amorphization and a partial defluorination on the compounds. The tribofilms are then composed of individual fluorinated nanoparticles embedded in a carbon matrix. The friction reduction mechanisms are attributed to the properties of this matrix rather than to individual nanoparticles.

The tested carbon nanostructures appear as interesting candidates to be used as new nanoadditives for lubricants. The control of the fluorination conditions allows one to achieve optimized friction properties. In addition to good intrinsic friction properties, the advantages of fluorinated nanocarbons are their sizes and geometries, well adapted for a good feeding of the sliding interface, their high chemical and temperature stability, and the possibility to obtain stable dispersions in liquid or gel lubricants.

Acknowledgments

The authors acknowledge the French Research Department, the Conseil Régional de la Guadeloupe, the Fond Social Européen (FSE) and the Fond Européen pour le Développement (FEDER) for their financial supports.

References

[1] D. Dowson, History of Tribology, Longmans, London, 1979.

[2] G. Amontons, Mémoires de Mathématiques et de Physique de l'Académie Royale des Sciences, 1699, pp. 206−257.

[3] C.A. Coulomb, Mémoires de Mathématiques et de Physique de l'Académie Royale des Sciences, 1785, pp. 161−332.

[4] H. Allamier, C. Priestner, D.E. Sander, F.M. Reich, Friction in automotive engines, CH 9, in: H. Pihtili (Ed.), Tribology in Engineering, 2013, ISBN 978-953-51-1126-9.

[5] R. Stribeck, Z. Verein, Deut. Ing. 46 (1902) 1341−1348.

[6] W.B. Hardy, Collected Works, University Press, Cambridge, 1936.

[7] W.A. Zisman, Friction and Wear, in: R. Davies (Ed.), Elesevier, Amsterdam, 1959, pp. 110−148.

[8] A.S. Akhmatov, Molecular Physics of Boundary Friction, IPST, Jerusalem, 1966.

[9] S.N. Postnikov, Electrophysical and Electrochemical Phenomena in Friction, Cutting and Lubrication, Van Nostrand Reinhold Comp., London, 1978, pp. 113−114.

[10] J.L. Mansot, Aspect microscopique de l'action des reducteurs de frottement en lubrification limite (Docteur Ingenieur thesis), Ecole Centrale de Lyon, 1982.

[11] J.L. Mansot, J.M. Martin, A. Gagnaire, J. Joseph, Wear 84 (1983) 115.

[12] J.M. Georges, T. Mathia, J. Mech. Appl. 2 (1978) 1−34.

[13] J.M. Martin, Contribution à la tribologie: Etude du mécanisme d'action d'un additif anti-usure en regime de lubrification limite. Aspects chimiques dans le cas des organo-dithiophosphates métalliques, Thèse d'état nO 7827, Université Claude Bernard, Lyon, 1978.

[14] J.M. Martin, J.L. Mansot, I. Berbezier, H. Dexpert, Wear 93 (1984) 117−126.

[15] J.M. Martin, J.L. Mansot, I. Berbezier, M. Belin, Wear 107 (1986) 355−366.

[16] J.M. Martin, M. Belin, J.L. Mansot, ASLE Trans. 49 (1986) 523−531.

[17] J.M. Martin, C. Grossiord, T. Lemogne, S. Bec, A. Tonck, Tribol. Int. 31 (2001) 627−644.

[18] J.L. Mansot, M. Hallouis, J.M. Martin, Colloids Surf. A 75 (1993) 25−31.

[19] S. Giasson, T. Palermoa, T. Buffeteau, B. Desbat, J.M. Turlet, Thin Solid Films 252 (1994) 111−119.

[20] J.L. Mansot, V. Golabkan, L. Romana, Ph Bilas, E. Alleman, Y. Bercion, Coll. Surf. A 243 (2004) 67−77.

[21] J.L. Mansot, J.M. Martin, Y. Bercion, L. Romana, Brazil. J. Phys. 39 (2009) 53−61.

[22] J.M. Martin, in: J.M. Martin, O. Ohmae (Eds.), Nanolubricants, Wiley, New York, 2008.

[23] S. Cohen, L. Rapoport, E.A. Poonomarev, H. Cohen, T. Tsirlina, R. Tenne, C. Levy-Clement, Thin Solid Films 324 (1998) 190−197.

[24] R. Tenne, M. Homyonfer, Y. Feldman, Chem. Mater. 10 (1998) 3225−3238.

[25] L. Rapoport, Y. Feldman, M. Homyonfer, H. Cohen, J. Sloan, J.L. Hutchison, R. Tenne, Wear 975 (1999) 225−229.

[26] L. Rapoport, Y. Bilik, Y. Feldman, M. Homyonfer, S.R. Cohen, R. Tenne, Nature 387 (1997) 791−793.

[27] R.L. Fusaro, H.E. Sliney, NASA Tech. Note D-5097, 1969.

[28] R.L. Fusaro, H.E. Sliney, ASLE Trans. 13 (1970) 56−65.

[29] R.L. Fusaro, H.E. Sliney, ASLE Trans. 16 (1973) 189−196.

[30] W. Zhang, M. Dubois, K. Guérin, P. Bonnet, H. Kharbache, F. Masin, A.P. Kharitonov, A. Hamwi, Phys. Chem. Chem. Phys. 12 (2010) 1388−1398.

[31] Y. Ahmad, E. Disa, K. Guérin, M. Dubois, E. Petit, A. Hamwi, P. Thomas, J.L. Mansot, J. Fluor. Chem. 168 (2014) 163−172.

[32] F. Chamssedine, M. Dubois, K. Guérin, J. Giraudet, F. Masin, D.A. Ivanov, L. Vidal, R. Yazami, A. Hamwi, Chem. Mater. 19 (2007) 161−172.
[33] W. Zhang, K. Guérin, M. Dubois, Z. El Fawal, D.A. Ivanov, L. Vidal, A. Hamwi, Carbon 46 (2008) 1010−1016.
[34] W. Zhang, K. Guérin, M. Dubois, A. Houdayer, F. Masin, A. Hamwi, Carbon 46 (2008) 1017−1024.
[35] W. Zhang, M. Dubois, K. Guérin, A. Hamwi, J. Giraudet, F. Masin, J. Solid State Chem. 181 (2008) 1915−1924.
[36] W. Zhang, M. Dubois, K. Guérin, P. Bonnet, E. Petit, N. Delpuech, D. Albertini, F. Masin, A. Hamwi, Carbon 47 (2009) 2763−2775.
[37] Y. Ahmad, E. Disa, M. Dubois, K. Guérin, V. Dubois, W. Zhang, P. Bonnet, F. Masin, L. Vidal, D.A. Ivanov, A. Hamwi, Carbon 50 (2012) 3897−3908.
[38] Y. Ahmad, M. Dubois, K. Guérin, A. Hamwi, W. Zhang, Carbon 94 (2015) 1061−1070.
[39] A.M. Kannan, A. Meghal, I.V. Barsukov, Electrochem. Commun. 8 (2006) 887−891.
[40] S.N. Naess, A. Elgsaeter, G. Helgesen, K.D. Knudsen, Sci. Technol. Adv. Mater. 10 (2009) 065002.
[41] S. Lynum, J. Hugdhal, K. Hox, R. Hildrum, M. Nordvik, Production of Micro Domain Particles by Use of a Plasma Process, Patent EP1017622.
[42] A.M. Panich, Synth. Metals 100 (1999) 169−185.
[43] M. Dubois, J. Giraudet, K. Guérin, A. Hamwi, Z. Fawal, P. Pirotte, et al., J. Phys. Chem. B 110 (2006) 11800−11808.
[44] J. Giraudet, M. Dubois, K. Guérin, C. Delabarre, A. Hamwi, F. Masin, J. Phys. Chem. B 111 (2007) 14143−14151.
[45] J. Giraudet, M. Dubois, A. Hamwi, W.E.E. Stone, P. Pirotte, F. Masin, J. Phys. Chem. B 109 (2004) 175−181.
[46] T. Mallouk, B.L. Hawkins, M.P. Conrad, K. Zilm, G.E. Maciel, N. Bartlett, Philos. Trans. R Soc. London A 314 (1985) 179−187.
[47] P. Thomas, K. Delbé, D. Himmel, J.L. Mansot, F. Cadoré, K. Guérin, M. Dubois, C. Delabarre, A. Hamwi, J. Phys. Chem. Solids 67 (2006) 1095−1099.
[48] A.M. Rao, A.W.P. Fung, S.L. di Vittorio, M.S. Dresselhaus, G. Dresselhaus, M. Endo, K. Oshida, T. Nakajima, Phys. Rev. B 45 (1992) 6883−6892.
[49] V. Gupta, T. Nakajima, B. Zemva, J. Fluor. Chem. 110 (2001) 145−151.
[50] V. Gupta, T. Nakajima, T. Ohzawa, B. Zemva, J. Fluor. Chem. 120 (2003) 143−150.
[51] F. Withers, M. Dubois, A.K. Savchenko, Phys. Rev. B 82 (2010) 073403.
[52] F. Tuinstra, J.L. Koenig, J. Chem. Phys. 53 (1970) 1126−1130.
[53] F. Chamssedine, D. Claves, Chem. Phys. Lett. 454 (2008) 252−256.
[54] T. Nakajima, N. Watanabe, Graphite Fluorides and Carbon − Fluorine Compounds, CRC Press, Boston, 1991.
[55] P. Thomas, J.L. Mansot, A. Molza, F. Bégarin, M. Dubois, K. Guérin, Tribol. Lett. 56 (2014) 259−271.
[56] H. Hertz, J. Reine Angew. Math. 92 (1881) 156−171.
[57] P. Bilas, L. Romana, B. Kraus, Y. Bercion, J.L. Mansot, Rev. Sci. Instrum. 75 (2004) 415−421.
[58] P. Bilas, L. Romana, F. Bade, K. Delbe, J.L. Mansot, Tribol. Lett. 34 (2009) 41−48.
[59] J.E. Sader, J.W.M. Chon, P. Mulvaney, Rev. Sci. Instrum. 70 (1999) 3967−3970.
[60] L.B. Ebert, J.I. Brauman, R.A. Huggins, J. Am. Chem. Soc. 96 (1974) 7841−7842.
[61] H. Touhara, K. Kadono, Y. Fujii, N. Wanatabe, Z. Anorg. Allg. Chem. 544 (1987) 7−20.
[62] H. Fujimoto, Carbon 35 (1997) 1061−1065.
[63] J.C. Charlier, X. Gonze, J.P. Michenaud, Phys. Rev. B 47 (1993) 16162−16168.

[64] S.S. Han, T.H. Yu, B.V. Merinov, A.C.T. van Duin, R. Yazami, W.A. Goddard III, Chem. Mater. 22 (2010) 2142–2154.
[65] P. Lazar, E. Otyepková, F. Karlický, K. Čépe, M. Otyepka, Carbon 94 (2015) 804–809.
[66] P. Bilas, P. Thomas, L. Romana, T. Cesaire, J.L. Mansot, C. Delabarre, M. Dubois, Society of Tribologists and Lubrication Engineers Annual Meeting and Exhibition, 2010, pp. 638–640.
[67] S.H. Cheng, K. Zou, F. Okino, H.R. Gutierrez, A. Gupta, N. Shen, P.C. Eklund, J.O. Sofo, J. Zhu, Phys. Rev. B 81 (2010) 205435.
[68] S. Kwon, J.H. Ko, K.J. Jeon, Y.H. Kim, J.P. Young, Nano Lett. 12 (2012) 6043–6048.
[69] P. Thomas, D. Himmel, J.L. Mansot, M. Dubois, K. Guérin, W. Zhang, A. Hamwi, Tribol. Lett. 34 (2009) 49–59.
[70] P. Thomas, D. Himmel, J.L. Mansot, W. Zhang, M. Dubois, K. Guérin, A. Hamwi, Tribol. Lett. 41 (2011) 353–362.

Perfluoropolyether-Functionalized Carbon-Based Materials and Their Applications

15

M. Sansotera[1,2], M. Gola[1,2], W. Navarrini[1,2]
[1]Politecnico di Milano, Milano, Italy; [2]Consorzio INSTM, Firenze, Italy

Chapter Outline

15.1 Introduction

In fluorine chemistry, the study of C—F bonds over the last two centuries has allowed to deepen progressively the knowledge about how the physical properties and chemical reactivities of carbon-based substrates could be affected by fluorination. The extreme electronegativity of the fluorine atom ensures that it is inductively electron withdrawing when bonded to carbon atoms: therefore, C—F bonds are strongly polarized with a relatively high ionic character and are stronger than other C—X bonds [1]. The bond dissociation energy of C—F bonds can be as high as 545 kJ/mol and it explains the typical chemical, thermal, and photochemical stability of perfluorinated materials [2,3].

New Fluorinated Carbons: Fundamentals and Applications. http://dx.doi.org/10.1016/B978-0-12-803479-8.00015-2

The first fluorinated carbon-based material was the graphite fluoride, $(CF_x)_n$, reported by Ruff et al. in 1934 [4]. It was followed by the discovery of fluorine–graphite intercalation compounds (GIC) and the development of several preparation techniques for fluorinated carbon blacks (CBs) [5–10]. The commercial interest on these materials started in the late 1960s and now they are considered as specialty products which found applications in lithium batteries, lubrication, and imaging processes [6,11].

For a long time graphite and CB represented the only carbon materials suitable for fluorination. However, in 1985 a new era of synthetic carbon allotropes began after the discovery of fullerenes, and it continued with the synthesis of carbon nanotubes (CNTs) and the rediscovery of graphene [12]. As a consequence of the expected remarkable physical properties of these new carbon allotropes, the study of their fluorination reactions became appealing along with the development of their applications [13–15]. In particular, the nanostructuration of the allotropic forms of carbon and the development of new fluorination routes for such kind of materials emerged as two key parameters which would generate new materials with improved performances [16]. Besides fluorination, other approaches were also considered and perfluoroalkylation resulted as one of the most promising reaction. For example, fluorinated fullerenes resulted unstable with respect to hydrolysis by losing readily fluoride ions and generating HF, while perfluoroalkylation of fullerenes provided the necessary stability against hydrolysis, in addition to chemical stability and hydrophobicity of the material surface [17]. Perfluoroalkylated fullerenes were prepared by reaction with perfluoroalkyl radicals which were generated by various approaches [17–20]. Since the late 1990s, the sidewall addition chemistry of single-walled carbon nanotubes (SWCNTs) was also developed and fluorination as well as fluoroalkylation strategies were studied accordingly [21–25]. The chemical derivatization of SWCNTs was particularly desirable for the improvement of their solubility and their processibility [26]. Perfluoroalkyl iodide, diazo-perfluorocompounds, and perfluorodiacyl peroxides were considered as source of perfluorinated alkyl radicals for the addition reaction [23,24,27]. Other carbon-based materials, such as diamond-like carbon (DLC) and graphitic CB, were also functionalized with perfluorocarbon chains by exploiting the chemistry of perfluoroalkyl radicals on the unsaturated moieties of such substrates [28,29].

Interestingly, perfluoropolyethers (PFPEs) represent an alternative class of compounds belonging to fluorinated materials suitable for the functionalization of carbon-based materials. PFPEs are fluid polymeric materials which exhibit exceptional thermo-oxidative stability and chemical resistance, high gas permeability, low surface energy (SE), good lubricity characteristics, and extended bioinertness [30]. They can be synthesized by photooxidative polymerization of perfluoroolefins through the formation of stable oligomeric mixtures of random peroxidic perfluorinated polyethers as intermediates [31–33]. The homolytic cleavage of peroxide bonds can be induced by thermal or photochemical treatments, generating alkoxy radicals with PFPE structure in a concentration depending on the peroxide content (usually expressed as weight percentage of peroxidic oxygen) and temperature [30]. Carbon-centered PFPE radicals can be also obtained by rapid β-scission reaction of the early formed PFPE alkoxy radicals [34]. Thus, the addition of these electrophilic PFPE radicals to carbon-based materials with available unsaturated moieties appeared as a suitable alternative route for the preparation of a new class of fluorinated materials with improved performances: the PFPE-functionalized carbon-based materials [35].

The aim of this chapter is to describe the functionalization of several carbon-based materials with PFPE moieties, the properties of these materials, and their recognized applications. High thermal and chemical stability might be expected for such materials because both PFPEs and carbon-based materials are exceptionally thermally as well as chemically stable. In addition, it can be argued that typical properties of PFPEs, like high gas permeability and low SE, were stably conferred to the carbonaceous surface of these materials by the covalent linkage of PFPE chains. On the basis of these characteristics, several applications of PFPE-functionalized carbon-based materials have been developed in the field of electrochemical devices like fuel cells and lithium batteries, in self-lubricating surfaces and in polymer formulation. Other applications in the field of separation techniques are based on the fluorous effect ascribable to PFPE chains.

15.2 Functionalization With Perfluoropolyether Moieties

Peroxidic PFPEs are a family of perfluorinated polymers with structures characterized by the alternation of oxygen and carbon-based units, where the oxygen units can be both ether (O) and peroxidic (O—O). Particularly, in linear peroxidic PFPEs the carbon-based units can be perfluoromethylene (CF_2) or perfluoroethylene (CF_2CF_2). A general extended representation of these structures is shown in Formula (i) below:

$$TO(CF_2CF_2O)_m(CF_2O)_n(O)_vT' \qquad (i)$$

where T and T′ can be perfluorinated or acyl fluoride chain end groups and the average molecular weight (MWA) is usually comprised between a few thousand to tens of thousands; the peroxidic units are generally less than the ether units [30,33]. Differently, in branched PFPE peroxides the polymeric backbone can be constituted by perfluoro-iso-propylene oxide (($(CF_3)CFCF_2O$ and $CF_2CF(CF_3)O$ randomly distributed), perfluoro(methyl)methylene oxide ($CF(CF_3)O$), and perfluoromethylene oxide (CF_2O) units, as reported in Formula (ii):

$$TO[CF_2CF(CF_3)O]_q[CF(CF_3)O]_p(CF_2O)_n(O)_vT' \qquad (ii)$$

The MWA of branched PFPE peroxides typically falls in the range 700—50,000 and the peroxide content normally varies in the range 0.75—1.25% by weight [30]. Peroxidic PFPEs are synthesized by oxidative photopolymerization of perfluoroolefins at cryogenic temperatures in the range −80 to −40°C: branched and linear macromolecules derive from photooligomerization of hexafluoropropene (HFP) and tetrafluoroethylene (TFE), respectively. This synthetic route has been also extended to mixtures of HFP/TFE and to a variety of other fluorinated olefins, including perfluorocyclobutene, perfluorobuta-1,3-diene, and various substituted olefins and higher homologs of HFP [30].

Thermal treatments at the appropriate temperatures (150–200°C) allow the cleavage of peroxidic units present along the polymeric backbone, generating PFPE alkoxy radicals; hence reactive carbon-centered PFPE radicals are quickly obtained by the subsequent cleavage of a bond in β-position to the oxygen-centered radicals, ie, β-scission reactions [30]. In these conditions, such PFPE radicals can directly bond to carbonaceous substrate, creating covalent C–C bonds between PFPE chains and unsaturated moieties on the surface of the carbon-based material (Scheme 15.1) [35,36].

The perfluorinated moieties are grafted without any hydrogenated spacer, which usually decreases both thermal and chemical stabilities of the resulting materials. Carbonyl fluoride and primary PFPE alcohols are formed as main side-products of the chemical treatment [34]. Primary perfluorinated alcohols are thermodynamically unstable by an estimated 80–160 kJ/mol and instantaneously rearrange to acyl fluorides by eliminating HF (Scheme 15.2) [37].

The perfluorinated radicals react preferentially on the aromatic rings but, when benzylic positions are available, low to medium amounts of by-product deriving from the hydrogen abstraction from the methyl group can be detected [38].

The O–O bond energy is influenced by the structure of the units near the peroxidic moiety. Particularly, the energies of peroxidic bonds between two perfluoromethylene units and between two perfluoroethylene units are around 197 and 168 kJ/mol, respectively [30]. After the homolytic cleavage of the first peroxidic O–O bond, the resulting PFPE radicals initially maintain a peroxidic content and the sum of the peroxidic units in the resulting PFPE oxyradicals is one less than in the starting chain (Scheme 15.3). Perfluoromethylene (b and $e = 1$) or perfluoroethylene units (b and $e = 2$) can be randomly found as adjacent groups to the radical centers on the oxygen atoms in the PFPE oxyradicals (Scheme 15.3). Since the reactivity of oxyradicals is weakly influenced by the PFPE chain, the subsequent reaction steps for the two separated PFPE radical residuals can be described analogously even if they could have different molecular weights.

The PFPE oxyradicals activate a single electron transfer propagation mechanism where the unsaturated moieties on the carbonaceous surfaces are active as radical scavengers (Scheme 15.4). In this step, the reactivity of the oxyradicals is mainly determined by the two successive units adjacent to the radical center: if the neighboring group is a C_2 unit ($b = 2$), a β-scission reaction occurs, by forming a carbon-centered radical with the elimination of a molecule of carbonyl difluoride, COF_2; if the neighbor group is a C_1 unit ($b = 1$) and the successive unit is a C_2 unit ($a = 2$), two consecutive β-scission reactions occur and a carbon-centered radical is obtained with the elimination of two COF_2 molecules; the case of two successive C_1 units ($b = 1, a = 1$) adjacent to the radical center is quite unlikely, because C_1 units are typically less than C_2 units. Overall, the reactivity of the PFPE oxyradicals determines the stoichiometry of COF_2 production.

The PFPE carbon-centered radicals are generated by the β-scission reactions and are the reactive species responsible of functionalization of the carbon-based substrates, as demonstrated by ^{19}F solid state MAS-NMR and DFT calculations at PBE0 level [39].

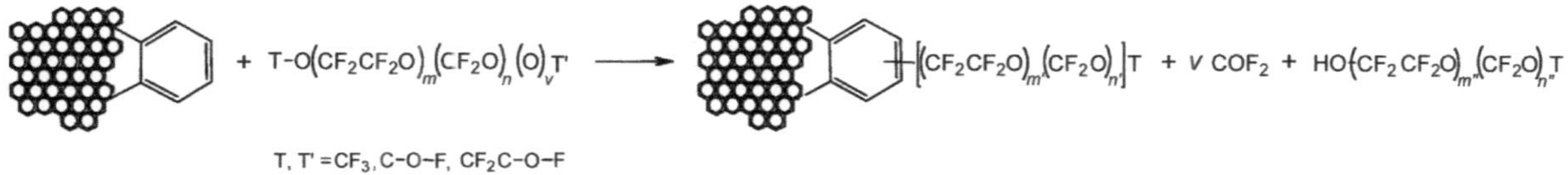

Scheme 15.1 Reaction between a carbonaceous substrate with graphitic moieties and linear perfluoropolyether peroxide.

$$HO(CF_2CF_2O)_{m''}(CF_2O)_{n''}T \longrightarrow F{-}C({=}O){-}CF_2O(CF_2CF_2O)_{m''-1}(CF_2O)_{n''}T + HF$$

Scheme 15.2 Rearrangement of primary perfluoropolyether alcohols to acyl fluorides.

$$TO(CF_2CF_2O)_m(CF_2O)_n(O)_vT' \longrightarrow TO(C_iF_{2i}O)_{p_i}(O)_yC_aF_{2a}OC_bF_{2b}O\cdot + T'O(C_iF_{2i}O)_{q_i}(O)_zC_dF_{2d}OC_eF_{2e}O\cdot$$

$m + n = \Sigma p_i + \Sigma q_i + 4$ $\quad$ i = 1,2

$v = y + z + 1$ $\quad$ a = 1,2 ; b = 1,2

d = 1,2 ; e = 1,2

Scheme 15.3 Homolytic cleavage of O—O bond in perfluoropolyether (PFPE) peroxide and formation of two separated PFPE oxyradicals.

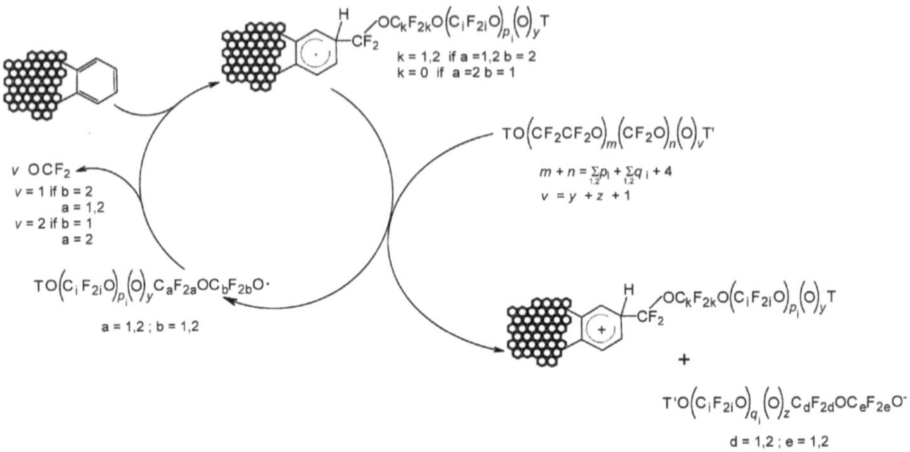

Scheme 15.4 Covalent linkage of perfluoropolyether carbon-centered radicals to the carbon-based substrate.

As a consequence of the chemical linkage of the PFPE carbon-centered radicals to the carbonaceous substrate, an aryl radical adduct with a linked PFPE chain is formed as intermediate (Scheme 15.4). It is subsequently oxidized by an additional molecule of the starting PFPE peroxide that splits in two parts (Scheme 15.4): a PFPE primary alkoxide and a new PFPE oxyradical, which is structurally similar to the PFPE oxyradicals generated by the initial O—O homolysis and continues the propagation process. Then, a proton is removed from the aryl oxidized adduct and reacts with the PFPE alkoxide, by forming an unstable primary alcohol and restoring the unsaturation on the substrate (Scheme 15.5).

The PFPE residuals maintain some nondecomposed peroxidic moieties inside the chain, even after the linkage with the CB. However, the peroxidic content in all the PFPE residuals tends to zero during the treatment, because the number of peroxidic groups is lowered during each step and the process proceeds until all the peroxidic moieties are decomposed. In fact, as the reaction continues, all the nondecomposed peroxidic moieties react through either homolytical cleavage of O—O bonds or SET-induced decomposition, leaving PFPE portions linked to the substrate or as bridges between two reactive sites. Consequently, the thermal decomposition of PFPE peroxide in the presence of a carbon-based substrate leads overall to the

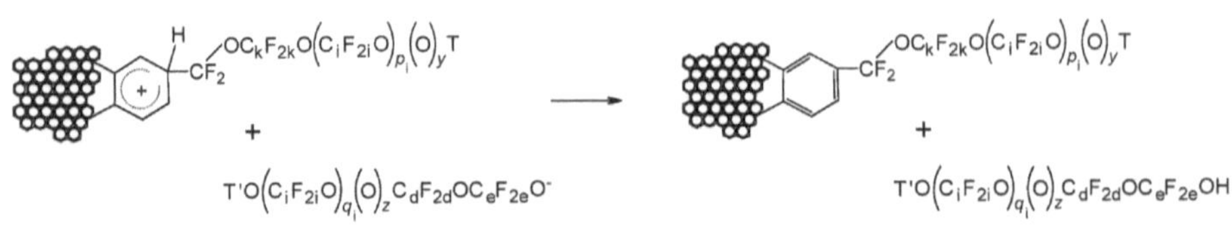

Scheme 15.5 Formation of the perfluoropolyether primary alcohol by proton abstraction from the aryl carbocation.

formation of a nonperoxidic PFPE layer which is covalently linked on the carbonaceous surface. The C_1 and C_2 content ratio in the linked PFPE chains remains approximately equal to that of the starting chain. The homocoupling reactions of PFPE radicals can competitively generate a portion of PFPE fluids with analogous C_2/C_1 ratio as side-products which are not linked to the carbon-based substrates [30].

15.3 Perfluoropolyether-Functionalization of Carbon-Based Materials

15.3.1 Graphite

Graphite is a crystalline form of carbon and the most diffuse carbon-based allotrope. Graphite has a layered planar structure and in each layer carbon atoms are arranged in a honeycomb lattice with C—C bond length of 1.421 Å [40]. Graphitic layers are alternated with a separation distance between the planes of 3.35 Å and carbon atoms are stacked so that half the atoms (A) in one layer are directly above those in the layer below and the other half (B) are above the holes, as described by ABAB Bernal structure [41]. Bonding between layers is via van der Waals forces of 0.7 kJ/mol and this weak interaction is at the basis of the sliding of graphitic layers and of graphite exfoliation [40].

Functionalization of carbon-based materials is nowadays one of the most important topics in research due to the wide variety of possible applications, and graphite is usually functionalized for comparison as commercially available well-defined material with model sp^2 surface [42,43]. Thus, chemical functionalization of graphite by means of a treatment with PFPE peroxides has been performed and the characterization of PFPE-functionalized graphite shows that the bulk remains mainly unaltered and the surface is modified by chemical linkage of PFPE chains [44]. The morphology is overall influenced by the functionalization and, even if the surface area of graphite is typically low depending on defects (ie, below 10 m^2/g), at low functionalization degrees a decrease in specific surface area is noticed due to the filling of graphite defects by the polymeric PFPE chains; at high functionalization degrees, the surface area usually increases back because of the opening of graphite layers by exposing inner pores or creating new mesopores [45,46]. In addition, graphite becomes highly hydrophobic and a water contact angle (CA) of around 145 degree can be measured [44]. Gravimetric analysis during PFPE functionalization of graphite measures a decrease in weight, ascribable to graphite exfoliation, and solubilized particles of highly PFPE-functionalized graphite have been detected by TEM analysis in the fluorinated solvent at the end of the treatment (Fig. 15.1). It is worth to notice that graphite exfoliation with the exposure of new bare layers occurs at a PFPE functionalization degree corresponding to a fluorine content giving an F/C ratio around 0.05, as measured by X-ray photoelectron spectroscopy (XPS) analysis [44].

PFPE functionalization induces modifications on graphite surface by varying chemical composition, morphology, and wettability; the ordered stacking of graphitic layers is progressively altered until graphite exfoliation and solubilization in fluorinated

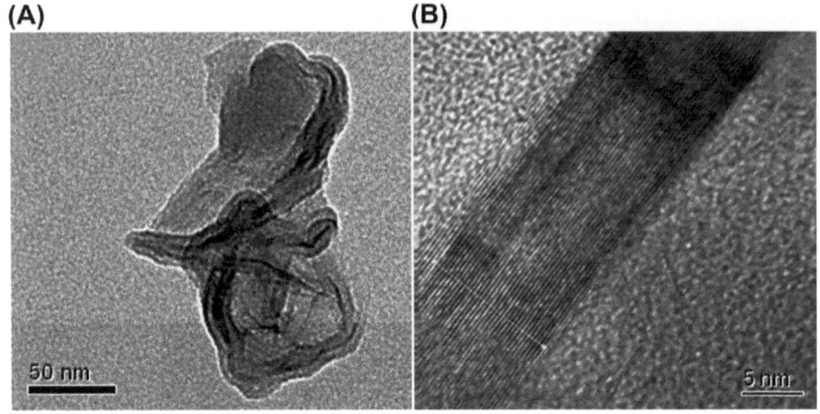

Figure 15.1 Transmission electron microscopy (TEM) image of a perfluoropolyether-functionalized graphite particle solubilized in perfluorinated solvent (A) and high TEM magnification showing a detail of the residual stacked graphitic layers (B).

solvents. Thus, the presence of turbostratic disorder in carbonaceous materials promotes the functionalization at high rates and inhibits the exfoliation, opening a new route for the design of PFPE-functionalized materials with a wide palette of properties.

15.3.2 Highly Oriented Pyrolytic Graphite

The highly oriented pyrolytic graphite (HOPG) is a pyrolytic form of graphite with a high degree of crystallographic orientation along the *c*-axis that is perpendicular to the surface of the substrate [47]. HOPG is used mainly as a growth support for CNTs, nanohorns, grapheme, and other oriented carbon-based materials [48]. The HOPG has an anisotropic structure and it is characterized by properties, such as electrical resistivity, thermal conductivity, and chemical reactivity, that differ on the edges along the *c*-axis and on the planes of graphitic layers [49,50]: HOPG surface consists predominantly of extraordinary large graphene sheets with a few defects; HOPG edges can be idealized as terminal zones of graphene sheets with conjugated systems, which reactivity is characterized by a large number of irregularities.

The chemical treatment of HOPG with PFPE peroxide provides fundamental insights about the functionalization mechanism of carbon-based materials by means of PFPE radicals, due to HOPG high crystallographic orientation: EDS analyses on the surface as well as on the edges of a PFPE-functionalized HOPG measure the corresponding elemental composition and a threefold higher F/C ratio can be determined on the edges [44]. Thus, HOPG edges result more reactive toward PFPE radical attack than the ordered aromatic systems of graphene-like structure on HOPG surface.

Scanning electron microscopy (SEM) images show that HOPG morphology is not affected by the functionalization with PFPE chains: in particular, the smoothness of HOPG surface (Fig. 15.2) and the roughness of the edges (Fig. 15.2) are mainly preserved. PFPE functionalization confers highly hydrophobic properties to the entire

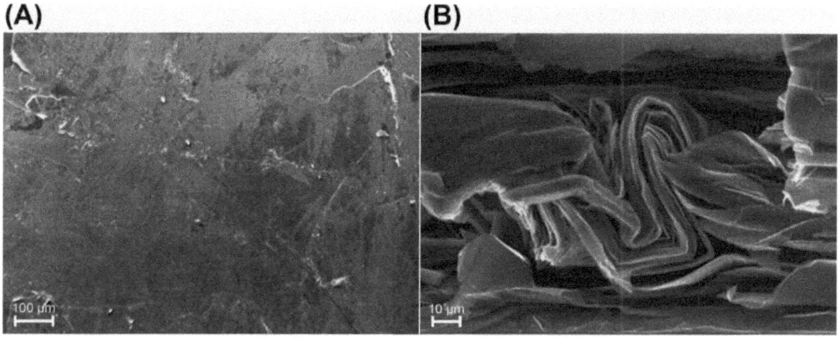

Figure 15.2 Scanning electron microscopy magnifications of perfluoropolyether-functionalized highly oriented pyrolytic graphite: surface view (A 330×) and detail of an edge (B 2500×).

HOPG: CAs higher than 120 degree can be measured on both surface and edges: 125 and 140 degree, respectively [44]. However, the wettability of HOPG surface is substantially different from that of its edges: in fact, the hydrophobic behavior of HOPG surface follows the Wenzel model which describes a homogeneous wetting regime; the hydrophobicity of HOPG edges can be described by the Cassie-Baxter model which includes the influences on CAs due to the roughness and the irregularities of the edges.

The functionalization of HOPG with PFPE chains shows a general reactivity that can be extended to all carbon-based materials: PFPE radical attack is achieved more easily on defected carbonaceous edges than on a regular graphitic surface and this difference is at the basis of diverse behaviors in macroscopic properties.

15.3.3 Diamond-Like Carbon

DLCs are formed by an amorphous mixture of sp^3 and sp^2 hybridized carbons with hydrogen atoms incorporated in the structure (a-C:H DLC). DLC films are remarkable protective coatings for tools and other devices because of their excellent properties such as hardness, low coefficient of friction, chemical inertness, and infra-red transparency [51]. The properties of DLC coatings are normally used in tribological applications, where the SE is a discriminating factor.

The SE of DLC coatings can be significantly reduced by the introduction of fluorine atoms into the matrix or by the deposition of fluorinated compounds on the surface. Only in recent years chemical modifications of DLC films have attracted the attention of researchers; in particular Nakamura et al. first reported the use of perfluoroazooctane as reagent for the linkage of perfluorooctyl radicals on DLC surface [52].

Direct perfluoroalkylation strategies for unsaturated substrates are convenient tools for the introduction of perfluoroalkyl groups with carbon—carbon bond formation [53—57]. Perfluorinated radicals can directly bond to the sp^2 sites of the DLC structure avoiding any spacer, which usually decreases both thermal and chemical stabilities of the resulting fluorinated materials [58,59]. PFPE peroxides decompose generating PFPE oxygen- and carbon-centered radical species: the former by homolytic cleavage of the O—O bonds and the latter by rapid β-scission of the resultant PFPE alkoxy radicals [19]. The half-life time of the PFPE peroxides depends on their structure and it is approximately 30 min at temperatures in the range between 140 and 250°C [20]. A comparison between MWA and equivalent molecular weight of PFPE peroxide reveals the average number of peroxidic units present in the macromolecular structure of the PFPE peroxide. Depending on the relative position in the PFPE chain, the peroxidic moieties show different reactivities and consequently quite different decomposition temperatures [20]. Therefore at the beginning of the thermal decomposition, the PFPE radicals maintain some undecomposed peroxidic moieties inside the chain. After the PFPE radicals reacted with the DLC surface, these peroxidic moieties can decompose generating PFPE radicals linked to the DLC surface as well as free PFPE radicals. Considering the PFPE radicals linked to the DLC surface, they can follow different reaction pathways: they can react with neighbor sites on the DLC surface (bridging) or with free PFPE radicals (coupling) reducing the peroxidic content in the PFPE chain (Scheme 15.6). Overall the treatment of a DLC film by dip coating in a solution of PFPE peroxide, followed by the thermal decomposition of the PFPE peroxide, leads to the formation of a non-peroxidic PFPE layer covalently linked to the DLC surface.

XPS measurements on DLC film coated by means of PFPE peroxides revealed that the amount of fluorine atoms approximately can equate that of carbon atoms (F/C ratio ≈ 1), due to the macromolecular structure of the PFPE chains [28]. High-resolution XPS spectrum in C1s region of a PFPE-coated DLC visibly shows an

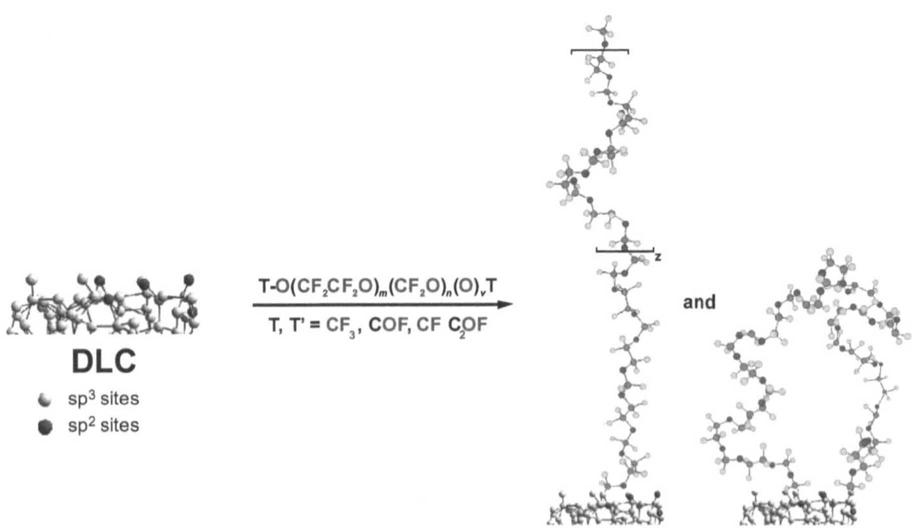

Scheme 15.6 Covalent linkage of perfluoropolyether (PFPE) chains on a diamond-like carbon (DLC) surface by the thermal decomposition of PFPE peroxides. PFPE radicals are generated by reduction of the peroxidic content in a PFPE peroxide and, once linked to the DLC surface, react with other free PFPE radicals (coupling) or with neighbor sites on the DLC surface (bridging).

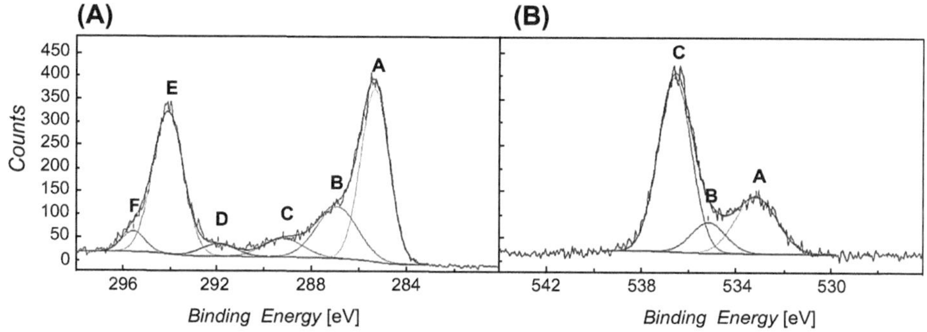

Figure 15.3 High-resolution X-ray photoelectron spectroscopy spectra of perfluoropolyether-coated diamond-like carbon: C1s region (A); O1s region (B).

intense peak at 293 eV (Fig. 15.3A) due to C—F bond in the PFPE chains, and the signals of carbon atoms in the carbonaceous matrix are clearly separated at lower binding energies, in the range between 284 and 290 eV. In PFPE-coated DLC the oxygen content increases roughly equating the amount stoichiometrically expected by the linkage of PFPE chains. In high-resolution spectrum in O1s region, PFPE oxygen atoms at 536 eV can be clearly distinguished by the oxygen of the DLC matrix at 532 eV due to hydroxylated/oxygenated moieties (Fig. 15.3B).

AFM three dimensional images of a bare a-C:H DLC sample and of a DLC sample after the chemical treatments with PFPE peroxides have been reported in Fig. 15.4 [28]: the uncovered DLC substrate is typically characterized by an irregular morphology with a root mean square (RMS) roughness of approximately 8 nm (Fig. 15.4A); the treatment with PFPE decreases the RMS roughness value up to value of around 4 nm (Fig. 15.4B). A significant change in the surface morphology is observed in the case of PFPE coating: in agreement with the XPS results, the functionalization produces a thick coating that covers valleys and peaks of the substrate.

The covalent linkage of PFPE chains on DLC thin films generates the permanent modification of their wettability. In particular, the PFPE coating increments both CAs with water (from 75 to 105 degree) and *n*-dodecane (from 5 to 60 degree),

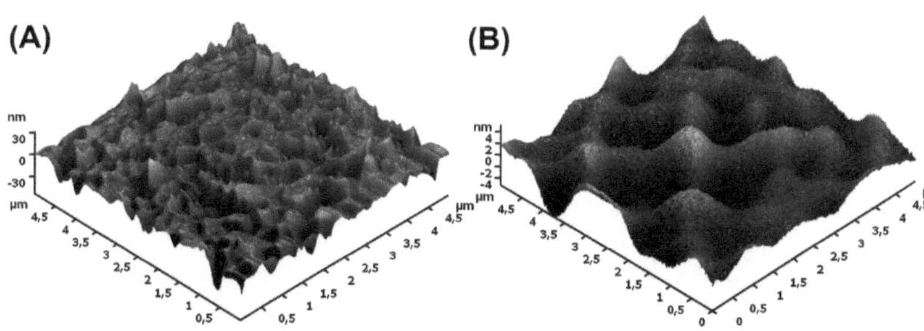

Figure 15.4 AFM images of bare diamond-like carbon (DLC) thin film (A) and of perfluoropolyether-coated DLC (B).

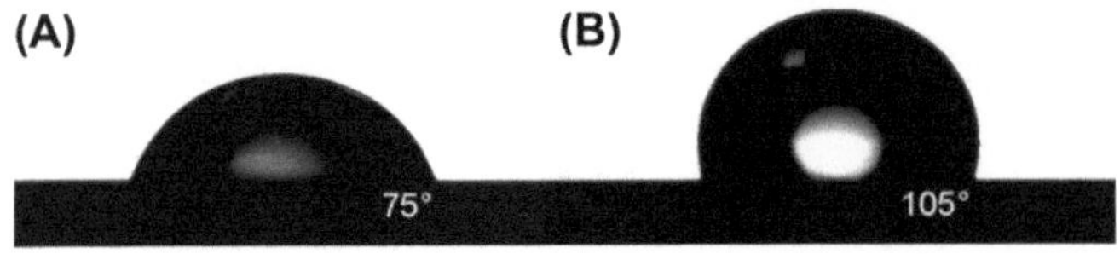

Figure 15.5 Water droplets on bare (A) and perfluoropolyether-coated diamond-like carbon films (B).

corresponding to a general decrease of SE (Fig. 15.5) [28]: a value of 34 mN/m has been measured on the uncovered a-C:H DLC, while the PFPE coating revealed 18 mN/m as values of SE. The length of the fluorinated polymeric chains and the thickness of the PFPE layer deeply modify the surface properties of DLC thin films, even if few milliequivalents of PFPE peroxide are used in the functionalization treatment.

The deposition of PFPE fluids on the DLC surface similarly generates a variation of the surface properties and CAs higher than 90 degree can be measured, revealing the typical hydrophobic properties of fluorinated coatings [60,61]. However, simple washings with fluorinated solvent can remove these physically deposited coatings, because no chemical linkage of the fluorinated chains occurs in this case. Conversely, the surface properties of the coatings obtained by functionalization with PFPE peroxides are permanently stable, even after several washings with pure fluorinated solvent.

15.4 Perfluoropolyether-Functionalization of Carbonaceous Materials

15.4.1 Graphitic Carbon Black

CB is generally regarded as paracrystalline carbon that is a variant of disordered carbon-based materials made up by graphene layers randomly translated to each other and rotated about the normal stacking by changing the interlayer spacing and/or the shape of the atomic layers [62]. The structure of graphitic CB is formed by polycyclic aromatic structured aggregates with several hydrogenated and hydroxylated/oxygenated moieties on the edge (Fig. 15.6) [63]. Thus, PFPE functionalization strategies for substrates with unsaturated moieties can be usefully applied for the covalent linkage of PFPE chains onto such carbonaceous materials [35,38,54−57,64]. In particular, PFPE peroxide reacts with graphitic CB through a radical pathway by linking PFPE chains to the carbonaceous matrix without any spacer, as described in Section 15.2.

Overall, the chemical treatment of CB with PFPE peroxide leads to the formation of a non-peroxidic PFPE layer covalently linked to the CB surface (see Scheme 15.1). Carbonyl fluoride and primary perfluorinated alcohols are generated as by-products. The unbounded PFPE moieties can be efficiently removed by several washing steps.

The most evident effect on CB due to its functionalization with PFPE chains is the modification of its hydrophilic surface properties, until the exceeding of the threshold

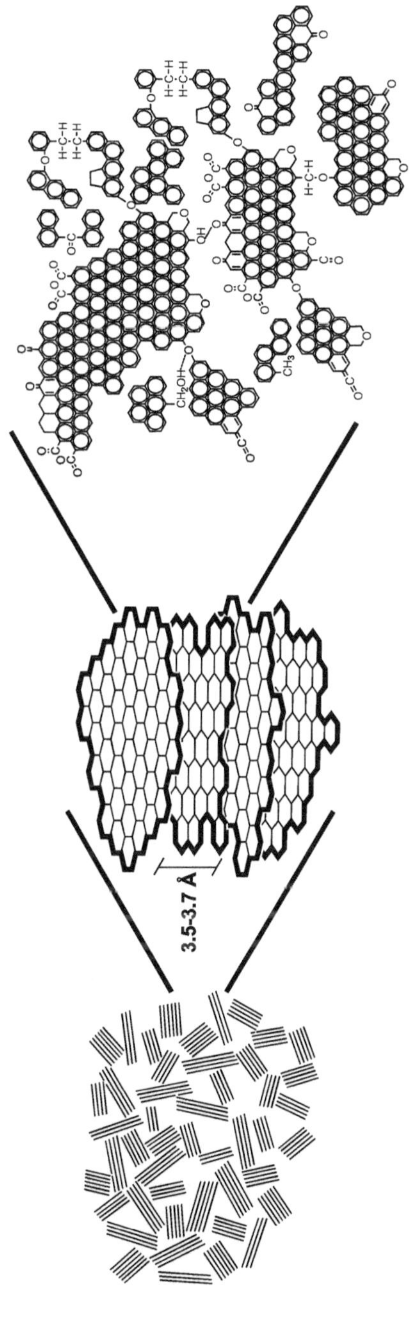

Figure 15.6 Scheme of graphitic CB: polycyclic aromatic structured aggregates with several hydroxyalted/oxygenated moieties on the edge.

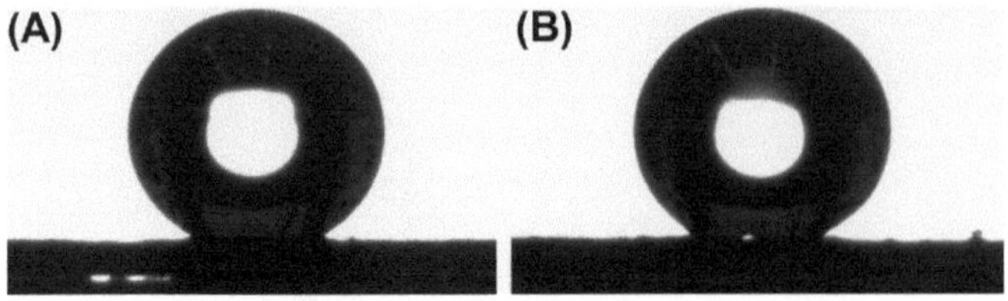

Figure 15.7 Static water droplets on CB showing contact angles of approximately 170 degree after functionalization with perfluoropolyether (PFPE) peroxide: 1:1 (A) and 2:1 (B) PFPE/CB weight ratio.

of the superhydrophobicity [65]. The CA measurement on pellets of bare CB can show that water droplets are adsorbed in few seconds (2−4 s) by the carbonaceous matrix [29]. Water droplets carefully placed on pellets made of PFPE-treated CB are enduringly stable: thus, static CAs over 150 degree and CA hysteresis below 10 degree can be obtained (Fig. 15.7) [65]. Therefore, the superhydrophobicity threshold (ie, 150 degree) can be exceeded in PFPE-functionalized CB.

In CB functionalized with PFPE chains, the polyperoxidic structure of PFPE peroxide allows the linkage in a row of several connected PFPE chains, gradually growing the entire PFPE layer on carbon surface. In addition, the higher the amount of PFPE peroxide in the chemical treatment, the more discernible are the effects due to the linked PFPE chains. The uncoiled average distance between two peroxidic moieties in PFPE peroxide with an average equivalent weight of around 1000 g/eq is approximately 3 nm. However, the lengths of the PFPE chains on the CB can be higher than 3 nm, thanks to the growing properties of the PFPE layer and depending on the amount of peroxide used during the chemical treatment. Thus, the PFPE chains are long enough to cover micro- and mesopores ($2 < d_p < 50$ nm), and it is reasonable to assume that macromorphology is preserved because the macroporous structure ($d_p > 50$ nm) is typically not modified by the chemical treatment with PFPE peroxides.

XPS analysis on samples of PFPE-functionalized CB revealed that the higher the load of PFPE peroxide in the chemical treatment, the higher the amount of fluorine detected on CB surface [65]. Coherently, a general increase of oxygen content on the CB surface was also observed, equating what stoichiometrically expected on the basis of chemical composition of PFPEs. This behavior can be explained on the basis of the polyperoxidic structure of the PFPE peroxide and the growing properties of PFPE chains linked on the CB surface. In fact, depending on the relative position in the PFPE chain, the peroxidic moieties show different reactivities and consequently different decomposition temperatures [30,34]. Therefore, at the beginning of the chemical treatment, PFPE radicals maintained some undecomposed peroxidic moieties inside the chain, even after linking the CB surface. Thus, these peroxidic moieties decomposed generating further PFPE radicals already linked to the CB. These radicals reacted with free PFPE radicals (coupling) and grew the PFPE layer linked on the CB surface or alternatively linked neighbor unsaturated sites on the CB surface (bridging). Overall, the chemical treatment with PFPE peroxide formed a non-peroxidic PFPE layer covalently linked to the CB surface. The portion of unlinked PFPEs can be efficiently removed by appropriate washing steps with fluorinated solvents.

Graphitic CB is widely used in electrochemical devices because of its typical conductivity and the determination of the resistivity in function of the applied pressure can be employed to characterize its electrical properties [66]. The electrical data are typically reported as resistivity and the electrical conductivity is easily evaluated as reciprocal. The compacting pressure is directly proportional to the contact pressure between CB particles and it is well known that the contact resistivity decreases by increasing the contact pressures for electrical contact [66,67]. Furthermore, the carbon volume fraction increases as the compacting pressure increases [68]. Thus, by increasing the loads on the piston, the electrical resistivity typically decreases and it is gradually leveled off, due to the leveling off of both contact resistivity and carbon volume fraction.

A typical graphitic CB such as Vulcan XC72R can be assumed as standard reference material for comparison. The resistivity of this carbon varies from 0.7 to 0.1 Ω cm and decreases under 0.2 Ω cm when the compacting pressure is around 4.0 MPa (Fig. 15.8A). After the chemical treatments with PFPE peroxides, the values of resistivity slightly increase, but even at high load of fluorinated peroxide (1:1 and 2:1 PFPE/CB weight ratio) they still remain in the range of conductive materials, exhibiting low resistivity around 1 Ω cm just under a load 4.4 MPa (Fig. 15.8A). The increase in resistivity can be ascribed to the typical nonconductive properties of fluorinated materials, but the liquid-like features of the PFPE layer permit the contact between CB particles and, therefore, the overall resistivity of PFPE-functionalized CB remains in the conductivity range [69]. As expected, a CB directly fluorinated with elemental fluorine loses its electrical conductivity and behaves as an insulator (Fig. 15.8A). In fact, the fluorination with elemental fluorine covalently bonds fluorine atoms directly to carbon atoms on CB surface and into the subsuperficial zone of the particles, changing the sp^2 state of the carbon layers of the π-electron systems in sp^3 state.

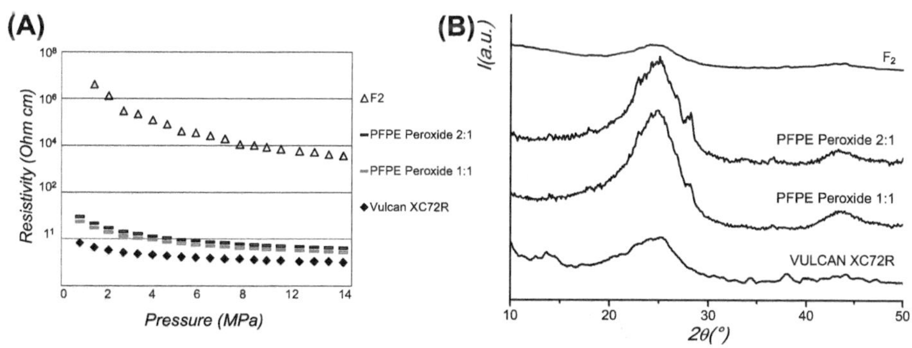

Figure 15.8 Electrical resistivity (in logarithmic scale) at different pressures (A) and X-ray diffraction patterns (B) of a graphitic CB (Vulcan XC72R) before and after chemical treatments with perfluoropolyether peroxide (1:1 and 2:1 weight ratio) and with elemental fluorine (F_2).

The X-ray diffraction (XRD) pattern of a graphitic CB shows a broad diffraction centered at $2\theta = 24$ degree and weak diffraction signals at around 10−15 degree and 44 degree (Fig. 15.8B). The diffraction at lower angles correspond to the 001 planes while the signals around 24 and 44 degree to the (002) and (100/101) reflections of the graphitic structure, respectively [70,71]. The delocalized π-electron systems of these graphitic contributes are responsible for the conductive properties of the CB [68]. In PFPE-functionalized CB, no shifts in the peak positions can be detected and (002) diffraction profiles become narrower with respect to the bare material, indicating minor changes in the graphite structure (Fig. 15.8B). Thus, while the interlayer distances (002 planes) does not change, a decrease in the full width at half maximum indicates an increase in crystallite size along the orthogonal direction to these planes, or apparent stack size, which is consistent with the chemical modification of 002 planes. Thus, the graphitic π-electron systems are not modified in CB bulk by the linkage of PFPE chains via carbon−carbon bond formation and, therefore, the electrical properties of PFPE-functionalized-CB is maintained even after the functionalization. Conversely, in the case of fluorination with elemental fluorine, only a residual part of the bulky π-electron systems is maintained and the CB surface as well as the subsuperficial zone of CB particles are extensively modified (Fig. 15.8B). Thus, the conductivity decline due to direct fluorination can be ascribed to the formation of a fluorinated insulating layer on CB particles.

The functionalization of graphitic CB with PFPE chains can transfer the typical hydrophobic properties of fluorinated materials to the carbonaceous surface, even exceeding the superhydrophobicity threshold, but the conductive properties are maintained despite the deep variations in the surface properties. Thus, PFPE-functionalized CBs are very promising materials in the field of electronic applications and devices for energy production and storage. In addition, they are also emerging as fillers in host matrices such as polymers and fluoropolymers.

15.4.2 Carbon Cloth

The term "carbon cloth" (CC) identifies woven carbon fibers obtained by polyacrylonitrile (PAN) pyrolysis [72]. CC is a macroporous carbonaceous material commonly used in electrochemical applications and its structure contains condensed aromatic cycles, suitable for the linkage of perfluorinated radicals formed in the decomposition of PFPE peroxide (see Section 15.2 for the reaction mechanism) [69,73].

CC functionalization with PFPE chains allows to obtain a thin and homogeneous perfluorinated layer covalently linked and wrapped on each fiber of the material. The maximum linkable PFPE is slightly over 1 wt% of CC, but this threshold is related to an almost complete coverage of carbon fibers by the PFPE layer [69]. In fact, the coverage of fiber surface has been evaluated by the detection depth of XPS (<10 nm) on a 1 wt% PFPE-functionalized CC and it resulted around 95 wt% [74]. Such a high surface coverage confers to CC fibers the substrate PFPE-like surface properties and PFPE-functionalized CCs have very high levels of superhydrophobicity: water CAs measured on functionalized CCs ranges from 166 to 172 degree on lowly (ie, 0.24 wt%) and highly functionalized (ie, 1 wt%) substrates, respectively

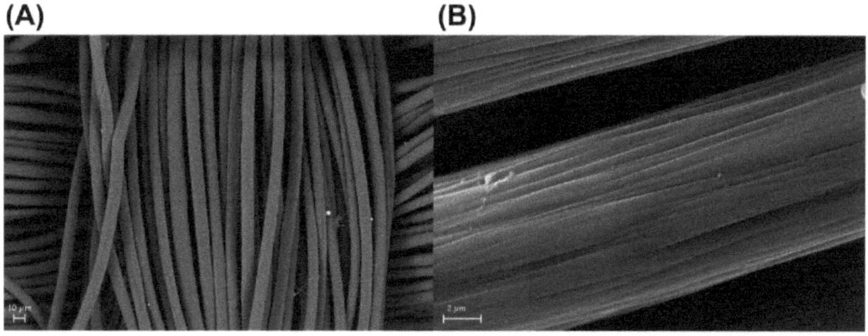

Figure 15.9 Scanning electron microscopy magnifications (A 300×; B 5000×) of perfluoropolyether (PFPE)-functionalized carbon cloth (1 wt% of linked PFPE chains). The fibers are evenly covered by a PFPE layer without any filling of the voids between the fibers.

[69]. In addition, electrochemical tests on PFPE-functionalized CCs evidenced that their conductive performance remains in an appropriate range for their application as conductive materials (see Section 15.6.1 for more details) [69,74].

PFPE chain length remains in the nanometric range during the functionalization and it is not sufficient to bridge-link different fibers: therefore, the voids between CC fibers (ie, ~1 μm) results unaltered by the reaction (Fig. 15.9) [39].

The chemical functionalization of CCs by means of PFPE peroxide is expected to work similarly on other substrate based on carbon fibers, like carbon papers or carbon foams. The linkage of PFPE moieties on carbon fibers, even with low loadings of perfluorinated polymer, confers very high hydrophobicity to the functionalized material by letting free interspaces between fibers. Thus, PFPE-functionalized carbon fiber-based materials can find application in the fields where gas transfer and conductivity are required without any water crossing.

15.5 Perfluoropolyether-Functionalization of Carbon-Based Nanomaterials

15.5.1 Carbon Nanotubes

CNTs are rolled-up graphite planes and, depending on their molecular structure, they can show conductive properties suitable for many applications ranging from microelectronics to conductive polymers or electromagnetic shielding materials [75,76]. Multi-walled carbon nanotubes (MWCNTs) are reported to be always electrically conductive and their physical as well as chemical modification has been widely studied by many researchers [27,76,77]. MWCNT are conveniently functionalized because the covalent linkage of specific groups on their external conjugated skeleton can occur selectively on the surface and the bulk can remain unchanged preserving the electric properties [78]. In particular, the functionalization with PFPE chains can confer to MWCNTs the peculiar characteristics of fluorinated material, such as low SE, high chemical stability, and UV resistance [3].

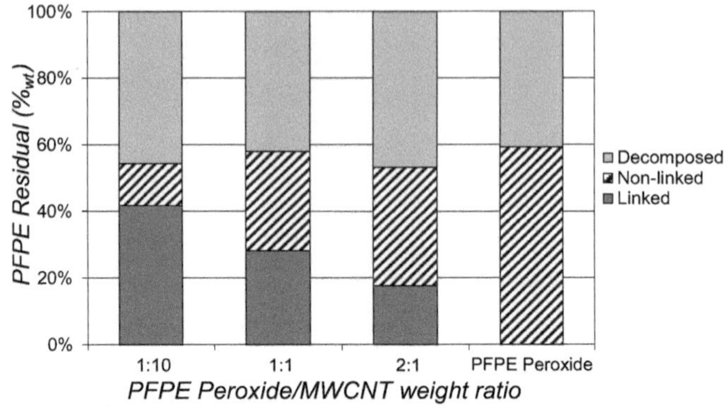

Figure 15.10 Mass balances of linked, nonlinked, and decomposed perfluoropolyether (PFPE) residuals after carbon nanotubes functionalization with 1:10, 1:1, and 2:1 PFPE peroxide/multi-walled carbon nanotube weight ratios (compared to the decomposition of pure PFPE peroxide). The percent weight of PFPE residuals is referred to the initial load of PFPE peroxide.

The functionalization degree of MWCNTs with PFPE chains can be evaluated by a mass balance that determines the portions of linked, nonlinked, and decomposed PFPE after the chemical treatment with PFPE peroxide. A comparison between mass balances referred to PFPE-functionalized MWCNTs with 1:10, 1:1, and 2:1 PFPE peroxide/MWCNT weight ratios can reveal some useful details of the chemical treatment (Fig. 15.10) [79]. The linked portion of PFPE gradually decreases as the amount of loaded PFPE peroxide increases, because the active sites on CNTs surface are increasingly occupied by PFPE chains and, thus, the functionalization reaction proceeds progressively less efficiently. The decomposed portion of PFPE is mainly related to the production of gaseous carbonyl difluoride, COF_2, due to β-scission reactions related to the chemistry of PFPE oxyradicals generated during the functionalization [34,39]: it is typically around 40–50 wt% of loaded PFPE peroxide. This amount is in agreement with the weight loss of pure PFPE peroxide (around 40 wt%) subjected to a thermal treatment similar to that of functionalization reaction, indicating that the decomposed portion of PFPE is mostly related to the reactivity of PFPE peroxide, slightly influenced by oxygen traces, moisture, or by the reductive effects of the carbonaceous substrate. As a consequence, the nonlinked portion of PFPE gradually increases with higher amounts of loaded PFPE peroxide. This trend is related to a more frequent coupling of PFPE radicals that leads to the formation of not-peroxidic PFPE fluids and it can be considered as a side-reaction competitive to the functionalization.

The morphology of MWCNTs is influenced by functionalization with PFPE chains and the surface area decreases as the amount of PFPE peroxide increases. In particular, the PFPE chains are usually long enough to progressively cover micropores ($d_p < 2$ nm) and mesopores ($2 < d_p < 50$ nm), and it is reasonable to assume that the macroporous structure ($d_p > 50$ nm) is weakly modified by the functionalization.

The functionalization of MWCNTs with PFPE chains changes their wettability, which is shifted from hydrophilic to superhydrophobic. The covalent linkage of fluorinated chains on the surface of MWCNTs has a fundamental role in this variation. In fact, pure MWCNTs usually exhibit a behavior related to intrinsic hydrophilicity of the graphitic network of sp^2 carbon: the water droplets are adsorbed in few seconds (2–4 s) by the carbonaceous matrix [80]. Conversely, functionalization with PFPE peroxide gradually increases the hydrophobic properties of the PFPE-functionalized MWCNTs: water droplets can be placed on MCNTs, are not adsorbed and results enduringly stable, allowing the measurement of CAs [79]. The enhancement of water repellency over the superhydrophobic threshold (ie, CAs >150 degree) can be obtained in samples with high degree of functionalization. The superhydrophobic properties can be also confirmed by the measurements of CA hysteresis, which results lower than 10 degree for the superhydrophobic samples. In addition, the wetting properties conferred to MWCNT surface by the functionalization with PFPE chains are permanently stable over time. In fact, the CA values with water are constant even if they are repeated after 10,000 h of exposure to air and moisture. On the contrary, fluorinated MWCNTs obtained by direct fluorination with elemental fluorine contains a part of fluorine atoms which are weakly grafted to the MWCNT, and these fluorinated moieties are easily hydrolyzed when exposed to air, in particular to atmospheric moisture [79].

The physical adsorption of a PFPE fluid onto MWCNTs generates a variation of their surface properties, like a decrement of surface area and an improvement of the hydrophobic behavior. However, washings with fluorinated solvent increase back the surface area and remove the hydrophobic properties, because no chemical linkage of the fluorinated chains occurs in this case.

The electrical behavior of PFPE-functionalized MWCNTs can be evaluated by the determination of the resistivity under different applied pressure (Fig. 15.11A). The compacting pressure causes an overall decrease in resistivity at increasing loadings, due to an increase of the local contact forces. In fact, the compression improves the contacts of nanotubes and averagely diminishes the distances between their contacts [81]. The resistivity of pure MWCNTs can range from 0.5 to 0.1 Ω cm and it is slightly increased by the functionalization with PFPE peroxide: resistivity rises, but still

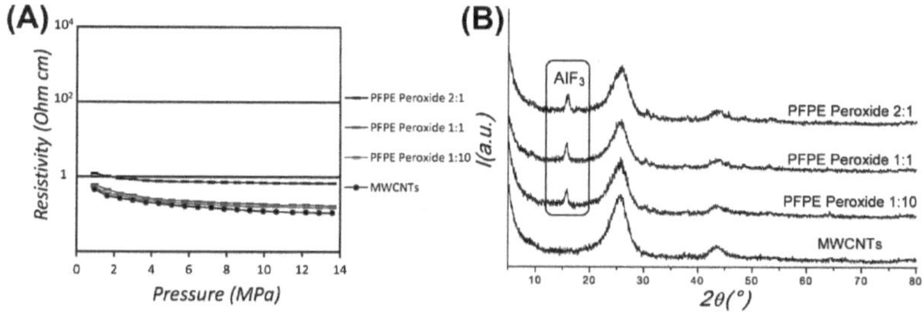

Figure 15.11 Electrical resistivity (in logarithmic scale) at different pressures (A) and X-ray diffraction patterns (B) of multi-walled carbon nanotubes (Nanocyl NC 7000) before and after chemical treatments with perfluoropolyether peroxide (1:10, 1:1, and 2:1 weight ratio).

remains lower than 1 Ω cm (Fig. 15.11A) [79]. This increment can be ascribed to the typical nonconductive properties of PFPE chains that wraps MWCNTs after the functionalization. In addition, the sp^2 state of the carbon atoms in the conjugated π-electron systems of MWCNTs is changed in sp^3 state where PFPE radicals attack the nanotubes during the functionalization. However, PFPE chains can covalently link and surround MWCNTs, but the core of MWCNTs remains unaltered by preserving the conductive properties of the material [79]. In fact, the persistence of 002 reflection peak at 2θ of 25.64 degree corresponding to a spacing of 3.47 Å in XRD patterns of pristine and PFPE-functionalized MWCNTs suggests that the bulky structure of the cylindrical concentric carbon layers is mainly preserved despite the covalent linkage of PFPE chains (Fig. 15.11B). Therefore, the functionalization of MWCNTs with PFPE chains occurs prevalently onto the outer tube and the inner tubes remain almost unaltered.

The combination of electrical conductivity and superhydrophobicity in PFPE-functionalized MWCNTs is highly desirable and these combined properties may contribute to the development of a wide range of applications, including display technologies, solar cells, new electrode materials, and flexible electronic devices.

SWCNTs are more pliable but harder to synthesize than MWCNTs. Due to their structure-depending electronic behavior that can vary from metallic to semiconducting, SWCNTs can be used in numerous applications, such as field-emission displays, nanocomposite materials, nanosensors, and logic elements. Many applications utilizing SWCNTs require the chemical modification of nanotubes to confer them properties for an easier handling and manipulation [27]. In addition, the tethering of chemical moieties with specific recognition sites on nanotube surface can exploit the triggering of a predictable response. Fluorination has a relevant role in surface covalent chemistry of SWCNTs because fluorinated nanotubes can be easily solubilized in various solvents. Radical perfluoroalkylation can be considered as alternative route for SWCNT functionalization with fluorinated moieties, and perfluoroalkyl iodide, diazo-perfluorocompounds, and perfluorodiacyl peroxides are typical sources of perfluorinated alkyl radicals [23,24,27]. Analogously, SWCNT functionalization with PFPE radicals obtained by thermal decomposition of PFPE peroxides can represent a suitable technique for the modification of surface and morphological properties as well as for the tuning of the electric behavior [82]. In particular, explorative studies revealed that PFPE functionalization of SWCNT improve their solubility and dispersibility in fluorinated solvents.

15.5.2 Fullerenes

Among carbon allotropes, fullerenes are largely studied due to their specific reactivity towards radicals and, nowadays, the functionalization of buckyball fullerenes can be obtained by several reactions leading to different fullerene-based molecules [83−91]. Among these reactions, perfluoroalkylation of fullerenes, such as reactions with perfluoroalkyl iodides (R_FI) or perfluorodiacyl peroxides ($[R_FCO_2]_2$), have a significant relevance because perfluoroalkylfullerenes (PFAFs) have enhanced electron affinities than the parent fullerenes [17−20,92,93]. This property is very important for their possible use in photovoltaic and other electronic devices [94].

Due to similarity between classical perfluoroalkylation techniques and functionalization based on PFPE peroxides [29]. Buckyball fullerenes can be considered as suitable substrate for the linkage of PFPE chains.

Preliminary tests on PFPE functionalization were already performed and provided promising results, but new challenges emerged [95]. In fact, product separation and structural characterization are particularly demanding due to the formation of several isomers and mixtures of compounds with different functionalization degrees. In addition, the PFPE residuals covalently linked on the fullerenes can have a wide distribution of chain lengths because of the polyperoxidic nature of PFPE peroxide. PFPE-functionalized fullerenes remain a very interesting new class of molecules because of synergic effects related to the presence of perfluoropolymeric chains on a buckyball core.

15.6 Applications

15.6.1 Gas Diffusion Layers for Fuel Cells

CC is usually employed in polymer electrolyte membrane fuel cells (PEMFCs) as single-layer gas diffusion layer (GDL) or as macroporous part of dual-layer GDLs [73].

In PEMFCs the GDL is the part of the cell that allows the gases to diffuse from the external feeding toward the catalyst located in direct contact with the ionomeric membrane. GDL works even as current collector and physical support for the catalyst layer (CL) that is usually composed by precious metals on carbon powder. Moreover, GDL regulates the water management in PEMFCs, that is a key factor of their functioning: a deficient water content reduces ionic conductivity in the membrane as well as in the CL and induces severe contact resistance between the membrane and the CL; excessive water reduces catalytic sites for electrochemical reactions and impedes reactant transport through the unreactive region [73]. Liquid water comes from condensed water vapor produced by oxygen reduction at the cathode and it can diffuse into the membrane toward the anode because a higher liquid water pressure at the cathodic interface between membrane and CL is formed by an electro-osmotic drag. In addition, liquid water can diffuse across the GDL and accumulate at the interface between CL and GDL. Thus, it can flow toward the gas-flow channel when its pressure exceeds a threshold determined by GDL pore geometry and hydrophobicity [96]. In PEMFC an effective water management is necessary to meet fast response to power demand.

CCs are suitable for their use as a GDL material, thanks to their high chemical stability, proper macroporosity, and electrical conductivity, but they lack in the hydrophobicity required for a correct water management [97]. Therefore, a hydrophobization treatment is necessary and the most common method is a coating with 10–40 wt% polytetrafluoroethylene (PTFE) sintered at 350°C. However, this approach presents some drawbacks that negatively influence the desired material properties, like electrical conductivity, porosity and gas permeability [69].

The functionalization of CC with PFPE chains achieves a more effective hydrophobization than the PTFE coating, with minor or absent influences on the properties

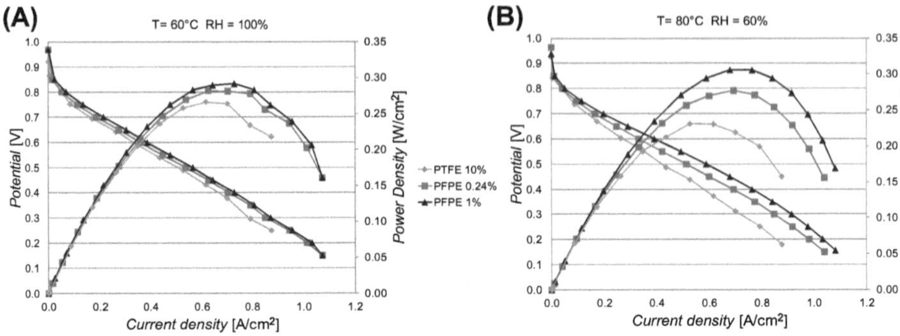

Figure 15.12 Polarization and power density curves of polymer electrolyte membrane fuel cells using perfluoropolyether- and polytetrafluoroethylene-functionalized single-layer gas diffusion layers in wet (A) and dry (B) conditions.

of the starting material (see Section 15.4.2 for more details): in particular, the hydrophobicity on a 0.24 wt% PFPE-functionalized CC (ie, CA 166 ± 3 degree) is greater than on a 10 wt% PTFE-coated CC (ie, CA 134 ± 3 degree), despite the amount of fluorinated polymer is more than one order of magnitude lower [98]. Thus, the presence of a lower amount of polymer induces lower ohmic resistances, as evidenced by differences in the resistive region of the polarization curves reported in Fig. 15.12 [74]. Moreover, the superhydrophobic effect due to the functionalization through PFPE peroxide is more durable than that can be achieved by the treatment with other perfluorinated polymers [74].

In the region of mass transport losses in PEMFC polarization curves, PFPE-functionalized CCs perform better than PTFE-coated CC, thanks to easier gas diffusion in the GDL (Fig. 15.12). In fact, PFPE chains are intrinsically highly permeable to gases and, because of their liquid-like properties, they do not affect gas diffusivity as solid barriers. On the contrary, PTFE is a solid polymer that wraps CC fibers, reduces the free volume, and increases the pressure drops for gas transport in the GDL. The superhydrophobic effect due to PFPE functionalization of CC surface improves also the water management, facilitating water removal, and gas transport. Tests on a working PEMFC in both dry and wet conditions (ie, 60 and 100% cathodic relative humidity, respectively) revealed that GDLs based on PFPE-functionalized CCs show the voltage drop at approximately 1.07 A/cm^2, while the PTFE-coated standard falls at 0.9 A/cm^2 (Fig. 15.12). Thus, PFPE-functionalized GDLs can increase the power density of the system up to 20%, by shifting the power density peak at 0.15 A/cm^2 up [74].

The dual-layer assembly of GDLs is made of a macroporous substrate coupled with a microporous layer (MPL) and it shows better performances than the single-layer GDLs, like improvements in electrical conductivity and water management [73]. MPL is usually made of CB powder and it is hydrophobized by a standard PTFE treatment; alternatively, PFPE functionalization can be applied for the preparation of superhydrophobic conductive CB [65,73].

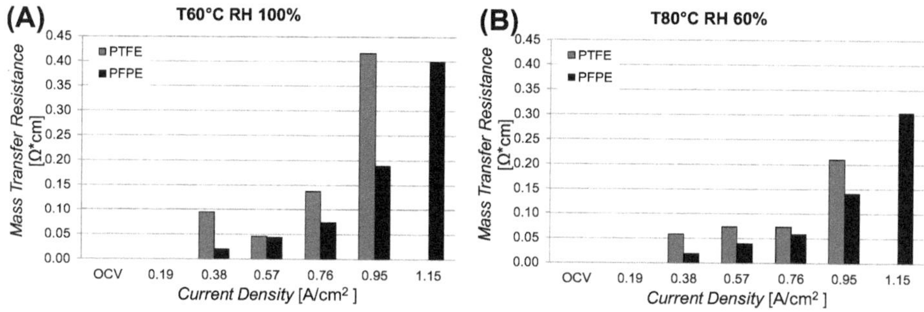

Figure 15.13 Mass transfer resistances of polymer electrolyte membrane fuel cells using perfluoropolyether-functionalized and polytetrafluoroethylene-hydrophobized dual-layer gas diffusion layers in wet (A) and dry (B) conditions obtained by electrochemical impedance spectroscopy.

Electrochemical impedance spectroscopy (EIS) analyses can discriminate at specific current densities the different contributions of resistances in PEMFC polarization curves: activation loss, ohmic polarization, and mass transport resistance. EIS on PFPE-functionalized and PTFE-hydrophobized dual-layer GDLs reveals that the former induces significant decrements in mass transfer resistance in both wet and dry PEMFC conditions (Fig. 15.13). In particular, the improvements are more evident at high humidification of the feeding gases (ie, 100% cathodic relative humidity) (Fig. 15.13A). These performances are mainly related to an increment in the flux of the inlet gases reaching the catalyst, due to the lower pressure drop and to the better water removal driven synergically by PFPE chains.

Polarization curves of operating PEMFCs assembled with dual-layer GDLs (Fig. 15.14) shows that PFPE-functionalized dual-layer GDLs can increase the current density output up to 30% more than PTFE-coated GDLs: in fact, in both dry and wet conditions, PFPE-functionalized and PTFE-coated dual-layer GDLs show the voltage drop at approximately 1.3 and at 1.0 A/cm^2, respectively [74]. In terms of power

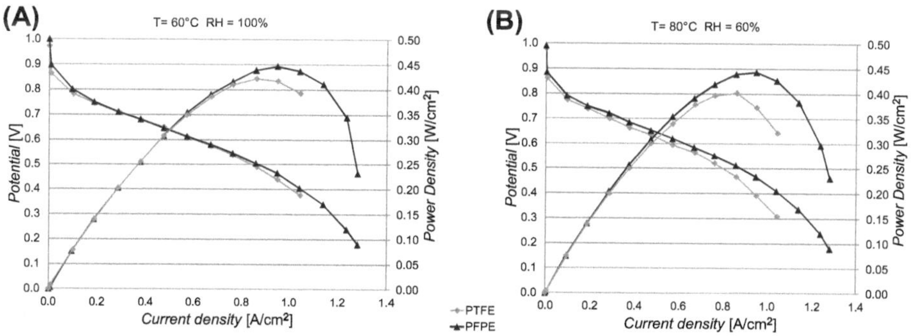

Figure 15.14 Polarization and power density curves of polymer electrolyte membrane fuel cells using perfluoropolyether- and polytetrafluoroethylene-functionalized dual-layer gas diffusion layers in wet (A) and dry (B) conditions.

density the improvement due to PFPE functionalization can be determined as up to 10% more than what obtained by standard PTFE sintering.

Chemical functionalization of GDL materials with PFPE chains is an improvement in GDL hydrophobization techniques and it can increase PEMFC current and power density outputs, moreover by inducing more durability and better thermo-oxidative stability if compared to the standard coating methods.

15.6.2 Self-Lubricating Surfaces

Recent studies propose the fabrication of fluorinated DLC films by using fluorocarbon gases activated by radiofrequency glow discharge, plasma and chemical vapor deposition techniques [99–101]. These approaches have some drawbacks mainly due to the chemical instability and the mechanical properties decline of the DLC coatings [102]. Progress in the technology of magnetic media has led to the application of liquid lubricants films on the DLC coatings. The hard surfaces are typically lubricated with fluorinated or perfluorinated organic compounds, such as PFPEs [103]. Many attempts have been made to obtain a layer of PFPE lubricant on the DLC surface by dip coating or by vapor deposition so as to minimize any friction and wear [104–106]. Systematic studies have been also performed on the effect of sliding speed on friction and wear performance of PFPE lubricants with different functional groups and molecular weights [60,61].

The functionalization of DLC through the covalent linkage of PFPE chains modifies permanently surface properties such as morphology and wettability (see Section 15.3.3 for more details). In addition, the linkage of PFPE chains transfers the typical lubricant features of PFPEs to DLC thin film. The friction forces on bare and PFPE-functionalized DLC films can be evaluated by means of lateral force microscopy on the basis of the ratio between lateral force signals and the normal forces: a reduction of friction on DLC is observed in the presence of a covalently linked PFPE coating (Fig. 15.15). Thus, PFPE-functionalized DLC films emerge as suitable substrates for

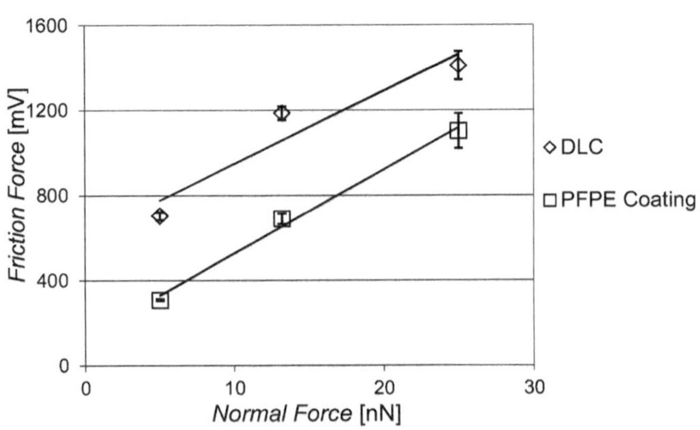

Figure 15.15 Lateral force signals versus normal forces on the native diamond-like carbon film and perfluoropolyether coating.

the development of self-lubricating surfaces with low SE and high chemical inertness. The design of functionalized surfaces can widen the application range of DLC also in other fields, like anti-fingerprint protective layers, self-cleaning surfaces, and functional coatings.

15.6.3 Fluorous Stationary Phases in Chromatography

A very recent branch of fluorine chemistry is the fluorous chemistry: the adjective fluorous refers to the characteristics of highly fluorinated saturated organic materials, molecules or molecular fragments; in a simplified definition it means "highly fluorinated" or "rich in fluorines" and based on sp^3-hybridized carbon [107].

Fluorous molecules show an affinity for other fluorous compounds (*fluorophilic*) and fluorous separation techniques are methods that separates fluorous or fluorous-labeled molecules from the other types of molecules [107]. Fluorous silica gel is a traditional silica functionalized by linking $-(CH_2)_nR_F$ chains and it has been developed for fluoruous analytical and preparation separations, such as column chromatography. Silanes with the general formula $X\text{-}Si(Me)_2(CH_2)_nR_F$ are used as "silylating" reagents for the linkage of $-(CH_2)_nR_F$ chains on silica gel and the length of R_F chains influences the retention versus fluorous molecules [108].

Chromatographic packings based on carbonaceous materials are commonly used in gas chromatography (GC) due to their high surface area, high thermal stability, and high chemical resistance. Typically, packed columns with carbonaceous packings are prepared by physical adsorption or impregnation of an unbound stationary phase. The functionalization of carbonaceous materials by means of PFPE peroxide (see Sections 15.2 and 15.4.1 for details) is a method to obtain a PFPE-based fluorous stationary phase that is covalently linked on the packing material. Chromatographic columns based on PFPE-functionalized packings can overcome some drawbacks of other standard columns for separation of fluorinated compounds, like a low thermal stability that limits the analysis below 70°C, the phenomenon of "bleeding" at temperatures above 40°C and a poor resistance towards aggressive chemicals, such as hydrofluoric acid, fluorophosgene, elemental fluorine, perfluoro-acyl-fluorides, and perfluorinated olefins.

Preliminary tests of GC columns packed with PFPE-functionalized carbonaceous materials have revealed different advantages due to their application in the separation of different fluorinated mixtures: high peak resolution; good separation efficiency; thermal stability that extends the temperature analysis up to 190°C; an effective dwindling of "bleeding" [109]. In addition, tests on aggressive mixtures show that PFPE-functionalized stationary phases can be applied even in the presence of elemental fluorine: eg, a GC column packed with a PFPE-functionalized graphitic CB adsorbent efficiently separates CO_2/F_2 mixtures as high as 190°C without any "bleeding" effect and with a high peak resolution [109].

Moreover, the chemical linkage of the PFPE-based stationary phase on the carbonaceous packing, instead of a standard physical adsorption, can play a fundamental role in future applications like liquid phase separations; in fact, the chemical linkage prevents the removal of the stationary phase by fluids with high solvating power, such as in heavy fluorous approaches.

References

[1] B.E. Smart, Characteristics of C–F systems, in: R.E. Banks, B.E. Smart, J.C. Tatlow (Eds.), Organofluorine Chemistry – Principles and Commercial Applications, Plenum Press, New York (USA) and London (UK), 1994.

[2] D.M. Lemal, Perspective on fluorocarbon chemistry, J. Org. Chem. 69 (2004) 1–11.

[3] W. Navarrini, M.V. Diamanti, M. Sansotera, F. Persico, M. Wu, L. Magagnin, S. Radice, Effect of amorphous fluorinated coatings on photocatalytic properties of anodized titanium surfaces, Prog. Org. Coat. 74 (2012) 794–800.

[4] O. Ruff, C. Bretschneider, F. Ebert, Die Reaktionsprodukte der verschiedenen Kohlenstoffformen mit Fluor II (Kohlenstoff-monofluorid), Z. Anorg. Allg. Chem. 217 (1) (1934) 1–18.

[5] W. Rudorff, G. Rudorff, Tetrakohlenstoffmonofluorid, eine neue Graphit-Fluor-Verbindung, Chem. Ber. 80 (5) (1947) 417–423.

[6] T. Nakajima, Synthesis, structure, and physicochemical properties of fluorine-graphite intercalation compounds, in: T. Nakajima (Ed.), Fluorine-carbon and Fluoride-carbon Materials: Chemistry, Physics, and Applications, Marcel Dekker, Inc., New York (USA), 1995.

[7] G. Nansé, E. Papirer, P. Fioux, F. Moguet, A. Tressaud, Fluorination of carbon blacks: an X-ray photoelectron spectroscopy study: I. A literature review of XPS studies of fluorinated carbons. XPS investigation of some reference compounds, Carbon 35 (1997) 175–194.

[8] G. Nansé, E. Papirer, P. Fioux, F. Moguet, A. Tressaud, Fluorination of carbon blacks. An X-ray photoelectron spectroscopy study. Part II. XPS study of a furnace carbon black treated with gaseous fluorine at temperatures below 100°C. Influence of the reaction parameters and of the activation of the carbon black on the fluorine fixation, Carbon 35 (1997) 371–388.

[9] G. Nansé, E. Papirer, P. Fioux, F. Moguet, A. Tressaud, Fluorination of carbon blacks: an X-ray photoelectron spectroscopy study: III. Fluorination of different carbon blacks with gaseous fluorine at temperatures below 100°C influence of the morphology, structure and physico-chemical characteristics of the carbon black on the fluorine fixation, Carbon 35 (1997) 515–518.

[10] T. Shirasaki, F. Moguet, L. Lozano, A. Tressaud, G. Nansé, E. Papirer, Fluorination of carbon blacks: an X-ray photoelectron spectroscopy study: IV. Reactivity of different carbon blacks in CF_4 radiofrequency plasma, Carbon 37 (1999) 1891–1900.

[11] K. Guérin, M. Dubois, A. Houdayer, A. Hamwi, Applicative performances of fluorinated carbons through fluorination routes: a review, J. Fluorine Chem. 134 (2012) 11–17.

[12] A. Hirsch, The era of carbon allotropes, Nat. Mater. 9 (2010) 868–871.

[13] J.H. Holloway, E.G. Hope, R. Taylor, G.J. Langley, A.G. Avent, T.J. Dennis, J.P. Hare, H.W. Kroto, D.R.M. Walton, Fluorination of buckminsterfullerene, J. Chem. Soc. Chem. Commun. 14 (1991) 966–969.

[14] H. Selig, C. Lifshitz, T. Peres, J.E. Fischer, A.R. McGhie, W.J. Romanow, J.P. McCauley Jr., A.B. Smith III, Fluorinated fullerenes, J. Am. Chem. Soc. 113 (14) (1991) 5475–5476.

[15] H. Tuohara, Electrochemical properties of fluorinated carbon nanotubes, in: T. Nakajima, H. Groult (Eds.), Fluorinated Materials for Energy Conversion, Elsevier, Amsterdam (NL), 2005.

[16] J.T. Robinson, J.S. Burgess, C.E. Junkermeier, S.C. Badescu, T.L. Reinecke, F.K. Perkins, M.K. Zalalutdniov, J.W. Baldwin, J.C. Culbertson, P.E. Sheehan, E.S. Snow, Properties of fluorinated graphene films, Nano Lett. 10 (8) (2010) 3001–3005.
[17] P.J. Fagan, P.J. Krusic, C.N. McEwen, J. Lazar, D. Holmes Parker, N. Herron, E. Wasserman, Production of perfluoroalkylated nanospheres from buckminsterfullerene, Science 262 (1993) 404–407.
[18] P.J. Fagan, P.J. Krusic, E. Wasserman, Fluoroalkylated Fullerene Compounds, 1994. US Patent 5,354,926.
[19] H.P. Fritz, R. Hiemeyer, Formation in situ of perfluoroalkylated fullerenes, Carbon 33 (11) (1995) 1601–1605.
[20] O.V. Boltalina, I.V. Kuvychko, N.B. Shustova, S.H. Strauss, Perfluoroalkylation of fullerenes, in: F. D'Souza, K.M. Kadish (Eds.), Handbook of Carbon Nano Materials – Volume 1: Synthesis and Supramolecular Systems, World Scientific Publishing Co. Pte. Ltd., Singapore (SGP), 2011.
[21] E.T. Mickelson, C.B. Huffmann, A.G. Rinzler, R.E. Smalley, R.H. Hauge, J.L. Mangrave, Fluorination of single-wall carbon nanotubes, Chem. Phys. Lett. 296 (1–2) (1998) 188–194.
[22] V.N. Khabashesku, W.E. Billups, J.L. Mangrave, Fluorination of single-wall carbon nanotubes and subsequent derivatization, Acc. Chem. Rev. 35 (12) (2002) 1087–1095.
[23] T. Nakamura, M. Ishihara, T. Ohana, A. Tanaka, Y. Koga, Sidewall modification of single-walled carbon nanotubes using photolysis of perfluoroazooctane, Chem. Commun. 11 (2004) 1336–1337.
[24] M. Holzinger, O. Vostrowsky, A. Hirsch, F. Hennrich, M. Kappes, R. Weiss, F. Jellen, Sidewall functionalization of carbon nanotubes, Angew. Chem. Int. Ed. 40 (21) (2001) 4002–4005.
[25] K.S. Coleman, S.R. Bailey, S. Fogden, M.L.H. Green, Functionalization of single-walled carbon nanotubes via Bingel reaction, J. Am. Chem. Soc. 125 (29) (2003) 8722–8723.
[26] E.T. Mickelson, I.W. Chiang, J.L. Zimmerman, P.J. Boul, J. Lozano, J. Liu, R.E. Smalley, R.H. Hauge, J.L. Mangrave, Solvation of fluorinated single-wall carbon nanotubes in alcohol solvents, J. Phys. Chem. B 103 (21) (1999) 4318–4322.
[27] S. Banerjee, T. Hemraj-Benny, S.S. Wong, Covalent surface chemistry of single-walled carbon nanotubes, Adv. Mater. 17 (1) (2005) 17–29.
[28] W. Navarrini, C.L. Bianchi, L. Magagnin, L. Nobili, G. Carignano, P. Metrangolo, G. Resnati, M. Sansotera, Low surface energy coatings covalently bonded on diamond-like carbon films, Diam. Relat. Mater. 19 (4) (2010) 336–341.
[29] M. Sansotera, C.L. Bianchi, G. Lecardi, G. Marchionni, P. Metrangolo, G. Resnati, W. Navarrini, Highly hydrophobic carbon black obtained by covalent linkage of perfluorocarbon and perfluoropolyether chains on the carbon surface, Chem. Mater. 21 (19) (2009) 4498–4504.
[30] D. Sianesi, G. Marchionni, R.J. De Pasquale, Perfluoropolyethers (PFPEs) from perfluoroolefins photooxidation – Fomblin® and Galden® fluids, in: R.E. Banks, B.E. Smart, J.C. Tatlow (Eds.), Organofluorine Chemistry – Principles and Commercial Applications, Plenum Press, New York (USA) and London (UK), 1994.
[31] D. Sianesi, A. Pasetti, C. Corti, Photooxidation of perfluoroolefins; polyperoxides and polyethers of hexafluoropropene, Makromol. Chem. 86 (1) (1965) 308–311.
[32] M. Malavasi, D. Sianesi, Novelties and prospects in the synthesis of perfluoropolyethers by oxidative polymerization of fluoroolefins, J. Fluorine Chem. 95 (1–2) (1999) 19–25.

[33] P.A. Guarda, S. Petricci, G. Fontana, G. Marchionni, Synthesis and characterization of poly(tetrafluoroethyleneperoxides), J. Fluorine Chem. 130 (11) (2009) 1011–1016.
[34] P.A. Guarda, E. Barchiesi, G. Fontana, S. Petricci, M. Pianca, G. Marchionni, Peroxidic perfluoropolyether from tetrafluoroethylene oxidation: micro structural analysis by NMR spectroscopy and mechanistic considerations, J. Fluorine Chem. 126 (1) (2005) 141–153.
[35] W. Navarrini, M. Sansotera, P. Metrangolo, P. Cavallotti, G. Resnati, Modification of Carbonaceous Materials, PCT Patent, 2009. WO 2009/019243 A1.
[36] M. Sansotera, W. Navarrini, L. Magagnin, C.L. Bianchi, A. Sanguineti, P. Metrangolo, G. Resnati, Hydrophobic carbonaceous materials obtained by covalent bonding of perfluorocarbon and perfluoropolyether chains, J. Mater. Chem. 20 (39) (2010) 8607–8616.
[37] R.D. Chambers (Ed.), Fluorine in Organic Chemistry, Blackwell Publishing Ltd, Oxford (UK), 2004, p. 254.
[38] C. Corvaja, A. Famulari, L. Franco, M. Galimberti, P. Metrangolo, W. Navarrini, G. Resnati, M. Sansotera, Perfluorodiacyl Peroxides: a class of free radical sources with a wide range of thermal decomposition temperatures, Chim. Oggi-Chem. Today 24 (3) (2006) 17–22.
[39] M. Sansotera, W. Navarrini, M. Gola, C.L. Bianchi, P. Wormald, A. Famulari, M. Avataneo, Peroxidic perfluoropolyether for the covalent binding of perfluoropolyether chains on carbon black surface, J. Fluorine Chem. 132 (12) (2011) 1254–1261.
[40] P. Delhaes (Ed.), Graphite and Precursors, Gordon & Breach Science Publishers, Inc, Philadelphia (USA), 2000.
[41] J.D. Bernal, The structure of graphite, Proc. Roy. Soc. A 106 (1924) 749–773.
[42] G. Zhang, P.M. Kirkman, A.N. Patel, A.S. Cuharuc, K. McKelvey, P.R. Unwin, Molecular functionalization of graphite surfaces: basal plane versus step edge electrochemical activity, J. Am. Chem. Soc. 136 (32) (2014) 11444–11451.
[43] P.-L. Girard-Lauriault, R. Illgen, J.-C. Ruiz, M.R. Wertheimer, W.E.S. Unger, Surface functionalization of graphite and carbon nanotubes by vacuum-ultraviolet photochemical reactions, Appl. Surf. Sci. 258 (2012) 8448–8454.
[44] M. Invernizzi, Functionalization of Graphene Nanoplatelets and HOPG Through Elemental Fluorine and Perfluoropolyether Peroxides (Master's thesis), Universitá degli Studi di Milano, Milano, Italy (2013).
[45] O.N. Shornikova, E.V. Kogan, N.E. Sorokina, V.V. Avdeev, The specific surface area and porous structure of graphite materials, Russ. J. Phys. Chem. A 83 (6) (2009) 1022–1025.
[46] B.-J. Kim, S.-J. Park, Preparation of nanoporous carbons from graphite nanofibers, Nanotechnology 17 (2006) 4395–4398.
[47] H. Chang, A.J. Bard, Observation and characterization by scanning tunneling microscopy of structures generated by cleaving highly oriented pyrolytic graphite, Langmuir 7 (1991) 1143–1153.
[48] M.R. Falvo, J. Steele, R.M. Taylor II, R. Superfine, Gearlike rolling motion mediated by commensurate contact: carbon nanotubes on HOPG, Phys. Rev. B 62 (16) (2000) R10665–R10667.
[49] D.-Q. Yang, E. Sacher, Carbon 1s X-ray photoemission line shape analysis of highly oriented pyrolytic graphite: the influence of structural damage on peak asymmetry, Langmuir 22 (3) (2006) 860–862.
[50] J. Barzola-Quiquia, J.-L. Yao, P. Rödiger, K. Schindler, P. Esquinazi, Sample size effects on the transport characteristics of mesoscopic graphite samples, Phys. Stat. Sol. A 205 (12) (2008) 2924–2933.

[51] B. Bhushan, Chemical, mechanical and tribological characterization of ultra-thin and hard amorphous carbon coatings as thin as 3.5 nm: recent developments, Diamond Relat. Mater. 8 (1999) 1985–2015.

[52] T. Nakamura, T. Ohana, M. Suzuki, M. Ishihara, A. Tanaka, Y. Koga, Chemical modification of DLC films with perfluorooctyl functionality, Diamond Relat. Mater. 14 (2005) 1019–1022.

[53] H. Sawada, Fluorinated peroxides, Chem. Rev. 96 (1996) 1779–1808.

[54] H.Y. He, Y. Zhao, K. Zhang, X.M. Meng, C.X. Zhao, Surface perfuorofuctionalization of polystyrene and styrene-butadiene-styrene block copolymer by perfuorodiacyl peroxides, J. Fluorine Chem. 106 (2000) 117–120.

[55] G. Resnati, I. Wlassics, M. Sansotera, P. Metrangolo, W. Navarrini, Perfluoroalkylation of aromatic compounds with perfluorodiacyl peroxides, Chim. Oggi-Chem. Today 25 (3) (2007) 23–25.

[56] A. Bravo, H.-R. Bjørsvik, F. Fontana, L. Liguori, A. Mele, F. Minisci, New methods of free-radical perfluoroalkylation of aromatics and alkenes. absolute rate constants and partial rate factors for the homolytic aromatic substitution by n-perfluorobutyl radical, J. Org. Chem. 62 (1997) 7128–7136.

[57] T. Nakamura, Y. Koga, Perfluorooctylation of aromatic compounds with perfluoroazooctane upon 185 nm irradiation in a two-phase system, J. Chem. Soc. Perkin Trans. 2 (3) (1998) 659–662.

[58] P.C. Hartmann, A. Collet, M. Viguier, C. Blanc, Acrylic monomers and polymers with perfluoroalkylated biphenyl side groups: synthesis and phase transitions, J. Fluorine Chem. 125 (2004) 1909–1918.

[59] S. Saidi, F. Guittard, C. Guimon, S. Geribaldi, Low surface energy perfluorooctyalkyl acrylate copolymers for surface modification of PET, Macromol. Chem. Phys. 206 (2005) 1098–1105.

[60] M. Minn, S.K. Sinha, S.-K. Lee, H. Kondo, High-speed tribology of PFPEs with different functional groups and molecular weights coated on DLC, Tribol. Lett. 24 (1) (2006) 67–76.

[61] Y. He, Y. Fujikawa, H. Zhang, K. Fukuzawa, Y. Mitsuya, Evaluations of tribological characteristics of PFPE lubricants on DLC surfaces of magnetic disks, Tribol. Lett. 27 (1) (2007) 1–11.

[62] Z.Q. Li, C.J. Lu, Z.P. Xia, Y. Zhou, Z. Luo, X-ray diffraction patterns of graphite and turbostratic carbon, Carbon 45 (2007) 1686–1695.

[63] H.P. Boehm, Some aspects of the surface chemistry of carbon blacks and other carbons, Carbon 32 (5) (1994) 759–769.

[64] I. Wlassics, V. Tortelli, M. Sala, D. Montrone, Perfluoropolyether (PFPE) (poly)diacyl peroxides: synthesis and applications, J. Fluorine Chem. 121 (1) (2003) 65–74.

[65] M. Sansotera, W. Navarrini, G. Resnati, P. Metrangolo, A. Famulari, C.L. Bianchi, P.A. Guarda, Preparation and characterization of superhydrophobic conductive fluorinated carbon blacks, Carbon 48 (2010) 4382–4390.

[66] J. Sanchez-Gonzalez, A. Marcias-Garcia, M.F. Alexandre-Franco, V. Gomez-Serrano, Electrical conductivity of carbon blacks under compression, Carbon 43 (4) (2005) 741–747.

[67] N. Probst, E. Grivei, Structure and electrical properties of carbon black, Carbon 40 (2) (2002) 201–205.

[68] D. Pantea, H. Darmstadt, S. Kaliaguine, C. Roy, Electrical conductivity of conductive carbon blacks: influence of surface chemistry and topology, Appl. Surf. Sci. 217 (1–4) (2003) 181–193.

[69] M. Sansotera, W. Navarrini, M. Gola, G. Dotelli, P. Gallo Stampino, C.L. Bianchi, Conductivity and superhydrophobic effect on PFPE-modified porous carbonaceous materials, Int. J. Hydrogen Energy 37 (7) (2012) 6277–6284.

[70] V.S. Babu, M.S. Seehra, Modelling of disorder and X-ray diffraction in coal-based graphitic carbons, Carbon 34 (10) (1996) 1259–1265.

[71] K.K.C. Ho, G. Beamson, G. Shia, N.V. Polyakova, A. Bismarck, Surface and bulk properties of severely fluorinated carbon fibres, J. Fluorine Chem. 128 (11) (2007) 1359–1368.

[72] M. Inagaki, F. Kang, M. Toyoda, H. Konno (Eds.), Advanced Materials Science and Engineering of Carbon, Butterworth-Heinemann, Oxford (UK), 2014.

[73] S. Park, J.-W. Lee, B.N. Popov, A review of gas diffusion layer in PEM fuel cells: materials and designs, Int. J. Hydrogen Energy 37 (7) (2012) 5850–5865.

[74] M. Gola, M. Sansotera, W. Navarrini, C.L. Bianchi, P. Gallo Stampino, S. Latorrata, G. Dotelli, Perfluoropolyether-functionalized gas diffusion layers for proton exchange membrane fuel cells, J. Power Sources 258 (2014) 351–355.

[75] L. Zhu, J. Xu, Y. Xiu, Y. Sun, D.W. Hess, C.P. Wong, Growth and electrical characterization of high-aspect-ratio carbon nanotube arrays, Carbon 44 (2) (2006) 253–258.

[76] Y.J. Kim, T.S. Shin, H.D. Choi, J.H. Kwon, Y.-C. Chung, H.G. Yoon, Electrical conductivity of chemically modified multiwalled carbon nanotube/epoxy composites, Carbon 43 (1) (2005) 23–30.

[77] B. Ameduri, From vinylidene fluoride (VDF) to the applications of VDF-containing polymers and copolymers: recent developments and future trends, Chem. Rev. 109 (12) (2009) 6632–6686.

[78] H. Yao, C.C. Chu, H.J. Sue, R. Nishimura, Electrically conductive superhydrophobic octadecylamine-functionalized multiwalled carbon nanotube films, Carbon 53 (2013) 366–373.

[79] S. Talaeemashhadi, M. Sansotera, C. Gambarotti, A. Famulari, C.L. Bianchi, P.A. Guarda, W. Navarrini, Functionalization of multi-walled carbon nanotubes with perfluoropolyether peroxide to produce superhydrophobic properties, Carbon 59 (2013) 150–159.

[80] A.K. Bhalchandra, K.P. Vijayamohanan, Tuning the wetting properties of multiwalled carbon nanotubes by surface functionalization, J. Phys. Chem. C 112 (9) (2008) 3183–3186.

[81] P. Slobodian, P. Riha, R. Olejnik, P. Saha, Electromechanical properties of carbon nanotube networks under compression, Meas. Sci. Technol. 22 (12) (2011) 1–7.

[82] S. Talaeemashhadi, Surface Modification of Carbon-based Materials by Fluorine-containing Reagents (Ph.D. dissertation), Politecnico di Milano, Milano, Italy, 2014.

[83] P.J. Krusic, E. Wasserman, B.A. Parkinson, B. Malone, E.R. Holler Jr., Electron spin resonance study of the radical reactivity of C_{60}, J. Am. Chem. Soc. 113 (1991) 6274–6275.

[84] J.R. Morton, K.F. Preston, P.J. Krusic, S.A. Hill, E. Wasserman, ESR studies of the reaction of alkyl radicals with C_{60}, J. Phys. Chem. 96 (1992) 3576–3578.

[85] M. Walbiner, H. Fischer, Rate constants for the addition of the benzyl radical to C_{60} in solution, J. Phys. Chem. 97 (1993) 4880–4881.

[86] N.M. Dimitrijević, P.V. Kamat, R.W. Fessenden, Radical adducts of fullerenes C_{60} and C_{70} studied by laser flash photolysis and pulse radiolysis, J. Phys. Chem. 97 (1993) 615–618.

[87] D.M. Guldi, H. Hungerbühler, E. Janata, K.-D. Asmus, Redox processes and alkylation reactions of C_{60} as studied by pulse radiolysis, J. Phys. Chem. 97 (1993) 11258–11264.

[88] R. Borghi, L. Lunazzi, G. Placucci, P.J. Krusic, D.A. Dixon, L.B. Kight Jr., Regiochemistry of radical addition to C_{70}, J. Phys. Chem. 98 (21) (1994) 5395–5398.

[89] O.V. Boltalina, J.M. Street, R. Taylor, Formation of triumphene, $C_{60}F_{15}Ph_3$: first member of a new trefoil-shaped class of phenylated [60]fullerenes, Chemm. Comm. 17 (1998) 1827–1828.

[90] M.P. Barrow, X. Feng, J.I. Wallace, O.V. Boltalina, R. Taylor, P.J. Derrick, T. Drewello, Characterization of fullerenes and fullerene derivatives by nanospray, Chem. Phys. Lett. 330 (2000) 267–274.

[91] R. Taylor, A.K. Abdul-Sada, O.V. Boltalina, J.M. Street, Isolation and characterisation of forty-nine fluorinated derivatives of [70]fullerene, J. Chem. Soc. Perkin Trans. 2 (5) (2000) 1013–1021.

[92] I.E. Kareev, N.B. Shustova, I.V. Kuvychko, S.F. Lebedkin, S.M. Miller, O.P. Anderson, A.A. Popov, S.H. Strauss, O.V. Boltalina, Thermally stable perfluoroalkylfullerenes with the skew-pentagonal-pyramid pattern: $C_{60}(C_2F_5)_4O$, $C_{60}(CF_3)_4O$, and $C_{60}(CF_3)_6$, J. Am. Chem. Soc. 128 (2006) 12268–12280.

[93] O.V. Boltalina, A.A. Popov, I.V. Kuvychko, N.B. Shustova, S.H. Strauss, Perfluoroalkylfullerenes, Chem. Rev. 115 (2015) 1051–1105.

[94] A.A. Goryunkov, I.V. Kuvychko, I.N. Ioffe, D.L. Dick, L.N. Sidorov, S.H. Strauss, O.V. Boltalina, Isolation of $C_{60}(CF_3)_n$ ($n = 2, 4, 6, 8, 10$) with high compositional purity, J. Fluorine Chem. 124 (2003) 61–64.

[95] M. Sansotera, S. Talaeemashhadi, M. Gola, C.L. Bianchi, P.A. Guarda, W. Navarrini, Chemical linkage of perfluoropolyether chains to carbon-based nanomaterials: recent results and new perspective, in: 17th European Symposium on Fluorine Chemistry, July, Paris, France, 2013, pp. 21–25.

[96] N. Zamel, X. Li, Effective transport properties for polymer electrolyte membrane fuel cells – with a focus on the gas diffusion layer, Prog. Energy Combust. Sci. 39 (1) (2013) 111–146.

[97] A.L. Dicks, The role of carbon in fuel cells, J. Power Sources 56 (2) (2006) 128–141.

[98] P. Gallo Stampino, C. Cristiani, G. Dotelli, L. Omati, L. Zampori, R. Pelosato, M. Guilizzoni, Effect of different substrates, inks composition and rheology on coating deposition of microporous layer (MPL) for PEM-FCs, Catal. Today 147S (2009) S30–S35.

[99] M. Hakovirta, R. Verda, X.M. He, M. Nastasi, Heat resistance of fluorinated diamond like carbon films, Diamond Relat. Mater. 10 (2001) 1486–1490.

[100] M. Rubio-Roy, E. Bertran, E. Pascual, M.C. Polo, J.L. Andújar, Fluorinated DLC deposited by pulsed-DC plasma for antisticking surface applications, Diamond Relat. Mater. 17 (2008) 1728–1732.

[101] G.Q. Yu, B.K. Tay, Z. Sun, L.K. Pan, Properties of fluorinated amorphous diamond like carbon films by PECVD, Appl. Surf. Sci. 219 (2003) 228–237.

[102] S.C. Trippe, R.D. Mansano, F.M. Costa, R.F. Silva, Mechanical properties evaluation of fluor-doped diamond-like carbon coatings by nanoindentation, Thin Solid Films 446 (2004) 85–90.

[103] C. Kajdas, Nanoscale boundary lubrication of diamond-like carbon coatings with fluorinated compounds, J. Synth. Lubrication 18 (2001) 17–38.

[104] A.M. Homola, L.-J.J. Lin, D.D. Saperstein, Process for Bonding Lubricant to a Thin Film Magnetic Recording Disk, 1990. US Patent 4960609.

[105] J. Gui, X. Ma, M.J. Stirniman, System and Method for Thin Film Protective Overcoat, 2003. US Patent Appl. 20030104158 A1.

[106] J. Gui, X. Ma, M.J. Stirniman, Method for Thin Film Protective Overcoat, 2004. US Patent 6767592 B2.
[107] J.A. Gladysz, D.P. Curran, I.T. Horvath (Eds.), Handbook of Fluorous Chemistry, Wiley-VCH Verlag GmbH & Co. KGaA, Weinheim (D), 2004.
[108] D.P. Curran, Z. Luo, Fluorous synthesis with fewer fluorines (light fluorous synthesis): separation of tagged from untagged products by solid-phase extraction with fluorous reverse-phase silica gel, J. Am. Chem. Soc. 121 (1999) 9069−9072.
[109] A. Beretta, A Study on the Functionalization of Carbonaceous Materials Through Perfluoropolyether Peroxides for Gas Chromatographic Applications (Master's thesis), Università degli Studi di Milano, Milano, Italy, 2015.

Nanoelectronics Based on Fluorinated Graphene

16

K.-I. Ho[1], C.-S. Lai[1], C.-Y. Su[2]
[1]Chang Gung University, Tao-Yuan, Taiwan; [2]National Central University, Chung-Li, Taiwan

Chapter Outline

16.1 Introduction

Since the discovery of 2D carbon materials of graphene in 2004 [1], it has attracted great attention due to its tremendous material properties, including the excellent mechanical strength [2], good chemical stability, high thermal conductivity (above 5000 W/mK) [3], high optical transparency, and the extremely high carrier mobility ($\sim$200,000 cm^2/V) [4]. All these unique material properties make it potential for the applications in optical electronics and nanoelectronic devices, such as the photosensor and high-speed RF transistors [5,6]. However, it is noteworthy that the zero band gap of pristine graphene hinders its practical application in logic electronics. Previous works have been an effort on graphene band gap opening, including the formation of graphene nanoribbon (GNR), situational doping with N, graphene hybrid heterostructure, and substrate-induced band gap tuning (Fig. 16.1A) [7]. Most of them bring the undesired drawbacks such as complex and expensive procedures and degraded carrier mobility due to the increased defect density. Recently, the graphene functionalized with hydrogen (H) or oxygen (O), known as the hydrogenated graphene (graphane) and graphene oxide (GO), were reported to tailor the electronic structure of graphene. However, graphane rapidly lost H atoms from surface when it

New Fluorinated Carbons: Fundamentals and Applications. http://dx.doi.org/10.1016/B978-0-12-803479-8.00016-4

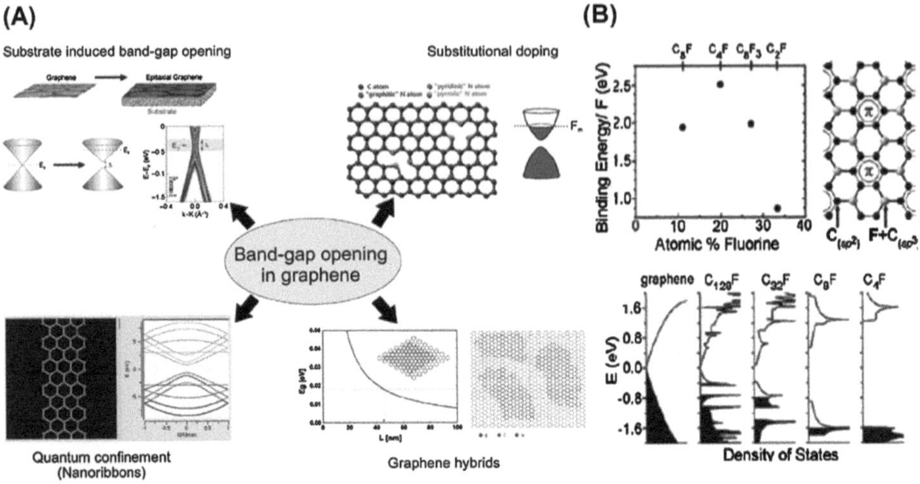

Figure 16.1 (A) The current reported methods for the graphene band gap opening. (B) The band gap tuning from semi-metallic to insulator by adjusting the stoichiometry of carbon-to-fluorine ratio of fluorinated graphene [7,16,17].

was subjected to the moderate temperature [8], thus the instability issue hinder its particle usage. On the other hand, the GO is the alternative derivative of graphene, where the basal plane and edges were randomly functionalized with many kinds of oxygen functional groups (hydroxyl, epoxy, carboxyl, etc.). The electronic structure highly depends on the stoichiometry of C/O ratio. However, the functional groups severely damage the graphene carbon networks, leading to the degradation of the carrier mobility.

Recently, more stable (~250°C) graphene derivatives, created by fluorine (F) functionalization on graphene surface, have been regarded as a promising approach to alter the graphene electronic structure. The so-called fluorinated graphene or fluorographene shows comparable mechanical strength to pristine graphene, while there are some distinct material properties such as the wide optical band gap, the highly fluorographene was found to be more transparent in the visible light region, and the optical absorption began from the energy of >3 eV [9]. Moreover, the electrical properties can be tuned from semi-metallic to insulator by simply adjusting the stoichiometry of carbon-to-fluorine ratio. According to the theoretical calculation, the stoichiometry of fluorographene from $C_{32}F$ to C_4F is able to modify the energy band gap from 0.8 to 2.9 eV, respectively (Fig. 16.1B). The recent progress by using selective fluorination can therefore create an in-plane heterojunction structure on a graphene film, which is attractive for versatile applications on the nanoelectronics or optical electronics [10–12]. Moreover, the highly fluorinated graphene exhibits the electrical property close to insulator, which was demonstrated to be a promising gate dielectric in the electronic devices, such as the resistive random access memory (RRAM) and thin film transistors. This kind of 2D insulator can also be regarded as a building block

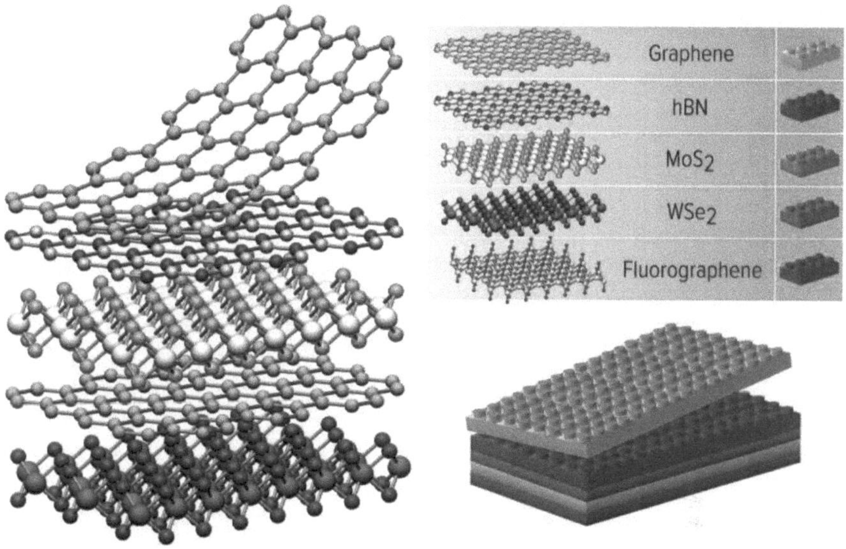

Figure 16.2 Fluorographene was regarded as the building block for the emerging 2D devices, where the various 2D layered materials (graphene, MoS_2, WS_2, etc.) assembled together to achieve an artificial superlattice and could be applied as a functional electronics [13].

and integrated in the emerging 2D devices, where various 2D layered materials (graphene, MoS_2, WS_2, etc.) are assembled by stacking together to achieve a functional electronics (Fig. 16.2) [13].

The aforementioned material properties of fluorinated graphene shed the light on next-generation architectures of electronics. Prior to this, the development for large-scale fabrication process is of great importance. Many fluorinated processes have been reported to fluorinate carbon materials. The earlier works to obtain the fluorinated graphite or single-walled carbon nanotubes were frequently processed by using CF_4, SF_6, F_2, and HF as reactive gasses [14]. The work in 2004 employed XeF_2 as the fluorinating agent for the fluorination of monolayer graphene. In brief, the samples were exposed to XeF_2 at temperature 70°C and were standing for several hours. The XeF_2 gas reacted with graphene, resulting in the fluorine content of ~20% (C_4F) and ~50% (CF) on single side and both sides of graphene, respectively. A more recent work reported a more controllable approach based on the CF_4 plasma treatment on graphene [15]. However, the high energy radicals dissociated from the plasma system normally generate lots of defect sites on graphene film. To further address this issue, the low-damaged fluorination process, based on CF_4 plasma, was demonstrated, where the filter was designed to screen out the high energy ion bombardments and UV photons. The fluorination process based on the plasma technique was considered as the promising approach to obtain high quality and controllable fluorinated graphene, which allows for its integration with nowadays integrated circuit (IC) fabrication processing.

16.2 The Synthesis of Fluorinated Graphene

16.2.1 Vapor Phase Fluorination

To fluorinate graphene, XeF_2 has been chosen as the fluorine source. Robinson et al. has demonstrated graphene fluorination on different substrates like silicon-on-insulator (SOI) and Cu [17]. As shown in Fig. 16.3A, the degrees of fluorination are significantly distinct. 50% fluorination could be readily achieved for the graphene supported on SOI substrate for 200 s exposure, whereas the maximum realizable fluorination of graphene lying on Cu is 20%. Fig. 16.3A clearly illustrates the mechanism for these results. The XeF_2 gas would etch the underlying Si through the pinhole or defects of graphene and lead to the double-side fluorination. With 100 s exposure, graphene/SOI only achieved 20% fluorination, where the Si/SiO_2 fraction was not varied, implying that the XeF_2 gas does not etch with the Si in the initial stage, thus limited fluorination degree of 20% was implemented by single-side fluorination. When Si was started to be etched by XeF_2 for over 100 s exposure, graphene was double-side exposed to XeF_2, resulting in 50% of fluorination. In contrast, graphene/Cu sample could only achieve 20% fluorination on top side graphene, while the bottom side graphene preserves its intrinsic properties due to the underlying Cu protection.

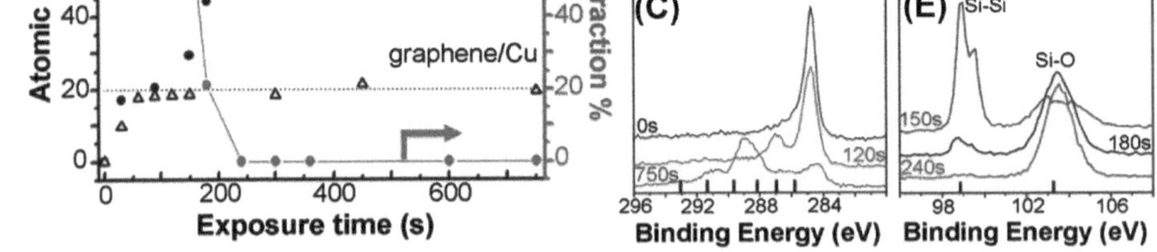

Figure 16.3 X-ray photoelectron spectroscopy (XPS) analysis for fluorinate graphene during XeF_2 exposure. (A) Fluorine fraction for single (graphene/Cu) and double (graphene/SOI) side fluorination when it was subjected to XeF_2. The inset illustration shows that the underlying Si was etched due to the reactive agent of XeF_2 through a pinhole and edges in the graphene film. (B) The XPS carbon spectra (C 1s) for the sample of graphene/Cu after XeF_2 exposure. Several C—F bonding states are indicated. (C) The XPS carbon spectra for sample of graphene/SOI after XeF_2 exposure. (D) XPS carbon spectra clearly show the fact of defluorination when it was exposed to hydrazine vapor. The inset reveals the fluorine spectra before and after 24-h hydrazine treatment. (E) XPS silicon spectra from (A), showing the evolution of Si—Si peak during XeF_2 etching for different exposure durations [17].

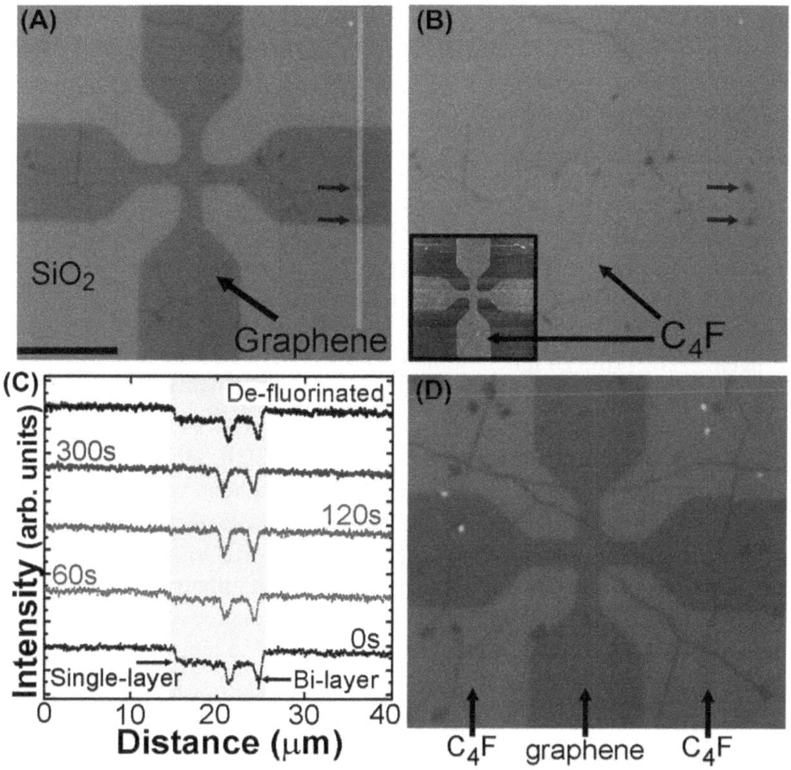

Figure 16.4 The optical changes of graphene upon fluorination: (A) and (B) were the optical micrograph on graphene that before and after single-side fluorination, respectively (scale bar = 10 μm). The result shows that the fluorinated graphene regime become optical invisible. The inset in (B) shows the AFM height image for single-side fluorination. The height for a patterned and fluorinated graphene is about 1.0−1.5 nm. (C) The green component of the RGB signal, recorded by a CCD camera, taken along the *green line (light grey in print versions)* as shown in (A). The *black arrows* indicated in (A) and (B) mark the location of the bilayered graphene regions. The profiles were taken at various XeF_2 exposure times and the sample after defluorination. (D) A different graphene film selectively patterned with fluorine (note: image is contrast enhanced; *dark central cross* is graphene; *light regions* are partially fluorinated graphene) [17].

Moreover, the optical properties for fluorinated graphene were first investigated by Robinson et al. Fig. 16.4A−B shows the optical photography of graphene before and after fluorination, respectively. It was found out that the high optical contrast between graphene and SiO_2 becomes invisible after fluorination. It is remarkable that few wrinkles and dots of graphene were still visible after XeF_2 exposure. A comprehensive explanation was proposed in Fig. 16.2C, which displays the optical absorption along the green line in Fig. 16.2A. Single and bilayer graphene exhibited significantly distinct absorption (ie, the amount of adsorption of bilayer graphene was twice than the amount of single-layer graphene). After 300 s XeF_2 exposure, bilayer graphene exhibits the same amount of absorption with single-layer graphene, indicating the fluorination only occurred on the surface layer while the second layer preserves the pristine property of graphene.

16.2.2 Chemical Fluorination

In the previous approach, XeF_2 plays the role of fluorine source to react with the intrinsic graphene, leading to the formation of fluorinated graphene. However, this method lacks cost-effectiveness and compatibility to integrate this procedure with industrial process, especially for the device manufacturing. In this regard, Wang et al. attempted to establish a solution-based fluorination technique by a frequently used reactive agent—HF [18]. Graphene oxide was employed as the starting material. In brief, the 40 mL graphene oxide dispersion (2 mg/mL) and 0.5 mL HF (40 wt%) were mixed by ultrasonication for 1 min. The mixture was then transferred to a 50 mL Teflon-lined autoclave and kept at 180°C for 30 h, leading to the formation of fluorinated graphene. The collected fluorinated graphene products were purified by washing and drying. The approach shows the advantages of cost-effective and rapid synthesis for producing fluorinated graphene. It is worth to note that the oxygen-contained functional groups in graphene oxide played an important role for converting the structure to the fluorinated graphene. Moreover, the degree of fluorination could be easily tuned from 9.5% to 32.3% by adjusting the reaction temperature, times and the amounts of HF during the reaction.

16.2.3 Plasma-Assisted Fluorination

Vapor and chemical fluorinations were still doubtful regarding the concerns of time-consuming and compatibility with modern solid-state device fabrication. Time-effective and complementary metal-oxide semiconductor (CMOS)-compatible approaches for graphene fluorination were desired. Therefore, plenty of efforts have been devoted to investigate the plasma-assisted fluorination process on graphene, where fluorine precursor with stable gas phase was used. Shen et al. employed CF_4 plasma as the fluorine resource, the result demonstrated the significant thinning effect on graphene layers and in situ fluorine functionalization. The process was carried out at 0.8-Torr when the CF_4 plasma was ignited under the radio frequency of 20 W with different exposure times ranging from 1 to 5 s. It is worthy to note that the employed relatively lower power and shorter treatment time, compared to traditional plasma for dry etching, allowed the authors to gradually reduce the numbers of layers of graphene sheets and gently functionalized graphene surface with fluorine atoms. However, the etching effect was still observed (Fig. 16.5) when it was exposed to longer treatment during the fluorination process, suggesting that it was difficult to achieve well-controlled fluorination on graphene based on conventional plasma treatment system [15].

As shown in Fig. 16.6A, Baraket et al. applied the electron beam generated plasma as a fluorination route to suppress the commonly observed ion-bombardment issue, which occurs in the conventional plasma system/atmosphere. Here, the electron beams (2 KeV) were injected to a diluted argon atmosphere to produce large area ($60 \times 25\ cm^2$) plasma sheets, whereas the diluted SF_6 and O_2 were used as the

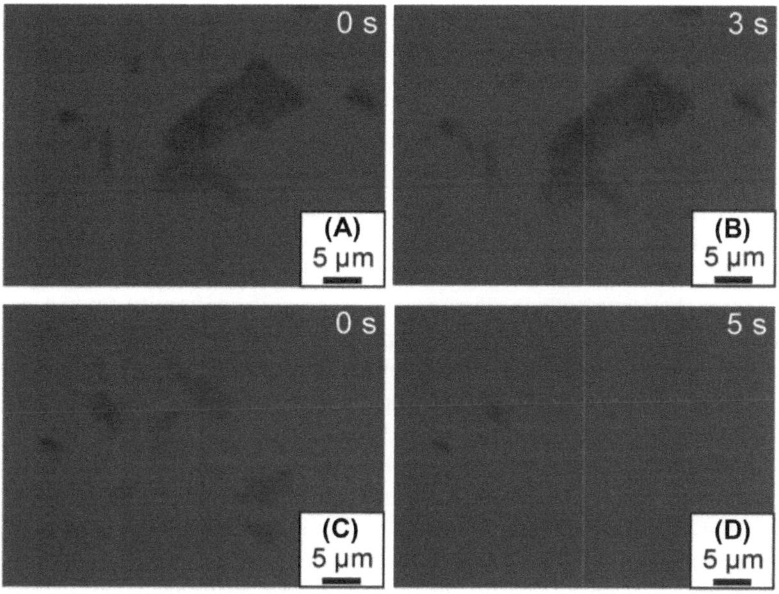

Figure 16.5 Optical images of samples before and after CF_4 plasma treatment. (A) and (C) Optical images demonstrating the morphologies of as-prepared samples before CF_4 plasma treatment, (B) optical image of the sample in (A) after 3 s of CF_4 plasma treatment, (D) optical image of the sample in (C) after 5 s of CF_4 plasma treatment [15].

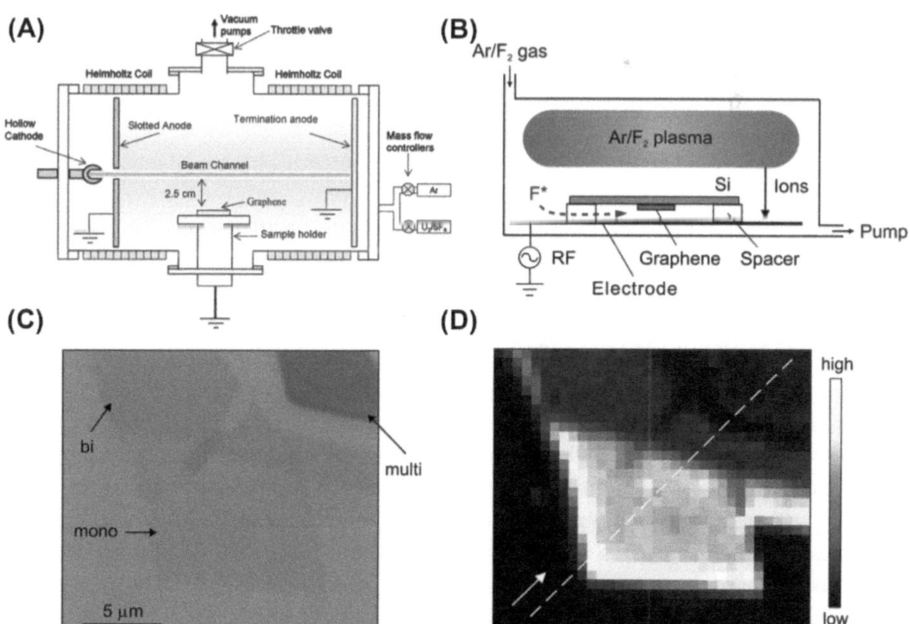

Figure 16.6 (A) Schematic illustration of the plasma processing systems in the work of Baraket et al., (B) schematic structure of the "face down" fluorination process in the reaction ion etching system, (C) optical photo of 3 min fluorinated graphene, mono-, bi-, and multi-layer regions are included, (D) Raman mapping of the integrated D peak (1300–1410 cm^{-1}) intensity [19,20].

doping resources during this procedure [19]. Moreover, Tahara et al. also proposed a "face down" architecture for the graphene fluorination as shown in Fig. 16.4B [20]. The approach could significantly reduce the ion-bombardment, which was induced from the Ar/F_2 plasma. The results from the Raman mapping according to the feature peak of D band indicate the fluorination of graphene was spatially homogeneous within the 10 μm^2 size as shown in Fig. 16.4D. However, an issue of the fluorination homogeneity might be raised for this technique due to the different diffusing length of fluorine radicals from edge to center of the substrate, resulting in spatial nonuniformity of the fluorine content on graphene sheets.

To address the issues of ion damage and large area uniformity for the graphene fluorination, Ho et al. introduced a low-damage plasma for the graphene fluorination process [12], which could fit the requirement for achieving large area fluorinated graphene. In this work, the CF_4 plasma was chosen as the fluorine source due to the highly compatible process in semiconductor manufacturing. The graphene on oxidized Si was heated to 200°C to facilitate the efficiency of fluorination, which was a definitely acceptable processing temperature in CMOS fabrication. The results demonstrate that the C/F fraction ranging from 27.4 to 5.6 could be adjusted by varying the treatment time (from 5 to 20 min). From the material characterization, the XPS C1s spectrum of a representative case for 10 min exposure time is shown in Fig. 16.7A, where the typical components such as strong peak of C−C (284.5) and other peaks of C−CF (285.8 eV), C−CF_2 (292.1 eV), and CF_3 (293.5 eV) were observed. In addition, the trend of increased sheet resistance on fluorinated graphene according to various exposure times were observed, in which resistances at 10 k, 100 k, and 1 Mohm/sq were corresponded to 5, 10, and 20 min fluorination as shown in Fig. 16.5B, respectively. Regarding the homogeneity of fluorination, optical microscope image and Raman mapping in Fig. 16.5C and D clearly reveal the excellent uniformity of fluorination on graphene in terms of optical contrast and the mapping from characters of D' band intensity.

16.3 Fluorinated Graphene on the Nanoelectronic Devices

16.3.1 Graphene-Based Nanoelectronics

Graphene has been an interesting material in the application of nanoelectronics due to its outstanding properties, as we mentioned in Section 16.2.1. In this section, the novel strategies by introducing graphene for upgrading or replacing current electronic technologies will be reviewed. To be more specific, the ways to replace indium tin oxide (ITO) as the conductive layer in touch screen and compete with III-IV compounds, as well as the Si-based materials as the substrates in radio-frequency and logic transistor are discussed below.

ITO was the most frequently used material as the conductive layer in touch screen and many optical electronics devices. However, ITO is a brittle material, where the electrical conductivity was easily degraded when it was subjected to the higher degree

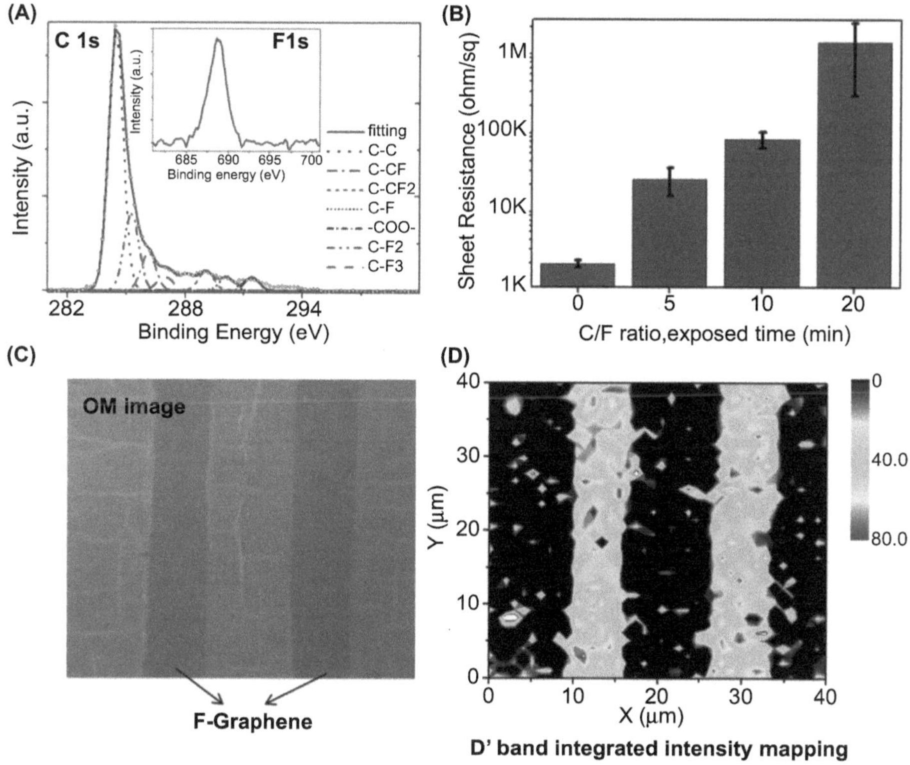

Figure 16.7 (A) The XPS results of the C1s core level for a fluorinated graphene sample (exposure time of 10 min). Several features peaks of carbon–fluorine bonding states are indicated. The inset shows the fluorine (F1s) peak at 688.8 eV. (B) The statistics of sheet resistance extracted from graphene with different fluorination time. (C) The optical microscope image of graphene with 1 min CF_4 plasma exposure. (D) Raman mapping according to the feature peak of D' band on a selectively patterned fluorinated graphene sample [12].

of mechanical bending. Also, the high cost due to the limited amounts of rear element of indium, making it difficult for widespread and versatile applications. These factors therefore limit the ITO for next-generation nanoelectronics, especially for these flexible and wearable devices. In this regards, graphene could be a more promising candidate material for replacing ITO as a novel conductive electrode. Sukang Bae et al. demonstrated a novel roll to roll technique for transferring large-area graphene films on flexible substrate [21]. The results show that graphene exhibits comparable transparency and conductivity to ITO (eg, ~125 Ω/sq for single-layer graphene and ~30 Ω/sq for four layers graphene), indicating their potential for replacing ITO as frontier electronics.

Regarding the application in field effect (FET) devices, graphene exhibits a symmetry gate-controlled current (ambipolar characters), first demonstrated by Novoselov et al. [1]. Graphene was reported to exhibit carrier mobility up to 10,000 cm^2/V and its mean free path is approximately 0.4 μm as well as it has ballistic transport and

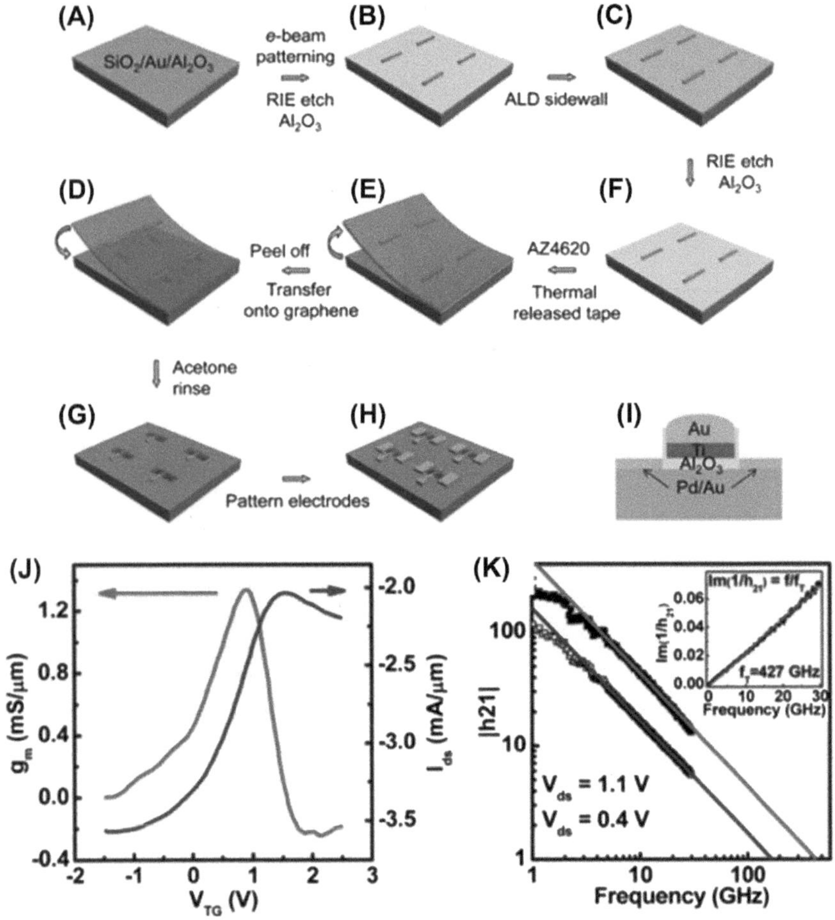

Figure 16.8 (A–I) Schematic illustration of self-aligned transistors with transferred gate stacks. (J–K) Room-temperature dc and ratio-frequency characteristics of the self-aligned peeled graphene transistor with transferred gate stacks. The cutoff frequency is 427 GHz for 1.1 V bias (*solid block*) and 169 GHz for 0.4 V bias (*open block*). (Inset) Extraction of fT by Gummel's method [22].

extremely high carrier current density (>10^8 A/cm^2), all these unique properties making it as a promising candidate for transistor in post-Si electronic material.

In the past decade, scientists made great efforts on practical electronic applications, such as ratio-frequency (RF) and logic transistor. Lin et al. established the graphene-based RF-transistor which had cutoff frequency (f_T) above 100 GHz with 240 nm channel length [6]. Furthermore, Cheng et al. enhanced the f_T up to 427 GHz with a self-aligned technique, which achieves the highest record [22] of graphene-based FET as shown in Fig. 16.8. The performance is even competitive to the state-of-the-art on GaAs high electron mobility transistor devices (Fig. 16.9A) [23].

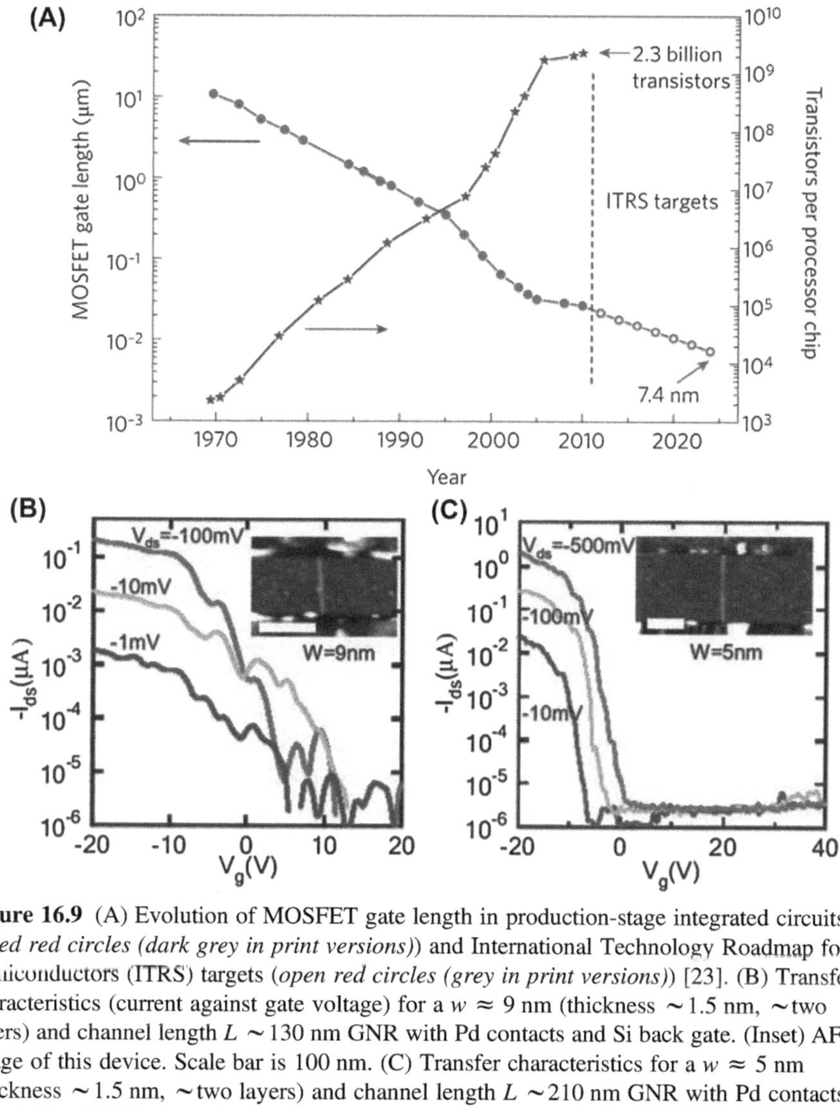

Figure 16.9 (A) Evolution of MOSFET gate length in production-stage integrated circuits (*filled red circles (dark grey in print versions)*) and International Technology Roadmap for Semiconductors (ITRS) targets (*open red circles (grey in print versions)*) [23]. (B) Transfer characteristics (current against gate voltage) for a $w \approx 9$ nm (thickness ~1.5 nm, ~two layers) and channel length L ~130 nm GNR with Pd contacts and Si back gate. (Inset) AFM image of this device. Scale bar is 100 nm. (C) Transfer characteristics for a $w \approx 5$ nm (thickness ~1.5 nm, ~two layers) and channel length L ~210 nm GNR with Pd contacts. (Inset) The AFM image of this device. Scale bar is 100 nm [24].

On the other hand, the lack of band gap in graphene obscures its application in logic transistors (ie, one could never switch off graphene-based transistor to represent inlogical signal "0"). In this regard, the transistor made by graphene nanoribbon (GNR) with atomic smooth edge was predicted to open the band gap while preserving the high carrier mobility under room temperature operation [24]. The current on–off ratios of 10^7 were realized in the width of sub-10 nm as shown in Fig. 16.9B–C. However, the undesired mobility degradation came along with the excellent on–off ratio, the advantage for replacing Si is therefore lost (ie, mobility below 200 cm^2/V hardly competed with 1350 cm^2/V of conventional Si-based transistor). Nonideal metallic contact and relatively long channel length were the main reasons for degradation of mobility on graphene-based FETs [24]. Also, Yang et al. pointed out the fact that the line-edge scattering from GNR edges began to lower the mobility when the ribbon width is scaled below 60 nm, and the mobility was significantly lower than 200 cm^2/V when its width is down scaled to 20 nm [25]. Thus, GNR still suffers from the fact of mobility degradation. Moreover, the latest COMS technology of device scaling is around 16–22 nm. The sub-10 nm level lithography technology might be achieved after 2025 or even later [23], which implies the challenge to implement GNR in CMOS technology.

In addition, the bilayer graphene exhibits a parabolic shape of valence band near the K point. From the theoretically simulated [26] and experimentally verified [27] studies, it was found out that the bilayer graphene shows moderate band gap when it was subjected to a perpendicular electric field. The on–off ratio of 10^2 and band gap of >130 meV are achieved [27]. It has been one of the candidates for practical logic transistor application. However, the practical on–off ratio in logic transistor is 10^6 or higher, whereas ratio of 10^2 in bilayer gFET has not fitted the current requirement. Furthermore, dual gate architecture causes the difficultly for large-scale manufacturing. The performance of top-gated graphene transistors is still limited by the poor quality of gate dielectric during the deposition process (ie, low quality or yield of insulator deposition leads to unstable electrical characteristics) [28,29].

To realize the graphene-based logic device, fluorinated graphene could be a promising strategy for band gap opening/tuning and could also improve the gate dielectric deposition when it was integrated to a nanoelectronics. In the following sections, we will reveal the approaches to address the mentioned issues.

16.3.2 In-Plane Heterostructure for the Graphene-Based Field-Effect Transistors

To achieve high on–off ratio, Moon et al. first demonstrated a heterostructure, graphene(Gr)/fluorographene(GrF)/graphene(Gr), as shown in Fig. 16.10A [10]. The diode performance was shown in Fig. 16.3D, the turn-on voltage significantly depended on the GrF channel length due to the barrier constructed by graphene since it was determined from the amounts of tunneling current. The shorter channel length was created, the lower turn-on voltage was achieved. In addition, it was found out that the diode current increases significantly when reducing the GrF channel length.

The gate-controlled resistance of Gr/GrF/Gr was shown in Fig. 16.11. Resistance varied from off-state of 3 GΩ down to on-state of 30 kΩ. The inset revealed the mechanism of carrier tunneling. As the gate voltage was achieving turn-on voltage, the band of GrF was bended and provided a narrowed band barrier, implying the higher possibility of carrier tunneling through the channel.

However, the GrF represents the fully fluorinated graphene exhibiting a completely different band structure than pristine graphene (ie, instead of the zero effective mass of

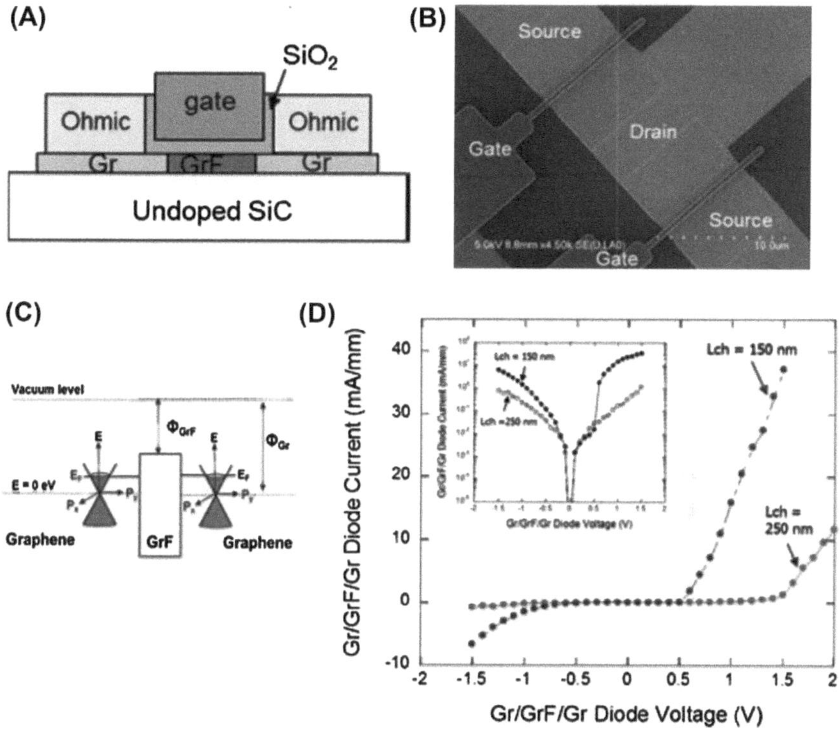

Figure 16.10 (A) Schematic illustration of Gr/GrF/Gr transistor; (B) scanning electron microscopy image of transistor; (C) band diagram of Gr/GrF/Gr; (D) diode characterization of Gr/GrF/Gr [10].

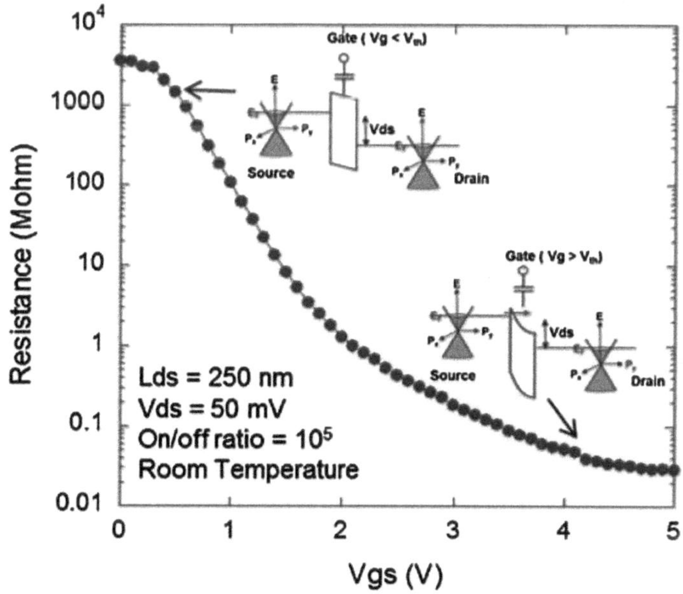

Figure 16.11 The electrical characterization of gate-controlled Gr/GrF/Gr [10].

carrier in graphene, the heavy effective mass degrades the carrier transport and reduces carrier mobility.) Therefore, the partially fluorinated graphene could be an efficient method to keep the high mobility of graphene.

The second application of fluorinated graphene in nanoelectronics is the feasible patterning approach, which was employed by a novel chemical isolation concept to reduce the edge scattering effect which occurs when graphene was patterned from geometric insulation method (ie, removal of unwanted graphene region by O_2 plasma etching) [11]. In Fig. 16.12, Lee et al. showed a novel probe lithography technique to define patterns of GNR, where the defined regime out of channel was exposed and followed by the formation of highly fluorinated graphene. Almost nondegraded mobility was represented in such chemically isolated GNR transistors. The GNR with geometric isolation shows mobility at 692 cm^2/V while pristine graphene and GNR with chemical isolation show a higher mobility above 900 cm^2/V. The result

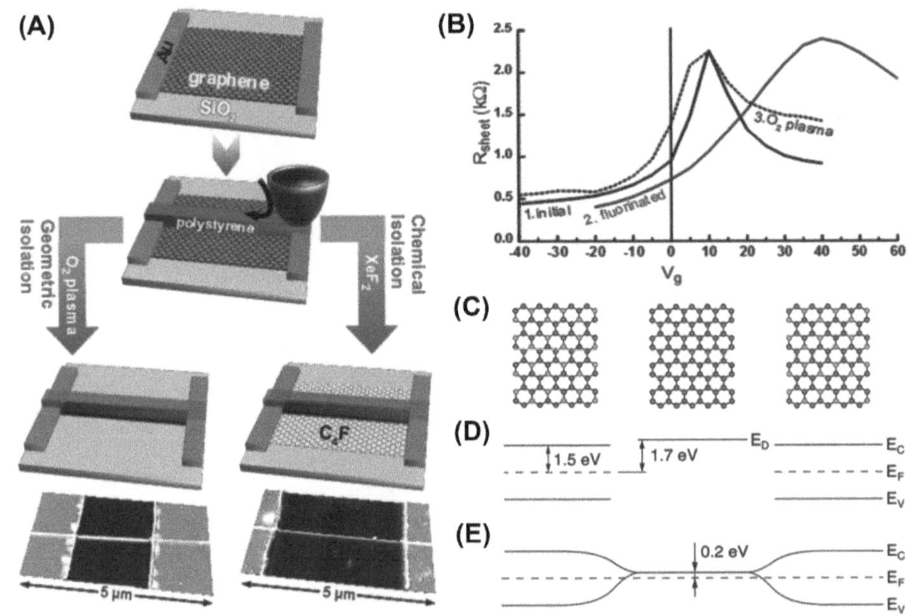

Figure 16.12 (A) Fabrication processes of geometric and chemical isolation for patterning graphene nanoribbon (GNR) by O_2 plasma and XeF_2 exposure, respectively. (B) The characteristics of sheet resistance (R_{sheet}) versus gated voltage (V_g) for GNR devices after fluorination and subsequent O_2 plasma etch. (1) Initial device with the Dirac point at 10 V ($\mu_h = 908$ cm^2/V). (2) PS/GNR device after fluorination ($\mu_h = 858$ cm^2/V). (3) Additional O_2 plasma etch of the same device ($\mu_h = 692$ cm^2/V). After etching C_4F surrounding the GNR, the Dirac point was recovered to the initial state. Note that exposure to the oxygen plasma degrades the mobility. (C–E) Structure and energy diagram across the graphene ribbon. (C) The graphene ribbon is surrounded by large areas of C_4F. Schematic energy diagrams before (D) and after (E) the system has equilibrated. E_D is the Dirac energy in graphene, E_C (E_V) the conduction (valence) band energy in C_4F, and E_F the Fermi level [11].

indicates that the fluorinated graphene brings a new concept for creating and designing the high performance nanoelectronics by selectively modifying material's electrical properties rather than removing it by etching or damaging.

It is remarkable that the V_{Dirac} of GNR was strongly p-doped due to the work function difference between pristine and fluorinated graphene. Electrons near the GNR's Fermi level will be flowing into extended states of C_4F. That results in 0.2 eV energy separation between Dirac point and the Fermi level. Additionally, V_{Dirac} could be recovered to the initial state after removing the laterally surrounded C_4F by O_2 plasma etching as shown in Fig. 16.12B, curve 3 (ie, converting the same GNR from the chemical isolation to geometric isolation). That evidenced the surrounding C_4F does act as the p-doping factor for GNR.

Ho et al. developed a novel technique to fabricate pristine, partially fluorinated and fully fluorinated graphene in a one-step formation process [12]. Au buffer layer was deposited above channel graphene to reduce the fluorination degree in channel graphene, whereas the naked graphene outside the channel was highly fluorinated as an insulator as shown in Fig. 16.6A.

The eventual electrical property in semi-conductive graphene transistor was demonstrated in Fig. 16.13B–C. On–off ratio of drain current could be enlarged to 10.1, which is higher than that of pristine graphene device ($\sim$3). Although the on–off ratio is not as excellent as that showed by Moon et al. at 10^5. This strategy opens up a new way to modulate current by semi-conductive graphene with chemical isolation method. There is plenty of room to improve the electrical properties for the logic circuit application by optimizing the contact resistance and device downscaling.

16.3.3 Fluorinated Graphene as Novel Dielectric Materials in Nanoelectronics

It is difficult to directly deposit dielectric film on graphene with uniform thickness and high quality crystal by atomic layer deposition (ALD) method, due to the hydrophobic nature of graphene surface, where the nonuniform formation of oxygen–carbon covalent bonds on graphene results in failed growth or porous surface of dielectric film [28,30,31]. High quality dielectric could grow on modified graphene while one can raise concerns regarding formation of defects in graphene or unexpected disturbances by external seed layers. Regarding the PVD system, defects are definitely introduced into graphene because of tremendous physical kinetic energy [29].

To overcome this issue, fluorinated graphene was applied as dielectric and investigated in capacitor and graphene-based FET [32]. Here the fundamental dielectric properties of fluorinated graphene are shown in Fig. 16.14. Capacitance density is significantly voltage independent and the corresponding dielectric constant is kept at 1.2. Frequency dispersion of fluorinated graphene and graphene oxide are displayed in Fig. 16.7B for comparison. The capacitance degradation of fluorographene is around 10% in the range of 1–100 kHz, implying the reliability of both graphene-based insulators. While employing a gate insulator, the investigation of its ability for resisting leakage current is essential. Here, the 5 nm thickness of fluorinated

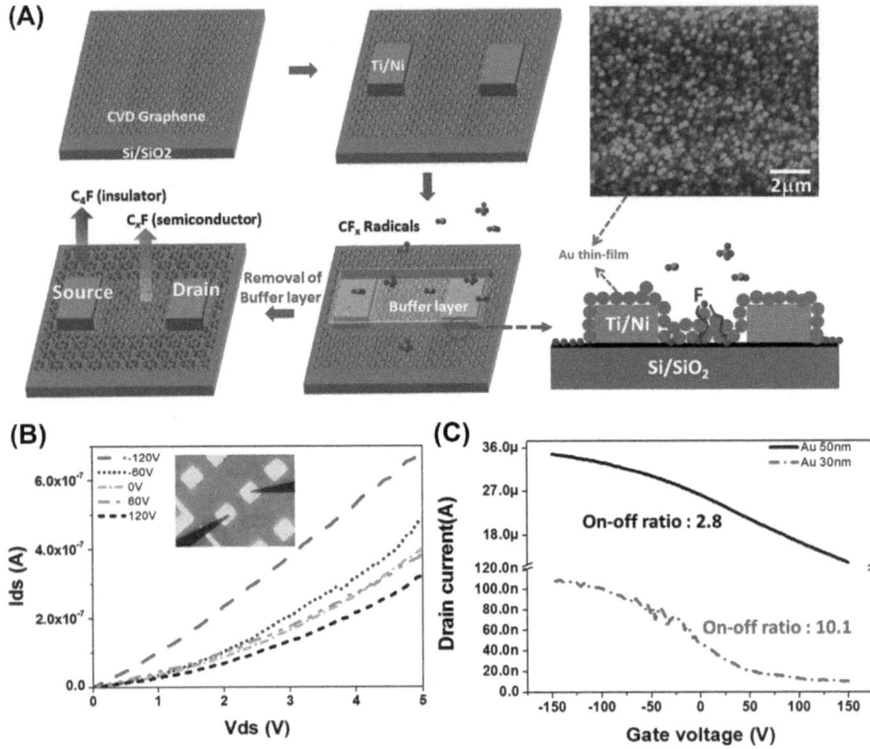

Figure 16.13 (A) One-step formation of semiconductor/insulator lateral heterostructure implemented by gold buffer layer to decrease the fluorination degree within channel (B) Drain current (I_{DS}) versus drain voltage (V_{DS}) under different gate voltage. The inset shows the optical image of the transistor. (C) The drain current versus gate voltage curves of graphene channel deposited with varied thickness of Au buffer layer [12].

graphene as dielectric exhibits excellent breakdown electric field above 10 MV/cm. The gate leakage current density as a function of electric field was shown in Fig. 16.14D. Fluorinated graphene is highly insulating and performing a low electric field-dependent leakage current of 46.2 $\mu A/cm^2$ at 10 MV/cm, indicating the promising application for atomic layered insulator in nanoscale electronics.

A multilayer graphene fluorination process was established to prevent randomly created defects through transferring fluorinated graphene, which often lead to current leakage paths for dielectric application. With appropriate fluorination time, multilayer graphene of top few layer graphene could be readily converted to fluorinated graphene, whereas keeping the pristine graphene at the bottom regime. Thus, a unique vertical heterostructure of insulator/semiconductor could be readily fabricated. In Fig. 16.15, Ho et al. demonstrated the proof-of-concept FET device with fluorinated graphene as dielectric and pristine graphene as channel. A representative transport curve of graphene was demonstrated, which indicated the electrical properties of pristine graphene were definitely maintained.

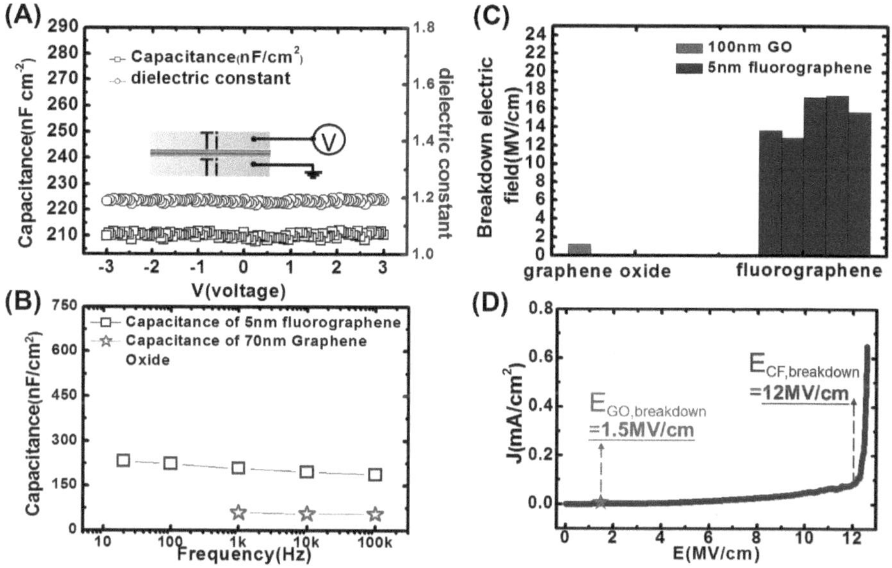

Figure 16.14 (A) Capacitance value and dielectric constant of fluorographene. The inset shows a schematic of a MIM capacitor. (B) Frequency dependence of capacitance degradation for fluorographene and GO. (C) Dielectric breakdown of 10 L of fluorographene and 100 nm of graphene oxide. (D) The current density of the fluorographene capacitor in (C) [32].

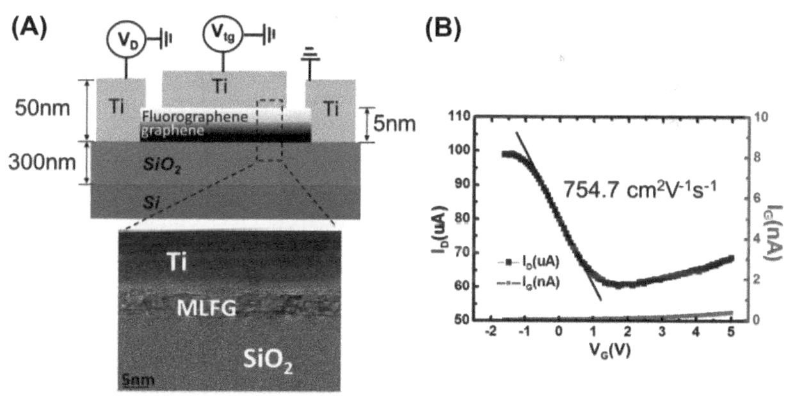

Figure 16.15 (A) An illustration of a graphene-based field effect composed of a fluorographene gate dielectric and the corresponding transmission electron microscopy cross-sectional image. (B) The transfer characteristics and the monitored gate current (I_G) of these transistors, which demonstrate hole mobility of 760 cm^2/V at $V_D = 0.5$ V [32].

As a result, high FET mobility of 754.7 cm/V was observed in transistors, which was competitive to typical mobility of CVD-graphene. Therefore, forming fluorinated graphene was a suitable route to fabricate top-gated graphene-based transistor and definitely compatible with conventional IC technology.

Besides, fluorinated graphene could also act as a passivation layer for screening charged impurity on the surface of substrate (eg, SiO_2). Ho et al. report that the graphene device supported by the fluorinated graphene film, results in the significant enhancement in its electrical properties of graphene transistors in terms of carrier mobility and current swing [33]. The fluorinated graphene could be an excellent supporting buffer layer for devices assembly of emerging 2D layered materials, such as the MoS_2-based electronics, to improve its carrier mobility.

16.4 Conclusion

Fluorinated graphene shows unique atomic configurations and material properties when compared with its counterpart of pristine graphene, especially the tunability of electronic properties by adjusting its fluorine content on graphene surface, thus allowing for tailoring the electrical conductance from semi-metallic to an insulator. Also, the optical properties can be transformed to be a wide band gap material when it was highly fluorinated. The high degree of graphene fluorination controllability makes it a versatile and potential material for next-generation nanoelectronics and optical electronics. The selective fluorination of graphene by the chemical isolation for patterning graphene nanoribbon allows one to create the atomic in-plane heterojunction, the integrated transistor exhibits higher on−off ratio while preserving its high carrier mobility (900 cm^2/V). Moreover, highly fluorinated graphene has been demonstrated to be durable dielectric material, where the 5 nm fluorinated graphene exhibits a breakdown of electric field above 10 MV/cm, indicating the potential application for atomic layer dielectric layer in nanoscale electronics. The recent progress by using scalable fluorination methods, such as plasma techniques, has brought the benefits for cost-effective, large-area, and IC compatible processing, which make it more feasible for practical device integration. For example, the mechanical strength and bendability of fluorinated graphene comprising the high performance electronic behavior allows it for the application in future flexible electronics.

References

[1] K.S. Novoselov, A.K. Geim, S. Morozov, D. Jiang, Y. Zhang, S.V. Dubonos, I. Grigorieva, A. Firsov, Science 306 (2004) 666−669.
[2] C. Lee, X. Wei, J.W. Kysar, J. Hone, Science 321 (2008) 385−388.
[3] A.A. Balandin, S. Ghosh, W. Bao, I. Calizo, D. Teweldebrhan, F. Miao, C.N. Lau, Nano Lett. 8 (2008) 902−907.
[4] K.I. Bolotin, K. Sikes, Z. Jiang, M. Klima, G. Fudenberg, J. Hone, P. Kim, H. Stormer, Solid State Commun. 146 (2008) 351−355.
[5] F. Bonaccorso, Z. Sun, T. Hasan, A. Ferrari, Nat. Photonics 4 (2010) 611−622.
[6] Y.-M. Lin, C. Dimitrakopoulos, K.A. Jenkins, D.B. Farmer, H.-Y. Chiu, A. Grill, P. Avouris, Science 327 (2010) 662.

[7] D. Jariwala, A. Srivastava, P.M. Ajayan, J. Nanosci, Nanotechnology 11 (2011) 6621–6641.
[8] D. Elias, R. Nair, T. Mohiuddin, S. Morozov, P. Blake, M. Halsall, A. Ferrari, D. Boukhvalov, M. Katsnelson, A. Geim, Science 323 (2009) 610–613.
[9] R.R. Nair, W. Ren, R. Jalil, I. Riaz, V.G. Kravets, L. Britnell, P. Blake, F. Schedin, A.S. Mayorov, S. Yuan, Small 6 (2010) 2877–2884.
[10] J.S. Moon, H.-C. Seo, F. Stratan, M. Antcliffe, A. Schmitz, R.S. Ross, A. Kiselev, V.D. Wheeler, L.O. Nyakiti, D.K. Gaskill, IEEE Electron Device Lett. 34 (2013) 1190–1192.
[11] W.-K. Lee, J.T. Robinson, D. Gunlycke, R.R. Stine, C.R. Tamanaha, W.P. King, P.E. Sheehan, Nano Lett. 11 (2011) 5461–5464.
[12] K.I. Ho, J.H. Liao, C.H. Huang, C.L. Hsu, W. Zhang, A.Y. Lu, L.J. Li, C.S. Lai, C.Y. Su, Small 10 (2014) 989–997.
[13] A. Geim, I. Grigorieva, Nature 499 (2013) 419–425.
[14] J.S. Burgess, J.W. Baldwin, J.T. Robinson, F.A. Bulat, B.H. Houstona, Fluorinated carbon nanomaterials: XeF_2 fluorination of graphene, in: ACS Symposium Series, Oxford University Press, 2011, pp. 11–30.
[15] C. Shen, G. Huang, Y. Cheng, R. Cao, F. Ding, U. Schwingenschlögl, Y. Mei, Nanoscale Res. Lett. 7 (2012) 1–8.
[16] S. Zhou, G.-H. Gweon, A. Fedorov, P. First, W. De Heer, D.-H. Lee, F. Guinea, A.C. Neto, A. Lanzara, Nat. Mater. 6 (2007) 770–775.
[17] J.T. Robinson, J.S. Burgess, C.E. Junkermeier, S.C. Badescu, T.L. Reinecke, F.K. Perkins, M.K. Zalalutdniov, J.W. Baldwin, J.C. Culbertson, P.E. Sheehan, Nano Lett. 10 (2010) 3001–3005.
[18] Z. Wang, J. Wang, Z. Li, P. Gong, X. Liu, L. Zhang, J. Ren, H. Wang, S. Yang, Carbon 50 (2012) 5403–5410.
[19] M. Baraket, S. Walton, E. Lock, J. Robinson, F. Perkins, Appl. Phys. Lett. 96 (2010) 231501.
[20] K. Tahara, T. Iwasaki, A. Matsutani, M. Hatano, Appl. Phys. Lett. 101 (2012) 163105.
[21] S. Bae, H. Kim, Y. Lee, X. Xu, J.-S. Park, Y. Zheng, J. Balakrishnan, T. Lei, H.R. Kim, Y.I. Song, Nat. Nanotechnol. 5 (2010) 574–578.
[22] R. Cheng, J. Bai, L. Liao, H. Zhou, Y. Chen, L. Liu, Y.-C. Lin, S. Jiang, Y. Huang, X. Duan, Proc. Natl. Acad. Sci. U.S.A. 109 (2012) 11588–11592.
[23] F. Schwierz, Nat. Nanotechnol. 5 (2010) 487–496.
[24] X. Li, X. Wang, L. Zhang, S. Lee, H. Dai, Science 319 (2008) 1229–1232.
[25] Y. Yang, R. Murali, IEEE Electron Device Lett. 31 (2010) 237–239.
[26] P. Gava, M. Lazzeri, A.M. Saitta, F. Mauri, Phys. Rev. B. 79 (2009) 165431.
[27] F. Xia, D.B. Farmer, Y.-M. Lin, P. Avouris, Nano Lett. 10 (2010) 715–718.
[28] B. Lee, S.-Y. Park, H.-C. Kim, K. Cho, E.M. Vogel, M.J. Kim, R.M. Wallace, J. Kim, Appl. Phys. Lett. 92 (2008) 203102.
[29] Z.H. Ni, H.M. Wang, Y. Ma, J. Kasim, Y.H. Wu, Z.X. Shen, ACS Nano 2 (2008) 1033–1039.
[30] X. Wang, S.M. Tabakman, H. Dai, J. Am. Chem. Soc. 130 (2008) 8152–8153.
[31] O.M. Nayfeh, T. Marr, M. Dubey, IEEE Electron Device Lett. 32 (2011) 473–475.
[32] K.-I. Ho, C.-H. Huang, J.-H. Liao, W. Zhang, L.-J. Li, C.-S. Lai, C.-Y. Su, Sci. Rep. 4 (2014).
[33] K.-I. Ho, M. Boutchich, C.-Y. Su, R. Moreddu, E.S.R. Marianathan, L. Montes, C.-S. Lai, Adv. Mater. 27 (2015) 6519.

Index

'*Note*: Page numbers followed by "f" indicate figures and "t" indicate tables.'

H

L

M

N

FSC
www.fsc.org
MIX
Papier | Fördert gute Waldnutzung
FSC® C083411

Kolibri 360 GmbH
Ferdinand-Jühlke-Straße 7
99095 Erfurt, Deutschland
produktsicherheit@kolibri360.de